T0222587

MECHANISMS IN SCIENCE

In recent years what has come to be called the 'New Mechanism' has emerged as a framework for thinking about the philosophical assumptions underlying many areas of science, especially in sciences such as biology, neuroscience, and psychology. This book offers a fresh look at the role of mechanisms, by situating novel analyses of central philosophical issues related to mechanisms within a rich historical perspective of the concept of mechanism as well as detailed case studies of biological mechanisms (such as apoptosis). It develops a new position, Methodological Mechanism, according to which mechanisms are to be viewed as causal pathways that are theoretically described and are underpinned by networks of difference-making relations. In contrast to metaphysically inflated accounts, this study characterises mechanism as a concept-in-use in science that is deflationary and metaphysically neutral, but still methodologically useful and central to scientific practice.

STAVROS IOANNIDIS is Assistant Professor of Philosophy of Natural Sciences at the National and Kapodistrian University of Athens. He is principal investigator of the project MECHANISM, funded by the Hellenic Foundation for Research and Innovation.

STATHIS PSILLOS is Professor of Philosophy of Science and Metaphysics at the National and Kapodistrian University of Athens. He is the author of *Scientific Realism: How Science Tracks Truth* (1999) and *Causation and Explanation* (2002), and editor (with Henrik Lagerlund and Ben Hill) of *Reconsidering Causal Powers* (2021).

MECHANISMS IN SCIENCE

Method or Metaphysics?

STAVROS IOANNIDIS

University of Athens

STATHIS PSILLOS

University of Athens

Shaftesbury Road, Cambridge CB2 8EA, United Kingdom

One Liberty Plaza, 20th Floor, New York, NY 10006, USA

477 Williamstown Road, Port Melbourne, VIC 3207, Australia

314–321, 3rd Floor, Plot 3, Splendor Forum, Jasola District Centre, New Delhi – 110025, India

103 Penang Road, #05–06/07, Visioncrest Commercial, Singapore 238467

Cambridge University Press is part of Cambridge University Press & Assessment,
a department of the University of Cambridge.

We share the University's mission to contribute to society through the pursuit of
education, learning and research at the highest international levels of excellence.

www.cambridge.org
Information on this title: www.cambridge.org/9781009011495

DOI: 10.1017/9781009019668

First published 2022
First paperback edition 2024

A catalogue record for this publication is available from the British Library

Library of Congress Cataloging-in-Publication data
NAMES: Ioannidis, Stavros, author. | Psillos, Stathis, 1965– author. Title: Mechanisms in science:
method or metaphysics? / Stavros Ioannidis, University of Athens, Greece, Stathis Psillos,
University of Athens, Greece.
DESCRIPTION: Cambridge, United Kingdom; New York, NY, USA: Cambridge University
Press, 2022. | Includes bibliographical references and index.
IDENTIFIERS: LCCN 2021056613 (print) | LCCN 2021056614 (ebook) | ISBN 9781316519905
(hardback) | ISBN 9781009011495 (paperback) | ISBN 9781009019668 (epub)
SUBJECTS: LCSH: Mechanism (Philosophy) | Biology–Philosophy. Classification: LCC BD553 .I53
2022 (print) | LCC BD553 (ebook) | DDC 146/.6–dc23/eng/20211202
LC record available at https://lccn.loc.gov/2021056613
LC ebook record available at https://lccn.loc.gov/2021056614

ISBN   978-1-316-51990-5   Hardback
ISBN   978-1-009-01149-5   Paperback

To my parents, Panagiota and Giorgos, with gratitude. —S.I.
To the memory of those two who made this exciting journey possible for me, my father Dimitris (1927–2001) and my mother Maria (1934–2019). —S.P.

Contents

Figures

Preface

This book is the product of genuinely collaborative work (the names appear in alphabetical order). We have known each other for almost two decades now. We met for the first time when S.I., still at high school, walked into S.P.'s office to ask advice concerning degrees in philosophy. S.I. became an undergraduate student in S.P.'s dept; then he did graduate studies in the philosophy of biology in the University of Bristol and joined the History and Philosophy of Science Department of the University of Athens as a postdoc in 2012.

Over the years, we have developed a philosophical partnership that extends from issues in the history of philosophy (mostly during the seventeenth century) to issues in contemporary metaphysics of science. Key to this partnership is mutual respect and tolerance as well as a common philosophical outlook. Both of us agree that good philosophy should be conceptually clear and historically sensitive. In fact, it seems that each and every philosophical problem is better illuminated if it is subjected to rigorous conceptual dissection; yet when the various parts are synthesised again, treating them in their historical concreteness enhances our understanding of their trajectory in time and space.

The project that led to this book started, like all of our joint ventures, with discussions over coffee on an early Saturday morning in a café in the centre of Athens. At the beginning of 2015, one of us (S.P.) received a kind invitation from Phyllis Illari and Stuart Glennan to contribute a piece on mechanisms and counterfactuals to the *Routledge Handbook of Mechanisms*. S.P. invited S.I. to work with him on this project. We started as we always did, by drafting the table of contents and dividing up the writing. Dozens of meetings and some heated discussions later, we submitted the piece in August 2016.

At the time of the mechanism project, we were both involved in a long and thorough study of the relations between metaphysics and physics in the seventeenth century. We started with Descartes and moved on to

Newton and Leibniz, with seminars, reading groups and workshops. We focused in particular on the transition from an Aristotelian power-based ontology to the modern law-based account of the world in terms of matter in motion. These endeavours brought with them the question of the relation between the Old Mechanism of the seventeenth century and the New Mechanism of the twenty-first. In October 2016 S.P. presented this historical narrative at the annual conference of L'academie Internationale de Philosphie des Sciences in Dortmund. The conference was on mechanisms and was organised by Brigitte Falkenburg and Gregor Schiemann.

The New Mechanism started to become the focal point of our research. Searching for a mechanism to study in detail, we came across the case of apoptosis, a.k.a. programmed cell death. We soon realised that that is a very rich case of a mechanism and it became the subject of our study. The more we thought about apoptosis, the more it became a showcase of our own approach to mechanism. This approach, which we called Methodological Mechanism, was aired first at conferences, most notably at the conference on Mechanisms in Medicine at the University of Kent at Canterbury in July 2017, organised by the gurus of mechanisms in the United Kingdom (the members of the 'Evaluating Evidence in Medicine' project (https://blogs.kent.ac.uk/jonw/projects/evaluating-evidence-in-medicine/).

The reception was mixed. John Worrall, our good friend, had to leave right after S.P. delivered the paper and on his way out he whispered in S.P.'s ear, 'You are back on the straight and narrow; you've become a logical positivist.' The hosts (Jon Williamson and Phyllis Illari) were more critical, arguing that the position is too thin, while María Jiménez Buedo (a young talented philosopher from Madrid) was enthusiastic. The result was a kind invitation from her to UNED in Madrid in May 2018, and a couple of talks in her research group (together with Jesús Zamora, Mauricio Suarez and David Teira) on the notion of mechanism in biology and the social sciences. In the meantime, our paper on apoptosis and Methodological Mechanism had appeared in print in the journal *Axiomathes*. Our main thesis, that the concept of mechanism in use in the sciences is mainly or exclusively methodological and not metaphysical, started to acquire some traction.

At roughly the same time, S.P. had finished a book review of Glennan's *The New Mechanical Philosophy* for the *Australasian Journal of Philosophy*. This book came to us as manna from heaven. Stuart presented very eloquently and forcefully the view that we wanted to oppose – the metaphysics-first view, as it were – and made us think harder about how best to defend our own practice-first approach. In the end, the review grew longer and longer and only a small part of it appeared in the *Australasian Journal of Philosophy*. The rest of it, focused as it was on activities qua a new ontic category, was destined to go into another paper, which was invited by the journal *Teorema*. In this paper we took on the concept of activity and developed our own difference-making account of the workings of a mechanism.

At the beginning of 2019, we received a kind email from our good friend Orly Shenker inviting us to a star-studded workshop in Jerusalem on the levels of reality, towards the end of May 2019. We took this opportunity to sharpen our thoughts on the issue of levels of mechanisms, and in particular on the relation of constitution that is supposed to hold between a mechanism and other mechanisms as its parts. That is an idea we reject in favour of causation. At the end of the day only S.I. managed to go as S.P.'s mother fell terminally ill and passed away a couple of weeks later.

The last station of this journey was in Geneva in September 2019, at the EPSA 19 Conference, where S.I. presented work concerning our main thesis, that mechanisms are causal pathways described in theoretical language. A few months later, just before the COVID-19 pandemic and the first lockdown, a good chunk of manuscript was submitted to Hilary Gaskin and Cambridge University Press. We should thank Hilary for her patience, care and support throughout the occasionally very demanding and tiring health-related issues that we both faced during the period of the completion of the manuscript. Two anonymous readers for Cambridge University Press made wonderfully detailed, critical but positive comments on the first draft. Without them the book would have been philosophically poorer.

Philosophy is essentially a communal enterprise. The book has benefitted a lot from a number of individuals who cared enough to make oral and written comments, ask questions and pose various challenges. The list of all those we would like to thank wholeheartedly is a lot bigger than the list of those mentioned by name. Hence our deepest thanks go to all those who asked questions and made comments, but whose names we don't know. But also to Konstantina Antiochou, Ken Binmore, Diderik Batens, Craig Callender, Nancy Cartwright, Paul Churchland, Peter Clark, Lindley Darden, Mauro Dorato, Jan Faye, Alexander Gebharter, Mania Georgatou, Michel Ghins, Donald Gillies, Olav Gjelsvik, Alan Hajek, Haris Hatziioannou, Chris Hitchcock, Carl Hoefer, Ilhan Inan, Gürol Irzik, Philip Kargopoulos, Patricia Kitcher, Buket Korkut, Daniel Kostić,

Vassilis Livanios, Peter Machamer, Vincent Müller, Daniel Nolan, Panagiotis Oulis (RIP), Kostas Pagondiotis, Demetris Portides, Nils Roll-Hansen, Pavlos Silvestros, Mauricio Suárez, Javier Suárez, David Teira, Amalia Tsakiri, Eric Watkins, Erik Weber and Jim Woodward. Special mention goes to the following for giving us venues to present our work (as well as constructive criticism): Brigitte Falkenburg, Stuart Glennan, Meir Hemmo, Phyllis Illari, Valeriano Iranzo, Maria Jimenes Buedo, Federica Russo, Orly Shenker, Gregor Schieman, Erik Weber and Jon Williamson. We are also thankful to our research group in Athens – the usual suspects – who patiently heard and relentlessly criticised various drafts of chapters of the book, to Marilina Smyrnaki for her help in compiling the bibliography, to Marios Ioannidis for preparing the illustrations and to Stephanie Sakson for her careful and valuable copyediting work.

Over the years, ideas that eventually formed parts of the book have been presented in seminars at the University of California San Diego, Caltech, American College of Thessaloniki, Bogazici University, University of Oslo, University of Ghent, UNED University in Madrid, Western University, Aristotle University of Thessaloniki and the University of Cyprus; also at the fourth Athens–Pittsburgh conference on Proof and Demonstration in Science and Philosophy, in Delphi (June 2003); at the workshop of the Metaphysics in Science Group in Athens (June 2003); at the Symposium on Mechanisms in the Sciences, APA Central Division, Chicago (April 2006); at the Conference on Causality in the Sciences, University of Kent (September 2008); at the Conference on Mechanisms and Causality in Science, University of Kent (September 2009); at the Workshop on the Metaphysics of Science, University of Warsaw (January 2010); at the Symposium on the Metaphysics of Science, College de France (May 2012); at ISHPSSB 2015 (Montreal, UQAM); at the AIPS Conference on Mechanistic Explanations, in Dortmund (October 2016); at the 'Mechanisms in Medicine' workshop, Centre of Reasoning, University of Kent (July 2017); at the Conference on the Multi-Level Structure of Reality (Israel Institute for Advanced Studies, Hebrew University of Jerusalem, and University of Haifa, May 2019); and at the EPSA19, Geneva (September 2019). S.P. would like to thank wholeheartedly the three women in his life, who help him get his bearings: Athena, Demetra and Artemis.

S.I.'s work has received funding from the Hellenic Foundation for Research and Innovation (HFRI) and the General Secretariat for Research and Innovation (GSRI), under grant agreement no. 1968.

Parts of this book have been based on reworked and expanded material that first appeared in journals and books. We thank the various publishers and editors for permission to use material in the book. Specifically:

Ioannidis, S. & Psillos, S. (2017). In defense of methodological mechanism: the case of apoptosis. *Axiomathes* 27 601–19. Reprinted by permission from Springer Nature, Copyright © 2017. (**Chapters 3, 4 and 10**)

Ioannidis, S. & Psillos, S. (2018). Mechanisms, counterfactuals and laws. In S. Glennan and P. Illari, eds., *The Routledge Handbook of Mechanisms and Mechanical Philosophy*, 1st ed. New York: Routledge, pp. 144–56. Copyright © 2018 by Routledge. Reproduced by permission of Taylor & Francis Group. (**Chapter 5**)

Ioannidis, S. & Psillos, S. (2018). Mechanisms in practice: a methodological approach. *Journal of Evaluation in Clinical Practice* 24: 1177–83. Copyright © 2018, John Wiley & Sons. (**Chapters 3 and 4**)

Psillos, S. (2004). A glimpse of the secret connexion: harmonizing mechanisms with counterfactuals. *Perspectives on Science* 12: 288–319. Copyright © 2004 The Massachusetts Institute of Technology. (**Chapters 6 and 7**)

Psillos, S. (2011). The idea of mechanism. In P. Illari, F. Russo, and J. Williamson, eds., *Causality in the Sciences*. Oxford: Oxford University Press, 771–88. This material has been reproduced by permission of Oxford University Press [http://global.oup.com/academic]. (**Chapter 2**)

Psillos, S. (2015). Counterfactual reasoning, qualitative: philosophical aspects. In J. Wright, ed., *International Encyclopedia of the Social & Behavioral Sciences*, 2nd ed. Oxford: Elsevier, vol. 5, pp. 87–94. Copyright © 2015, with permission from Elsevier. (**Chapter 7**)

Psillos, S. (2019). Review of Stuart Glennan, *The New Mechanical Philosophy*. *Australasian Journal of Philosophy* 97: 621–4. Copyright © Australasian Association of Philosophy, reprinted by permission of Taylor & Francis Ltd, http://www.tandfonline.com on behalf of Australasian Association of Philosophy. (**Chapter 6**)

Psillos, S. & Ioannidis, S. (2019). Mechanisms, then and now: from metaphysics to practice. In B. Falkenburg and G. Schiemann, eds., *Mechanistic Explanations in Physics and Beyond*. European Studies in Philosophy of Science. Cham: Springer Nature, pp. 11–31.

Reprinted by permission from Springer Nature, Copyright © 2019. **(Chapter 1)**

Psillos, S. & Ioannidis, S. (2019). Mechanistic causation: difference-making is enough. *Teorema* 38, no. 3: 53–75. www.unioviedo.es/Teorema **(Chapters 4 and 6)**

Introduction

When we think about mechanisms there are two general issues we need to consider. The first is broadly epistemic and has to do with the understanding of nature that identifying and knowing mechanisms yields. The second is broadly metaphysical and has to do with the status of mechanisms as building blocks of nature (and in particular, as fundamental constituents of causation). These two issues can be brought together under a certain assumption, which has had a long historical pedigree, namely, that nature is fundamentally mechanical.

That's a thought that was introduced by René Descartes and was popularised by Pierre Gassendi and Robert Boyle. Indeed, Descartes referred to his own theory of nature as 'Mechanica' since it considers sizes, shapes and motions, adding that it's a true theory of the world (Letter to Fromondus, 3 October 1637). Boyle called 'mechanical philosophy' his own corpuscularian theory which was based on matter and motion, two principles more 'catholic and universal' than any others. Mechanical philosophy was both a metaphysics of nature and a scientific theory. The point of contact was in the assumption that all worldly phenomena were ultimately the product of the mechanical affections of tiny corpuscles. That's the content of the old mechanist view that nature is *mechanical.*

And yet this very assumption has had no concrete ahistorical conceptual content. Rather, its content has varied according to the dominant conception of nature that has characterised each epoch. Nor has it been the case that the very idea of mechanism has had a fixed and definite content. Even if in the seventeenth century and beyond, the idea of mechanism had something to do with matter in motion subject to mechanical laws, *current* conceptions of mechanism have only a very loose connection with this assumption.

What kinds of commitments does more recent talk of mechanisms imply? Until the 1980s, the dominant views about mechanisms in the philosophy of science had been metaphysical. Mechanism has been seen as a view about causation: mechanisms were taken to provide the missing link (David Hume's 'secret connexion') between the cause and the effect. 'Mechanism', on this approach, is the very causal connection, and has been described in various ways as mark transmission (Salmon 1984), persistence, transference or possession of a conserved quantity (Mackie 1974; Salmon 1997; Dowe 2000).

In the 1990s, the New Mechanical Philosophy emerged, which is a view about the causal structure of the world: the world we live in is a world of mechanisms. A mechanism, nowadays, is virtually *any* relatively stable arrangement of entities such that, by engaging in certain interactions, a function is performed or an effect is brought about. Take a very typical characterisation of mechanism by William Bechtel and Adele Abrahamsen (2005, 423):

> (M) A mechanism is a structure performing a function in virtue of its component parts, component operations, and their organisation. The orchestrated functioning of the mechanism is responsible for one or more phenomena.

On this conception, a mechanism is *any* structure that is identified as such (i.e., as possessing a certain causal unity) via the function it performs. Moreover, a mechanism is a *complex* entity whose behaviour (i.e., the function it performs) is determined by the properties, relations and interactions of its parts. This priority of the parts over the whole – and in particular, the view that the behaviour of the whole is determined by the behaviour of its parts – is the distinctive feature of this broad account of mechanism.

Behind this broad understanding of mechanism, there is a certain metaphysics of nature. New mechanists take 'mechanisms' to be complex *entities* in their own right which are characterised by a certain ontological signature. This is supposed to be a signature that all mechanisms share in common, something that unifies all mechanisms and grounds their role in causation and explanation. The dominant view is that mechanisms are *structured wholes of entities and activities.* The latter are supposed to be the 'ontic correlate' of verbs; they are meant to ground the productive relations among entities and the productivity of the mechanism as a whole.

This metaphysics of mechanisms, a.k.a. 'new mechanical ontology' of entities, activities, organisation of parts into wholes and so on, invites a number of questions: What, in general terms, are the constituents of mechanisms? And what are their relations with more traditional metaphysical categories, such as objects, properties, powers and processes? So, fundamental ontology is brought to bear on how best to understand mechanisms in science.

A main claim of this book is that those philosophical views that offer metaphysically 'inflated' accounts of mechanisms are not necessary in order to understand scientific practice. And not just that. We shall also claim that there is no argument from the practice of using mechanisms in science to any metaphysics of mechanisms. That's mainly because we take it that the concept of mechanism as it is used in science, and in biology in particular, is methodological. We call our view Methodological Mechanism (MM). The main tenet of MM is that commitment to mechanism in science is first and foremost a methodological stance. The core of MM is a deflationary account of mechanism that is ontologically non-committal. According to what we call *Causal Mechanism*,

(CM) A mechanism is a causal pathway that is described in theoretical language.

Moreover, we claim that commitment to mechanism in science means adopting a certain methodological postulate, that is, that one should always look for the causal pathways producing the phenomena of interest. As such, it does not make any general ontological assertions about the 'deep' nature of causal processes.

On our deflationary account, talk of mechanisms in (biological) practice is talk about how causes (described in the language of theories) operate to bring about a certain effect. To identify a mechanism, then, is to identify a specific causal pathway that connects an initial 'cause' (the causal agent) with a specific result. Wherever there is a cause for a specific effect, there exists a mechanism that accounts for how the cause operates. The scientific task, then, is to identify the *mode of operation* of the cause, that is, the causal pathway. Identification of the causal pathway is crucial in order to establish that a causal link exists between a putative causal agent and a result (e.g., a disease state). Moreover, knowing the causal pathway makes interventions possible (and in the case of pathology, treatment). Our examination and defence of CM will unravel some limitations inherent in any attempt to extract metaphysical conclusions from scientific practice.

Admittedly, CM is thin. But it does *not* follow from CM that mechanisms are not 'things in the world'. After all, they *are* causal pathways! In characterising CM as deflationary or metaphysically agnostic, the point is that there is no need to say something 'deeper' than this in order to have a useful concept of mechanism that elucidates practice: a mechanism simply *is* a sequence of causal steps (or a process) that leads from an initial 'cause' to an end result. In sum: mechanisms in science and, in particular, in biology are stable causal pathways, described in the language of theory, where to identify a *causal* pathway is to identify difference-making relations among its components. The best way to introduce the main thesis of the book is by a parable.

A Parable

Imagine you've been a lifelong metaphysician trying to understand the fundamental building blocks of reality. You've read (more or less) everything written on the subject, which after an accident of classification, came to be called 'metaphysics'. You started with Aristotle's treatises that deal with the fundamental ontological categories, and which because they were placed after his φυσικά (physica; physics), were collectively dubbed μετά τά φυσικά (metaphysica; metaphysics). And in the fullness of time, you acquired views about all important topics: universals versus particulars; categorical versus dispositional properties; necessitarian versus non-necessitarian laws of nature; Humean versus non-Humean accounts of causation; and so on. You've been particularly excited by the concept of mechanism. You read about the mechanical account of nature that emerged in the seventeenth century, you reflected on the alternative, non-mechanical, way of explanation. You came to believe that causation is intimately connected with the presence of mechanism. And of course you lived through the revival of the mechanistic account of nature in the end of the twentieth century.

Metaphysics being what it is, that is, quite remote from ordinary and scientific experience, you had various doubts about the theories you entertained but at no time did you doubt that there are well-founded answers to the questions you grappled with. Now (that's the beginning of the parable), you are standing at the Pearly Gates about to meet your maker. After the usual introductions (with lots of rhetorical questions on God's part, 'Did you enjoy what you did for a living?,' etc.) God asks you the question you've been waiting and longing for: 'My child, is there anything you'd like to ask me because you haven't been satisfied with the answers you've hit upon yourself?' You gather all the strength you have and say: 'Yes, my Lord, there are indeed quite a few things that gave me sleepless nights and I'd die to find out the answer.' God being very busy replies candidly that only three short questions are allowed. Here is your first: Are there universals? God replies. (Unfortunately, there is no record of his answer.) Here is your second: 'Are there mechanisms in nature?' God reflects for a minute and then replies: 'Certainly! Light reflection, chemical bonding, mitosis but also economic crises and demonstrations are mechanisms. Where there is causation, there is mechanism!' Time for the third question, God says. You are puzzled. You collect all of your philosophical might and ask: 'Well, that's a list! Isn't there anything all these, and plenty of others like those in the list, share in common in virtue of which they are *mechanisms*? Isn't there something metaphysically deep that constitutes a mechanism?' God is somewhat upset since as he pointedly remarks these are *two* questions, but, knowing that his interlocutor is a philosopher he let it pass. 'So', you say with genuine aporia in your eyes. 'Well', he says, 'they are all causal pathways described in some theoretical language. That's what they have in common.' And before he parts company he winks and adds: 'I'm afraid you were barking up the wrong tree. "Mechanism" is everywhere as a piece in methodology but not in ontology.'

The Aim and Scope of Our Argument

The very aim of the book at hand is to explain the foregoing reply in detail and to show its plausibility. Lest we are misunderstood, that's not God's reply (who knows what this might be). But it's a possible reply to be assessed as any other philosophical theory.

We should be clear from the outset about the scope of our argument. Let us again distinguish between two questions one may ask about causation, one metaphysical, the other methodological. The metaphysical one concerns how exactly causation bottoms out: Does it bottom out in irreducible productive activities, in the manifestation of powers, in Humean regularities or (perhaps) in primitive difference-making relations? The methodological question is this: If we focus on how scientists discover mechanisms and how experiments are being done, are difference-making relations enough in order to understand scientific practice, or do we need also to say something about the ultimate nature of the truth-makers, if there are any, of these difference-making relations?

Concerning the question about methodology, we claim that if we focus on methodology and the epistemology of practice, difference-making relations should be enough (the case of the cause of scurvy, discussed in Chapter 4, will drive this point home). We do not have to say anything more about the truth-makers of causal claims. So, if one just focuses on the methodological question, one can remain agnostic about the metaphysics. It is in this sense that we described CM as a metaphysically agnostic position.

Concerning the question of metaphysics, one can, surely, give reasons to prefer some particular metaphysical picture over others. Our point is that whatever else these reasons are, they are not related to scientific practice. That is, what we emphatically reject is a particular kind of strategy of offering arguments in favour of a specific metaphysics of causation: that in order to understand the concept of mechanism *as it is used in the sciences* we need to be committed to a layer of thick metaphysical facts, for example, about activities or powers, as the truth-makers of causal claims.

In fact, in the subsequent chapters we will offer reasons to favour a particular metaphysical picture, namely, a Humean view that grounds difference-making relations in laws of nature understood as regularities, and to reject activities-based views that several mechanists accept. What is important to note however is that the argument against activities and in favour of Humean regularities will not be based on scientific practice. Rather, it will be based on an examination of the philosophical merits of the various specific metaphysical views. This means that if one is not inclined to pose the metaphysical question about causation and wishes to confine oneself to what is licensed by scientific practice, one can adopt CM for the concept-in-use and remain a metaphysical agnostic.

Given our complete view that combines a practiced-based CM with a particular approach to metaphysics, two opposite reactions are possible if one wants to reject our position. One can either opt for a stronger account about the metaphysics of causation, or an even more minimal metaphysical view. Let us be clear about each of them (although a fuller answer will be given in subsequent chapters).

When it comes to the search for stronger accounts that ground difference-making relations to something metaphysically robust, we take the central issue to be whether these accounts can tell illuminating stories about how mechanisms are used in science – we think that they do not. The point can be made as follows: when it comes to the practice of discovering difference-making relations, it does not matter whether we live in a Humean or a non-Humean world containing perhaps irreducible activities; in all cases, scientific practice would be the same and the search for mechanisms equally prevalent. This means, again, that such stronger accounts of causation do not help us understand scientific practice – of course, as noted already, one may have other kinds of reasons to favour a Humean metaphysics, for example, on grounds of conceptual economy.

Accounts of causation that may assume difference-making relations while not requiring an explicit stance on laws rely typically on a notion of variation among the values of relevant variables or invariance under interventions, which is less robust than regularity. Our differences with such accounts are less important than the similarities. Hence, they dismiss calls for metaphysically thicker views of causation and focus on how causal statements are established empirically. The key problem with such accounts is to avoid collapsing causation to correlations.

Conceptual Geography

In the book, we draw a number of distinctions and offer a number of characterisations of mechanism. To avoid confusion, we shall try here to offer a map of the conceptual landscape of mechanism.

The central distinction we draw is between the Old Mechanism and the New Mechanism, which is based on two pillars. The first is historical, while the second is conceptual. Old Mechanism emerged as a reaction to Aristotelian natural philosophy; it was foremost a theory of the natural world backed up by a metaphysics of nature. It was an attempt to leave behind the physics and the metaphysics of later scholasticism and to replace it with a new physics. The relation of the new physics with metaphysics was a matter of dispute. The clearest relation was in Descartes's metaphysical physics, to use Dan Garber's apt expression. By contrast, New Mechanism is, by and large, a philosophical re-interpretation of scientific practice and not a new science. As such, it is mostly a new metaphysics of nature. The second pillar of the distinction concerns the key features of the new metaphysics of science. The key feature of the metaphysics of Old Mechanism was that everything material was explainable by reference to shape, size and motion. A mechanism was any configuration of matter in motion subject to laws. By contrast to this 'flat' ontology, New Mechanism takes *any* kind of thing to be a mechanism provided it has a certain structure of entities engaging in activities, the entities and the activities being a lot more 'colourful' than the 'dull' entities (corpuscles) and the 'activities' (forces acting by contact) of Old Mechanism. A bit provocatively put, a mechanism of the new mechanical metaphysics is a matter of structure whereas Old Mechanism is a matter of content.

Occasionally we refer to 'Old Mechanism (narrowly understood)' aiming to draw a distinction *within* Old Mechanism between a narrow (Cartesian) understanding of mechanism as involving contact-action and a more liberal (post-Newtonian) account of mechanism which takes it that a mechanism is any system subject to Newton's laws, namely, the laws of *mechanics*. As Robert Schofield (1970, 15) has argued, post-Newtonian mechanists took it that 'causation for all the phenomena of nature was ultimately to be sought in the primary particles of an undifferentiable matter, the various sizes and shapes of possible combinations of these particles, their motions, and the forces of attraction and repulsion between them which determine these motions'. By the time of Henri Poincaré, this more liberal account of Old Mechanism was the dominant one.

This more liberal version of Old Mechanism we call 'mechanical mechanism' and we contrast it with what we call following A. C. Ewing 'quasi-mechanical' mechanism. That's a distinction that can, arguably, be traced to Immanuel Kant's *Third Critique* and is meant to introduce a conception of mechanism such that the properties of the whole are determined by the properties of its parts, but with no particular reference to mechanical properties and laws. Rom Harré has called these mechanisms *generative mechanisms*. They are taken to underpin causal connections: in virtue of them causes are supposed to produce their effects. As Harré aptly put it: 'not all mechanisms are mechanical' (1972, 118). This idea of quasi-mechanism or generative mechanism is a precursor of New Mechanism.

Finally, we draw a distinction between mechanism-of and mechanism-for, which relates to the role of mechanisms in causation. The dominant conception among new mechanists is that a mechanism is *for* a behaviour or function. The function/behaviour of a mechanism determines the boundaries of the mechanism and the identification of its components and operations. For us, however, the notion of a mechanism-for captures the functional notion of mechanism. A mechanism-of is any causal process, irrespective of whether or not a function is performed. We will argue that every mechanism-for is a mechanism-of, but not conversely. The mechanism-for/-of distinction serves to illustrate the fact that some causal pathways have a function and a functional role to play (they are mechanisms-for) while other causal pathways do not (they are mechanisms-of). The distinction then is important for driving home the point that there are mechanisms everywhere (where there are causal pathways) even if there is no function they perform.

The Road Map

Having drawn the conceptual map of mechanism, it's time to move to an orderly summary of the chapters of the book.

In Part I we look at some main aspects of the historical development of the concept of mechanism. This broader narrative is motivated by the explicit intention of the new mechanists to link the current mechanical philosophy with its older counterpart in the seventeenth century.

Chapter 1 examines the relationship between Old and New Mechanism and uses it to illuminate the relations between metaphysical and methodological conceptions of mechanism. This historical examination will directly motivate our new deflationary account of mechanism developed in the subsequent chapters. We start by focusing on the role of mechanistic explanation in seventeenth-century scientific practice, by discussing the views of Descartes, Christiaan Huygens, Gottfried Wilhelm Leibniz and Boyle, and the attempted mechanical explanations of gravity by Descartes and Huygens. We thereby illustrate how the metaphysics of Old Mechanism constrained scientific explanation. We then turn our attention to Isaac Newton's critique of mechanism. The key point is that Newton introduces a new methodology that frees scientific explanation from the metaphysical constraints of the older mechanical philosophy. Last, we draw analogies between Newton's critique of Old Mechanism and our critique of New Mechanism. The main point is that causal explanation in the sciences is legitimate even if we bracket the issue of what mechanisms or causes are as things in the world.

In Chapter 2, we continue our historical discussion of mechanistic explanation. The chief purpose of this chapter is to disentangle what we call *mechanical* and *quasi-mechanical* mechanism and point to the key problems they face. We begin by offering an outline of the mechanical conception of mechanism, as this was developed after the seventeenth century. We then present Poincaré's critique of mechanical mechanism in relation to the principle of conservation of energy. The gist of this critique is that mechanical mechanisms are too easy to be informative, provided that energy is conserved. We then advance the quasi-mechanical conception of mechanism and reconstruct G. W. F. Hegel's critique of the idea of quasi-mechanism, as this was developed in his *Science of Logic*. Hegel's problem, in essence, was that the unity that mechanisms possess is external to them and that the very idea that *all* explanation is mechanical is devoid of content. Finally, we bring together Poincaré's problem and Hegel's problem and argue that though mechanisms are not the building blocks of nature, the search for mechanism is epistemologically and methodologically welcome.

Part II develops our own science-first difference-making account of Causal Mechanism.

In Chapter 3 we present the first main part of our case for CM, by discussing in detail apoptosis, a central biological mechanism. We examine how Kerr and his co-workers first introduced apoptosis in 1972. We then present the most important stages in scientific research regarding apoptosis during the last decades that led to its identification as a central biological mechanism, explaining the shift from morphological descriptions to biochemical descriptions of the mechanism. We generalise the molecular definition of a pathway to arrive at a more general notion of a causal pathway. We also show that several distinctions used by biologists in order to differentiate between causal pathways and identify the genuine biological mechanisms (active vs passive, programmed vs non-programmed, physiological vs accidental) do not correspond to internal features of causal pathways, but concern an external feature, that is, the role those processes play within the organism.

In Chapter 4 we build on this discussion in order to argue that understanding mechanisms in the CM sense is all that is needed in order to understand biological practice. We clarify the main commitments of our view by presenting three theses that together constitute CM. (1) Mechanisms are to be identified with causal pathways; (2) causal relations among the components of a pathway are to be viewed in terms of difference-making; and (3) CM is metaphysically agnostic. A key point is that, in contrast to mechanistic theories of causation, for CM causation as difference-making is conceptually prior to the notion of a mechanism. We examine in some detail the discovery of the mechanism of scurvy in order to argue that difference-making is what matters in practice. We then turn to the main inflationary accounts of mechanism and contrast them with our deflationary view and its metaphysical agnosticism. We argue that CM offers a general characterisation of mechanism as a concept-in-use in the life sciences that is deflationary and thin, but still methodologically important.

In Chapter 5, we examine the relation between mechanisms and laws/counterfactuals by revisiting the main notions of mechanism found in the literature. We distinguish between two different conceptions of 'mechanism': *mechanisms-of* and *mechanisms-for*. We argue that for both mechanisms-of and mechanisms-for, counterfactuals and laws are central for understanding within-mechanism interactions. Concerning mechanisms-for, we claim that the existence of irregular mechanisms is compatible with the view that mechanisms operate according to laws. The discussion in this chapter, then, points to an asymmetrical dependence between mechanisms and laws/counterfactuals: while some laws and counterfactuals must be taken as primitive (non-mechanistic) facts of the world, all mechanisms depend on laws/counterfactuals.

In Chapter 6, we defend the *difference-making* thesis of CM, that is, the view that mechanisms are underpinned by networks of difference-making relations, by showing that difference-making is more fundamental than production in understanding mechanistic causation. Our argument is two-fold. First, we criticise Stuart Glennan's claim that mechanisms can be viewed as the truth-makers of counterfactuals and argue that counterfactuals should be viewed as metaphysically more fundamental. Second, we argue against the view that the productivity of mechanisms requires thinking of them as involving activities, qua a different ontic category. We criticise two different routes to activities: Glennan's top-down approach and Phyllis Illari and Jon Williamson's bottom-up approach. Given these difficulties with activities and mechanistic production, it seems more promising to start with difference-making and give an account of mechanisms in terms of it.

Given the centrality of counterfactual difference-making relations to the argument of the book, in Chapter 7 we say a few more things about the contrary-to-fact conditionals. We offer a primer on the logic and the semantics of counterfactuals, focusing on the two main schools of thought: the metalinguistic and the possible-worlds approach. We also present and examine James Woodward's interventionist counterfactuals and the Rubin-Holland model. We argue that the counterfactual approach is more basic than the mechanistic, but information about mechanisms can help sort out some of the methodological problems faced by the counterfactual account.

In Part III we show how our account of Causal Mechanism goes beyond New Mechanism and defend our own Methodological Mechanism.

In Chapter 8 we examine Carl Craver's well-known account of constitutive mechanisms, which takes the organised entities and activities that are the components of the mechanism to constitute the phenomenon to be explained. The main aim of the chapter is to criticise the adequacy of this view for illuminating mechanism as a concept-in-use in biological practice. We identify two main problems for the constitutive view: the problem of external components and the fact that some mechanisms can exist outside the entity the behaviour of which they underlie; we argue that both problems undermine the usefulness and appropriateness of viewing typical and paradigmatic cases of biological mechanisms in constitutive terms. The main claim of the chapter is that in order to understand the notion of mechanism as a concept-in-use, there is no need to posit a non-causal relation of constitution.

In Chapter 9 we present and defend a causal account of multilevel mechanistic explanation by examining various case studies from biology. We argue that two key consequences of Causal Mechanism are: (1) that levels and mechanisms are distinct notions and (2) that levels of multilevel explanations are levels of composition. This view is in stark contrast to Craver's account according to which levels in multilevel explanations are levels of mechanisms and multilevel explanations are instances of constitutive explanations. A key claim of the chapter is that whatever contributes to the phenomenon is part of the same pathway; but causal pathways can contain entities from multiple levels of composition. In order to motivate and illustrate our view, we use various examples from biology and medicine. We criticise some common views associated with the picture of a hierarchy of mechanistic levels and argue that our view allows for causation at higher levels.

In Chapter 10 we pick up the various threads and defend the main thesis of the book (Methodological Mechanism), namely, the claim that to be committed to mechanism is to adopt a certain methodological postulate, that is, to look for causal pathways for the phenomena of interest. We compare our view of Methodological Mechanism with an important discussion by Joseph Henry Woodger (1929) concerning the meaning of mechanism, which has been ignored in current discussions, as well as with the views of Robert Brandon. We then formulate a dilemma that new mechanists face; the dilemma arises from the unstable combination of two main tenets of New Mechanism, an ontological and a methodological one, both of which depend on the general characterisation of mechanism but that pull in opposite directions. We argue that CM is able to resolve the dilemma, by providing the best defence of the methodological tenet of New Mechanism, while at the same time preventing the adoption of a robust version of the ontological tenet.

In the last chapter, the Finale, we examine to what extent CM can be seen as a descendant of the original notion of mechanism developed in seventeenth century, by examining possible extensions of the seventeenth-century notion of mechanism and discussing whether they can be used to characterise mechanism as a concept-in-use.

In sum: Methodological Mechanism is mechanism enough.

PART I

Ideas of Mechanism

CHAPTER I

Mechanisms, Then and Now

1.1 Preliminaries

Let's call 'Old Mechanism' (or mechanical natural philosophy) the general conception about the nature of the world and of science that was developed in the seventeenth century by thinkers such as Descartes, Boyle, Huygens and Leibniz. The currently popular New Mechanism too constitutes a general framework for understanding science and nature, which emerged and spread in philosophy of science in the beginning of the twenty-first century. In this chapter, we look at some main aspects of the historical development of the concept of mechanism, by examining the differences and similarities between New and Old Mechanism. The motivation comes from the explicit intention of new mechanists to link the current mechanical philosophy with its older counterpart in the seventeenth century (see Machamer et al. 2000). However, there are several questions to be asked about the real relationship between the new and old mechanical philosophy.

A key similarity between Old and New Mechanism is that, for many old and new mechanists, Mechanism is both a metaphysical position about the structure of reality and a methodological thesis about the general form of scientific explanation and the methodology of science. The main aim of this chapter is to examine the relation between the metaphysics of mechanisms and the methodological role of mechanical explanation in the practice of science, by presenting and comparing the key tenets of Old and New Mechanism. We will use this historical examination to motivate our deflationary account of mechanism developed in the book.

1.2 Old versus New Mechanism

The mechanical world view of the seventeenth century was both a metaphysical thesis and a scientific theory. It was a metaphysical thesis insofar as it was committed to a reductionist account of all worldly phenomena to configurations of matter in motion subject to laws. In particular, it was committed to the view that all macroscopic phenomena are caused, and hence are accounted for, by the interactions of invisible microscopic material corpuscles. Margaret Wilson captured this view succinctly:

> The mechanism characteristic of the new science of the seventeenth century may be briefly characterised as follows: Mechanists held that all macroscopic bodily phenomena result from the motions and impacts of submicroscopic particles, or corpuscles, each of which can be fully characterised in terms of a strictly limited range of (primary) properties: size, shape, motion and, perhaps, solidity and impenetrability. (1999, xiii–xiv)

But this metaphysical thesis did, at the same time, license a *scientific theory* of the world, namely, a certain conception of scientific explanation and of theory construction. To offer a scientific explanation of a worldly phenomenon X was to provide a configuration Y of matter in motion, subject to laws, such that Y could cause X. A mechanical explanation then was (a species of) *causal explanation*: to explain that Y causes X was tantamount to constructing a mechanical model of how Y brings about X. The model was mechanical insofar as it was based on resources licensed by the metaphysical world view, namely, action of particles by contact in virtue of their primary qualities and subject to laws of motion.[1]

Nearly four centuries later, the mechanical world view has become prominent again within philosophy of science. It's become known as the 'New Mechanical Philosophy' or the 'New Mechanism' and has similar aspirations as the old one. New Mechanism, as Stuart Glennan puts it, 'says of nature that most or all the phenomena found in nature depend on mechanisms – collections of entities whose activities and interactions, suitably organized, are responsible for these phenomena. It says of science that its chief business is the construction of models that describe, predict, and explain these mechanism-dependent phenomena' (2017, 1).

So, New Mechanism too is a view about both science *and* the metaphysics of nature. And yet, in New Mechanism the primary focus has been on scientific practice, and in particular on the use of mechanisms in discovery, reasoning and representation (see Glennan 2017, 12). The focus on the metaphysics of mechanisms has emerged as an attempt to draw conclusions about the ontic signature of the world starting from the

---

[1] For the purposes of this chapter, we ignore issues of mind-body causation and we focus on body-body causation. We also ignore divisions among mechanists concerning the nature of corpuscles, the existence of vacuums, etc.

concept of mechanism as it is used in the sciences. According to Glennan, as the research into the use of mechanism in science developed, 'it has been clear to many participants in the discussion that metaphysical questions are unavoidable' (2017, 12). It is fair to say that New Mechanism aims to ground the metaphysics of mechanisms on the practice of mechanical explanation in the sciences.

1.3   Old Mechanism: From Metaphysics to Practice

A rather typical example of the interplay between the metaphysical world view and the scientific conception of the world in the seventeenth century was the attempted mechanical explanation of gravity.

1.3.1   Mechanical Models of Gravity

Let us start with Descartes. The central aim of the third and fourth parts of Descartes's *Principia Philosophiae*, published in 1644, was the construction of an account of natural phenomena. In Cartesian physics, the possible empirical models of the world are restricted from above by a priori principles which capture the fundamental laws of nature and from below by experience. Between these two levels there are various theoretical hypotheses, which constitute the proper empirical subject matter of science. These are mechanical hypotheses; they refer to configurations of matter in motion. As Descartes explains in Part III, 46, of the *Principia*, since it is a priori possible that there are countless configurations of matter in motion that can underlie the various natural phenomena, 'unaided reason' is not able to figure out the right configuration of matter in motion. Mechanical hypotheses are necessary but experience should be appealed to, in order to pick out the correct one: '[W]e are now at liberty to assume anything we please [about the mechanical configuration], provided that everything we shall deduce from it is {entirely} in conformity with experience' (III, 46; 1982, 106).

These mechanical hypotheses aim to capture the putative causes of the phenomena under investigation (III, 47). Hence, they are explanatory of the phenomena. Causal explanation – that is, mechanical explanation – proceeds via *decomposition*. It is a commitment of the mechanical philosophy that the behaviour of observable bodies should be accounted for on the basis of the interactions among their constituent parts and particles, hence, on the basis of unobservable entities. Descartes states (IV, 201; 1982, 283) that sensible bodies are composed of insensible particles. But to get to know these particles and their properties a *bridge principle* is necessary, that is, a principle that connects the micro-constituents with the macro-bodies. According to this principle, the properties of the minute particles should be modelled on the properties of macro-bodies. Here is how Descartes put it:

> Nor do I think that anyone who is using his reason will be prepared to deny that it is far better to judge of things which occur in tiny bodies (which escape our senses solely because of their smallness) on the model of those which our senses perceive occurring in large bodies, than it is to devise I know not what new things, having no similarity with those things which are observed, in order to give an account of those things [in tiny bodies]. {E. g., prime matter, substantial forms, and all that great array of qualities which many are accustomed to assuming; each of which is more difficult to know than the things men claim to explain by their means}. (IV, 201; 1982, 284)

In this passage Descartes does two things. On the one hand, he advances a *continuity thesis*: it is simpler and consonant with what our senses reveal to us to assume that the properties of micro-objects are the same as the properties of macro-objects. This continuity thesis is primarily *methodological*. It licenses certain kinds of explanations: those that endow matter in general, and hence the unobservable parts of matter, with the properties of the perceived bits of matter. It therefore licenses as explanatory certain kinds of unobservable configurations of matter, namely, those that resemble perceived configurations of matter. On the other hand, however, Descartes circumscribes mechanical explanation by noting *what it excludes*, that is, by specifying what does not count as a proper scientific explanation. He's explicit that the Aristotelian-scholastic metaphysics of substantial forms and powerful qualities is precisely what is abandoned as explanatory by the mechanical philosophy.[2]

All this was followed in the investigation of the mechanism of gravity and the (in)famous vortex hypothesis according to which the planets are carried by vortices around the sun. A vortex is a specific configuration of matter in motion – matter revolving around a centre. The underlying mechanism of the planetary system then is a system of vortices:

> [T]he matter of the heaven, in which the Planets are situated, unceasingly revolves, like a vortex having the Sun as its center, and . . . those of its parts

---

[2] Descartes accepts that scientific explanation does not require the truth of the claims about the micro-constituents of things (IV, 204; 1982, 286). In the next paragraph, however, he argues that his explanations have 'moral certainty' (IV, 205; 1982, 286–7).

which are close to the Sun move more quickly than those further away; and ... all the Planets (among which we {shall from now on} include the Earth) always remain suspended among the same parts of this heavenly matter. (III, 30; 1982, 96)

The very idea of this kind of configuration is suggested by experience, and by means of the bridge principle it is transferred to the subtle matter of the heavens. Hence, invisibility doesn't matter. The bridge principle transfers the explanatory mechanism from visible bodies to invisible bodies. More specifically, the specific continuity thesis used is the motion of 'some straws {or other light bodies} ... floating in the eddy of a river where the water doubles back on itself and forms a vortex as it swirls'. In this kind of motion we can see that the vortex carries the straws 'along and makes them move in circles with it'. We also see that 'some of these straws rotate about their own centers, and that those which are closer to the center of the vortex which contains them complete their circle more rapidly than those which are further away from it'. More importantly for the explanation of gravity, we see that 'although these whirlpools always attempt a circular motion, they practically never describe perfect circles, but sometimes become too great in width or in length.' Given the continuity thesis, we can transfer this mechanical model to the motion of the planets and 'imagine that all the same things happen to the Planets; and this is all we need to explain all their remaining phenomena' (III, 30; 1982, 96). Notably, the continuity thesis offers a heuristic for discovering plausible mechanical explanations.

Huygens (1690/1997) came to doubt the vortex theory, 'which formerly appeared very likely' to him (p. 32). He didn't thereby abandon the key tenet of mechanical philosophy. For Huygens too the causal explanation of a natural phenomenon had to be mechanical. He said, referring to Descartes: 'Mr Descartes has recognized, better than those that preceded him, that nothing will be ever understood in physics except what can be made to depend on principles that do not exceed the reach of our spirit, such as those that depend on bodies, deprived of qualities, and their motions' (pp. 1–2).

Huygens posited a fluid matter that consists of very small parts in rapid motion in all directions and which fills the spherical space that includes all heavenly bodies. Since there is no empty space, this fluid matter is more easily moved in circular motion around the centre, but not all parts of it move in the same direction. As Huygens put it 'it is not difficult now to explain how gravity is produced by this motion' (p. 16). When the parts of the fluid matter encounter some bigger bodies, like the planets, 'these bodies [the planets] will necessarily be pushed towards the center of motion, since they do not follow the rapid motion of the aforementioned matter'. And he added: 'This then is in all likelihood what the gravity of bodies truly consists of: we can say that this is the endeavor that causes the fluid matter, which turns circularly around the center of the Earth in all directions, to move away from the center and to push in its place bodies that do not follow this motion.' In fact, Huygens devised an experiment with bits of beeswax to show how this movement towards the centre can take place.

Newton of course challenged all this, along the lines that the very idea of causal explanation should be *mechanical*. But before we take a look at his reasons and their importance for the very idea of mechanical explanation, we should not fail to see the broader metaphysical grounding of the mechanical project. For, as we noted, in the seventeenth century mechanism offered the metaphysical foundation of science.

1.3.2   Mechanical versus Non-Mechanical Explanation

The contours of this endeavour are well known. Matter and motion are the 'ultimate constituents' of nature, or, as Boyle (1991, 20) put it, the 'two grand and most catholic principles of bodies'. Hence, all there is in nature (but clearly not the Cartesian minds) is determined (caused) by the mechanical affections of bodies and the mechanical laws. Here is Boyle again: '[T]he universe being once framed by God, and the laws of motion being settled and all upheld by his incessant concourse and general providence, the phenomena of the world thus constituted are physically produced by the mechanical affections of the parts of matter, and what they operate upon one another according to mechanical laws' (1991, 139).

The Boylean conception, pretty much like the Cartesian, took it that the new mechanical approach acquired content by excluding the then dominant account of explanation in terms of 'real qualities': the scholastics 'attribute to them a nature distinct from the modification of the matter they belong to, and in some cases separable from all matter whatsoever' (pp. 15–16). Explanation based on real qualities, which are distinct (and separable) from matter, is not a genuine explanation. They are posited without 'searching into the nature of particular qualities and their effects' (p. 16). They offer sui generis explanations: why does snow dazzle the eyes? Because of 'a quality of whiteness that is in it, which makes all very white bodies produce the same effect' (p. 16). But what is whiteness? No further story about its nature is offered, but just that it's a 'real entity' inhering in the substance. Why do white objects produce this effect rather than that? Because it is in their nature to act thus.

Descartes made this point too when, in his *Le Monde*, he challenged the scholastic rivals to explain how fire burns wood, if not by the incessant and rapid motion of its minute parts. In his characteristic upfrontness, Descartes contrasted two ways to explain how fire burns wood. The first is the Aristotelian way, according to which 'the 'form' of fire, the 'quality' of heat and the 'action' of burning' are 'very different things in the wood' (Descartes 2004, 6). The other is his own mechanistic way: when the fire burns wood, 'it moves the small parts of the wood, separating them from one another, thereby transforming the finer parts into fire, air, and smoke, and leaving the larger parts as ashes' (2004, 6).

This causal explanation, based as it is on matter in motion, is preferable precisely because it is explanatory of the burning; in contrast, the Aristotelian is not, precisely because it does not make clear the mechanism by which the fire consumes the wood: '[Y]ou can posit "fire" and "heat" in the wood and make it burn as much as you please: but if you do not suppose in addition that some of its parts move or are detached from their neighbours then I cannot imagine that it would undergo any alteration or change' (p. 6). To the then dominant account of real qualities, the new mechanical metaphysics juxtaposed a different view of qualities. For something to be a quality it should be determined by the mechanical affections of matter, that is, by 'virtue of the motion, size, figure, and contrivance, of their own parts' (Boyle 1991, 17). Hence, there can be no change in qualities unless there is a change in mechanical affections. Though 'catholic or universal matter' is common to all bodies (being, as Boyle [p. 18] put it, 'a substance extended, divisible, and impenetrable'), it is diversified by motion, which is regulated by laws.

The key point then is that the mechanical account of nature is both a metaphysical grounding of science *and* a (the) way to do science: offering mechanical explanations of the phenomena. It covers everything, from the very small to the very large. Here is Boyle again: 'For both the mechanical affections of matter are to be found, and the laws of motion take place, not only in the great masses and the middle-sized lumps, but in the smallest fragments of matter; and a lesser portion of it, being as well a body as a greater, must, as necessarily as it, have its determinate bulk and figure' (p. 143).

The metaphysical grounding of mechanical explanation renders it a distinct kind of explanation, which separates it sharply from rival accounts. Concomitantly, it becomes very clear what counts as a non-mechanical alternative. An explanation couched in terms of 'nature, substantial forms, real qualities, and the like' is 'unmechanical' (p. 142). But a sui generis chemical account of nature is unmechanical too. As Boyle put it:

> [T]hough chemical explications be sometimes the most obvious and ready, yet they are not the most fundamental and satisfactory: for the chemical ingredient itself, whether sulphur or any other, must owe its nature and other qualities to the union of insensible particles in a convenient size, shape, motion or rest, and contexture, all which are but mechanical affections of convening corpuscles. (p. 147)

The opposition to both of these non-mechanical accounts is weaved around a certain metaphysical account of the world as fundamentally mechanical and a reductive-decompositional account of scientific explanation itself.

1.3.3   Boyle on Mechanical Explanation

Boyle's discussion of the nature of mechanical explanation in his 'About the Excellency and Grounds of the Mechanical Hypothesis' deserves further analysis, as it is particularly relevant for our purposes in this book. In that essay, Boyle contrasts mechanical with other kinds of explanations and points out that only the former provide information about how exactly a result is produced. What is particularly interesting for us is that Boyle focuses on what can be broadly described as 'medical' examples. Consider the following passage:

> They that, to solve the phenomena of nature, have recourse to agents which, though they involve no self-repugnancy in their very notions, as many of the judicious think substantial forms and real qualities to do, yet are such that we conceive not how they operate to bring effects to pass – these, I say, when they tell us of such indeterminate agents as the soul of the world, the universal spirit, the plastic power, and the like, though they may in certain cases tell us some things, yet they tell us nothing that will satisfy the curiosity of an inquisitive person, who seeks not so much to know what is the general agent that produces a phenomenon, as *by what means, and after what manner, the phenomenon is produced.* (p. 144, emphasis added)

Here, Boyle points out that, in giving an explanation, it is not enough to state what the cause is; what is more important is to state how exactly a cause operates to bring about the effect. Failure to do this, Boyle thinks, is the main problem with explanations that merely state a causal agent without providing further information as to the manner that this agent acts. Boyle goes on to give an example of such an unsatisfactory medical explanation:

> The famous Sennertus and some other learned physicians tell us of diseases which proceed from incantation: but sure it is but a very slight account that a sober physician, that comes to visit a patient reported to be bewitched, receives of the strange symptoms he meets with and would have an account of, if he be coldly answered that it is a witch or the devil that produces them. (p. 144)

Similarly,

> it would be but little satisfaction to one that desires to understand the causes of what occurs to observation in a watch, and how it comes to point at and strike the hours, to be told that it was such a watchmaker that so contrived it; or to him that would know the true cause of an echo to be answered that it is a man, a vault, or a wood, that makes it. (p. 144)

The point that Boyle makes in these passages is that, quite apart from the accusation that notions such as substantial forms and real qualities are obscure, such explanations as well as others (e.g., in terms of plastic powers, which Boyle thinks are not as bad as the ones offered by the scholastics) do not fulfil what he takes as *a general adequacy condition* that an explanation has to satisfy, that is, to provide *information* as to how exactly a cause acts. Consider, for example, the following causal claim: administering of the poison led to the death of the person. We can interpret Boyle as saying that, qua explanation of death, such an explanation is incomplete; what is missing is how exactly administering of the poison led to death. Stating that the poison possessed the power to bring about death is tantamount to saying that it in fact produced death, that is, that it was the cause of death. What we need in addition to this, however, is *the way* it did so. A way to satisfy this demand is to provide information about the changes that the poison produced within the organism and explain how they eventually led to death. Here is how Boyle puts it:

> I consider that the chief thing that inquisitive naturalists should look after in the explicating of difficult phenomena is not so much what the agent is or does, as what changes are made in the patient to bring it to exhibit the phenomena that are proposed, and *by what means, and after what manner, those changes are effected*: so that, the Mechanical philosopher being satisfied that one part of matter can act upon another but by virtue of local motion or the effects and consequences of local motion, he considers that as, if the proposed agent be not intelligible and physical, it can never physically explain the phenomena, so, *if it be intelligible and physical, it will be*

reducible to matter and some or other of those only catholic affections of matter
already often mentioned. (p. 145, emphasis added)

According to Boyle, then, what an 'inquisitive person' should do in
offering an explanation of how an outcome such as death by poison comes
about is to describe the series of changes that led from the event that
counts as the cause to the resulting outcome. Moreover, such an explanation should be given in mechanical terms. The reason is that only by
means of local motion can we understand how a cause (i.e., a part of
matter) can operate on something (i.e., another part of matter). So, non-
physical causes cannot explain (since they do not act by means of local
motion), and for physical causes to be explanatory, they have to be reduced
to matter and to the 'catholic affections' of matter, including motion. The
requirement that an adequate explanation has to state the means by which
the cause acts is here supplemented by the further requirement that the
account of how exactly it acts has to be given in mechanical terms. We can
view the resulting account of explanation as a combination of a methodological thesis (i.e., that an explanation should state how exactly the cause
acts) with an ontological one (i.e., that it should be given in mechanical
terms). Boyle goes on:

> whatever be the physical agent, whether it be inanimate or living, purely
> corporeal or united to an intellectual substance, the above-mentioned
> changes, that are wrought in the body that is made to exhibit the phenom-
> ena, may be effected by the same or the like means, or after the same or the
> like manner ... And if an angel himself should work a real change in the
> nature of a body, it is scarce conceivable to us men how he could do it
> *without the assistance of local motion*, since, if nothing were displaced or
> otherwise moved than before (the like happening also to all external bodies
> to which it related), it is hardly conceivable how it should be in itself other
> than just what it was before. (p. 146, emphasis added)

Boyle argues here that since real change requires local notion, a mechanical
explanation can always be given no matter what the exact nature of the
agent is. Even if the agent is immaterial, to produce a certain outcome is to
produce a series of changes in local motion. This leads, finally, to the
following point:

> From the foregoing discourse it may (probably at least) result that if, besides
> rational souls, there are any immaterial substances (such as the heavenly
> intelligences and the substantial forms of the Aristotelians) that regularly are
> to be numbered among natural agents, their way of working being
> unknown to us, they can but help to constitute and effect things, but will
> very little help us to conceive how things are effected: so that, *by whatever*

principles natural things be constituted, it is by the Mechanical principles that
their phenomena must be clearly explicated. (p. 150, emphasis added)

Boyle here points out that even if substantial forms were to be accepted, they (as well as other immaterial substances) are useless if our aim is to understand how the phenomena are brought about. The thought here is that, in order to explain phenomena, we have to explain how they come about; but we can do this only by stating how the various causal agents produce changes in the properties of parts of matter by means of local motion; we cannot conceive how substantial forms can result in local motion; so, substantial forms are useless in offering explanations of how phenomena are produced.

In sum, Boyle's main thought is that a satisfactory causal explanation has to explain the way a cause acts in bringing about a certain phenomenon. To do this, a causal explanation has to be mechanical, where a mechanical explanation explains how the effect is brought about in terms of changes in the mechanical properties of parts of matter, including local motion. As we will explain in Section 1.4, in giving our account of mechanism and mechanistic explanation, we will be in agreement with Boyle's insights. But whereas we will keep his methodological thesis (i.e., that in giving a mechanistic explanation in science one has to explain how exactly a cause acts by describing the sequence of causal steps leading from the cause to the effect), we will reject the ontological thesis (i.e., that there exists a privileged description of this causal sequence, either in physico-chemical terms or in terms of one's favourite metaphysics of causal processes).

1.3.4 Newton against Mechanism

When Newton offered a non-mechanical account of gravity, he primarily challenged the idea that legitimate scientific explanation ought to be mechanistic, at least in the narrow sense of taking all action to be by *contact*. There is a sense in which Newton prioritised explanation by unification under laws and not by mechanisms. This is seen in the *Preface* to the second (1713) edition of the *Principia*, authored by Roger Cotes under the supervision of Newton. In this preface, Cotes presents Newton's method as a middle way (*via media*) between Aristotelianism and Mechanism. To be sure, the mechanical explanations offered by the Cartesians were an improvement over the Scholastic explanations because they relied on demonstrations on the basis of laws. Still, taking 'the foundation of their speculations from hypotheses', the mechanists are 'merely putting together a romance [i.e., fiction], elegant perhaps and charming, but nevertheless a romance' (Newton 2004, 43).

Thus put, the point sounds epistemic; it concerns the increased risk involved in hypothesising a mechanism which is supposed to underpin, and hence to causally explain, a certain phenomenon. Cotes adds:

> But when they [the mechanists] take the liberty of imagining that the unknown shapes and sizes of the particles are whatever they please, and of assuming their uncertain positions and motions, and even further of feigning certain occult fluids that permeate the pores of bodies very freely, since they are endowed with an omnipotent subtlety and are acted on by occult motions: when they do this, they are drifting off into dreams, ignoring the true constitution of things, which is obviously to be sought in vain from false conjectures, when it can scarcely be found out even by the most certain observations. (p. 43)

Still, it's fair to say that Newton's *via media* was based on a different understanding of scientific explanation: it should certainly look for causes – hence, scientific explanation should be causal – but the sought-after causes need not act by the principles of Mechanism. Newton's way, Cotes says, is to 'hold that the causes of all things are to be derived from the simplest possible principles', but unlike the mechanists' way, it 'assume(s) nothing as a principle that has not yet been thoroughly proved from phenomena'. The 'explication of the system of the world most successfully deduced from the theory of gravity' is the 'most illustrious' example of Newton's way (p. 32).

Newton emphatically denied feigning any hypotheses about the cause of gravity. For him, 'it is enough that gravity really exists and acts according to the laws that we have set forth and is sufficient to explain all the motions of the heavenly bodies and of our sea' (p. 92). Gravity according to Newton is a non-mechanical force since it 'acts not in proportion to the quantity of the *surfaces* of the particles on which it acts (as mechanical causes are wont to do) but in proportion to the quantity of *solid* matter, and whose action is extended everywhere to immense distances, always decreasing as the squares of the distances' (p. 92). He added that the very motion of the comets makes it plausible to think that the regular elliptical motion of the planets (as well as of their satellites) cannot 'have their origin in mechanical causes' (p. 90).

In his already mentioned *Discourse on the Cause of Gravity* (1690), Huygens expressed his dissatisfaction with Newton's failure to offer a *mechanical* explanation of the cause of gravitational attraction. Favouring his own explanation of gravity in terms of the centrifugal force of the subtle and rapidly moving matter that fills the space around the Earth and the other planets, Huygens noted that Newton's theory supposes that gravity is 'an inherent quality of corporeal matter'. 'But', he immediately added, such a hypothesis 'would distance us a great deal from mathematical or mechanical principles' (1690/1997, 35).

Yet Huygens had no difficulty in granting that Newton's law of gravity was essentially correct when it comes to accounting for the planetary system. As he put it:

> I have nothing against *Vis Centripeta*, as Mr. Newton calls it, which causes the planets to weigh (or gravitate) toward the Sun, and the Moon toward the Earth, but here I remain in agreement without difficulty because not only do we know through experience that there is such a manner of attraction or impulse in nature, but also that it is explained by the laws of motion, as we have seen in what I wrote above on gravity. (p. 31)

Explaining the fact that gravity depends on the masses and diminishes with distance 'in inverse proportion to the squares of the distances from the centre' (p. 37) was, for Huygens, a clear achievement of Newton's theory despite the fact that the mechanical cause of gravity remained unidentified.

Commitment to mechanical explanation was honoured by Gottfried Wilhelm Leibniz too. In a piece titled 'Against Barbaric Physics: Toward a Philosophy of What There Actually Is and against the Revival of the Qualities of the Scholastics and Chimerical Intelligences' (written between 1710 and 1716), he defended the mechanical view by arguing that corporeal forces should be grounded mechanically when it comes to their application to the natural world. Leibniz was very clear that though he allowed 'magnetic, elastic, and other sorts of forces', they are permissible 'only insofar as we understand that they are not primitive or incapable of being explained, but arise from motions and shapes' (Leibniz 1989, 313). So, forces are necessary, but a condition for their applicability to the natural world is that they are seen as 'arising from motions and shapes'. What he took it to be 'barbarism in physics' was to posit sui generis, that is non-mechanically grounded, 'attractive and repulsive' forces that act at a distance (pp. 314–15). Newton's gravity was supposed to be such a barbaric force!

In a letter he sent to Nicolaas Hartsoeker (Hanover, 10 February 1711), Leibniz makes it clear that the proper scientific explanation should be mechanical. It is not enough for scientific explanation to identify the law by means of which a certain force acts; what is also required is the specification of the mechanism by means of which it acts. The mechanism is, clearly, on top of the law and given independently of it. Without the mechanism the power is 'an unreasonable occult quality'. He says:

> Thus the ancients and the moderns, who own that gravity is an occult quality, are in the right, if they mean by it that there is a certain mechanism unknown to them, whereby all bodies tend towards the center of the earth. But if they mean that the thing is performed without any mechanism by a simple primitive quality, or by a law of God, who produces that effect without using any intelligible means, it is an unreasonable occult quality, and so very occult, that it is impossible it should ever be clear, though an angel, or God himself, should undertake to explain it. (Newton 2004, 112)

Newton couldn't disagree more. In an unsent letter written circa May 1712 to the editor of the *Memoirs of Literature*, Newton referred explicitly to Leibniz's letter to Hartsoeker and stressed that it is not necessary for the introduction of a power – such as gravity – to specify anything other than the law it obeys; no extra requirements should be imposed, and in particular no requirement for a mechanical grounding:

> And therefore if any man should say that bodies attract one another by a power whose cause is unknown to us, or by a power seated in the frame of nature by the will of God, or by a power seated in a substance in which bodies move and float without resistance and which has therefore no vis inertiae but acts by other laws than those that are mechanical: I know not why he should be said to introduce miracles and occult qualities and fictions into the world. For Mr. Leibniz himself will scarce say that thinking is mechanical as it must be if to explain it otherwise be to make a miracle, an occult quality, and a fiction. (Newton 2004, 116)

Note well Newton's point. The fact that an explanation does not conform to a certain mechanical framework does not make it fictitious, occult or miraculous. Non-mechanical explanations are legitimate insofar as they identify the law that covers or governs a certain phenomenon. Hence, Newton promotes a methodological shift: causal explanation without mechanisms but subject to laws.

Causal explanation then need not be mechanical to be legitimate and adequate. This is Newton's key thought. In breaking with a tradition which brought under the same roof a certain metaphysical conception of the world and a certain view of scientific explanatory practice, Newton distinguished the two and laid emphasis on the explanatory practice itself, thereby freeing it from a certain metaphysical grounding.

Though this is not the end of the story of Old Mechanism (more will be said in the next chapter), Newton's key thought, we shall argue, is of relevance in the current debates over New Mechanism, to which we shall now turn our attention.

1.4 New Mechanism: From Practice to Metaphysics

It is useful to differentiate between two ways to conceptualise mechanisms in the post-1970 literature. First, mechanism has been used as a primarily metaphysical concept, mostly aiming to illuminate the metaphysics of causation. Second, mechanism has been taken to be a concept used in science, and philosophical accounts of mechanism have aimed to elucidate this concept.

To be sure, some philosophical approaches to mechanism, most notably Glennan's (1996), blend these two conceptions (the metaphysical one and the concept-in-use). However, it's fair to say that there are two quite distinct points of origin of the recent philosophical accounts of mechanism: the first starts from metaphysics (as was the case for Descartes and other old mechanists), the second from scientific practice. Using this distinction between mechanism as a primarily metaphysical concept and as a concept-in-use in science, we can differentiate between two kinds of approaches to the metaphysics of mechanisms.

On the first approach, the aim is to show what the connection is between mechanism qua a metaphysical category and other central metaphysical concepts, notably, causation. In the context of the metaphysics of causation, 'mechanistic' accounts are theories about the link between cause and effect. Such theories are meant to be anti-Humean in that they view causation as a productive relation; that is, the cause somehow brings about or produces the effect. The aim of the mechanistic view of causation is to illuminate the productive relation between the cause and the effect by positing a mechanism that connects them and by explicating 'mechanism' in a suitable way such that causal sequences are differentiated from non-causal ones. The central thought, then, is that A causes B if and only if there is a mechanism connecting A and B.

Two kinds of views have become prominent: those that characterise the mechanism that links cause and effect in terms of the persistence, transference or possession of a conserved quantity (Mackie 1974; Salmon 1997; Dowe 2000) and those that connect a mechanistic account to causal production with a power-based one (see Harré 1970 for an early such view). Despite their differences, these views share in common the claim that mechanisms are the ontological tie that constitutes Hume's 'secret connexion'. We call such mechanisms *mechanisms-of*. Mechanisms-of are ontological items that underlie or constitute certain kind of processes, that is, those that can be deemed causal. We will deal with these accounts in more detail in Chapter 5 (see also Psillos 2002).

On the second approach, working out a metaphysics of mechanisms is not the starting point but rather the end point of inquiry. Starting with mechanism as a concept-in-use in science, one tries first to give a general characterisation of this concept and then to derive metaphysical conclusions, that is, conclusions about the (mechanistic) structure of the world. This kind of bottom-up inquiry has yielded several well-known general accounts of mechanisms as well as theses about the ontic signature of a mechanistic world.

1.4.1   The Metaphysics of New Mechanism

Here are three well-known general characterisations of a mechanism in recent mechanistic literature:

> Mechanisms are entities and activities organized such that they are productive of regular changes from start or set-up to finish or termination conditions. (Machamer et al. 2000, 3)

> A mechanism for a behavior is a complex system that produces that behavior by the interaction of a number of parts, where the interactions between parts can be characterized by direct, invariant, change-relating generalizations. (Glennan 2002, S344)

> A mechanism is a structure performing a function in virtue of its component parts, component operations, and their organization. The orchestrated functioning of the mechanism is responsible for one or more phenomena. (Bechtel & Abrahamsen 2005, 423)

The focus on mechanism as a concept-in-use is common to all three accounts; none of the three accounts can be viewed as falling under the rubric of mechanistic theories of causation. And yet, all these and similar accounts yield specific metaphysical commitments about what kind of things in the world mechanisms are. All these accounts are committed to the thesis that a general characterisation of mechanism must itself be cashed out in metaphysical terms.[3] Hence, talk of mechanisms in science is taken to have quite direct consequences about the kind of ontology

---

[3] As Bechtel and Abrahamsen are more interested in offering an account of explanation rather than saying what mechanisms are as things in the world, their characterisation of mechanism can be read in a metaphysically deflationary way.

presupposed by such talk. In order to substantiate this point, let us look at the three accounts mentioned earlier in some more detail.

Peter Machamer, Lindley Darden and Carl Craver's (henceforth MDC) account is perhaps the most ontologically inflated, as it is explicitly committed to both entities *and* activities as distinct and separate ontological categories. It is thus committed to a particular view about the metaphysics of causation: causation within mechanisms is to be characterised in terms of *production*, where the productive relation is captured by the various different kinds of activities identified by science.

Glennan's case is interesting, since in his 2002 article he refrains from taking mechanisms to entail a productive account of causation. Instead, within-mechanism interactions are characterised in terms of invariant, change-relating generalisations. As we will see below, however, Glennan has connected his account of mechanisms with a power-based understanding of causation. Hence, he is committed to causal powers as parts of the building blocks of mechanisms.

Last, Bechtel and Abrahamsen's account does not include a specific characterisation of what mechanistic causation amounts to at all. Here, however, as in the other two accounts, we have a series of general terms, the meaning of which needs to be unpacked. So, MDC include in their accounts 'entities' and 'organisation'; Glennan in his early formulations includes 'complex system' and 'parts'; and Bechtel and Abrahamsen talk about 'structure', 'function', 'parts' and 'organisation'.

All these accounts suggest the further need to explain what this 'new mechanical ontology' of entities, activities, organisation of parts into wholes and so on amounts to; what, in general terms, the constituents of mechanisms are and what their relations are to more traditional metaphysical categories, such as things, properties, powers and processes.

Notably, there has been a tendency recently to offer a more minimal general characterisation of a mechanism. For example, according to Illari and Williamson:

> A mechanism for a phenomenon consists of entities and activities organized in such a way that they are responsible for the phenomenon. (2012, 120)

Glennan's recent version is almost identical:

> A mechanism for a phenomenon consists of entities (or parts) whose activities and interactions are organised so as to be responsible for the phenomenon. (2017, 17)

Glennan calls this account *Minimal Mechanism*. The key motivation here is for a general characterisation of mechanism broad enough to capture examples of mechanisms in different fields, from physics to the social sciences. But even in this minimal mode, mechanisms, according to Glennan, 'constitute the causal structure of the world' (2017, 18).

This minimal account of mechanism might appear to fit the bill of capturing a concept-in-use in science. On closer inspection, however, it is committed to a rather rich metaphysical account of mechanism: the minimal account is not more minimal than the metaphysically inflated accounts noted above. The reason is that both of the foregoing minimal accounts still invite questions about the ontic status of mechanisms. For example, how exactly do entities and activities differ? What is the relation between activities and interactions? How should organisation be understood? Glennan (2017, 13) explicitly talks about a 'new mechanical ontology' as the upshot of the minimal account. The 'minimal mechanism', he adds, 'is an ontological characterization of what mechanisms are as things in the world' (p. 19).

New Mechanism, then, aims to provide a new ontology of mechanisms. We can identify three commonly accepted key theses concerning mechanical ontology:

(1)   The world consists of mechanisms.

Thesis 1 is a typical view among mechanists: mechanisms are taken to be *things in the world*, with (more or less) objective boundaries.[4] Ours is a mechanistic world. As Glennan puts it, '[t]hat is just how we have found the world to be' (2017, 240).

(2)   A mechanism consists of objects of diverse kinds and sizes structured in such a way that, in virtue of their properties and capacities, they engage in a variety of different kinds of activities and interactions such that a certain phenomenon P is brought about.

Thesis 2 (or something very similar) can be taken as the common core of the general characterisations of mechanism as a concept-in-use given by new mechanists. It identifies the components of a mechanism and the relations among them. As mechanisms are things in the world (thesis 1), their components are also particular things in the world. Besides, these parts engage in activities by being 'active, at least potentially' (Glennan 2017, 21). Activity is understood as a manifestation of the powers things have. Glennan is quite explicit that 'Activities manifest the powers

---

[4] The issue of boundaries is far from settled in the literature; we will come back to this issue in Chapter 8.

(capacities) of the entities involved in the Activity' (p. 31). Positing powers is supposed to explain why 'activities are powerful'; being powerful, activities are what 'an entity does, not merely something that happens to an entity' (p. 32). But activities are not enough. Interactions are needed too because 'there is no production without interaction' (p. 22). 'The fundamental point of ontological agreement among the New Mechanists', as Glennan (2017, 21 n. 6) puts it, is that entities cannot exist without activities or activities without entities. It's not hard to see that the minimal account of mechanism is taken to imply or suggest a rather substantive metaphysical conception of mechanism, which, until further notice at least, is broadly neo-Aristotelian.

(3)   The main way to explain a certain phenomenon P is to offer the mechanism that produces it.

Thesis 3 connects the previous theses with a claim about explanation (and more specifically, causal explanation): since in a mechanistic world phenomena are produced by mechanisms, the main task of scientific explanation is to identify the mechanism that produces a certain phenomenon, that is (by thesis 2), to identify the organised entities and activities that produce the phenomenon.

Despite their differences, there are important similarities between Old and New Mechanism (which justify viewing both positions as mechanistic). On the one hand, as we saw, new mechanists differ from their seventeenth-century predecessors in that they do not start their analysis with a metaphysical concept of mechanism; rather, they aim at giving a general characterisation of mechanism as a central concept of scientific practice. This characterisation is non-reductive in that it is not committed to the view that mechanisms are configurations of matter in motion subject to laws (and contact action). But, on the other hand, they are committed to mechanisms being configurations of powerful entities engaged in activities and interactions. As Glennan puts it: 'Mechanisms are particular and compound, made up of parts (entities) whose activities and interactions are located in particular regions of space and time' (2017, 57). Hence, New Mechanism is similar to seventeenth-century Mechanism, in that it is committed to a mechanical *ontology*. This ontology (theses 1 and 2 above), while not a global metaphysics in the sense of the seventeenth century, is still a thesis about the ontic signature of the world. Here is Glennan again: 'New Mechanist ontology is an ontology of compound systems. It suggests that the properties and activities of things must be explained by reference to the activities and organization of their parts' (p. 57). Instead of resulting in a 'flat' ontology where everything there is consists in matter in motion, this new mechanical metaphysics ends up with a hierarchy of particular things – mechanisms – which may contain a diverse set of entities and activities, rather than the limited set endorsed by the corpuscularians, and whose productivity is grounded in causal powers, rather than in a few fundamental laws of motion.

But we can ask: Are these ontological commitments really necessary in order to understand scientific practice? Are they licensed by the practice of science? Remember here that the primary aim of new mechanists is to give a general characterisation of mechanism as a concept-in-use. So, ideally, the general account of mechanism should capture as far as possible the extension of a concept-in-use in the various sciences. The minimal account of mechanism discussed so far, though broad enough to play this role, inflates the concept-in-use by making it amenable to a certain metaphysical description of its basic components.

Note that our claim is not that the metaphysical questions are not philosophically interesting questions to ask; they are, especially if we are interested in giving an account of the ontological structure of reality. Moreover, such a kind of project has to be informed by what science has to say about the world. If, however, our aim is to understand how a specific concept – *mechanism* – is used in scientific practice, these questions seem, at least prima facie, irrelevant, especially if a general characterisation of mechanism is possible that does not include such things.

1.4.2   Mechanism in Scientific Practice

A metaphysically deflationary view of mechanism as a concept-in-use that is broad enough to capture all examples of mechanisms that we find in science seems indeed possible. This is skinnier than those accounts of mechanism offered by Illari and Williamson and by Glennan. We nonetheless claim that this skinny account is enough to capture the concept-in-use. It will be the main aim of Chapter 4 to present this account in detail, and the subsequent chapters will further illuminate various features of the view. Here we will just introduce the basic idea behind our account. This skinny or, as we will prefer to describe it, *deflationary* account of mechanism is achieved by dropping the reference to activities and interactions and by understanding mechanism as the causal pathway of a certain phenomenon, described in the language of theory. According to this account that we call *Causal Mechanism* (CM), a mechanism in science just is a causal pathway described in theoretical language:[5]

(CM) A mechanism is a theoretically described causal pathway.

The central idea behind CM stems from the Boylean insight introduced in Section 1.3.3: when scientists talk about a 'mechanism', what they try to capture is the way (i.e., the causal pathway) a certain result is produced. Say, for example, that a pathologist tries to find out how a certain disease state is brought about. They will look for a specific mechanism, that is, a causal pathway that involves various causal links between, for example, a virus infection and changes in properties of the organism that ultimately lead to the disease state. In pathology such causal pathways are referred to as the 'pathogenesis' of a disease, and when pathologists talk about the 'mechanisms' of a disease, it is such pathways that they have in mind (see Lakhani et al. 2009).

According to CM, then, mechanisms and causation are closely related: when two events are causally connected, there is a mechanism (i.e., a causal pathway) that connects them and accounts for the specific way that the cause brings about the effect. Scientists succeed in identifying a mechanism, if they succeed in describing the relevant causal pathway in terms of the theoretical language of the particular scientific field. An especially clear example of the identification of a new mechanism is the case of the mechanism of cell death known as apoptosis; we will examine this case in detail in the next chapter.

The view of mechanisms as causal pathways differentiates CM from accounts that explicitly view mechanisms as complex systems (Glennan 1996), kinds of structures (Bechtel & Abrahamsen 2005) or more generally as organised entities of some sort; it doesn't differentiate it from more processual views, such as the MDC account. CM stresses that mechanisms are not systems, but causal processes. It is therefore closer to the older Salmon-Dowe view, as well as to the Boylean conception sketched above, than some more recent accounts.

There exists, however, a very important difference between Boyle's notion of mechanical explanation and CM. As we saw earlier, for Boyle

---

[5] We have first presented *Causal Mechanism* in Ioannidis and Psillos (2017), where we have called it *Truly Minimal Mechanism* to differentiate it from Glennan's *Minimal Mechanism*. But in this book we will use 'Causal Mechanism' to describe our account, as this indicates the close relationship between mechanism and causation, where causation is used to understand what a mechanism in science is, rather than vice versa (as in mechanistic theories of causation – we will examine these theories and the relations between mechanisms and causation in detail in Chapter 5).

and other mechanical philosophers of the seventeenth century, mechanical explanations had to be couched in very specific terms, that is, in terms of the changes produced by parts of matter to the 'mechanical affections' (including motion) of other parts of matter. Thus, the methodological claim of mechanical philosophers – that is, that to explain how the phenomena are produced one should identify the mechanisms that produce them – did incorporate a claim about the specific theoretical language that such explanations should be couched in. And the main justification of this latter claim was ontological: what really exists in the world is matter that behaves according to certain laws that govern its motion. Old Mechanism, then, combined a methodological claim about the preferred form of scientific explanation with an ontological claim, that is, a claim about how the world is constructed.

In contrast to this more restricted way to conceive of mechanistic explanation, according to CM there is no privileged theoretical language in terms of which the causal pathway that produces the phenomena has to be described. This is, in one sense, in agreement with the dominant views of what a mechanism is: as both MDC (Machamer et al. 2000) and Glennan (1996) stress in the papers that first offered general characterisations of the notion of mechanism, the contemporary concept is more general than its seventeenth-century counterpart, as the parts of a mechanism interact in various ways (e.g., by chemical interactions) and thus are not 'mechanisms' in the restricted seventeenth-century sense of the term.

In another sense, however, CM is different from current accounts, as it stresses that there is no privileged ontological description of a mechanism. So, while current accounts combine the methodological claim that science should discover the causal pathways that produce the phenomena with an ontological claim about the metaphysics of mechanisms, CM deflates the metaphysics and puts the methodological claim at the centre. We shall further examine this feature of CM by revisiting Newton's views. Before this, however, let us dispel a natural but pointed objection. Recall our main thesis that 'A mechanism is a theoretically described causal pathway.' Does this invite the interpretation that a mechanism does not exist before it is theoretically described? We, of course, do not believe this; our view is that mechanisms just are independently and objectively existing causal processes. The aim of the phrase 'theoretically described' is to highlight that the causal processes that constitute mechanisms are to be described in the theoretical terms of the relevant scientific domain and not in terms of general ontological categories.

1.5 Newton Revisited

What does Newton's critique of the Old Mechanism have to do with our understanding (and criticism) of New Mechanism? In a letter to Leibniz dated 16 October 1693, Newton challenged him to offer a mechanical explanation of 'gravity along with all its laws by the action of some subtle matter' and to show that 'the motion of planets and comets will not be disturbed by this matter'. If this were available, Newton said, he would be 'far from objecting'. But no such explanation was forthcoming and Newton was happy to reiterate his view that 'since all phenomena of the heavens and of the sea follow precisely, so far as I am aware, from nothing but gravity acting in accordance with the laws described by me; and since nature is very simple . . . all other causes are to be rejected' (Newton 2004, 108–9). Newton does not simply say that causal explanation might not be mechanical. His point is that causal explanation should be liberated from the tenets of (the narrowly understood) Old Mechanism. It would not be enough to offer a mechanical account of the cause of gravity; the laws that gravity obeys should be mechanically explicable, and, as Newton repeatedly stressed, this was not forthcoming. Though causal explanation matters, it doesn't matter if it is subject to various (old) mechanical constraints.

We noted already that the new mechanical conception of nature is far from the seventeenth-century conception that everything should be accounted for in terms of (configurations of) matter in motion. So it's far from us to tar New Mechanism with the same brush as Old Mechanism. For instance, the key ontology of the old mechanical picture was justified by and large a priori, whereas the key ontology of New Mechanism is grounded in scientific practice; in this case, it is practice that constrains metaphysics. Be that as it may, we are now going to argue that there exists a kind of Newtonian move against New Mechanism too.

What is clear from the present discussion is that, regardless of the main difference noted above, the new idea of mechanism is no less metaphysically loaded than the old one. Where the seventeenth-century mechanists looked for stable arrangements of matter in motion subject to laws, the twenty-first-century mechanists look for stable arrangements of powerful entities engaged in various activities and interactions. These mechanisms are supposed to be the building blocks of nature, and the scientific task is to unravel them. They underpin 'mechanistic explanations' which, as Glennan put it, show 'how the organized activities and interactions of some set of entities cause and constitute the phenomenon to be explained' (2017, 223). Mechanistic explanation 'always involves characterizing the activities and interactions of a mechanism's parts' (p. 223). Where the seventeenth-century mechanists saw 'action by contact' as a requisite for a proper mechanical explanation, new mechanists see powers and 'activities'. Why is Newton's key thought relevant to the modern debates about mechanisms? The key thought, to repeat, was that causal explanation should identify causes and the laws that govern their action irrespective of whether or not these causes can be taken to satisfy further (mostly metaphysically driven) constraints. In other words, Newton showed that certain causal explanations of phenomena (in terms of non-mechanical forces) are both legitimate and complete insofar as they identify the right causes and are empirically grounded.

We take it that the point CM stresses, is, *mutatis mutandis*, analogous to Newton's. The point of CM is that causal explanation need not be mechanistic in the new mechanists' ontic sense and that being couched in the way new mechanists propose, causal explanation is subjected to constraints unwarranted by scientific practice. Insofar as mechanism is a concept-in-use in science, it may well be seen referring to the causal pathway of the phenomenon to be explained, couched in the language of theories. Preserving the spirit of Newton's key thought, we might say that causal explanation is legitimate even if we bracket the issue of 'what mechanisms or causes are as things in the world' (Glennan 2017, 12) or the issue of what activities are and how they are related to powers and the like. The issue then is not 'an ontological characterization of what mechanisms are as things in the world' (p. 19), but a methodological characterisation of them as causal pathways described in the language of theories.

To press the analogy a bit more, questions such as 'If entities, activities, and the mechanisms they constitute are compounds, of what are they compounded? Where does one entity or activity or mechanism end, and when does another begin? And on what account do we decide that a collection of interacting entities is to count as a whole mechanism?' (p. 29) are pretty much like the questions concerning the cause of the properties of gravity that Newton thought need not be asked and answered for a scientifically legitimate conception of causal explanation.

We do not want to claim that questions such as the above are not connected to scientific practice. After all, even the question of the cause of gravity that Newton refrained from answering was connected to scientific practice. The point, rather, we take from Newton is that answering these questions is not required for offering adequate causal explanations of the phenomena under study. Similarly, for CM, answering questions such as the above is not required in order to have legitimate mechanistic explanations. In other words, the properties of mechanism over and above those that are required by its methodological use need not be specified; nor is there an explanatory lacuna if they are not.

According to CM, the concept of mechanism as used in practice need not, and should not, be understood in a metaphysically inflated sense. Hence, new mechanists, in offering such metaphysically inflated accounts, need to show that such accounts are indeed indispensable for doing good mechanistic science. To conclude, as Newton remained agnostic about the underlying mechanism of gravity, so CM remains agnostic about the metaphysical ground of any particular causal pathway. As in the case of gravity, *it is enough that mechanisms qua causal pathways really exist and act as they do.*

CHAPTER 2

Extending Mechanism beyond the Two 'Most Catholic Principles of Bodies'

2.1 Preliminaries

In the previous chapter we offered a sketchy but hopefully intriguing account of the history of the mechanical world view. Now, similar stories have been told by the new mechanists themselves. A story about Newton occurs in Glennan (1992), and an account of the relationship between the austere Old Mechanism and richer New Mechanism shows up in a historical introduction to a journal issue by Craver and Darden (2005). Where does our story so far differ from theirs? Though all these accounts are very interesting, ours is better, we think, for three reasons. First, it is a lot more detailed than other accounts; second, and relatedly, we pay a lot more attention to historical accuracy compared with other accounts; third, and more importantly, we focus on an issue that is not discussed in detail in other accounts, that is, the reasons that led to the abandonment of Old Mechanism. Our analysis why Old Mechanism was abandoned leads to a fresh look at the differences and similarities between Old and New Mechanism and examines what the lessons are for current accounts.

Our view then is that history is important not so much because of the common elements between Old and New Mechanism, or as an 'origin story' of New Mechanism, but because of the lessons one can learn from a close examination, at both the metaphysical and the methodological levels, of the limitations and shortcomings of Old Mechanism when applied to the problem of explaining gravity. So, while for old mechanists such as Descartes and Leibniz scientific explanations of gravity (and scientific explanations in general) had to be couched in mechanical terms (i.e., in terms of matter in motion), with Newton this view was superseded. In fact, we have argued that according to Newton causal explanation should be liberated from the search for mechanisms. Would that be too strong a conclusion since, one might argue, for Newton the motion of the planets and the tides was mechanistic, even if there was not a mechanistic explanation (indeed any explanation at all) of the gravitational force itself? Answering this potential objection requires delving into the rest of our story of the fate of the mechanical conception of the world. This will be part of the present chapter. Put in a nutshell, however, our reply is that the issue is partly terminological and partly substantive. It is substantive insofar as in the case of explanations that refer to forces acting at a distance (whether in the case of gravity or anything else), and for which no mechanical explanation can be given (i.e., their action cannot be explained in mechanical terms), we have causal explanations without mechanisms in the sense that natural philosophers like Leibniz or Huygens understood the notion. As Leibniz argues in his essay 'Against Barbaric Physics' (which we discussed briefly in Section 1.3.4), such explanations go against basic mechanistic tenets. So, it is clear to us that with Newton there is a fundamental change in what counts as a legitimate scientific explanation; consequently, if we want to designate Newton's explanations in terms of forces acting at a distance as 'mechanical', we have here a new and more liberal notion of mechanism. And that's why the issue is partly terminological. For as we shall show in the present chapter, there is a sense in which Newton did modify and expand the notion of mechanism prevalent in seventeenth century, which was further modified and extended by Kant and others after him.

2.2   Mechanical versus Quasi-Mechanical Mechanism

It will be helpful and accurate to distinguish between two concepts of mechanism – or, if you like, between two ideas of mechanism. We may call the first *mechanical* mechanism and the second *quasi-mechanical* mechanism. The first conception of mechanism is narrow: mechanisms are configurations of matter in motion subject to mechanical laws. It is this conception that has been associated with the rise and dominance of the mechanical conception of nature in the seventeenth century.

The second conception of mechanism is broader. A quasi-mechanical mechanism is *any* arrangement of parts into wholes in such a way that the behaviour of the whole depends on the properties of the parts and their mutual interactions. Harré (1972, 116) has called this kind of mechanism *generative* mechanism. The focus is not on the mechanical properties of the parts, nor on the mechanical principles that govern the behaviour of the parts and determine the behaviour of the whole. Instead, the focus is on the causal relations there are between the parts and the whole. Generative mechanisms are taken to be the bearers of causal connections. It is in virtue of them that the causes are supposed to produce the effects. There is a concomitant conception of mechanical explanation as a kind of decompositional explanation: an explanation of a whole in terms of its parts, their properties and their interactions. This second conception is, arguably, associated with Kant's idea of mechanism in his third critique.

2.2.1   Mechanical Mechanism

In the seventeenth century, as we explained in detail in the previous chapter, the mechanical conception of nature was taken to be a weapon against the Aristotelian view that each and every explanation was not complete unless some efficient *and* some final cause were cited. The emergent mechanical philosophy placed in centre-stage the new science of mechanics and left Aristotelian physics behind. Accordingly, the call for a mechanical explanation of phenomena has had definite content: all natural phenomena are *produced* by the mechanical interactions of the parts of matter according to mechanical laws.

The broad contours of the mechanical conception of nature were not under much dispute, at least among those who identified themselves as mechanical philosophers. The key ideas were that all natural phenomena are explicable mechanically in terms of matter in motion; that efficient causation should be understood, ultimately, in terms of *pushings* and *pullings*; and that final causation should be excised from nature.[1] Though definite, this conception was far from monolithic. As Marie Boas (1952) has explained in detail, there had been different and opposing conceptions as to the structure of matter (atomistic vs corpuscularian), the reality of the void (affirmation of the existence of empty space vs the plenum) and the primary qualities of matter (solely extension vs richer conceptions that include solidity, impenetrability and other properties). And yet the unifying idea was that all explanation is mechanical explanation and proceeds in terms of matter and motion.

With Newton, the content of the scientific conception of nature was altered and broadened. To the eyes of most of his contemporaries (notably to Leibniz and Huygens) Newton had just abandoned the principles of

---

[1] To be sure, most mechanical philosophers did find a role for final causation via God's design of the world, but crucially, this design was precisely that of a *mechanism*. More specifically, mechanical philosophers denied the presence in nature of immanent final causes such as Aristotelian forms. Indeed, an important characteristic of the mechanical conception of nature was its denial of *forms* as part of the acceptable ontology.

mechanical philosophy, especially in light of the admission of action at a distance. Attractive and repulsive forces that act a distance were deemed by Leibniz 'barbarism in physics'. But this view led him to a delicate position. On the one hand, he severely criticised his predecessors (including Descartes and Gassendi) for purging forces from nature. On the other hand, he criticised Newton and his followers for introducing sui generis forces in nature. His own supposed *via media* was that though force must be added to mass (against the Cartesians), 'that force is exercised only through an impressed impetus' (against the Newtonians). Leibniz, to be sure, was also opposed to the sui generis powers of traditional medieval-Aristotelian explanations of natural phenomena, what were deemed occult qualities, on the grounds that they 'lead us back to the kingdom of darkness'. For him the claim that all derivative forces in nature – that is, all natural forces – should be forces exerted in the collision between bodies (no matter whether they are visible or invisible), that is, forces acting by contact, was a condition of intelligibility of dynamical explanation: contact action offered an *intelligible* mechanism which underscores all natural forces. Leibniz was quick to reprimand Locke for abandoning the view that 'no body is moved except through the impulse of a body touching it'.

Leibniz's 'principles of mechanism' entailed that neither Descartes (and the Cartesians) nor Newton (and the Newtonians) offered proper mechanical explanations of natural phenomena. The culprit in Descartes's case was the lack of a notion of force (the Leibnizian *vis viva*), whereas in Newton's case it was the non-impulsive nature of gravity.

The point here is that the category of *force* was firmly introduced alongside the traditional mechanical categories (the 'most catholic principles') of *matter* and *motion*. Though Newton insisted that his concept of force was mathematical (see *Principia*, Book I, Definition VIII), he actually set it in a mechanical framework in which it was measured by the *change* in the quantity of motion it could generate. And yet mechanical interactions were enriched to include attractive and repulsive forces between particles. Mechanical explanation was taken to consist in the subsumption of phenomena under *Newton's laws*.

That's why we stressed in Section 2.1 that there is partly a terminological issue here: we might as well designate Newton's explanations in terms of forces acting at a distance 'mechanical', provided we keep in mind that we have here a new and more liberal notion of mechanism: mechanical is a system subject to Newton's laws, namely, the laws of *mechanics*.

Capitalising on Gregor Schiemann's enlightening book (2008), it can be argued that even within what we have called the mechanical conception of mechanism, there have been two distinct senses of mechanism, one wide and another narrow. The wide sense takes it that matter in motion is the ultimate cause of all natural phenomena. As such, mechanism covers everything, but its content is quite unspecific, since there is no commitment to specific laws or principles that govern the workings of the mechanism. The narrow sense of mechanism, on the other hand, has it that mechanisms are governed by the *laws of mechanics*, as enunciated paradigmatically by Newton and Lagrange. Mechanics becomes privileged because it offers universal structural principles. But then, the form of the mechanical conception of nature depends on the details of the principles of mechanics, and the content of the concept of mechanical mechanism is specified by the historical development of mechanics.

Schiemann draws an important distinction between monistic and dualistic conceptions of mechanics and, consequently, of mechanisms. On the monistic conception, there is only one fundamental mechanical category; on the dualistic conception, there are two fundamental categories. The monistic conception is further divided into two sub-categories: one takes matter to be the fundamental mechanical concept (called 'materialist' by Schiemann), while the other takes force to be the single fundamental mechanical category (called 'dynamic' by Schiemann). Huygens and Descartes had materialist conceptions of mechanical mechanism, while Leibniz had a dynamic conception. The dualist conception of mechanical mechanism admits two distinct fundamental mechanical concepts – matter *and* force. Newton was a dualist in this sense and so was Helmholtz, according to Schiemann. Helmholtz's case is particularly instructive since he proved the principle of conservation of energy. It is precisely this principle that, as we shall see in the next section, holds the key to the very possibility of a mechanical explanation of all phenomena.[2]

With the emergence of systematic theories of heat, electricity and magnetism, one of the central theoretical questions was how these were related to the theories of mechanics. In particular, did thermal, electrical and magnetic phenomena admit of *mechanical* explanations?

This question was addressed in two different ways. One, developed mostly in Britain, was by means of building of mechanical *models*. These models were meant to show (1) the realisability of the system under study

---

[2] As Schiemann (2008, 90) notes, what made the principle of conservation of energy special, at least for Helmholtz, was that energy can 'be used directly for measuring things (particularly mechanical work and heat) and their conserving properties can be examined experimentally in physical processes'.

(e.g., the electromagnetic field) by a mechanical system and (2) the possible inner structure and mechanisms by means of which the physical system under study operates. The other way was developed mostly in continental Europe and was the construction of abstract mechanical *theories* under which the phenomena under study were subsumed and explained. These theories were mechanical because they started with principles that embodied laws of mechanics and offered explanation by deductive subsumption. This tradition scorned the construction of mechanical models (especially of the wheels-and-pulleys form that many British scientists of the time were fond of). But even within this model-building tradition, especially in its mature post-Maxwellian period, mechanical models were taken to be, by and large, heuristic and illustrative devices – the focus being on the development of systematic theories (mostly based on abstract theoretical principles such as those of Lagrangian mechanics) under which the phenomena under study were subsumed and explained. Joseph Larmor (1894, 417) drew this division of labour clearly when he noticed

> [t]he division of the problem of the determination of the constitution of a partly concealed dynamical system, such as the aether, into two independent parts. The first part is the determination of some form of energy-function which will explain the recognised dynamical properties of the system, and which may be further tested by its application to the discovery of new properties. The second part is the building up in actuality or in imagination of some mechanical system which will serve as a model or illustration of a medium possessing such an energy function.

Indeed, as one of us has argued in detail elsewhere (Psillos 1999, 132–4), this distinction, in effect between an abstract mechanical (Lagrangian) theory and the various concrete configurations of matter in motion, became commonplace among the Maxwellians after James Clerk Maxwell's mature dynamical theory of electromagnetic field which rested on the general principles of Lagrangian dynamics and was independent of any particular model concerning the carrier of light waves (see Maxwell, 1873, vol. 2, chapters 5–9; Klein 1972, 69–70). Maxwell made clear that the principles of Lagrangian mechanics allowed him to pursue the most general laws of behaviour of the electromagnetic field, whose constitution is unknown. He stressed:

> We know enough about electric currents to recognise, in a system of material conductors carrying currents, a dynamical system which is the seat of energy, part of which may be kinetic and part potential. The nature of the connexions of the parts of this system is unknown to us, but as we have

dynamical methods of investigation which do not require a knowledge of the mechanism of the system, we shall apply them to this case. (vol. 2, p. 213)

Hence, he formulated the kinetic and potential energies of the system in terms of electric and magnetic magnitudes and proceeded – by means of Lagrangian principles – to the derivation of the laws of motion of this system, thereby deriving the equations of the EM field (vol. 2, p. 233).[3]

2.3 Poincaré's Problem

This liberalisation of the conception of mechanical explanation, together with conceptual issues in the foundations of mechanics, brought with it the following question, which came under sharp focus towards the end of the nineteenth century: How exactly was the idea of a mechanical explanation to be rendered? The problem here was not so much related to the nature of explanation as to what principles count as *mechanical*. In 1900, Poincaré addressed the International Congress of Physics in Paris with the paper 'Relations entre la physique expérimentale et de la physique mathématique' (1900; this paper was reproduced as chapters nine and ten of his *La science et l'hypothése* of 1902). He did acknowledge that most theorists had a constant predilection for explanations borrowed from mechanics. Historically, these attempts had taken two particular forms: either they traced all phenomena back to the motion of molecules acting at a distance in accordance with central force laws or they suppressed central forces and traced all phenomena back to the contiguous actions of molecules that depart from the rectilinear path only by collisions. 'In a word', Poincaré said, 'they all [physicists] wish to bend nature into a certain form, and unless they can do this they cannot be satisfied' (ibid.). And he immediately queried: 'Is nature flexible enough for this?'

The answer is positive, but in a surprising way. Poincaré's groundbreaking contribution to this issue was the proof of a theorem that a necessary and sufficient condition for a complete mechanical explanation of a set of phenomena is that there are suitable experimental quantities that can be identified as the kinetic and the potential energy such that they

---

[3] For a brief account of Maxwell's derivation of the equations of the field, see Andrew Bork (1967). For a detailed historical account of the derivation of the equations in the symmetrical form known today, see Hunt (1991, 122–8 and 245–7).

satisfy the principle of conservation of energy.[4] Given that such energy functions can be specified, Poincaré proved that there will be *some* configuration of matter in motion (i.e., a configuration of particles with certain positions and momenta) that can underpin (or model) a set of phenomena. As he put it:

> In order to demonstrate the possibility of a mechanical explanation of electricity, we do not have to preoccupy ourselves with finding this explanation itself; it is sufficient to know the expressions of the two functions T and U which are the two parts of energy, to form with these two functions the equations of Lagrange and, afterwards, to compare these equations with the experimental laws. (1890/1901, viii)

Poincaré presented these results in a series of lectures on light and electromagnetism – delivered at the Sorbonne in 1888 and published as *Électricité et optique* in 1890 – which primarily aimed to deliver Maxwell's promise, that is, to show that electromagnetic phenomena could be subsumed under, and represented in, a suitable mechanical framework. As Poincaré put it, he aimed to show that 'Maxwell does not give a mechanical explanation of electricity and magnetism; he confines himself to showing that such an explanation is possible' (p. iv). In effect, Poincaré noted that once the first part of Larmor's foregoing division of labour is dealt with, the second part (the construction of configurations of matter in motion) takes care of itself. Maxwell's achievement, according to Poincaré, was precisely this and he 'was then certain of a mechanical explanation of electricity' (1902/1968, 224).

The irony was that Poincaré's demonstration had the following important corollary: if there is one mechanical explanation of a set of phenomena – that is, if there is a possible configuration of matter in motion that can underpin a set of phenomena – there are an *infinity* of them. And not just that. Another theorem proved by the French mathematician Gabriel Königs suggested that for any material system such that the motions of a set of masses (or material molecules) is described by a system of linear differential equations of the generalised coordinates of these masses, these differential equations (which are normally attributed to the existence of forces between the masses) would be satisfied even if one replaced all forces by a suitably chosen system of *rigid connections* between these masses. Indeed, Heinrich Hertz (1894/1955) had made use of this result to develop a system of mechanics that did away with forces altogether.

---

[4] The details of the proof (as well as further discussion of Poincaré's conception of mechanical explanation) are given in Psillos (1995).

Poincaré thought that these formal results concerning the multiplicity of mechanical configurations that could underpin a set of phenomena described by a set of differential equations were natural. They were only the mathematical counterpart of the well-known historical fact that in attempting to form potential mechanical explanations of natural phenomena, scientists had chosen several theoretical hypotheses, for example, forces acting at a distance, retarded potentials, continuous or molecular media, hypothetical fluids and so on. Poincaré was sensitive to the view that even though some of these attempts had been discredited in favour of others, more than one potential mechanical model of, say, electromagnetic phenomena was still available (see 1900, 1166–7).[5]

So, the search for a *complete* mechanical explanation of electromagnetic phenomena was heavily underdetermined by possible configurations of matter in motion. Different underlying mechanisms could all be taken to give rise to the laws of electromagnetic phenomena. By the same token, though the possibility of a mechanical explanation of electromagnetic phenomena is secured, the empirical facts alone could not dictate any choice between different mechanical configurations that satisfy the same differential equations of motion. The choice among competing underlying mechanisms (possible configurations of matter in motion) was heavily underdetermined by the empirical facts. How then can one choose between these possible mechanical configurations? How can one find the correct complete mechanical explanation of electromagnetic phenomena? For Poincaré this was a misguided question. As he said, 'The day will perhaps come when physicists will no longer concern themselves with questions which are inaccessible to positive methods and will leave them to the metaphysicians' (1902/1968, 225). His advice to his fellow scientists was to content themselves with the possibility of a mechanical explanation of all conservative phenomena and to abandon hope of finding the true mechanical configuration that underlies a particular set of phenomena. He

---

[5] The turning point in Poincaré's thinking about mechanics is in his review of Hertz's (1894) book for *Revue Générale des Sciences*. Concerning the 'classical system', which rests on Newton's laws, Poincaré agreed with Hertz that it ought to be abandoned as a foundation for mechanics (see 1897, 239). Part of the problem was that there were no adequate definitions of force and mass. But another part was that Newton's system was incomplete precisely because it passed over in silence the principle of conservation of energy (cf. 1897, 237). Like Hertz, Poincaré was more sympathetic to the 'energetic system', which was based on the principle of conservation of energy and Hamilton's principle that regulates the temporal evolution of a system (cf. 1897, 239–40). According to Poincaré (1897, 240–1) the basic advantage of the energetic system was that in a number of well-defined cases, the principle of conservation of energy and the subsequent Lagrangian equations of motion could give a full description of the laws of motion of a system.

stressed: 'We ought therefore to set limits to our ambition. Let us not seek to formulate a mechanical explanation; let us be content to show that we can always find one if we wish. In this we have succeeded' (1900, 1173).

According to Poincaré, the search for mechanical explanation (i.e., for a configuration of matter in motion) of a set of phenomena is of little value not just because this search is massively underdetermined by the phenomena under study but mainly because this search sets the wrong target. What matters, for Poincaré, is not the search of mechanism per se, but rather the search for *unity* of the phenomena under laws of conservation. Understanding is promoted by the unification of the phenomena and not by finding mechanical mechanisms that bring them about. As he said, 'The end we seek ... is not the mechanism. The true and only aim is unity' (p. 1173).

One may question the status of the law of conservation of energy as a mechanical principle. But that's beside the point. For the point is precisely that there is no fixed characterisation of what counts as mechanical. It may well be that Poincaré's notion of mechanical explanation is too wide from the point of view of physical theory, since it hardly excludes any phenomena from being subject to mechanical explanation. Still, and this is quite important, it does block certain versions of vitalism that stipulate new kinds of forces. As is well known, in the twentieth century, the search for mechanisms and mechanical explanations was taken to be a weapon against vitalism. One key problem with vitalist explanations (at least of the sort that C. D. Broad has dubbed *Substantial Vitalism*) is that they are in conflict with the principle of conservation of energy, and in *this* sense, they cannot be cast, even in principle, as mechanical explanations.

The significance of Poincaré's problem for the mechanical conception of mechanism can hardly be overestimated. But we should be careful to note exactly what this problem is. It is not that mechanical mechanisms are unavailable or non-existent. It is not that nature is *not* mechanical. Hence, it is not that mechanical explanation – that is, explanation in terms of mechanical mechanisms – is impossible. On the contrary, Poincaré has secured its very possibility, thereby securing, as it were, the victory of traditional mechanical philosophy over Aristotelianism. Rather, the problem for the mechanical conception of mechanism that Poincaré has identified is that mechanical mechanisms are *too* easy to get, provided nature is conservative. Under certain plausible assumptions that involve the principle of conservation of energy, the call for mechanical explanation is so readily satisfiable that it ceases to be genuinely informative.

2.4   Quasi-Mechanical Mechanisms

Ewing (1969, 216) drew a distinction between two conceptions of mechanical necessity in Kant's *Third Critique*. The first is related to what we have called the mechanical conception of mechanism: a determination of the properties of a whole by reference to matter in motion, and in particular by the mechanical properties of its parts and the mechanical laws they obey. The second, which Ewing calls 'quasi-mechanical', is still a determination of the properties of the whole by reference to the properties of its parts, but with no particular reference to mechanical properties and laws. This quasi-mechanical conception of mechanism is broader than the mechanical conception since there is no demand that the laws that govern the behaviour of the parts, or the properties of these parts, are mechanical – at least in the strict sense associated with the mechanical conception.

Peter McLaughlin (1990) has developed a similar account of Kant's conception of mechanical explanation, according to which the mechanism of nature is a form of causation, whose differentia is that it takes it that the whole is determined by its parts. Thus understood, a mechanical explanation is a kind of decompositional (or modular) explanation: an explanation of a whole in terms of its parts, their properties and their interactions. McLaughlin bases his account on the following point made by Kant in his *Critique of Judgement* (1790/2008, 408): 'Now where we consider a material whole, and regard it as in point of form a product resulting from the parts and their powers and capacities of self-integration (including as parts any foreign material introduced by the co-operative action of the original parts), what we represent to ourselves in this way is a mechanical generation of the whole.' Accordingly, what renders a structure a mechanism is the fact that it possesses a reductive unity: its behaviour is determined by the properties its parts have 'on their own, that is independently of the whole' (McLaughlin 1990, 153).

This is not the place to discuss in any detail whether this was indeed Kant's own conception.[6] The key point is that *if* this conception is viable at all (and, as the current mechanistic turn demonstrates, it *is*), then the concept of mechanism is not tied to mechanics, nor to the operation of specifically mechanical laws, nor to the ultimate determination of the

---

[6] There are competing views on this. Hannah Ginsborg takes it that Kant's conception of mechanism is closely tied to his account of forces and mechanical laws. For her, according to Kant, 'we explain something mechanically when we explain its production as a result of the unaided powers of matter as such' (2004, 42). For an attempted synthesis of Ginsborg's and McLaughlin's views, see Breitenbach (2006).

behaviour of mechanism by reference to mechanical properties and inter-actions. Rather, the mechanism is *any* complex entity that exhibits reduc-tive stability and unity in the sense that its behaviour is determined by the behaviour of its parts.

Kant, to be sure, contrasted mechanical explanation to teleological explanation. In its famous antinomy of the teleological power of judge-ment, he contrasted organisms to mechanisms. *Qua* material things, organisms (like all material things) should be generated and governed by merely mechanical laws. And yet, some material things (*qua* organisms, and hence natural purposes, as Kant put it) 'cannot be judged as possible according to merely mechanical laws (judging them requires an entirely different law of causality, namely that of final causes)' (1790/2008, 387). The defining characteristics of an organism – that is of a non-mechanism – are two: (1) the whole precedes its parts and, ultimately, determines them; and (2) the parts are in reciprocal relations of cause and effect. Famously, Kant claimed that the very idea of non-mechanism (organism) is regulative and not constitutive – we have the right to proceed *as if* there were organisms (non-mechanisms) but this is not something that can be known or proved, though Kant did think that this regulative principle is a *safe* presupposition, not liable to refutation by the progress of science.

This contrast of mechanism and non-mechanism suggests that the key feature of mechanism – what really sets it apart from organism – is the priority of the parts over the whole in the constitution of the mechanism and the determination of its behaviour.[7] It is also worth noting that it is precisely this contrast that C. D. Broad (1925) had in mind in his own critique of mechanism.

Broad mounted an attack on what he called 'the ideal of Pure Mechanism'. This is an extreme and purified version of what we have called the mechanical conception of mechanism. Broad's Pure Mechanism is a world view, which he (1925, 45) characterises thus:

---

[7] Ginsborg (2004) takes it that *qua* natural purposes, organisms are non-machine-like (and hence mechanically inexplicable) in the sense that 'they are not assemblages of independent parts, but that they are instead composed of parts which depend for their existence on one another, so that the organism as a whole both produces and is produced by its own parts, and is thus in Kant's words 'cause and effect of itself'' (p. 46). This way to read Kant's account of organism distinguishes it from mechanism in two senses. (1) Organism cannot be explained in terms of the powers of matter as such, and (2) organism is such that its parts depend on the whole and cannot 'exist independently of the whole to which they belong' (p. 47). Hence, what renders mechanism distinctive is precisely the fact that its unity and behaviour are determined by its parts, as they are independent of their presence in the whole. For a useful attempt to synthesise Kant's antinomy in the light of modern evolutionary biology, see Walsh (2006).

The essence of Pure Mechanism is

    a.   a single kind of stuff, all of whose parts are exactly alike except for differences of position and motion;

    b.   a single fundamental kind of change, viz, change of position . . .;

    c.   a single elementary causal law, according to which particles influence each other by pairs; and

    d.   a single and simple principle of composition, according to which the behaviour of any aggregate of particles, or the influence of any one aggregate on any other, follows in a uniform way from the mutual influences of the constituent particles taken by pairs.

The gist of Pure Mechanism is that it is an ontically reductive thesis and in particular a reductive thesis with a very austere reductive basis of a single kind of fundamental particle, a single kind of change and a single causal law governing the interaction of the fundamental particles. Broad contrasted this view with two others. The first is what he called *Emergent Vitalism*. This is the view that living organisms and their behaviour cannot be fully and exhaustively determined by the properties and behaviour of their component parts, as these would be captured by the ideal of Pure Mechanism. Emergent Vitalism is also opposed to a view we have already noted in Section 2.3, namely, Substantial Vitalism: that living organisms are set apart from mechanism by an extra element (a kind of life-conferring force) that they share while pure mechanisms do not. In denying Substantial Vitalism, Emergent Vitalism puts emphasis on the structural arrangement of the whole vis-à-vis its parts and on the interaction among the parts when they are put together in a whole. A certain whole W may consist of constituents A, B, C placed in a certain relation R(A, B, C). There is emergence – emergent properties – when A, B, C cannot determine, even in principle, the properties of R(A, B, C).

Broad (1925, 61) put this point in terms of the lack of an in-principle deducibility of the properties of R(A, B, C) 'from the most complete knowledge of the properties of A, B, and C in isolation or in other wholes which are not of the form R(A, B, C)'. This way to put the matter might be unfortunate, since what really matters is the metaphysical determination (or its lack thereof) of the whole by its parts and not deducibility per se – which is dependent on the epistemic situation we might happen to be in. But what matters for our purposes is Broad's thought that the denial of Pure Mechanism need not lead to the admission of spooky forces and mysterious powers, associated with Substantial Vitalism.

Still, our main concern here is not the opposition of Pure Mechanism to Emergent Vitalism, but rather its opposition to what Broad rightly took to be a milder form of mechanism. This form, which Broad associated with what he called *Biological Mechanism*, is committed to the view that the behaviour of a whole (and of a living body in particular) is determined by its constituents, their properties and the laws they obey, but relies on a broader conception of what counts as a constituent and what laws are admissible. As Broad put it:

> Probably all that he [a biologist who calls himself a 'Mechanist'] wishes to assert is that a living body is composed only of constituents which do or might occur in non-living bodies, and that its characteristic behaviour is wholly deducible from its structure and components and from the chemical, physical and dynamical laws which these materials would obey if they were isolated or were in non-living combinations. Whether the apparently different kinds of chemical substance are really just so many different configurations of a single kind of particles, and whether the chemical and physical laws are just the compounded results of the action of a number of similar particles obeying a single elementary law and a single principle of composition, he is not compelled as a biologist to decide. (p. 46)

This is, clearly, what we have called a quasi-mechanical conception of mechanism, and as Broad rightly notes, this kind of conception is enough to set the mechanist biologist apart from the emergent vitalist. The controversy need not be put, nor is it useful to be put, in terms of the ideal of Pure Mechanism.[8]

Enough has been said, we hope, to persuade the reader that there is a distinct quasi-mechanical idea of mechanism, which – to recapitulate – proclaims a form of determination of a whole by its parts, their properties and interactions, as these would occur independently of their presence in the whole. With this in mind, let us now see what the key problem of this quasi-mechanical conception of mechanism is.

2.5   Hegel's Problem

Seeing the heading above, the reader might wonder: Has something gone awfully wrong? What can Hegel possibly have to do with mechanisms and

---

[8] In his very useful (2005) paper, Garland Allen notes that 'operative, or explanatory mechanism refers to a step-by-step description or explanation of how components in a system interact to yield a particular outcome' (p. 261). He contrasts this with what he calls 'philosophical mechanism', which he takes to assert that living things are material entities. He then offers an instructive historical account of approaches to biological mechanism in the early twentieth century (and their opposition to vitalism), emphasising that 'the form that Mechanistic thinking took in the early twentieth century ... differed from earlier (eighteenth- and nineteenth-century) mechanistic traditions. It was physico-chemical not merely mechanical' (p. 280).

contemporary mechanistic explanation in the sciences? Well, it turns out not only that there is a quite different conception of mechanism that goes back (at least) to Hegel, but that he identified a problem that all accounts of mechanism (especially those mechanisms which we shall call mechanisms-for) should meet.

Long before Poincaré's critique of mechanical mechanism, Hegel had, in his *Science of Logic*, attacked the idea that all explanation must be mechanical. According to James Kreines (2004), Hegel argued that making mechanism an absolute category – applicable to everything – obscures the distinction between explanation and description and hence undermines itself.

Hegel's writings on mechanism are rather cryptic (and, perhaps, obscure). Essentially, he took the characteristic of mechanism to be that it possesses only an external unity. Its constituents (the objects that constitute it) retain their independence and self-determination, although they are parts of the mechanism. As he put it in his *The Encyclopaedia Logic* (1832/1991, 278), 'the relation of mechanical objects to one another is, to start with, only an external one, a relation in which the objects that are related to one another retain the semblance of independence'. And in his *Science of Logic* (2002, 711) he stressed: 'This is what constitutes the character of *mechanism*, namely, that whatever relation obtains between the things combined, this relation is one *extraneous* to them that does not concern their nature at all, and even if it is accompanied by a semblance of unity it remains nothing more *than composition, mixture, aggregation* and the like.' The determinant of the unity of a mechanism or, as Hegel put it, 'the *form* that constitutes [its] difference and combines [it] into a unity' is 'an external, indifferent one; whether it be a *mixture*, or again an *order*, a certain *arrangement* of parts and sides, all these are combinations that are indifferent to what is so related' (p. 713). And elsewhere, he stressed that, being external, the unity of the mechanism 'is essentially one in which no *self-determination* is manifested' (p. 734).

On Kreines's reading of Hegel's critique of mechanism, Hegel raised a perfectly sensible and quite forceful objection to the view that *all* explanation is mechanical explanation, that the only mode of explanation is mechanical and that to explain X is to offer a mechanical explanation of it.

Hegel's argument against the idea of mechanism – *qua* an all-encompassing explanatory concept – goes like the following. Mechanistic explanation proceeds in terms of breaking an object down to its parts and of showing its dependence on them and their properties and relations. Explanation, then, amounts to a certain decomposition of the explanandum, namely, of a composite object whose behaviour is the result of the properties of, and interactions among, its parts. But there are indefinitely many ways to decompose something to parts and to relate it and its behaviour to them. For the call for explanation to have any bite at all, there must be some principled distinction between those decompositions that are merely descriptions of the explanandum and those decompositions that are genuinely explanatory. In particular, some decomposition – that which offers the mechanical *explanation* – must be privileged over the others, which might well reflect only pragmatic criteria or subjective interests. But how is this distinction to be drawn within the view that all explanation is mechanical? If all explanation is indeed mechanical, and if mechanical explanation amounts to decomposition, no line can be drawn between explanation and description – no particular way to decompose the explanandum is privileged over the others by being mechanical – mechanical as opposed to what? All decompositions will be equally mechanical and equally arbitrary. Hence, there will be no difference between explanation and description.

Hegel was pushing this line of argument in order to promote his own organic view of nature and, in particular, to reinstate a teleological kind of explanation – one that explains the unity of a composite object in terms of its internal purposeful activity.[9] But the point he makes is very general. In essence, Hegel's problem is that something external to the mechanism (considered as an aggregate of parts) is necessary to understand how mechanistic explanation is possible. His general point is that the unity of a mechanism is not just a matter of arranging a set of elements into a whole; nor is it just a matter of listing their properties and mutual relations. Nor is it determined by the parts of the mechanism, as they are independent of their occurrence within the mechanism. There are indefinitely many ways to arrange parts into wholes or to decompose wholes into parts. Most of them will be arbitrary since they will *not* be explanatorily relevant. The unity of the mechanism comes from something external to it, namely, from its function – from what it is meant to be a mechanism *for*. The function that a mechanism performs is something external to the description of the mechanism. It is the function that fixes a criterion of explanatory relevance. Some descriptions of the mechanism are explanatorily relevant while others are not because the former and not the latter explain how the mechanism performs a certain function.

---

[9] For an informative and intelligible account of Hegel's organic world view, see Beiser (2005, chapter 4).

Let us illustrate this point with a couple of examples. Consider a toilet flush – a very simple mechanism indeed. What confers unity to it *qua* mechanism is the function it performs. As a complex entity, it can be decomposed into elements in indefinitely many different ways. Actually, in all probability, there is, in principle, a description of it in terms of the interactions of the atoms of water and their collisions with the walls of the tank and so on. What fixes the explanatorily relevant description is surely the function it performs. Or consider a telephone conversation through which some information is passed over from one end to the other – a very simple social mechanism. What confers unity to it *qua* mechanism is its function, namely, to transfer information between two ends. In all probability, there is a description of this mechanism in terms of the interactions of sound waves, collisions of particles, triggering of nerve-endings and so on. But this description is explanatorily irrelevant when it comes to explaining how this simple social mechanism performs its function. Notice that a point brought out by these examples, and certainly a point that Hegel had in mind, is that the truth of a description (supposing that it is to be had) does not necessarily render this description explanatorily relevant.

We could sum up Hegel's problem like this: first the function, then the mechanism.[10] The functional unity of the mechanism determines, ultimately, which of the many properties that the constituents of the mechanism have are relevant to the explanation of the performance and function of the whole. Hegel (1832/1991, 275) did think that mechanism is a form of objectivity, claimed that it is applicable to areas other than 'the special physical department from which it derives its name' but denied that it is an 'absolute category' that is constitutive of 'rational cognition in general'.

2.6   Bringing Together the Two Problems

Qua thinkers, Hegel and Poincaré were as different as chalk and cheese. Yet they both point – with different philosophical arguments – towards a decline of the mechanistic world view. It's not that there are no mechanisms. Actually, mechanisms, in the broad sense of stable arrangements of matter in motion, are ubiquitous. But it does not follow from this that nature has a definite mechanical structure (or, if that's too strong, that we cannot know which definite mechanical structure is the one actually characterising nature). This is, in essence, Poincaré's problem. How the

[10] This is indeed something that many new mechanists have come to accept.

mechanisms are individuated is a matter external to them: what counts as a mechanism, where it starts and where it stops, what kind of parts are salient and what kind of properties are relevant depend on the function they are meant to perform. The unity of the mechanisms is not intrinsic but extrinsic to them. This is, in essence, Hegel's problem. But even after a function has been determined, there are indefinitely many ways to config-ure mechanical mechanisms that perform it, that is, to offer a mechanical model (a configuration of matter in motion) that performs it. This is a corollary of Poincaré's problem.[11] Nature, even if it is mechanical, does not fix the boundaries of mechanisms. When it comes to the search for mechanisms, *anything* can count as a quasi-mechanism provided it per-forms a function that it is meant to explain. This is a corollary of Hegel's problem.

Does the search for mechanisms improve understanding? The answer is unequivocally positive. The description of a mechanism is a theoretical description and, as such, tells a story as to how the phenomenon under study is brought about – if the story is true, our understanding of nature is enhanced. Insofar as mechanisms are taken to be functionally individuated stable explanatory structures (whose exact content and scope may well vary with our best conception of the world) which enhance our understanding of how some effects are brought about or are the realisers of certain functions, they can play a useful role in the toolkit of explanation.

What does all this imply about the ontic status of mechanisms? Given that the world is governed by conservation laws, it allows descriptions of configurations of matter in motion subject to the laws of mechanics; that is, it admits of (a multitude of) mechanical mechanisms. But that's all spoils to the victor, the latter being the fact of the law of conservation of energy. Even then, however, mechanisms are functionally individuated; the spoils do not come to much – there is more than one way to peel an orange!

---

[11] Hegel was confident that 'not even the phenomena and processes of the physical domain in the narrower sense of the word (such as the phenomena of light, heat, magnetism, and electricity, for instance) can be explained in a merely mechanical way (i.e., through pressure, collision, displacement of parts and the like' (1832/1991, 195). Poincaré proved him wrong on this.

PART II

Causation and Mechanism

CHAPTER 3

Mechanisms in Scientific Practice
The Case of Apoptosis

3.1 Preliminaries

Talk of mechanisms is widespread within the life sciences. Pathologists talk about 'mechanisms of disease' (Lakhani et al. 2009) and pharmacologists about the 'mechanisms of action' of drugs (Rang et al. 2016). Molecular biologists talk about the 'fundamental mechanisms of life' (Alberts et al. 2014, 22), such as DNA replication and protein synthesis; developmental biologists talk about 'genetic mechanisms of animal development' (Alberts et al. 2014, 39), 'mechanisms for specifying the germ layers' (Wolpert et al. 2002, 89) and 'mechanisms of axis determination' (Wolpert et al. 2002, 143); and there exist 'morphogenetic mechanisms' (Wolpert et al. 2002, 254) and the 'mechanism of programmed cell death' (Slack 2005, 214).

Searching for mechanisms and mechanical explanations has been viewed as a main aim of life sciences, and science in general (see Craver & Tabery 2015). There exists a widespread consensus among philosophers of science that an adequate philosophical account of the practice of current sciences must be structured around this basic notion (see Glennan & Illari 2018a). However, philosophers have not yet reached a consensus about what a mechanism is. This lack of a generally accepted account may seem surprising to the outsider, given the prominence of the concept in scientific practice.

3.2 The Case of Apoptosis

3.2.1 Why Apoptosis?

If we want to understand what the things that scientists identify as mechanisms are, we had better examine how mechanisms are identified in practice. So, we take it that the best argument for CM is that precisely this conception of mechanism is the one that we find in biological (as well as in other scientific) contexts where the language of mechanisms is used. In order to substantiate this argument from scientific practice, we will examine in depth a central example of a biological mechanism, the mechanism of cell death known as apoptosis. We will use apoptosis as a benchmark to develop a bottom-up argument (i.e., one based on a case study of a prominent case of a biological mechanism) in favour of the adequacy of CM. Our discussion in this chapter will establish that CM is a typical use of 'mechanism' in life sciences. In the next chapter and the chapters to follow, we will generalise CM and apply it to other cases.

There are three main reasons why apoptosis is a particularly good example for our purposes. The term 'apoptosis' was introduced in the biological literature by John F. R. Kerr, Andrew H. Wyllie and Alastair R. Currie in a seminal 1972 paper as the name of a newly discovered mechanism, and in particular of 'a hitherto little recognised mechanism of controlled cell deletion' (Kerr et al. 1972, 239). It is thus a particularly instructive case that can be used to draw lessons about how a new mechanism is identified in biology, and hence to provide insights about what exactly a mechanism in life sciences is taken to be. Subsequent research has identified apoptosis as a ubiquitous mechanism for the regulation of cell populations with important clinical relevance and has offered various levels of description of its workings. As we will see, this provides further insights into the nature of mechanisms in biology. Last, the study of apoptosis (and of mechanisms of programmed and physiological cell death in general) transcends particular biological fields and has involved cytologists, developmental biologists, pathologists and molecular biologists, among others. Because of its broad role, this case offers a nice test case of the concept of mechanism as it is used in the life sciences. Apoptosis is described in all these biological disciplines as a 'mechanism' of cell death. The common concept of mechanism at play here, we will argue, is that a mechanism is just a causal pathway. The case of apoptosis will then provide strong reasons for the view that CM is the common denominator of (at least) many uses of the concept of mechanism in biology.

3.2.2   Early History of Cell Death

The history of cell death can be traced back to the mid-nineteenth century, when biologists were already aware that there exist processes that lead to the death of cells (see Clarke & Clarke 1996). Cytologists of the time, such as Walther Flemming in 1885, had even observed cells undergoing what we now consider as *apoptosis*. However, in the following decades and during the 1960s and even later, there was not much interest from biologists in cell death. To explain why this was so, Richard A. Lockshin notes that biologists at that time 'tended to think that death was accidental and that mitosis was the active homeostatic process' – cell death was not yet viewed as a 'biological process' (Lockshin 2008, 1092).

In the 1960s, new technological developments (e.g., electron microscopy, improved histological techniques) started a new era in the study of cell death. In 1964, Lockshin and Carroll Williams published a paper with the title 'Programmed Cell Death' (Lockshin & Williams 1964). According to Lockshin and Zahra Zakeri, this new expression ('programmed cell death') was meant to capture the fact that 'cells followed a sequence of controlled (and thereby implicitly genetic) steps towards their own destruction' (Lockshin & Zakeri 2001, 546).

That such a controlled process existed was well known to developmental biologists. John W. Saunders had already noticed that there exist 'reproducible patterns of cell death in chick embryos' (Lockshin & Zakeri 2001, 546) and the same was the case concerning metamorphic cell deaths in insects. As he put it, 'abundant death, often cataclysmic in its onslaught, is a part of early development in many animals; it is the usual method of eliminating organs and tissues that are useful only during embryonic or larval life or that are but phylogenetic vestiges' (Saunders 1966, 154). Characteristically, Saunders wrote about a 'death clock' intrinsic in cells: cells that normally die during chick development would also die 'on schedule' in culture. If, however, they were transplanted to a different area of other chicks, they survived. It was evident from this that cell death is a controlled and regulated process.

To be sure, the idea of 'programmed cell death' was metaphorical. As Lockshin explains in a recent review on the history of the subject, it was

> a felicitous turn of phrase designed to exploit the trendiness of the then-nascent computer era. The intent was to focus attention on what was relatively obvious: that cell deaths in developing and metamorphosing animals occurred at predictable developmental stages and in specific locations. They must be 'programmed' into the genetics of the organisms, in the same sense that the differentiation and growth of an organ, tissue, structure, or pigment would be considered to be fundamentally determined by the interplay of specific genes. (Lockshin 2016, 10–11)

3.2.3   Identifying Apoptosis

John F. R. Kerr, who had been working on the processes of cell death since the 1960s, notes that at the time most researchers were 'equating cell death with cell degeneration' (Kerr 2002, 472). So, an early hypothesis was that cell death was the result of damaged lysosomes, which were viewed as 'suicide bags'. Kerr, however, had discovered a certain type of cell death that was 'non-degenerative in nature' (2002, 472) – he first named it 'shrinkage necrosis' (Kerr 1971). This was no accident; the initial thought was that this process was a type of *necrosis*. But soon Kerr noted that it was a different kind of mechanism – what he and his collaborators called 'apoptosis'.

Shrinkage necrosis was identified by studying ischaemic liver injury. It was a type of cell death that differed from classical necrosis both morphologically, in that it involved scattered cells that were converted in small round bodies, rather than groups of cells as in classical necrosis, and also chemically, in that during shrinkage necrosis lysosomes were preserved, again in contrast to classical necrosis where they ruptured. In his 1971 paper, Kerr concluded: 'Shrinkage necrosis is a distinct and important type of cell death, which has received relatively little attention in the past. It probably results from noxious stimuli that are insufficiently severe to produce coagulative necrosis' (p. 19).

The concept of apoptosis was introduced in a seminal paper in 1972 by Kerr, Wyllie and Currie as 'a hitherto little recognized mechanism of controlled cell deletion' (1972, 239). What did Kerr et al. do to identify the mechanism of apoptosis? They described it in the language of theory as a 'vital biological phenomenon', which is 'complementary to mitosis in the regulation of animal cell populations' (p. 241). This was mainly a description of the specific causal pathway of the deletion of 'scattered single cells' (p. 241). (See Figure 3.1.)

There are two main stages in the apoptotic process as shown by electron microscopy: first, so-called apoptotic bodies are formed; second, apoptotic bodies are phagocytosed and degraded by other cells. Apoptotic bodies are small, spherical, membrane-bound structures that contain condensed, but otherwise intact and functional, cell organelles and fragments of nuclei. During the formation of apoptotic bodies the nucleus and the cytoplasm condense, and nucleus fragments and protuberances are formed on the surface of the cell. The cell then breaks apart and from the protuberances the apoptotic bodies are formed. Within the cells that phagocytose them, apoptotic bodies show changes that are 'very similar to ischaemic coagulative necrosis'. But, in contrast to coagulative necrosis, the absence of inflammation in apoptosis results in a process of cell death with minimal disruption of the tissue. So, apoptosis is distinct from necrosis and, as Kerr et al. put it, 'is well suited to a role in tissue homoeostasis' (p. 250).

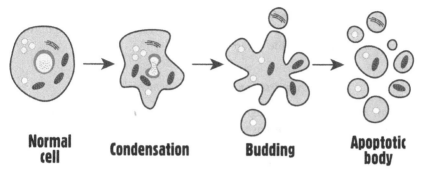

Normal cell   **Condensation**   **Budding**   **Apoptotic body**

Figure 3.1   Apoptosis as described by Kerr et al. (1972).

Crucially, the morphological changes that occur during apoptosis were 'essentially the same' (p. 244) in various types of circumstances studied by the authors, both physiological and pathological. Establishing this point was important, as a mistake that had previously been made in various cases was to confuse apoptotic bodies that have been phagocytosed with, for example, autophagic vacuoles (which often appear similar under the electron microscope). For example, apoptosis was found by Kerr and his collaborators to occur spontaneously in both treated and untreated malignant tumours and was involved in cases of pathological atrophy but also of normal involution of tissues, in normal development (e.g., during the development of digits) and in general in cellular turnover in normal adults. It was then taken to be a ubiquitous mechanism: 'a distinctive morphological process ... which plays a complementary but opposite role to mitosis in the regulation of animal cell populations' (pp. 255–6).

By describing the morphological pattern (causal pathway) of the apoptotic process and the various circumstances where it occurs, the authors reach the crucial conclusion that apoptosis has a regulatory role within the organism: it 'subserves a general homeostatic function' (Kerr 2002, 471). This regulatory role is the most important difference with coagulative necrosis, the classical type of cell death that had already been described morphologically in detail and which Kerr et al. contrasted to apoptosis. Coagulative necrosis does not regulate cell populations, as it is brought about when homeostatic mechanisms are irreversibly disturbed and is always caused by 'noxious stimuli' (Kerr et al. 1972, 239), whereas apoptosis is caused by both pathological and physiological stimuli. The triggering of apoptosis by *physiological* stimuli is the reason why apoptosis is considered of great importance by the authors. Summing up the events

that led to the 1972 paper, Kerr writes in a later review that '[a] serendipitous confluence of ideas thus made the formulation of the apoptosis
concept virtually inevitable' (Kerr 2002, 473).

3.2.4   Reconstruction of the Argument in Kerr et al.

We can reconstruct the strategy used by Kerr et al. in order to introduce
the mechanism of apoptosis as follows. First, Kerr et al. offered a *theoretical
description* of what they saw as a distinctive causal process. This was a kind
of process that had not been described in detail before and had very specific
morphological features. A main aim of their paper was to describe those
features. Electron microscopy revealed that this process involved the
following morphological changes: condensation of cytoplasm and nucleus,
fragmentation of the nucleus, formation of protuberances on the surface of
the cell, subsequent breaking of the cell and formation of spherical
structures that are membrane-bound and contain cell organelles and nuclei
fragments that are condensed but otherwise functional (named 'apoptotic
bodies' by Kerr et al.) and finally the phagocytation and degradation of
apoptotic bodies by other cells.

Second, the authors specified the *ubiquitous character* of the new process: the morphological changes associated with it are 'essentially the same'
in many circumstances, both physiological and pathological (e.g., in
malignant tumours, in cases of pathological atrophy, in normal development and in cellular turnover in normal adults).

Third, they noted a distinctive feature concerning the new process,
namely, its *non-disruptive* nature. In particular, whereas it was a process
resulting in the death of the cell, it did not produce inflammation. This
enabled them to discriminate it from necrosis, the process that results in
the death of the cell following a toxic stimulus. However, apoptosis was
triggered also by physiological stimuli. It thus seemed to have a specific
role within the organism. This immediately gives rise to the question:
What is a main function of apoptosis?

In order to identify such a function, Kerr et al. noted that a particular
kind of process has to exist, namely, some form of 'physiological cell death'
that is at work balancing divisions in cell populations (1972, 239).
However, necrosis, due to its disruptive nature, cannot play that role.
Apoptosis, however, precisely because it (1) is non-disruptive, (2) can be
triggered by physiological stimuli and (3) is ubiquitous, is a particularly
well-suited candidate to play that role. So, apoptosis 'is well suited to a role
in tissue homoeostasis, since it can result in extensive deletion of cells with

little tissue disruption' (p. 250). The conclusion, then, is that apoptosis plays a crucial regulatory role in tissue homeostasis.

To sum up, Kerr et al. argued as follows. Since apoptosis is

i)   a distinctive morphological process
ii) ubiquitous
iii) non-disruptive, in contrast to necrosis
iv) triggered by physiological stimuli; and given that
v)   a form of 'physiological cell death' that is at work balancing divisions in cell populations must exist; and since
vi) necrosis cannot play that role, while apoptosis is well suited to play that role; therefore,
vii) apoptosis is the mechanism with exactly this regulatory role across animals.

3.2.5   Extraction of Key Features

We can now extract from this case some salient features that are sufficient for the introduction and the characterisation of a new mechanism in biology. First of all, mechanisms are processes or, as we prefer to call them, borrowing a terminology widespread in biology, *causal pathways*. So, the first task that Kerr et al. had to accomplish in order to introduce the mechanism of apoptosis was to offer a theoretical description of a certain process, which is extended over time and is characterised by specific features. The process can be seen as *causal* as it is characterised by a regular sequence of causal steps and difference-making relations among its constituents: the formation of apoptotic bodies is dependent on the fragmentation of the nucleus, which depends on the condensation of cytoplasm and nucleus. The recurrence of this succession of stages under a variety of conditions offered evidence that this is indeed a causal sequence, with a possible genetic basis. As Kerr et al. put it, '[t]he ultrastructural features of apoptosis and its initiation and inhibition by a variety of environmental stimuli suggest to us that it is an active, controlled process' (p. 256).

Second, the description of the mechanism of controlled cell deletion makes clear that the full knowledge of the causal pathway is not necessary for the identification of the mechanism, as the identification of this new type of pathway did not require a full understanding (and hence a full description) of its workings. So, Kerr et al. noted that the mechanism of condensation, a main stage in the apoptotic process, was unknown. As they put it, 'the condensation is presumably a consequence of the extrusion of water, but its mechanism is still unknown' (p. 244). Additionally, in 1972 there was a lot that was not yet known 'of the factors that initiate apoptosis or of the nature of the cellular mechanisms activated before the appearance of the characteristic morphological changes' (p. 255). And of course, nothing was yet known of the biochemical processes underlying apoptosis or its genetic basis. Nevertheless, enough of its causal pathway was known to conclude that it is an 'active, controlled process' (p. 256). This shows that it is one thing to identify a mechanism for a certain phenomenon and it is quite another thing to acquire a full description of its workings. Hence, even with limited knowledge about the various causal details, it is possible to identify and initially describe a potentially important new causal pathway.

Descriptions of a mechanism can be made richer by offering more detailed characterisations. This can involve only the 'horizontal' dimension, as we may call it, of a process, for example, offering further cytological details of apoptosis; more interestingly, it can involve the 'vertical' dimension. Describing the 'mechanisms of condensation', for example, can be done at a cytological level by offering further cytological details, perhaps more fine-grained ones, or by offering details in the vertical dimension, for example, by giving a biochemical description of what is going on. Descriptions of biological mechanisms then are always couched in theoretical language and can be enriched in various ways. In particular, there can be alternative theoretical descriptions of the same mechanism; for example, we now have a cytological and a biochemical description of the apoptotic pathway (see below). Hence, at a minimum, a mechanism is *a certain theory-described causal pathway*.

Third, the regulatory role of apoptosis was crucial for its identification as a distinctive kind of mechanism. As we saw, Kerr himself in earlier work had already observed the process that was to be called apoptosis, but he did not regard it as a new kind of mechanism. Instead, he viewed it as a type of necrosis, that is, as a type of pathological cell death, that was 'nondegenerative in nature' compared with classical necrosis (Kerr 2002, 472). But shrinkage necrosis was not viewed as a new kind of mechanism.

It was the fact that apoptosis seemed to have a basic regulatory function within the organism that led, in the 1972 paper, to its identification as a new kind of mechanism. There, the apoptotic process is not taken to be a type of necrosis anymore; it is contrasted with necrosis and constitutes a new and very important kind of regulatory mechanism of cell death. Whereas the various microscopic observations described in the paper are used to establish a new causal pathway, the regulatory argument offered by

Kerr establishes that the new causal pathway constitutes a new kind of mechanism, thereby introducing a new taxonomy of types of cell death.

So, the function of a particular theoretically described process, that is, its role within the organism, is important when introducing a new biological mechanism.[1] The function serves to further distinguish the process from other similar processes, and it is crucial for establishing not only a distinct but also a new kind of mechanism. So, in our example, both necrosis and apoptosis are mechanisms of cell death. But they are distinct mechanisms since, first, they are characterised by different cytological features, and, second, they play different roles within the organism.[2]

The lesson from our discussion so far: what the identification of a new mechanism in the life sciences requires, as evidenced by the case study of apoptosis, is *a theoretical description of a causal pathway with a certain function.*

3.2.6   Apoptosis after 1972

By the mid 1970s it had already been recognised that 'cell death was as much a part of cell biology as mitosis, extension of an axon, the enzymatic sequence of glycolysis, or secretion' (Lockshin & Zakeri 2001, 547). This was a crucial conceptual breakthrough compared with the older way of thinking about cell death; before the 1970s cell death had not been seen as a phenomenon on a par with mitosis or glycolysis, which constitute fundamental biological mechanisms. However, the important point here is that even after the introduction and the wide recognition of apoptosis in the 1970s, research on apoptosis did not gain the importance it has today.

In the 1990s, however, the field was transformed into a fundamental research area within the life sciences. A major reason for this transformation was the realisation of its central role in many functions within the organism and the fact that the mechanism was understood to such an extent that it was seen as a phenomenon that was medically central. This reflects, then, a second major breakthrough in the history of apoptosis. By

---

[1] When we talk about biological functions of processes of causal pathways here and elsewhere, we mean the role a process plays in an organism that is the result of selection (cf. Millikan 1984; Neander 1991); apoptosis has been selected for regulating cell populations and has thus a homeostatic function, while necrosis does not. Artifacts, as products of design, can also have functions.

[2] New mechanists take mechanisms to be always mechanisms *for* a phenomenon, but typically this is not taken to entail that the phenomenon is some function. See Garson (2018) for an overview of the discussion. We agree that not all mechanisms have functions.

the mid 1990s, cell death 'was recognized as an interesting and biological event'; it was seen not just as 'an incidental part of life' but as 'a highly controlled and medically important element of existence' (Lockshin & Zakeri 2001, 545).

More specifically, there are three main reasons that can explain why the field of apoptosis was transformed from a modest topic to a central field of biological research. First, it was discovered that apoptosis was much more common than was initially thought by the development of techniques that made it easier to identify instances of apoptosis. Second, conserved genes that control cell death were identified, starting with the genes that determine the developmental pathway for programmed cell death in *Caenorhabditis elegans* (see Ellis & Horvitz 1986). Unravelling the molecular genetic mechanisms regulating cell death by using *C. elegans* as a model that began in late 1970s was a breakthrough in the study of regulated cell death. It was thus established that cell death was genetically based, and various important genes that are involved in the regulation of apoptosis were identified (e.g., bcl-2, fas, p53, ced-3; see Lockshin & Zakeri 2001). For example, in the 1990s ced-3 in *C. elegans* was sequenced and it was discovered that it was related to a mammalian protease and that a family of such proteases exist. Known as caspases, these proteases are central components of the apoptosis pathway and their sequence is widely conserved among animals.

The third reason which explains why the field of cell death increased in importance around 1990 was the recognition of the medical importance of apoptosis, which was in turn based on the discovery that apoptosis played a crucial role in several organismic functions. Central for this realisation were the discoveries that apoptosis is very common, is genetically based, is intimately related to the immune system and is important in cancer research. For example, it has been discovered that it is regulated by the p53 tumour suppressor gene. This established a relation between apoptosis and 'differentiation and maintenance of the immune system' (Lockshin & Zakeri 2001, 549). Thus, in the last decades the clinical relevance of apoptosis became evident.

3.2.7   Biochemical Pathways of Apoptosis

The causal pathway of apoptosis can nowadays be characterised not only morphologically, but also biochemically. To illustrate the kind of understanding of the mechanism of apoptosis that we now possess and to compare it with the cytological description outlined earlier, let us briefly

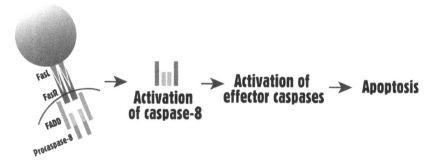

Figure 3.2   The extrinsic pathway of apoptosis.

review the description of the mechanism at the biochemical level, focusing on the mammalian apoptosis pathway (for a more detailed description, see, e.g., Shiozaki & Shi 2004; Cairrão & Domingos 2010).

A central part of the pathway is a process called the 'caspase cascade', which is a positive feedback cascade that involves the activation of caspases, a type of enzyme that when activated performs proteolysis. Caspases exist in the cytoplasm under normal conditions, but not in an active form. Apoptosis occurs when some caspases are activated. Active caspases in turn activate more caspases that eventually break down the cell, producing the morphological changes of apoptosis mentioned earlier.[3] There are two distinct signalling apoptotic pathways: the intrinsic pathway, where the initial apoptotic signal that ultimately leads to the activation of the caspase cascade comes from inside the cell, and the extrinsic pathway, where the initial signal comes from outside.

Here is a brief description of the biochemical mechanism underlying the extrinsic pathway of apoptosis (see Figure 3.2). The initial signal of the extrinsic pathway is the binding of an extracellular ligand to a death receptor, which is an intermembrane protein that has a domain within the cell. For example, T-lymphocytes have a Fas ligand that can bind to the Fas receptor, a protein located on the surface of the cell. The binding of, for example, the Fas ligand to the Fas receptor is the signal for the cell to commit suicide. The Fas receptor has a domain within the cell (Fas-associated death domain [FADD]); when the ligand binds to the Fas

---

[3] Apoptosis can also be produced without a caspase cascade. The central component of this pathway is the apoptosis-inducing factor (AIF), which is located in the intermembrane space in mitochondria. If the cell is damaged, AIF released into the cytoplasm and moves into the nucleus; it then binds to DNA and destroys it (e.g., neurons can commit apoptosis via this process).

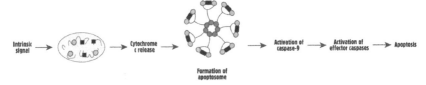

Figure 3.3   The intrinsic pathway of apoptosis.

receptor, FADD is activated and recruits the FADD adaptor protein. Next, procaspase-8 or procaspase-10 binds to the adaptor protein. The formation of this complex, known as the death-inducing signalling complex (DISC), is the signal for the caspase cascade. The procaspases are cleaved and form active caspases 8 and 10. These are the initiator caspases that lead to the activation of effector caspase-3, and so the positive feedback loop of the caspase cascade is activated, which breaks down intracellular proteins.

The biochemical mechanism underlying the intrinsic pathway of apoptosis can be summarised as follows (see Figure 3.3). A central component of the intrinsic pathway is the bcl-2 family of proteins, which include pro-apoptotic and anti-apoptotic proteins. Normally, in the cell there is equilibrium between pro-apoptotic bcl-2 proteins and anti-apoptotic ones, and apoptosis via the intrinsic pathway is prevented. This equilibrium can be disrupted when, for example, the DNA of the cell is irreparably damaged, or the cell stops receiving survival signals. When this happens, other pro-apoptotic bcl-2 proteins are synthesised (so-called BH3-only proteins), which bind to anti-apoptotic bcl-2 proteins. But then, the pro-apoptotic proteins that normally exist within the cell are not inhibited by the anti-apoptotic ones, and can activate the caspase cascade, by causing a central event in the intrinsic pathway, that is, the release of cytochrome c, a protein that normally exists in the intermembrane space of mitochondria, into the cytosol.

This release happens when, because of the disruption of the equilibrium between pro-apoptotic and anti-apoptotic proteins, pro-apoptotic bcl-2 proteins (Bax, Bak) aggregate and form channels in the outer mitochondrial membrane. In normal circumstances where there is a balance between anti-apoptotic and pro-apoptotic proteins, anti-apoptotic proteins (like Bcl-2 and Bcl-xL) bind to the pro-apoptotic ones, thereby stopping them from forming the channels. But when, for example, DNA is damaged and BH3-only proteins are synthesised, they bind to the anti-apoptotic ones, thereby inhibiting the inhibitors and so causing the channels to form. This in turn leads to the release of cytochrome c into the cytoplasm, where it binds to apoptotic protease factor 1 (Apaf-1), which causes Apaf-1 proteins to form a complex called the apoptosome. This activates initiator caspase-9, which then activates effector caspase-3, thereby activating the positive feedback loop of the caspase cascade that breaks down intracellular proteins.[4]

These biochemical descriptions of the signalling apoptotic pathways illustrate the point that there may exist different theoretical descriptions of one and the same mechanism. But another important point here is that the description at the biochemical level is richer than the cytological one in a crucial sense: it provides specific causal information; that is, it identifies difference-making relations among various components of the pathway. Knowing these difference-making relations is important for discovering types of interventions that can be made in order to control the outcome of the process. It is easy to see that the more detailed the description of the causal pathway, the more options for such interventions one has. So, knowing the biochemical description of the apoptosis pathway is important for discovering ways one can intervene to induce apoptosis for therapeutic purposes (for the relations between apoptosis and cancer treatment, see, e.g., Wong 2011). This focus on the potential intervention on causal pathways that is central in biomedicine, then, leads to an important desideratum that theoretical descriptions of causal pathways should have: they should be such so as to provide specific causal information that makes interventions possible (for an account of mechanism in medicine along similar lines, see also Gillies 2017, 2019).

3.2.8   *Causal Mechanism as a Generalised Pathway Concept*

Thus far we have examined the identification of the mechanism of apoptosis, and, based on this example, we have claimed that mechanisms in biology are causal pathways that are theoretically described.[5] There is

---

[4] Of course, this is a simplified version of the molecular pathways that are nowadays known, which are much more complicated. For example, we now know about the existence of inhibitors of apoptosis proteins (IAPs) that inhibit the activity of caspases by binding to them (via a zinc binding domain). Other mitochondrial proteins, however, can inhibit IAPs (e.g., SMAC, DIABLO) by binding to them when they are released from mitochondria (in case of mitochondrial injury). Other mitochondrial proteins (e.g., Htra2/Omi, AIF, endonuclease G) can cleave IAPs when released (cf. Wong 2011).

[5] Again, this should not be taken to mean that mechanisms are some kind of theoretical description; mechanisms for us are causal pathways and it is possible for the same causal pathway to have multiple descriptions. We add that mechanisms are theoretically described so as to highlight that they are not described in metaphysically loaded terms, as in the standard general characterisations that new mechanists offer.

another route to arrive at this notion, this time using the biochemical descriptions of the mechanism of apoptosis discussed above.

The intrinsic and extrinsic signalling pathways of apoptosis are examples of *biological pathways*. Biological pathways are central in molecular biology. According to the National Human Genome Research Institute, 'a biological pathway is a series of actions among molecules in a cell that leads to a certain product or a change in the cell' (www.genome.gov/about-geno mics/fact-sheets/Biological-Pathways-Fact-Sheet). Biological pathways are, for example, involved in metabolism (metabolic pathways), in the regulation of genes (gene-regulation pathways) and in transmitting signals (signal transduction pathways).

Metabolic pathways synthesise various molecules by utilising energy or release energy by breaking down complex molecules. Glycolysis, the pathway that converts glucose into pyruvate, is a central example. Another simpler example is the pathway in *Escherichia coli* that synthesises the amino acid isoleucine. This pathway has five steps that are catalysed by enzymes and, as is typical in biological pathways, exhibits feedback inhibition: if the amount of isoleucine produced by the cell is more than is needed, the accumulated isoleucine inhibits the activity of the enzyme that catalyses the first step in the pathway, and so the production of isoleucine decreases. Metabolic (and other) pathways are of course not discrete, but interact in complex ways with one another. As Nelson et al. put it,

> although the concept of discrete pathways is an important tool for organizing our understanding of metabolism, it is an oversimplification.... Metabolism would be better represented as a meshwork of interconnected and interdependent pathways. A change in the concentration of any one metabolite would have an impact on other pathways, starting a ripple effect that would influence the flow of materials through other sectors of the cellular economy. (2008, 28)

Signal transduction pathways enable cells to perceive and respond to changes in their environment. In signalling pathways, cells receive signals from the exterior environment or from the interior of the cell (e.g., when DNA is damaged), which lead to a series of chemical reactions that ultimately produce a cellular response to the initial stimulus. The intrinsic and extrinsic pathways of apoptosis we have seen are examples of signalling pathways. Last, gene-regulation pathways involve the regulation, that is, the turning on and off, of genes and can form complex gene-regulatory networks. These networks govern expression levels of mRNA and proteins and are central in ontogeny.

To illustrate, let us consider a part of the p53 signalling pathway. This example also shows the interactions between various pathways, as it is one of the ways the intrinsic apoptotic pathway can be initiated. When DNA is damaged, proteins (e.g., ATM, ATR) that can sense DNA damage become activated. These proteins are serine/threonine kinases; that is, they add phosphate groups to serine and threonine residues of other proteins. In particular, they phosphorylate two other serine/threonine kinases (ChK1 and ChK2), which become active and which in turn phosphorylate p53. Also known as the 'guardian of the genome', p53 is a tumour suppressor protein that is continuously produced within cells. In normal circumstances, p53 is prevented from acting: MDM2 binds to p53, deactivating it and also targeting it for ubiquitination; a ubiquitin group binds to the MDM2-p53 complex, and the complex is transferred to the proteasome where it is destroyed. However, when p53 is phosphorylated, MDM2 cannot bind and so p53 can form tetramers and can act as a transcription factor increasing the expression of certain genes, such as genes that code for proteins involved in DNA repair. Also, it can lead to the expression of p21, a tumour suppressor protein that arrests the whole cell cycle (thus preventing mitosis). If the levels of p53 remain high, this means that the cell cannot repair the damage; the only solution is then for the cell to commit suicide. So, p53 triggers apoptosis via the expression of pro-apoptotic proteins that cause the activation of the intrinsic signalling pathway of apoptosis.

This suffices to illustrate the central importance of biological pathways in all biological functions. The question that we want to pose now is the following: Are biological pathways mechanisms? We think that it is clear from how biologists use the notion that pathways are taken to be kinds of mechanisms. Let us illustrate this claim with a couple of examples from various biological fields.

A first example is, again, apoptosis: the pathways of apoptosis are also referred to as 'mechanisms' by biologists. As a second example, consider the following passage from Rang et al.'s pharmacology textbook:

> Cyclic AMP regulates many aspects of cellular function including, for example, enzymes involved in energy metabolism, cell division and cell differentiation, ion transport, ion channels and the contractile proteins in smooth muscle. These varied effects are, however, all brought about by a *common mechanism*, namely the activation of protein kinases by cAMP – primarily protein kinase A (PKA) in eukaryotic cells. Protein kinases regulate the function of many different cellular proteins by controlling protein phosphorylation (2016, 33; emphasis added).

Here, the activation of protein kinases by cAMP, characterised as a mechanism, is part of the cAMP-dependent pathway, which is a signalling pathway. Further examples are phrases such as 'metabolic mechanisms' and 'transduction mechanisms' to refer to metabolic and signal transduction pathways, respectively.

Here is another example, this time from developmental biology. According to Slack et al. (2005, 132),

> Vertebrate embryos are more or less bilaterally symmetrical, and so for many years the nature of the mechanism producing asymmetry was mysterious. In 1995 it was shown that some key genes (NODAL, sonic hedgehog (SHH), and activin receptor IIa (ACVR2A)) have asymmetrical expression patterns in the early chick, and regulated each other to form a pathway linking an initial symmetry-breaking event to the final morphological asymmetry of the heart and viscera.

Here, the mechanism that produces left-right asymmetry is said to be a pathway linking an initial symmetry-breaking event to the final outcome that is to be explained, that is, morphological asymmetry.

We think that these examples suffice to show that pathways are treated in biology as kinds of mechanisms. But does the converse also hold; that is, are all mechanisms in biology biological pathways? It may be argued that the notion of mechanism in biology is broader than the notion of a biological pathway. Thus, consider again the previous definition of a biological pathway, according to which a pathway is defined as 'a series of actions among molecules in a cell that leads to a certain product or a change in the cell'. Such a definition may imply that a biological pathway is a specifically *biochemical* concept. In contrast, the notion of mechanism can be viewed as a more general notion. For example, when Alberts et al. write about the 'fundamental mechanisms of life – for example, how cells replicate their DNA, or how they decode the instructions represented in the DNA to direct the synthesis of specific proteins' (2014, 22); the 'universal genetic mechanisms of animal development' (p. 39); or the 'fundamental mechanisms of cell growth and division, cell–cell signaling, cell memory, cell adhesion, and cell movement' (p. 1154) – it may be argued that it is not a specifically biochemical conception they have in mind. In addition, if a pathway is viewed as a biochemical concept, then physiological or pathological mechanisms cannot be biological pathways. Last, if a biological pathway only involves a series of changes within a cell, then mechanisms that involve many cells (e.g., morphogenetic mechanisms) cannot be identical to pathways.[6]

Even if this is the case and the notions of a biological pathway and a biological mechanism are not identical, we think that the biochemical notion can be straightforwardly generalised to arrive at a general notion of a causal pathway that, as we are going to argue throughout the book, is present in biology (and elsewhere). According to this more general notion, which we can call the 'generalised pathway concept', a mechanism is a sequence of causal steps (and not just of actions among molecules) that lead from an initial stimulus or cause to an effect (and not just to a cellular response) and is not confined within a cell, but can involve many cells, and more generally can include components from all levels of biological organisation.[7] Such pathways, then, can be described in molecular, cytological or other (even non-biological, e.g., psychological or sociological) terms. It is in this more generalised sense that the cytological descriptions of apoptosis by Kerr et al. are causal pathways. Some justification for the presence of a generalised pathway concept is the fact that pathway-talk is used in biology even in areas outside molecular biology. For example, biologists talk about 'cytological pathways', 'ecological pathways' and 'evolutionary pathways'. Our claim, then, which will be further corroborated in subsequent chapters that will discuss various other examples of mechanisms, is that mechanisms in biology can be identified with causal pathways in this more general sense.[8] According to this generalised pathway concept, then, or *Causal Mechanism*, as we prefer to call our account:

(CM) a mechanism = a theoretically described causal pathway.

---

[6] Such considerations may underlie the claim made by Giovanni Boniolo and Raffaella Campaner (2018) that mechanisms and pathways in molecular biology are not identical notions. See also Ross (2021) for a treatment of the notion of pathway in biology that shares some similarities with our account. Ross claims that pathways and mechanisms are different notions. In part, she arrives at this conclusion because she uses a notion of mechanism with which we do not agree. She writes: 'When [biologists] use the mechanism concept they often suggest that some biological phenomenon can be understood as a kind of machine or mechanical system.... This machine analogy encourages thinking of biological phenomena as having component parts that are spatially organized and that causally interact to produce some behavior of the system' (134). While we tend to agree that this is how many new mechanists think about mechanisms, we do not think that this is the best way to characterise the notion of mechanism as used by biologists. Ross also suggests that pathways abstract from causal detail, while mechanisms do not. We disagree, as many mechanisms abstract from causal detail too. Our view is that no clear distinction is to be found in biology between mechanism-talk and pathway-talk. Notions such as mechanisms, pathways, cascades and mediators all are instances of the same general concept of causal pathway.

[7] We discuss the important issue of the relationship between causal pathways and levels of organisation in detail in Chapter 9.

[8] Although in biology mechanisms as causal pathways often have function, sometimes they do not. See Section 3.4 and Chapter 4, Section 4.5.2.2, for details.

The advantage of this strategy to arrive at CM is that we start from a concept that is clearly central in biology and for which there even exists a definition, that is, the notion of a biological pathway, and we generalise it, in order to arrive at the more general notion of CM. We then use CM to illuminate the concept of a biological mechanism, for which no explicit definition exists in the biological literature (which, incidentally, suggests that mechanisms and biological pathways in the narrow sense are not viewed as identical concepts by biologists). We can thus illuminate the more general concept (i.e., the concept of a mechanism), which we take to be a concept-in-use central in biological practice, without using ordinary notions of what a mechanism is (such as the notion of a 'machine' as used in everyday contexts) or introducing metaphysical categories that seem to be extraneous to biological practice, but by using concepts already present in biology (i.e., the biochemical concept of a pathway).

Let us now turn to some objections to CM, focusing again on the example of apoptosis. It might by objected that the features of mechanisms in biology identified in Section 3.2.5 – that is, that mechanisms are causal pathways theoretically described – are not in fact exemplified in the case of the signalling pathways of apoptosis, as well as in other cases of biological pathways. That is, it might be claimed that apoptosis (1) is not linear, (2) has no obvious start and end points, (3) involves homeostasis and (4) is such that spatiotemporal organisation is crucial. Collectively, it might be argued, these features cannot be accounted for without a more metaphysically inflated account of mechanisms, such as those introduced in Chapter 2.

By way of reply, it should be noted that all the above features are compatible with our account of mechanisms as theoretical descriptions of causal pathways. First, pathways are processes; but it does not follow that they have to be 'linear' processes. Apoptosis as characterised cytologically might seem 'linear', that is, as a stepwise process with no negative or positive feedback loops. However, its biochemical description is far more complex, with multiple pathways and complex causal cycles, as is typically the case in biological pathways. But in any case, we still have, ultimately, a web of sequences of causal steps related by difference-making relations. Second, we do not take it as a general requirement that a process has to have well-defined starting and end points; this is something to be determined by biological practice. In the case of the extrinsic pathway of apoptosis, for example, the starting point of the process is specified by the binding of a ligand, such as the Fas ligand, to a death receptor. Third, homeostasis, which itself involves negative feedback loops, can also be described in terms of difference-making relations among the components of the pathways underlying the homeostatic process.[9] Fourth, the spatio-temporal organisation is indeed crucial for the functioning of any causal pathway, and because of this it is typically included in the theoretical description of the pathway: a detailed theoretical description of the causal pathway will state the spatiotemporal relations among the various components of the pathway that are required for the proper functioning of the mechanism.

To sum up, the example of apoptosis shows that some salient features of mechanisms in biology are the following: mechanisms are causal pathways described in theoretical language that have certain functions; these descriptions can be enriched by offering more detailed or fine-grained descriptions; the same mechanism can then be described at various levels using different theoretical vocabularies (e.g., cytological vs biochemical descriptions in the case of apoptosis);[10] last, the descriptions of biomedically important mechanisms are often such that they contain specific causal information that can be used to make interventions for therapeutic purposes.

3.3   Mechanisms of Cell Death

An important issue that crops up here for CM is how the various causal pathways are identified and distinguished from each other. Could it be the case that some causal pathways are *mechanisms* in a more robust sense while others are *merely* causal pathways? In the case at hand, the issue is how apoptosis can be distinguished from other cell-death processes. As we have noted already, the activation of caspases underlies the morphological changes that occur during apoptosis. However, not all processes of cell death are apoptotic. That is, not all processes that lead to cell death feature the particular sequence of morphological changes associated with apoptosis, nor do all involve caspase activation. This is why in the literature on cell death there are more general terms that are used to refer to the various

---

[9] See Woodward (2002) for a description of the *E. coli* lac operon regulatory mechanism in terms of difference-making; as Woodward notes, difference-making better captures the causal relationships present in negative regulation that is involved in homeostatic mechanisms, which involves cases of so-called double prevention that present difficulties to alternative characterisations of within-mechanism interactions.

[10] It is a widely held view in the philosophical literature on mechanisms that mechanisms form hierarchies (see Povich & Craver 2018 for a discussion of mechanistic levels). While we agree that a mechanism such as apoptosis can be described at various levels, we do not derive any ontological consequences from such talk (see Chapter 9 for our views on levels).

types of processes of cell death: on the one hand, there is 'physiological' or 'regulated' cell death; on the other, 'accidental cell death'. Lockshin and Zakeri explain the difference between the two by noting that when cells die as a result of a process of physiological cell death, 'such deaths are part of the normal function of the organism'; also, in physiological cell death genetic regulation is central. This is not the case in 'necrosis or oncosis which is accidental and in which the cell has no active role' (Lockshin & Zakeri 2001, 545).

Could it then be the case that there is a genuine difference between apoptosis and necrosis such that apoptosis is a mechanism in a more robust sense than just the causal pathway for cell deletion, whereas necrosis is *merely* a causal pathway? To address this issue, let us note first that, as Guido Majno and Isabelle Joris (1995) have pointed out, apoptosis and necrosis should not be juxtaposed: apoptosis is a process that leads to cell death, but necrosis should not be used to refer to such a process. As they stress, there is a distinction between cell death and necrosis: cell death comes about before necrotic changes can be observed. Necrotic changes (e.g., karyolysis, karyorhexis, loss of structure in the cytoplasm) 'are the features of a cell's cadaver, whatever the mechanism of the cell's death, be it ischaemia, heat, toxins, mechanical trauma, or even apoptosis' (Majno & Joris 1995, 11). This has actually been the traditional meaning of necrosis; that is, it refers to changes in tissues that are visible even without a microscope and as such occur after cell death.[11]

This point has been emphasised also by other authors. So, according to Kanduc et al. (2002, 167), it is 'scientifically unjustified' and 'unsound' to compare apoptosis to necrosis, as apoptosis is a *process* that leads to cell death, whereas necrosis refers to changes that occur to cells *after* they die. In general, then, we should distinguish between the processes that a dying cell undergoes (e.g., apoptosis) and the end result of these processes, which is the dead cell. 'Necrosis' should then refer to already dead cells and tissues and the changes that occur *after* cell death (see also Fink & Cookson 2005). This usage of the term 'necrosis' is precisely the one suggested by the Committee on the Nomenclature of Cell Death chartered by the Society of Toxicologic Pathologists to make recommendations 'about the use of the terms "apoptosis" and "necrosis" in toxicity studies' (Levin et al. 1999, 484). In line with the above, this committee recommended that

---

[11] For example, here is how Rudolf Virchow in the nineteenth century describes necrosis in *Cellular Pathology*: 'In necrosis we conceive the mortified [gangrenous] part to be preserved more or less in its external form' (quoted in Majno & Joris 1995, 3–4).

'when dead cells or tissues are observed in a histological lesion, "necrosis" is the appropriate morphological diagnosis, regardless of the pathway by which the cells or tissues died' (p. 486). They conclude: 'This Committee believes that returning to the long-established histopatholog-ical standard wherein the word necrosis denotes dead cells in a living tissue (regardless of their phenotype) should help to alleviate the confusion attendant on the notion, held by many, that a dichotomy exists between apoptosis and necrosis' (p. 489). Notably, Robert Sloviter (2002, 22) goes as far as to note that 'the term necrosis is now virtually meaningless because "necrotic" means nothing more than "dead"'.

According to Majno and Joris, 'the major sore spot in the nomenclature of cell death is precisely the lack of a suitable name for cell death that occurs not by apoptosis but by some external agent' (Majno & Joris 1995, 11). And this is precisely the point: cell death might have different causal pathways, and the difference between them is not that one (or some) of them counts as a mechanism while the other does not; rather, the differ-ence is in *how they are described.*

Majno and Joris's own suggestion is to characterise apoptosis in contrast with a specific process of cell death they call 'oncosis' (p. 12). A common non-apoptotic causal pathway that leads to cell death is when groups of cells die of ischaemia (*ischaemic necrosis*). During the causal pathway that leads to ischaemic necrosis the cell swells, and it is in order to capture this swelling process that the authors propose 'oncosis' as a term to refer to this mechanism. This causal pathway can nowadays be characterised in detail: reduced supply of oxygen and nutrients leads to ATP depletion, which ultimately results in protein denaturation, enzymatic digestion due to damaged lysosomes and loss of integrity of the plasma membrane, result-ing in influx of water and calcium into the cell, leading to swelling and ultimately rupture of the cell. Also, we know that ischaemia typically activates the causal pathway, but toxic agents can also initiate it. We can then talk about the mechanism of oncosis in the CM sense of the term. Oncosis and apoptosis, then, are two causal pathways (and hence mech-anisms) of cell death – by swelling and by shrinkage, respectively.

Can there be a way to distinguish apoptosis from oncosis such that only the former counts as a mechanism? We will consider three distinctions used by researchers of cell death that might be used in order to do this: (1) processes of physiological versus accidental cell death, (2) processes of programmed versus non-programmed cell death and (3) active versus passive processes.

A term widely used to describe a non-apoptotic, non-physiological type of cell death is 'accidental cell death'. By describing a process of cell death as 'accidental', biologists try to capture the idea that, in contrast to apoptosis, this is not a process that occurs under normal conditions, nor does it serve a general homeostatic function within the organism. However, as Majno and Joris (1995) note, apoptosis can also be induced by a variety of 'accidental' causes (e.g., heat, chemical agents, viruses). Levin et al. (1999) also note that 'dead cells having the cytological features of apoptosis can occur in large numbers as a pathological change, e.g., "single cell necrosis" in the liver and lymphocyte necrosis in the thymus, and that these changes can be induced by exogenous events such as exposure to toxicants' (p. 485). So, the very fact that apoptosis can be initiated by 'accidental causes' shows that the right contrast here is not between mechanism and non-mechanism, but rather between physiological cell death and accidental cell death, where 'physiological cell death' refers to a process of cell death that was initiated by physiological stimuli.

What about the programmed/non-programmed distinction? An important point here is that 'programmed cell death' and apoptosis should not be identified; thus, the former cannot be used to distinguish the latter. Programmed cell death is the phenomenon where cells die 'on schedule'; that is, they are programmed to die at a specific time. During development of the chick, for example, the morphology of the wing is produced as a result of the death of groups of cells; there is a 'genetic clock' that determines when the cells will die. But when the time comes for the cell to die, the specific programme that determines the form that cell suicide will take is triggered. The particular form of cell suicide can be apoptosis (indeed, very frequently it is), but it need not be. As Majno and Joris stress: 'The genetic programme of programmed cell death is a clock specifying the time for suicide, whereas the genetic programme of apoptosis specifies the weapons (the means) to produce instant suicide' (1995, 11): the weapon is precisely what we call the causal pathway. Again, the point here is that one cannot use this distinction to distinguish apoptosis as a genuine mechanism.

Biologists, we noted above, have characterised the contrast between processes of cell death along the active/passive lines: apoptosis was described as 'active' from the very beginning of its introduction, and contrasted with the 'passive' necrosis.[12] Perhaps what makes apoptosis a mechanism, then, is precisely that it is an active process. However, what does this distinction really mean? The idea here seems to be that in the case

---

[12] However, see Proskuryakov and Gabai (2010) for the argument that necrosis can in certain cases be considered an active and well-regulated process.

of apoptosis the cell is itself involved in its own demise ('cell suicide'), whereas in the case of oncosis the cell dies as a result of some exogenous influence ('cell murder'). That is, apoptosis involves a 'suicide programme' that is initiated under various circumstances and that is genetically based, in the sense that there exist specific genes that code for various components of the biochemical pathway underlying the apoptotic process (the distinctions regulated vs non-regulated, ordered vs unordered, controlled vs non-controlled appear to be used in a similar way).

To make this clear, Sloviter (2002, 23), after describing the two causal pathways as 'active cell death' (ACD) and 'passive cell death', notes that ACD is active in the sense that it requires 'active intracellular processes for death to result', whereas in passive cell death 'the cell plays no role in its own demise'; that is, cell death is 'immediate and involves no cellular activity', the cause being exogenous to the cell such as 'rapid freezing, aldehyde fixation, heat denaturation, and catastrophic physical destruction'. As such, passive cell death is of little interest since being immediate it 'offers no therapeutic window'.

The important point for us is that this difference between 'active' and 'passive' is merely a difference concerning the details of each causal pathway. Hence, there is no intrinsic difference between the two causal pathways as such: there is nothing particularly active in apoptosis and particularly passive in oncosis. The significant difference from a biological point of view is that because apoptosis involves a 'suicide programme', it can serve a homeostatic function, as argued by Kerr et al. (1972).

To avoid the insinuation that the use of 'active' and 'passive' processes might lead to views about the ontology of causation and mechanisms (e.g., to the distinction between entities and activities, as in Machamer, Darden & Craver 2000), let us see how biologists view the active/passive distinction. Kanduc et al. (2002) say:

> It is frequently assumed that the death of cells can be passive. This non-biological point of view on cell death ignores the role of cell death in cell development and adaptation. It cannot be assumed that 'ordinary' cell death or 'necrosis' is a passive process while the presumed special form of cell death, 'apoptosis' is active. *Both the ante-mortem and postmortem changes are active since both are enzyme-catalysed biochemical reactions.* (pp. 167–8, emphasis added)

So, to call a biological process 'active' should not commit us to a specific metaphysical view about the metaphysics of causation. In particular, it is not a reason to adopt an activities-based understanding of mechanisms as opposed to a difference-making account (see Chapter 6 for a detailed examination of activities as an ontic category).

3.4   Is Mechanism More than the Causal Pathway?

Hence, apoptosis and oncosis can both be considered mechanisms in the CM sense: they are both causal pathways that produce a result (apoptotic and ischaemic necrosis, respectively). However, the history of programmed cell death and apoptosis during the last 60 years might be used to argue that in biological practice what counts as a *biological* mechanism cannot just be a matter of identifying a specific causal pathway. Apoptosis seems to be a special kind of causal process with distinctive features that deserves to be labelled a mechanism. This can even be seen in biologists' description of apoptosis as a 'mechanism of cell death', whereas necrosis or oncosis are not usually described as 'mechanisms'.

Note that the reason apoptosis became a central biological mechanism is not due to some feature internal to the sequence of causal steps that constitutes the apoptotic pathway, but rather due to features that are external to the pathway itself; that is, because it is a key process that controls homeostasis. As we have seen, it is its role within the developing and adult organism, that is, its homeostatic function, that led to the formulation of the concept by Kerr et al in 1972; similarly, it is the discovery of its highly controlled nature and conservation of the genetic sequences of the components of the apoptotic pathway across animals (which shows its central functional importance in animals), as well as the realisation of the close relation between apoptosis and the immune system and cancer that followed the molecular genetic discoveries of the 1980s and 1990s, that gave it the central prominence it deservedly has today as a biological phenomenon.

To put the point differently, what we think the story of apoptosis shows is the following: in the world there are causal pathways for various phenomena; all causal pathways can be deemed mechanisms in the deflationary sense of CM; but not all those causal pathways are *biologically* interesting or significant, even if they occur frequently within organisms. Biologically interesting or significant causal pathways are those pathways that subserve a central *function* within the organism; that is, whether a causal pathway is biologically interesting has to do with features *external* to the pathway itself.

This does not imply that what is biologically interesting is something subjective. Rather, it implies that it is directly related to biological practice: what the community of biologists regards as the basic phenomena that must be explained in order to have both biological understanding and apply our knowledge clinically. Thus, we could say that what makes a causal pathway a specifically biological mechanism is not something internal to the pathway itself; rather, it concerns the functional role of that pathway within the organism. In other words, whether a causal pathway is considered a biological mechanism by biologists has to do with a relational property of the pathway. This relational property concerns how the pathway is related to the other processes that occur within the organism so as to subserve a certain function. In the case of apoptosis, this relational property concerns the homeostatic function that it subserves. This relational property is the difference between a causal pathway like apoptosis and a causal pathway such as oncosis.

However, suppose one were to argue as follows. We should certainly let biological practice itself decide what we should mean by a 'mechanism' in a biological context. If practice has it that a causal pathway is deemed a mechanism by an appeal to external features of the pathway, so be it. CM (the point would be) is false, since

(P-CM) A mechanism is a theoretically described causal pathway +X, where X is some biologically significant external feature of the causal pathway (like its function).

Now, if one were to argue like this, we would not seriously object. We are ready to accept that there may well exist features external to a particular causal pathway and, in particular, features that can be established by looking at biological practice, which determine whether a specific theory–described causal pathway counts as a biological *mechanism*. But, in our view, this attribution of 'mechanism' to certain causal pathways and perhaps not to others would entail that 'mechanism' is an honorific term attached to causal pathways that have certain (external) features. The further scientific question, then, is whether there is evidence that a causal pathway is a 'mechanism' in this sense or not.

In our view, the choice between CM and P-CM is not particularly significant: to adopt P-CM is to claim that we allow a distinction between a causal pathway for a phenomenon P and a specifically *biological* mechanism, where the difference between the two concerns an external feature of the respective causal pathways. Be that as it may, the important point is that both views are licensed by Methodological Mechanism: to be committed to either option, one need not be committed to some metaphysical view about causation or the ontology of mechanisms. As we will argue in Chapter 4, there is no need to do this in order to understand scientific practice.

Could someone insist that there is some other feature that differentiates a causal pathway from a mechanism? A possibility here is to adopt the requirement of causal modularity (cf. Woodward 2002). Causal modularity may be seen as the criterion that determines whether a process counts as machine-like or not; so, perhaps 'mechanism' should be used only for causal processes that exhibit causal modularity. While modularity can be important in many cases as a requirement for a causal representation of a system, the major disadvantage of this view is that many instances of 'mechanisms' in biology turn out not to be such, since they are not modular; apoptosis is a case in point (cf. Cassini 2016). So, adopting this view necessitates abandoning taking scientists' talk of mechanisms at face value.

Could someone insist that there is some other feature that differentiates a causal pathway from a mechanism? A possibility here is to adopt the requirement of causal modularity (see Woodward 2002). Causal modularity may be seen as the criterion that determines whether a process counts as machine-like or not; so, perhaps 'mechanism' should be used only for causal processes that exhibit causal modularity. While modularity can be important in many cases as a requirement for a causal representation of a system, the major disadvantage of this view is that many instances of 'mechanisms' in biology turn out not to be such, since they are not modular; apoptosis is a case in point (see Casini 2016). So, adopting this view necessitates abandoning taking scientists' talk of mechanisms at face value.

3.5   What Does the Case of Apoptosis Show?

What can we learn from the case of apoptosis? In our examination of this case we have use two strategies. The first strategy was to examine the cytological description of the mechanism of apoptosis in order to extract some main features of mechanisms in biology. The most important such feature that we found is that mechanisms are processes, that is, sequences of causal steps that we called causal pathways. The second strategy was to examine the biochemical pathway concept and to generalise it, in order to arrive at a generalised pathway concept. This strategy enabled us to argue that all causal pathways are mechanisms and all mechanisms are causal pathways. Both strategies lead to a general notion of a causal pathway (or mechanism) that we think is the typical notion of mechanism in the life sciences and elsewhere.

But of course apoptosis, it can be argued, is just one example. A full defence of the claim that mechanisms are identical to causal pathways would require the examination of many more cases. In a sense, this is correct. In the subsequent chapters of the book, we will examine various other cases of mechanisms and argue that they just are causal pathways. By the end of the book, our thesis that mechanisms are causal pathways will have become much more convincing.

However, the example of apoptosis is not just one case study among others. We take this case to be paradigmatic. First, the pathways of apoptosis illustrate a central concept in molecular biology, the concept of a molecular pathway. There are many specific examples of molecular pathways in current biology that are similar to the molecular pathways of apoptosis, some of them mentioned in Section 3.2.8. The second strategy, in particular, involved not just the particular case of the signalling pathways of apoptosis, but a general biochemical pathway concept that is prevalent in molecular biology. It is this general molecular concept that we generalised further in order to arrive at our generalised pathway concept and finally argue that all pathways are mechanisms, and vice versa. Nothing in this strategy depends on the particular case of molecular pathway chosen.

A second reason why we take apoptosis to be paradigmatic is that (as has been mentioned in Section 3.2.1) it has some very important features, compared with other cases. Apoptosis nicely illustrates how a *new* mechanism was first introduced, and thus it clearly shows what is involved in first identifying and describing a biological mechanism. Moreover, it illustrates several other features of mechanisms, such as that they can be described at many different levels, their relation with functions and how they are differentiated from similar mechanisms (such as necrosis). Being a mechanism that is central in various biological fields, apoptosis also illustrates a concept of mechanism that is common across biology. Last, it is a fundamental biological mechanism, but relatively unknown among new mechanists. In general, then, the example of apoptosis is for us a lot more than a case study.

It is also important to note that the case of apoptosis, as well as other cases of pathways and mechanisms described in this chapter and throughout the book, are not intended as premises of an inductive argument, the conclusion of which is that mechanisms are to be equated with causal pathways. If our argument were an inductive one, a lot more examples would be required to establish its conclusion. However, we do not view the generalisation from apoptosis to CM as a general concept of mechanism that applies widely within biology, as an induction; we take the apoptosis case as a paradigmatic illustration rather than as the starting point of an inductive argument. So, we think that the important issue is not just to include more cases but to include important and paradigmatic cases that illustrate our view.

We think, then, that the discussion in this chapter shows that the generalised pathway concept is prevalent in biology (subsequent chapters will further corroborate this claim). What remains to be shown is whether this concept suffices in order to capture all typical uses of 'mechanism' in biology. In other words, are there mechanisms that are not just causal pathways (nor pathways with a certain function), but something more? To answer this question, more cases should be examined. It is also important to discuss cases that are disciplinarily distinct, in order to be able to argue that mechanisms as causal pathways are to be found across the life sciences and elsewhere. In the subsequent chapters we will have the opportunity to apply the generalised pathway concept to various cases. For example, we think that this concept is easily applied to mechanisms in evolutionary biology, which have been thought to be problematic cases for other general characterisations of mechanism such as the MDC account (see Skipper & Millstein 2005; we will come back to this issue in Chapter 10).

Last, does the analysis in this chapter provide reasons to prefer CM to other accounts of mechanism? One could argue that if Minimal Mechanism or the MDC account can capture this case equally well, then the example of apoptosis does not help us very much. Let us first note that the main purpose of this chapter is not to claim that CM captures the case of apoptosis better than other accounts of mechanism. Our main aims here, and the reasons why we discuss at such length the apoptosis case, are, first, to examine the notion of a biological pathway – a systematic examination of this notion is absent from the literature and part of our aim in this chapter is to provide such an analysis – and, second, to use it in order to examine the concept of a biological mechanism. We will further develop CM and provide some reasons to prefer it to metaphysically inflationary accounts in Chapter 4. In Chapters 5 and 6 we will criticise inflationary accounts in detail and in Chapters 8 and 9 we will provide reasons why CM is preferable to Craver's account. So, we have various other reasons to prefer CM to other accounts of mechanism that are only indirectly related to our analysis of the apoptosis case. However, let us close this chapter by briefly mentioning two main reasons why the analysis in this chapter can be used to single out CM (we will come back to these points in Chapter 4). First, the examples of mechanisms examined in this chapter strongly suggest that typical and paradigmatic mechanisms in biology are causal pathways, as opposed to complex systems (Glennan 1996, 2002) or Craver's (2007a) 'constitutive' mechanisms. Second, while accounts that put more emphasis on the processual nature of mechanisms such as the

MDC account are consistent with the apoptosis case, if CM is also consistent with this example and CM is the more minimal view, this surely must be a reason to prefer CM over ontologically more inflated accounts. The important issue here is to identify the features of the concept-in-use. In particular, we take our case study to show that, as will become clearer in Chapter 4, the excess ontological content of the MDC and other accounts does not play a role *within scientific practice*.

Mechanisms as Causal Pathways

4.1 Preliminaries

Our key thesis, which we call *Causal Mechanism*, is that a mechanism is a causal pathway that is described in theoretical language, where the pathway is underpinned by networks of difference-making relations. In advancing CM, we will show:

1. that there is indeed a common notion of mechanism in virtue of which different kinds of mechanisms (at least in life sciences) count as such;
2. that this general account is ontologically minimal (in particular, it incorporates only a notion of causation as difference-making); and
3. that non-causal constitutive relations are not required to understand what a mechanism is in scientific practice.

Since we are offering a causal account of mechanism, we take it that the concept of causation is central in characterising a mechanism. We argue, however, that causation is a relation of robust dependence among particulars and that stronger accounts of causation are not required to understand the concept of a causal pathway.

4.2 Causal Mechanism: Three Theses

According to current mechanistic approaches to causation, causes *produce* their effects, where production is cashed out in terms of mechanisms. Mechanisms are taken to be complex systems or, in general, acting entities, which are composed of parts and have internal structure or organisation and certain spatiotemporal locations.[1] The mechanism has a characteristic

---

[1] The term 'complex system' is used by Glennan (1996, 2002), but then he drops it. MDC (2000) and Craver (2007a) give more processual accounts. See also Krickel (2018, chapter 2), who distinguishes between the 'Complex System Approach' (Bechtel & Abrahamsen, early Glennan) and the 'Acting

Figure 4.1 A mechanism as causal pathway.

behaviour in virtue of the properties, dispositions or capacities of its parts as well as in virtue of how these parts are organised and interact with each other. *What* the mechanism is doing (its characteristic activity, its behaviour or its output) is caused and explained by the details of *how* it is doing it. These details include the internal workings of the mechanism.

We shall critically discuss the mechanistic approach in some detail in Chapter 5. Suffices it to say that CM is not a version of a mechanistic account of causation; rather, it takes causation (as difference-making) to be conceptually prior to the notion of a mechanism (as a causal pathway). Let us now examine the main commitments of CM in more detail.

We have used the case of apoptosis as a benchmark to develop a characterisation of the concept of mechanism as used in biological practice. The core idea of this practice-based account is that mechanisms in biology are equivalent to stable causal pathways, described in the language of theory, which bring about a certain effect and (at least in some cases) perform a certain function (see Figure 4.1). We have called this account of mechanism *Causal Mechanism*:

(CM) A mechanism is a theoretically described causal pathway.

By way of further clarifying CM, it is useful to distinguish between three theses that together constitute CM. The core thesis of CM is that mechanisms are causal pathways. But, moreover, for CM causal relations among the components of the pathway are to be viewed in terms of difference-making. We will call this thesis the *difference-making* thesis of CM. In addition, according to CM the causal pathway has to be described in theoretical terms and not in terms of one's favourite metaphysics. CM then remains agnostic about the underlying metaphysics of mechanisms. We will call this thesis the *metaphysical agnosticism* thesis of CM. Let us examine these three theses in more detail.

The mechanisms-as-pathways thesis differentiates CM from most accounts of mechanism found in the literature. As already noted, some

---

Entities Approach' (Machamer et al.). In saying that mechanisms in biology are to be viewed as causal pathways, we disagree with both of these approaches.

new mechanists take mechanisms to be complex systems. Even if this is not explicitly incorporated in a general characterisation of mechanism, mechanisms are often viewed as kinds of organised entities or structures. The underlying intuition here seems to be the everyday notion of a mechanism or a machine: an organised structure made of parts that interact. But if the aim is to identify the notion of mechanism implicit in biology, we think that this is misleading, since mechanisms and systems, for biologists, are different things: mechanisms, we think, are best viewed as causal pathways, and it is strange to call a causal pathway a 'system'. In contrast to the view of mechanisms as complex systems, then, CM takes mechanisms to be processes. The main reason for viewing mechanisms in biology in terms of processes rather than in terms of systems, entities or machine-like structures, then, is that this is the notion that is at work in central examples of biological practice. The case of apoptosis examined in Chapter 3 provided support for this view.

It is important also to note, in relation to this core thesis of CM, that although we call our position *Causal Mechanism* in order to stress the close relationship between mechanism and causation, as well as the fact that causation as a concept is prior to mechanism, on our view a mechanism just is a causal pathway and every causal pathway is a mechanism. In other words, CM *identifies* mechanisms with causal pathways, both ontologically and conceptually; mechanisms and pathways are not distinct notions. Hence, it is not the case that some mechanisms are *causal* whereas others are not.

The view that comes closer to CM regarding the core thesis of mechanisms as causal pathways is the recent account of Donald Gillies about mechanisms in medicine. He takes mechanisms in medicine to be causal pathways connecting a cause with a particular effect. Gillies sums up his account as follows: 'Basic mechanisms in medicine are defined as finite linear sequences of causes ($C_1 \rightarrow C_2 \rightarrow C_3 \rightarrow \ldots \rightarrow C_n$), which describe biochemical/physiological processes in the body. This definition corresponds closely to the term "pathway" often used by medical researchers. Such basic mechanisms can be fitted together to produce more compli-cated mechanisms which are represented by networks' (Gillies 2017, 633).

In discussing the biochemical descriptions of the signalling pathways of apoptosis in Chapter 3, we have claimed that causal relations within a pathway are best understood in terms of difference-making. This is then a basic tenet of CM, which can now be expanded as follows: mechanisms are causal pathways involving difference-making relations among the compo-nents of the pathway. The *difference-making thesis* of CM differentiates

CM from accounts of mechanism that characterise within-mechanism interactions in terms of activities or the manifestation of powers. This second thesis of CM is shared by other accounts of mechanism. For Woodward (2002), difference-making is required to account for within-mechanism interactions. As he puts it, 'components of mechanisms should behave in accord with regularities that are invariant under interventions and support counterfactuals about what would happen in hypothetical experiments' (Woodward 2002, 374). In his 2002 paper, Glennan has adopted this view. Peter Menzies (2012) has used this interventionist approach to causation to give an account of the causal structure of mechanisms. Gillies also views causation in terms of difference-making. From those accounts, Gillies's view more closely resembles CM.[2] For Glennan, in particular, it is not the case that causation is conceptually prior to mechanism, since he uses mechanisms to offer a theory of causation, as we will see in more detail in Chapter 6.

The thesis of *metaphysical agnosticism* is a novel feature of CM that differentiates it from all extant accounts. Because of this thesis, the description of the pathway, according to CM, has to be given in the theoretical terms of the specific scientific field (or fields) and not in terms, for example, of entities and activities. Moreover, it is because of this thesis that CM puts the methodological role of mechanism at the centre; this will lead to the general framework that we will call Methodological Mechanism, which will be developed in Chapter 10.

It is important to stress that the theses of *metaphysical agnosticism* and *difference-making* are logically independent, as one can hold one, but not necessarily the other. For example, one can understand within-mechanism causal relations in terms of difference-making, but at the same time use metaphysical categories to understand what a mechanism is, or subscribe to a specific view about the truth-makers of causal relations. For example, one can hold a neo-Aristotelian view according to which causal relations supervene on the powers of entities, and provide an analysis of mechanisms in neo-Aristotelian terms. In advancing CM, we will reject this option. We

---

[2] Woodward's (2011) account of mechanisms also shares similarities with CM. Woodward is skeptical that there exists a common notion of mechanism present across diverse scientific fields. He writes: 'I believe it is a mistake to look for an account of "mechanism" that will cover all possible interpretations or applications of this notion. A better strategy is to try to elucidate some core elements in some applications of the concept and to understand better why and in what way those elements matter for the goals of theory construction and explanation' (p. 410). We think that mechanisms as causal pathways capture a core notion of mechanism present in many different scientific contexts.

will argue that in giving a characterisation of mechanism as a concept-in-use, one need not be committed to a specific view concerning the metaphysics of mechanisms: mechanism in our sense is a concept used in scientific practice and as such it is primarily a methodological concept. This will pose a challenge to those philosophers who adopt metaphysically inflationary accounts of mechanism: given the adequacy of CM to capture the concept-in-use, the burden is on the defender of a particular metaphysical characterisation of mechanism to say why a methodological account such as CM is not enough and why it should be inflated with metaphysical categories such as entities and activities – we will come back to this point in Section 4.5.

Conversely, one may accept *metaphysical agnosticism* by remaining neutral on the underlying metaphysics, without subscribing to a difference-making account of within-mechanism interactions. This shows that the thesis of *difference-making* needs independent justification. Our examination of the discovery of the mechanism of scurvy in Section 4.3 will start to provide such a justification and will show that *difference-making* is central for understanding mechanism as a concept-in-use.

It is also possible to hold *metaphysical agnosticism* (and perhaps also *difference-making*) and reject the core thesis of CM that mechanisms are causal pathways, in favour of the view that mechanisms are complex entities or, more generally, kinds of organised entities. That is, one could remain neutral on how to understand entities, activities and organisation from a metaphysical point of view. An important point here is that commitment to the minimal characterisation of organised entities and activities is more natural if one holds the complex systems view of mechanisms and not the view that mechanisms are causal processes. But, moreover, 'activities', for example, are typically introduced as a novel ontological category; hence, the elements of the minimal characterisation are too metaphysically loaded and very hard to read in such a neutral way. In Section 4.4, where we focus on the inflationary accounts of mechanisms, we will examine this important point in more detail. First, however, let us examine a second case study that will serve to further illuminate the view of mechanisms as causal pathways and the thesis of *difference-making*, that is, the case of scurvy.

4.3   The Case of Scurvy

In this section we look in some detail at the history of the discovery of the mechanism of scurvy. We use this case to further illustrate CM and, in particular, the view that difference-making suffices to capture causal relations among the components of the pathway. The key point will be that difference-making is what is important in biological practice when a new pathway is identified. Moreover, we will use the example of the discovery of the mechanism of scurvy in order to argue that mechanistic evidence in science is evidence about difference-making relations.[3]

4.3.1 The Discovery of the Mechanism of Scurvy

Scurvy, we now know, is a disease resulting from a lack of vitamin C (ascorbic acid). The absence of vitamin C in an organism causes scurvy, which starts with relatively mild symptoms (weakness, feeling tired, sore arms and legs) and if it remains untreated it may lead to death. If we take seriously the thought that absences, qua causes, are counterexamples to mechanistic causation, we should conclude that there is no mechanistic explanation of scurvy. But this would be clearly wrong. What is correct to say is that the lack of vitamin C disrupts various biosynthetic causal pathways, that is, mechanisms, for example, the synthesis of collagen. In the latter process, ascorbic acid is required as a cofactor for two enzymes (prolyl hydroxylase and lysyl hydroxylase) that are responsible for the hydroxylation of collagen. Some tissues such as skin, gums, and bones contain a greater concentration of collagen and thus are more susceptible to deficiencies. But ascorbic acid is also required in the enzymatic synthesis of dopamine, epinephrine and carnitine. Now, humans are unable to synthesise ascorbic acid, the reason being that humans possess only three of the four enzymes needed to synthesise it (the fourth enzyme seems to be defective). Hence humans have to take vitamin C through their diet.[4]

The disrupted causal pathways that prevent scurvy can be easily accommodated within the difference-making account of causation. Had vitamin C been present in the organism X, X wouldn't have developed scurvy. In fact, the very causal pathway can be seen as a network of relations of dependence (or difference-making). Abstractly put, had vitamin C been present in human organism X, X's lack of working L-gulonolactone oxidase (GULO) enzyme would not have mattered; enzymes prolyl

---

[3] Thagard (1999) has a brief discussion of the case of scurvy; although Thagard's 'official' account of mechanism in that work is that a mechanism is 'a system of parts that operate or interact like those of a machine' (p. 106), in discussing scurvy as well as other cases from medicine, he seems to work with a notion of mechanism similar to CM in that he takes mechanisms to be kinds of pathways.
[4] For a useful survey, see Magiorkinis et al. (2011).

hydroxylase and lysyl hydroxylase would have been produced and scurvy would have been prevented.[5]

The history of scurvy is quite interesting. During the Age of Exploration (between 1500 and 1800), it has been estimated that scurvy killed at least two million seamen. Although there were hints that scurvy was due to dietary deficiencies, it was not until 1747 that it was shown that scurvy could be treated by supplementing the diet with citrus fruits. In what is taken as the first controlled clinical trial reported in the history of medicine, James Lind, naval surgeon on HMS *Salisbury*, took 12 patients with scurvy 'on board the Salisbury at sea' (Lind 1753, 149). As he reported, 'Their cases were as similar as I could have them.' The patients were kept together 'in one place, being a proper apartment for the sick' and had 'one diet in common to all'. He then divided them into six groups of two patients each, and each group was allocated to six different daily treatments for a period of 14 days. One group was administered two oranges and one lemon per day for six days only, 'having consumed the quantity that could be spared' (p. 150). The other groups were administered cyder, elixir vitriol, vinegar, seawater, and a concoction of various herbs, all of which were supposed to be anti-scurvy remedies. As Lind put it: 'The consequence was, that the most sudden and visible good effects were perceived from the use of the oranges and lemons; one of those who had taken them being at the end of six days fit for duty.... The other was the best recovered of any in his condition' (p. 150). Lind's experiments provided evidence that citrus fruits could cure scurvy. He said that oranges and lemons are 'the most effectual and experienced remedies to remove and prevent this fatal calamity' (p. 157).

Though Lind had identified a difference-maker, he was side-tracked by looking for the cause of scurvy, which he found in the moisture in the air, though he did admit that diet may be a secondary cause of scurvy (see Bartholomew 2002; Carpenter 2012). But in 1793 his follower, Sir Gilbert Blane, who was the personal physician to the admiral of the British fleet, persuaded the captain of HMS *Suffolk* to administer a mixture of two-thirds of an ounce of lemon juice with two ounces of sugar to each sailor on board. As Blane reported, the warship 'was twenty-three weeks and one day on the passage, without having any communication with the land ... without losing a man' (quoted by Brown 2003, 222). To be sure, scurvy did appear, but it was quickly relieved by an increase in the lemon juice

---

[5] For a description of the causal pathways of the synthesis of vitamin C in the mammals that can synthesise it, see Linster and Van Schaftingen (2007).

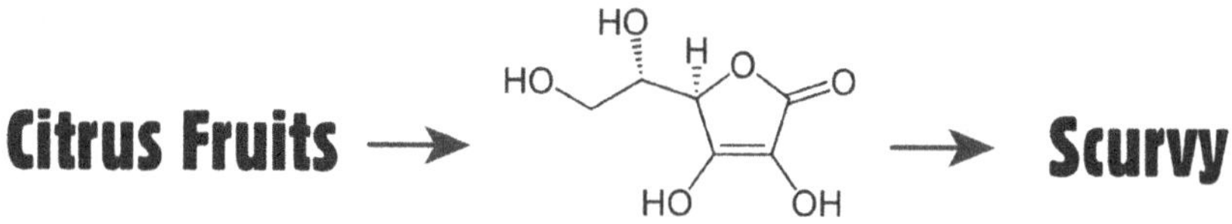

Citrus Fruits → [structure] → Scurvy

Figure 4.2 The mechanism of scurvy.

ration. When in 1795 Blane was appointed a commissioner to the Sick and Hurt Board, he persuaded the Admiralty to issue lemon juice as a daily ration aboard all Royal Navy ships. He wrote: 'The power [lemon juice] possesses over this disease is peculiar and exclusive, when compared to all the other alleged remedies' (Brown 2003, 222). But even when it was more generally accepted that citrus fruits prevent scurvy, it was the acid that was believed to cure scurvy.

The first breakthrough took place in 1907 when two Norwegian physicians, Axel Holst and Theodor Frølich, looked for an animal model of beriberi disease. They fed guinea pigs with a diet of grains and flour and found out, to their surprise, that they developed scurvy. They found a way to cure scurvy by feeding the guinea pigs with a diet of fresh foods. This was a serendipitous event. Most animals are able to synthesise vitamin C, but not guinea pigs. In 1912, in a study of the aetiology of deficiency diseases, Casimir Funk suggested that deficiency diseases (such as beriberi and scurvy) 'can be prevented and cured by the addition of certain preventive substances'. He added that 'the deficient substances, which are of the nature of organic bases, we will call "vitamins"; and we will speak of a beri-beri or scurvy vitamine, which means, a substance preventing the special disease' (Funk 1912, 342). By the 1920s, the 'anti-scurvy vitamine' was known as 'C factor' or 'anti-scorbutic substance' (see Hughes 1983). In 1927, Hungarian biochemist Szent-Györgyi isolated a sugar-like molecule from adrenals and citrus fruits, which he called 'hexuronic acid'. Later on, Szent-Györgyi showed that the hexuronic acid was the sought-after anti-scorbutic agent. The substance was renamed 'ascorbic acid'. In parallel with Szent-Györgyi's work, Charles King and W. A. Waugh identified, in 1932, vitamin C. The suggestion that hexuronic acid is identical with vitamin C was made in 1932, in papers by King and Waugh and by J. Tillmans and P. Hirsch (see Hughes 1983).

The breakthrough in scurvy prevention occurred when scientists started to look for what has been called the 'mediator' (see Figure 4.2), which is a code word for the 'mechanism', which 'transmits the effect of the treatment to the outcome' (Pearl & Mackenzie 2018, 270). As Baron and

Kenny (1986) put it, mediation 'represents the generative mechanism through which the focal independent variable is able to influence the dependent variable of interest' (p. 1173). This mechanism, however, is nothing over and above a network of difference-makers: citrus fruits → vitamin C → scurvy. One such difference-maker, citrus fruits, was identified by Lind and later on by Blane. This explains the success in preventing scurvy after citrus fruits were administered as part of the diet of sailors. It is noteworthy, however, that Lind and the early physicians did not look for the mediating factor in the case of scurvy. As Bartholomew (2002, 696) notes, Lind did not try to isolate a single common constituent in citrus fruits in particular and in fruit in general which makes a difference to the incidence of scurvy. Instead, he was trying to find out the contribution of difference sorts of vegetable to the relief from scurvy. Still, even without knowing the mediating variable (vitamin C), the intake of citrus in a diet did make a difference to scurvy relief.

In order to find the difference-maker in the case of vitamin C deficiency, it was necessary to find a model (animal) that does not synthesise its own vitamin C. In the late 1920s, Szent-Györgyi and his collaborator J. L. Svirbely used hexuronic acid, recently isolated by Szent-Györgyi, to treat guinea pigs in controlled experiments. They divided the animals into two groups. In one the animals were fed with food enriched with hexuronic acid, while in the other the animals received boiled food. The first group flourished, while the other developed scurvy. Svirbely and Szent-Györgyi decided that hexuronic acid was the cause of scurvy relief and they renamed it ascorbic acid. Ascorbic acid was the sought-after mediating variable: the difference-maker (see Schultz 2002).

We take the case of scurvy to illustrate very clearly that what matters in biological practice when a new pathway is identified are the difference-making relations among the components of the pathway. New mechanists who opt for some theory of causal production in order to ground the causal relations among the components of the mechanism need not, of course, deny the presence of difference-making, nor that difference-making is central in mechanism discovery. The main disagreement between us and new mechanists who adopt causal production is whether difference-making is enough for causation or whether we need to say something more about what *grounds* difference-making. The case of scurvy, while it illustrates our views about difference-making and Causal Mechanism, is not enough to establish the primacy of difference-making. In view of this, why not adopt a productive view of causation? We will come back to this crucial question in Section 4.5.1. In the next section, and to further clarify

CM, we will discuss the consequences of Causal Mechanism for the so-called Russo-Williamson thesis.

4.3.2 Causal Mechanism and the Russo-Williamson Thesis

According to Federica Russo and Jon Williamson (2007), in the health sciences, for a causal connection between A and B to be established, one needs evidence for the existence of both a difference-making relation between A and B *and* a mechanism linking A to B. The thought behind what has been called the Russo-Williamson thesis (RWT) seems straightforward: to accept a causal claim, evidence that a putative cause makes a difference to an effect is not enough, if we do not also know *how* the putative cause brings about the effect, that is, the mechanism (recall Boyle's views on mechanical explanation in Chapter 2).

To illustrate this, consider the following example used by Russo and Williamson to support their thesis. In 1833 it was observed at the Vienna Maternity Hospital that the death rates due to puerperal fever after childbirth in two different clinics of the hospital were different. Ignaz Semmelweis, who was an assistant physician at the hospital, hypothesised that the cause of the difference in the death rates was contamination by 'cadaverical' particles; doctors and medical students examined patients after having carried out autopsies and so they infected them. Semmelweis suggested that handwashing with a solution of 'chlorinated lime' would eliminate the contamination and decrease the death rates. Indeed, after handwashing was introduced, death rates decreased considerably. However, the hypothesis that contamination by 'cadaverical' particles was the cause of puerperal fever was accepted 'only after the germ theory of disease was developed and the Vibrio cholerae had been isolated by Robert Koch, establishing a viable mechanism' (Russo & Williamson 2007, 163). This case, according to Russo and Williamson, shows that 'causal claims made solely on the basis of statistics have been rejected until backed by mechanistic or theoretical knowledge' (p. 163).

Williamson (2011) relies on RWT to raise a problem for both mechanistic and difference-making theories of causation. The problem is supposed to be that these theories, taken on their own, are not compatible with the causal epistemology adopted in biomedicine and other scientific fields, which conforms to RWT. This means that, in biomedicine, as Williamson puts it, 'a causal relation typically signifies the existence of both a mechanistic and a difference-making relation, and evidence of the existence of both the mechanistic relation and the difference-making relation is typically required to establish the causal claim' (p. 435). This argument seems to raise a problem for difference-making accounts of mechanism (and thus for CM). If A causes B in virtue of a mechanism linking A to B, where a mechanism involves a chain of events linked by difference-making relations, it seems that evidence of difference-making is enough to establish a causal claim, contrary to what RWT asserts. In other words, for CM, 'mechanistic' evidence need not be different in kind from difference-making evidence.

So, in claiming that there is a difference between difference-making and mechanistic evidence, Russo and Williamson seem to implicitly reject a difference-making account of mechanism. To be sure, in their 2007 paper, Russo and Williamson do not commit to a particular conception of what a mechanism is. They argue, however, that in the biomedical sciences 'two different types of evidence – probabilistic and mechanistic – are at stake when deciding whether or not to accept a causal claim' (p. 163). In his 2011 paper, Williamson takes cases of absences and double prevention to constitute counterexamples to mechanistic causation, since such cases show that 'a causal relation need not be accompanied by a mechanistic relation' (p. 435). It seems, then, that Williamson had a specific view in mind about what a mechanism is, that is, that mechanistic causation should be viewed in terms of some kind of causal production (in a difference-making account of mechanism, cases of double prevention, at least, would not be problematic; see Woodward 2002).

A difference-making account of mechanism would suggest an alternative picture: while a difference-making relation between A and B is not enough to establish a causal relation between a putative cause A and effect B, a series of difference-making relations between factors causally intermediate between A and B offer further support for the causal claim. But, according to CM, such a causal pathway is exactly what a mechanism is. So, it seems that there is no difference between difference-making and mechanistic evidence for CM, at least of the kind that would lead us to talk about two different *kinds* of evidence.

Williamson and Michael Wilde (2016) deny this view. They assume that there is a distinction between two kinds of evidence, claiming that

> in order to establish that A is a cause of B there would normally have to be evidence both that (i) there is an appropriate sort of difference-making relationship (or *chain of difference-making relationships*) between A and B – for example, that A and B are probabilistically dependent, conditional on B's other causes – , and that (ii) there is an appropriate mechanistic connection (or chain of mechanisms) between A and B – so that instances of B can be explained by a mechanism which involves A. (p. 38; emphasis added)

In contrast to this, the case of scurvy shows that looking for mechanistic evidence is just looking for a special kind of 'difference-making' evidence and not for a different kind of evidence. This special difference-making evidence involves looking for the 'mediator'. As we have seen, Lind's experiments provided evidence for a difference-making relationship between citrus fruits and scurvy, but no evidence about how exactly citrus fruits acted so as to prevent scurvy. When it was realised by Funk that scurvy is a 'deficiency disease' – that is, it was produced because of the lack of a particular substance – it became obvious that citrus fruits acted to prevent scurvy by providing that preventive substance. So, scientists started looking for this preventive substance that was the mediating factor between citrus fruits and scurvy. As we have already seen, however, what was required for finding the mediator and establishing the pathway citrus fruits → vitamin C → scurvy was the isolation of a substance (hexuronic acid) from citrus fruits that was such as to prevent scurvy in controlled experiments with guinea pigs by Svirbely and Szent-Györgyi. So, the evidence for identifying the mediator was not evidence about particular entities engaging in activities, or some sui generis type of mechanistic evidence, as one would have believed if the activities-based account of mechanism were true; it was evidence about more difference-making relations, this time between the two initial variables (citrus fruits and scurvy) and the mediating variable vitamin C.

In view of the case of scurvy, we think that RWT can be accepted, but without being committed to the existence of a special type of 'mechanistic' evidence over and above difference-making relations.[6] Moreover, acceptance of RWT does not automatically lead to rejecting a difference-making account of causation. Given a difference-making account of mechanisms, RWT can be understood as follows: typically, to establish a causal connection between A and B, we have to have both evidence for a difference-making relation between A and B and evidence for one or more mediators; but all this evidence is, ultimately, evidence for difference-making relations.[7]

---

[6] Gillies (2011) offers a similar formulation for RWT. He suggests: 'In order to establish that A causes B, observational statistical evidence does not suffice. Such evidence needs to be supplemented by interventional evidence, which can take the form of showing that there is a plausible mechanism linking A to B' (p. 116).

[7] Hill's influential article (1965) has been viewed as offering a version of RWT (cf. Russo & Williamson 2007; Clarke et al. 2014). Note, however, that Hill does not talk explicitly about mechanisms in his paper. He offers 'plausibility' as a criterion for establishing causal claims, which can be understood as the existence of a biologically plausible mechanism; but he does not regard it as particularly important, since '[w]hat is biologically plausible depends upon the biological knowledge

4.4 Inflationary Accounts of Mechanism

Since the emergence of New Mechanism, there had been three dominant accounts of mechanism, offered by Machamer, Darden and Craver (2000); Glennan (1996; 2002) and Bechtel and Abrahamsen (2005). We have presented these accounts in Chapter 2. Other similar general characterisations of a mechanism have appeared in the recent literature, but the following (which Glennan 2017 has called *Minimal Mechanism*) represents a broad consensus:

> A mechanism for a phenomenon consists of entities (or parts) whose activities and interactions are organised so as to be responsible for the phenomenon. (p. 17)

What is important to stress for our purposes is that despite the fact that all these accounts are quite far from traditional mechanistic accounts of causation, they still offer what we might call a 'metaphysically inflated account of mechanism'. In spite of their differences in detail, they are all committed to a certain metaphysics of mechanisms and, in particular, to a certain 'new mechanical ontology', as Glennan has put it (2017, 48). This 'new ontology' of entities, activities, interactions, organisation of parts into wholes and the like creates the further philosophical need – which mechanists try to meet – to explain what they are and how they relate, if at all, with more traditional metaphysical categories.

Recall from Chapter 2 that the following are the key claims of the new mechanists:

1. The world consists of mechanisms.
2. A mechanism consists of objects of diverse kinds and sizes structured in such a way that, in virtue of their properties and capacities, they engage in a variety of different kinds of activities and interactions such that a certain behaviour B or a certain phenomenon P is brought about.
3. The main way to explain a certain behaviour B or a certain phenomenon P in science is to offer the mechanism of it.

---

of the day' (p. 298). As 'strongest support' for causation Hill takes experimental evidence, for example, whether some preventive action does in fact prevent the appearance of a disease. Last, his 'Coherence' criterion involves, among others, establishing a mediator; his example is 'histopathological evidence from the bronchial epithelium of smokers and the isolation from cigarette smoke of factors carcinogenic for the skin of laboratory animals' (p. 298), which was important in establishing a causal connection between smoking and lung cancer.

So, though the new mechanists do not commit themselves to a certain global metaphysics of mechanisms – they do not, for instance, align with the seventeenth-century view of mechanism as configurations of matter in motion subject to laws – their project is not much less metaphysical when they try to offer global accounts of what a mechanism is: the entities might be diverse but they are organised into a mechanism in virtue of their powers, capacities or activities and/or in virtue of their being subjected to laws or at least to invariant generalisations. Mechanisms are typically taken to be things in the world, with more or less objective boundaries, with causally interacting parts bringing about a certain phenomenon P or manifesting a certain behaviour B. Moreover, the blueprint of a mechanistic explanation is decomposition: the behaviour of a system is explained by the interactions/activities of its parts.

Illari and Williamson (2012) have offered their account of mechanism in an attempt to offer a less metaphysically committed view of what a mechanism is, and we have already noted that Glennan's *Minimal Mechanism* is a very similar account. Glennan takes it that an advantage of his minimal account is that mechanisms are everywhere constituting 'the causal structure of the world' (2017, chapter 2). Though we think that both Illari & Williamson and Glennan move the issue in the right direction, we take it that, as it stands, such a minimal account still invites a number of metaphysical questions that, we think, are not relevant to scientific practice. For instance, they invite various metaphysical questions as to the status of entities, their difference from activities, the need to introduce both activities and interactions, the role of the organisation in the performance of the function or behaviour, and so on. These might be philosophically legitimate questions to ask, but, we want to claim, they need not be asked and answered for an understanding of the role of mechanism as a concept-in-use in science.

Could one perhaps read the recent minimal general characterisations of mechanisms in a way that does not emphasise the ontological aspect? We take such characterisations to have two general aims, a metaphysical and a methodological. The metaphysical aim is to specify what exactly mechanisms are *as things in the world*. The methodological aim is to account for the notion of mechanism as this is present in scientific practice. We take it that these two aims are in potential conflict: what is important methodologically in the practice of science may not have clear metaphysical implications, and vice versa. One could choose to emphasise the methodological aspect more than the metaphysical. It's hard, however, to think of the current resurgence of the mechanist world view, even under the 'minimal' characterisations proposed by Glennan and Illari & Williamson, as not having a substantive metaphysical element in it. The very notion of activity as used in the post-MDC literature on mechanisms is heavily metaphysical (being the 'ontic correlate of verbs', as they say). But even if one chose to treat the standard 'minimal' characterisation in an ontologically thinner way (e.g., as not being committed to activities as an irreducible ontological item), it is still the case that even *this* account does exclude certain metaphysical views while fostering others (even if not entailing a specific one); for example, it excludes, on various grounds, Humean accounts of the underlying metaphysics. Reasons such as these make us claim that current accounts of mechanism have significant metaphysical content.[8]

To be more specific, we take the problem with the above accounts of mechanism to be the following. Such accounts invite a number of questions concerning the basic building blocks of mechanisms. For example, consider the claim (which is part of many accounts) that a mechanism consists of two distinct kinds of building blocks: entities (organised in a stable way into a spatiotemporal pattern) *and* activities. But what grounds the difference between entities and activities? Activities are supposed to ground the causal efficacy of mechanisms (according to MDC, mechanisms require 'the productive nature of activities' [2000, 4]); they are the ontic correlates of (transitive) verbs. At issue here, then, is the ontological structure that underlies and unites the scientific theoretical descriptions of mechanisms. Significantly, there are alternative, and competing, ways to characterise this ontological structure: instead of having both entities and activities, one can characterise activities in terms of entities and their causal powers (as, e.g., does Glennan 2017). The question here is: *What is added to scientific practice by insisting that a description of a mechanism has to be couched in some preferred philosophical categories, for example, entities and activities, powers or whatnot?*

New mechanists might reply here that it is unfair to take their metaphysical discussions as intending to inform our understanding of scientific practice. What new mechanists claim, it may be argued, is that the language of mechanisms, entities and activities is descriptively superior compared with the language of objects, properties and laws (or even difference-making relations) in order to characterise the practice of biology. So, MDC write about the 'descriptive adequacy' of their account of

---

[8] A possible exception is Bechtel and Abrahamsen's characterisation of mechanism, which, unlike those offered by Glennan, Illari & Williamson and MDC, is more focused on the methodological aspect and can be read in a metaphysically deflationary way.

mechanisms. The descriptive superiority of the language of entities and activities is then taken to imply a corresponding ontology of entities and activities. Thus, the main question new mechanists ask is something like 'What does scientific practice tell us about ontology?' and not 'What can the new mechanical ontology tell us about scientific practice?'

Concerning the first question, we think it is, by and large, premature and inadequate to infer ontology from language. As a rule, scientific language is compatible with various different metaphysical views (we will come back to this issue in Section 6.8). But even if we let the question from practice to ontology to pass, it would still be imperative to ask the second question too, namely, the question from ontology to practice. To see this let us consider the case of causation. If treating causation in terms of difference-making is enough to understand practice (as the case of scurvy illustrates), then any metaphysically inflated account of causation will add nothing by way of understanding practice further. This means that scientists do not need to commit themselves to any particular account about the metaphysics of causation. So, if the concept of causation that is sufficient for practice is difference-making, then a practice-based view of causation must focus on this (we will come back to this point about causation in Section 4.5.1).

Analogously, a practice-based approach to understanding what a mechanism is ought to focus on concepts that are central in practice, and certainly the notion of mechanisms as causal pathways is central in practice. If new mechanists think that inflationary accounts of mechanisms are central in practice, they ought to show how exactly these metaphysically loaded concepts function within practice, and in particular that they illuminate practice better than the concept of causal pathway. Otherwise, these concepts are not really grounded in practice; they are philosophical additions that have no real function within science.

We do not think, however, that these additions offer further illumination of the practice. Take the case of apoptosis. What clarity (or extra information) would be added to the cytological characterisation of apoptosis offered by Kerr et al. if we were to add that, for example, in the case of the condensation of cytoplasm and nucleus, condensation is an *activity* (whatever that means) and cytoplasm and nucleus are *entities* (whatever that means)? And are the protuberances that form on the surface of the cell and give rise to apoptotic bodies entities or activities? Or, what is the added value of the claim that apoptotic bodies have the *power* to degrade? Similar questions can be asked for the biochemical description of the apoptotic pathway.

In contrast to such metaphysically inflated accounts of mechanisms, our discussion of apoptosis showed that for the identification of a biological mechanism, a theoretical description of the causal pathway (and the function performed) is enough; there is no reason to characterise it further in terms of a preferred ontology. But then, what one can say about what a mechanism in general is, without giving an ontologically inflationary account, is to characterise it as a causal pathway that is theoretically described. Philosophers who characterise mechanisms in general in terms of a preferred ontological inventory have the burden to justify what this further characterisation adds to this deflationary account (i.e., Causal Mechanism) that can be extracted from biological practice.

4.5 Causal Mechanism as a Deflationary Account

The focus on biological practice, then, motivates a deflationary approach to mechanistic talk in biology and biomedicine: such talk does not have to be interpreted in a manner that leads to inquiry into how best to characterise the mechanical ontology of the world. Rather, commitment to mechanism is essentially a methodological (as opposed to an ontological) stance, and mechanism is primarily a methodological concept. We call this stance that incorporates CM as a general characterisation of mechanism *Methodological Mechanism* (MM). In Chapter 10 we will examine MM in detail, taking into account the main results reached in the book. Here we will discuss the main ideas behind CM.

CM can be viewed as based on two main claims, one negative and one positive. The negative claim is that the concept of mechanism as used in scientific practice need not (and should not) be characterised in abstract ontic terms aiming at ontic unity. The positive claim gives a generic characterisation of mechanism as a concept of practice present in various scientific fields. Let's examine these two claims in more detail.

The negative claim differentiates CM from many prevalent philosophical accounts of mechanism. CM remains agnostic (or non-committal) concerning the precise ontology underlying a causal process: it does not commit itself to the existence of activities, powers and the like as distinct from entities and their properties. This deflationary stance is, we think, a decisive advantage of CM over its rivals. First, as a generic account of mechanism applicable to various scientific fields, it retains the advantages of the recent accounts over older reductive approaches to mechanism (cf. Salmon 1997; Dowe 2000). But, second, unlike many recent accounts, it does not read off from scientific practice any views about the ontology of causation (views that do not seem to be supported by the concept of mechanism as used in practice). As noted already, this ontological agnosticism is agnosticism about a 'deeper' ontological description of the mechanism, that is, 'deeper' than the one offered by the relevant theory. As such it is far from implying that mechanisms as causal pathways are not things in the world. According to CM, in elucidating 'mechanism' as a concept of practice we need not go further and commit ourselves to a certain ontological ground of the difference-making relations. Third, by doing so, it avoids the need to answer several metaphysical questions that seem not important in scientific practice. In sum, it takes mechanism to be primarily a methodological, rather than an ontological, concept.

CM is thus opposed to views that, by the very fact of offering mechanistic explanations, biological practice yields specific commitments to a 'mechanistic' ontology. For example, it has been claimed that only a 'local' metaphysics of activities or powers can capture the sense in which mechanisms are both 'real and local' to the phenomenon they produce (Illari & Williamson 2011). While we take it to be fully consistent with CM to view mechanisms as causal pathways that are both 'real and local', in that the identification of a mechanism involves the localisation of the components of the pathway, we resist the further move to a 'local' (a.k.a. lawless) metaphysics. We have two main objections to such a move: first, what one may mean by a 'local' metaphysics is far from clear or uncontroversial (see Chapter 6). But, second, arguments such as those in Illari and Williamson (2011) take place in the context of a discussion on the ontology of mechanisms, and do not directly aim to characterise mechanism as a concept of practice. At the same time, of course, CM is compatible both with the view that the fundamental ontology of the world is broadly neo-Aristotelian as well as with a more Humean view (we will come back to the claim that the local character of mechanisms points to a lawless and powerful mechanical metaphysics in Chapter 6).

The main idea behind the positive claim of CM is that the search for mechanisms improves our understanding of natural phenomena. When scientists look for mechanisms that produce the phenomena, they seek to describe the causal pathways that lead from the initial event of the pathway to the resulting state. A mechanism, then, is some process that shows how exactly the effect is produced.[9] However, to say this is not to make an ontological claim about the structure of the world, but to stress that one of

---

[9] As we have seen in Chapter 2, the idea that mechanistic explanation enhances understanding by identifying a process that connects cause and effect was central in the context of the emergence of

the aims of science should be the discovery of pathways connecting events, which are regular enough so that they can be relied upon to enhance our understanding of how some effects are brought about, and of how we can intervene in order to prevent unwanted outcomes or to treat diseases. Mechanisms, then, can be viewed as theoretically described stable explanatory structures, whose exact content and scope may well vary with our best conception of the world. In particular, a pathway does not have to be described in some privileged (and maybe reductionist) language; pathways in science are described using the theoretical language of the relevant scientific field. Such a description is enough for the identification of a new mechanism, as we have seen in the case of apoptosis. Hence, viewing mechanism in methodological terms licenses our preferred account of mechanism: a mechanism in the biological sciences is a theoretically described causal pathway producing the phenomena of interest.

In the remainder of this section we will first look more closely at how we should understand the metaphysical agnosticism of CM. A main point that we stress is that metaphysical agnosticism is compatible with realism about causal pathways (Section 4.5.1). We will then turn to our claim that CM can serve as a general characterisation of mechanism in the life sciences and argue that it fulfils four important adequacy conditions: it is (1) practice-based, (2) common across fields, (3) topic-neutral and (4) diversifiable (Section 4.5.2).

4.5.1   Causal Mechanism and Metaphysics

We want to resist the temptation to offer a metaphysically inflated account of the causal pathway, in terms of an explicit specification of the mechanical ontology. A key reason for this is that the causal pathway is described in the *theoretical language* of a specific scientific field, and not in some privileged language of ontological categories. This suggests that the form of the description of the mechanism cannot be decided beforehand and in advance of how the concept of mechanism is used. What counts, each time, as a legitimate description of a causal pathway is something that has to be decided by scientific practice. Instead of imposing various ontological categories as those that constitute a general legitimate description of a mechanism, it should be left to the scientists themselves to decide how best to describe mechanisms using the theoretical language they employ to

mechanistic philosophy in the seventeenth century (see, e.g., Boyle's essay 'About the Excellency and Grounds of the Mechanical Hypothesis' in Boyle 1991).

understand and describe the world. CM has the consequence that a series of questions that new mechanists have been concerned with need not concern us if our aim is to understand scientific practice.[10]

However, what about causation? According to CM, causation as a difference-making relation is prior to the notion of mechanism as causal pathway.[11] Does this mean that, in order to identify a mechanism with a causal pathway, it is required to make a commitment about what the ontological nature of causation is? This does not seem necessary for understanding the concept-in-use. As we saw very clearly in the case of scurvy, scientific practice establishes robust causal connections, which can be used for understanding and manipulation, without necessarily being committed to a single and overarching ontic account of causation. Ultimately, whatever fundamental ontological theory of what causation is one might have (e.g., in terms of causal powers or regularities), the identification of causal relationships is based on theory-described difference-making relations; this is what scientists look for when establishing causal relations and causal pathways. In this sense, the causal pathway by means of which a phenomenon Y is brought about by a cause X, given that X initiates a chain of events that leads to Y, is the very network of theory-described difference-making relations among the various intermediaries of X and Y. It is a further question what the truth-makers of these difference-making relations are, and, according to us, this question need not be answered in order to discover and use causal relations in scientific practice. Hence, the point here is that in order to understand what a causal pathway (and hence a mechanism) as a concept-in-use is, and to identify mechanisms, we do not need a theory about the metaphysics of causation: CM is really, ontologically speaking, a deflationary view.[12]

Could perhaps one insist here that the distinction between difference-making and causal production can be found within scientific practice? The argument could go like the following. In biology, inhibition is a feature of many mechanisms. For example, proteins called repressors can bind to DNA and inhibit gene expression. Inhibition, now, is causation by

---

[10] These questions concern, for example, the components and boundaries of mechanisms (cf. Kaiser 2018), the metaphysics of causation (cf. Matthews & Tabery 2018) and the nature of mechanistic levels (cf. Povich & Craver 2018). We will come back to all these issues in subsequent chapters.

[11] It will be the aim of Chapters 5 and 6 to fully substantiate this claim.

[12] There are various other questions that can be raised concerning CM; for example, do we take pathways to be types or tokens? Here again, we defer to practice. Causal pathways, qua things in the world that produce an effect, are concrete particulars. But what is described theoretically in the language of theory is a type of causal pathway.

disconnection, which is considered a problem for productive accounts of causation. The productivity theorist, then, owes us an account of such cases. Similarly, we find in biology and elsewhere *continuous* causes; for example, the cytological description of apoptosis describes a continuous series of stages. It might be argued, with some plausibility, that such cases require a production view of causation, and thus causation as production is properly grounded in scientific practice. It would be a small step from here to arrive at a practice-based metaphysically inflationary account of mechanism.

Yet it's not hard to see that the difference-making account can accommodate difference-makers that are connected by some continuous process. This does not eo ipso imply that difference-making is the ground for productivity; this might well be true but more argument would be required. Rather, the point is that it is one thing to take causation to be production, and quite another to take causation to be a continuous relation among the components of a pathway. Humean accounts can satisfy continuity without being productive.

Our claim that mechanisms are theoretically described causal pathways should not be taken to imply that mechanisms are not things in the world; certainly, causal pathways are as real as anything. Rather, it is meant to highlight that the best description of the causal pathway (and hence of the mechanism) is given by the relevant theory and not by an abstract metaphysical account. Does it follow that every theoretically described pathway is a mechanism? Not at all. It should be a correctly described causal pathway. Not all theoretical descriptions of causal pathways are on a par. First, a theoretical description might fail to capture the actual causal dependencies in the world – such descriptions cannot pick out mechanisms. Second, some theoretical descriptions of a particular causal pathway might be less detailed compared with others, as shown by cytological and biochemical descriptions of apoptosis. How detailed a theoretical description can be is something that will be determined by scientific practice itself. In the case of apoptosis, for example, the initial description was quite detailed: it involved the careful description of various cytological features as well as situating the pathway within the overall functioning of the organism. A less detailed description such as 'cells die on their own', for example, is shown by practice to be too meagre to be of interest. Last, the actual initial identification of a pathway should be seen as the first step that leads to the further elucidation of the various causal dependencies that are involved in the pathway. The end result, as we saw in the case of the biochemical descriptions of the signalling pathways of apoptosis, can be a highly informative theoretical description that embeds the pathway within the known physiological and biochemical functions of the organism.

4.5.2 *The Generality of Causal Mechanism*

Let us now turn to our claim that CM offers a general characterisation of mechanism as a concept-in-use that is common across fields in biology and elsewhere. There are four important adequacy conditions that any such characterisation has to satisfy. First, it has to identify a concept that is actually at work within scientific practice. So, the first condition is that the characterisation has to be *practice-based*. Second, it has to be shown that the notion is *common* across various scientific fields. Third, to be common across many fields, in biology but also in non-biological fields, it has to be as *topic-neutral* as possible. Fourth, it has to be easily *diversifiable* in order to be adapted to more specific contexts without losing the common features that allow us to talk about a shared and common notion. In sum, a general characterisation of mechanism has to be (1) practice-based, (2) common across fields, (3) topic-neutral and (4) diversifiable.[13]

Our examination of the cases of scurvy and, especially, apoptosis and other biological pathways provides ample justification that CM identifies a practice-based notion. As we saw in Chapter 3, an advantage of CM over other accounts is that it can be viewed as a generalisation of a notion that is undoubtedly central in biological practice, that is, the notion of a biochemical pathway. In what follows, we will then focus on conditions (2)–(4).

4.5.2.1 *CM Is Common across Fields*

A possible objection against viewing CM as identifying a notion common across fields is that not all mechanisms are causal pathways. If this is true, then perhaps occurrences of the term 'mechanism' cannot all be accounted for in terms of CM and one should adopt a pluralist stance about what a mechanism in scientific practice is.

So, one might argue that not all mechanisms can be causal pathways, since (at least some) mechanisms are arrangements of entities capable of *implementing* a causal pathway, whether or not the causal pathway is activated; that is, there may exist 'inactivated mechanisms'. Can we

---

[13] Later (in Chapter 10), we will add the condition that such a characterisation has to be non-trivial. But, as this condition is best discussed in the context of our general framework of *Methodological Mechanism*, we will postpone examining it until Chapter 10.

properly describe something as a mechanism, however, even when it is not acting? If, it might be thought, the mechanism of apoptosis can be said to function properly, without apoptosis being initiated, this would seem to be a problem for the present view. Hence, should mechanisms be thought as always acting, or can there be mechanisms waiting-to-act?

Among new mechanists, there is no consensus on this issue. Illari and Williamson (2012), for example, think that a stopped clock would be such a non-acting mechanism; they suppose that examples of cases of 'mechanisms without activities' are cases of 'mechanisms' that instead of producing a change maintain stability (see Illari & Williamson 2012).[14] For Glennan (2017), in contrast, mechanisms are always acting. So, Glennan distinguishes between a mechanical system and a mechanism, where a mechanical system is 'a system that regularly engages in or is disposed to engage in mechanistic processes' (p. 26). A mechanical system, then, is not strictly speaking the mechanism; 'it is rather a thing in which mechanisms act' (p. 26). When a system S does something ψ, 'the mechanism is the organized activities and interactions of entities within the system that is responsible for that ψ-ing' (p. 26) and '[i]f mechanisms truly are understood to be entities acting, the mechanism persists only for the duration of the action' (p. 55). While we disagree with Glennan's general metaphysics of mechanism, we think that this distinction between mechanisms and systems is essentially correct and concordant with biological practice. In the case of apoptosis, the cell (or some part of the cell) can be taken as the system within which the process of apoptosis occurs. But apoptosis itself, qua a mechanism of regulated cell death, is there only when the relevant causal pathway is acting.

In reply, then, to the present objection, we should note that the causal pathway, as we understand it, and as was made vivid with the case of apoptosis, does involve entities, since it is entities that are causally connected by the processes of the causal pathway that leads to the required effect. But to call an arrangement of entities *mechanism*, in our view, is to make a claim about the causal pathway, which according to the theory does yield a certain effect. This does not mean that there are no mechanisms that maintain a stable state instead of producing change. But homeostatic mechanisms in biology are not like stopped clocks, chimneys

---

[14] The clock example was first discussed by Darden (2006), who argues that a stopped clock is a machine and not a mechanism, since mechanisms are 'inherently active'. A 'crystallised form of a molecule' is another example that according to her is not a mechanism, at least in the MDC sense (p. 281).

or pillars supporting roofs, which are examples that Illari and Williamson use to illustrate their view. For instance, the mechanisms within cells that maintain an equilibrium between apoptotic and anti-apoptotic proteins so that apoptosis is prevented involve various causal pathways. In the equilibrium state, anti-apoptotic proteins bind to pro-apoptotic ones, for example; when apoptosis is initiated, the pathways that maintain equilibrium are blocked, and different pathways are activated. In general, homeostatic regulation involves a dynamic equilibrium and so involves change. Thus, a causal pathway need not result in a specific change; its end result may well be the maintaining of a stable state. Of course, given some entities, a causal pathway involving these entities need not be activated; but for the mechanism to cause anything (either a change or a stable state) the causal pathway should be initiated.

Here is a related worry concerning the commonality criterion: Is it justified to generalise from the examples that we have been examining, the mechanisms of cell death (and other biological pathways) and the case of scurvy, to other mechanisms (perhaps all cases of mechanisms) in biology? While further examples will certainly be useful in order to show that the features of mechanisms identified here are typical in the life sciences (and in the following chapters we will discuss various other examples), we think that such a generalisation is (to say the least) very plausible and that cases such as apoptosis and scurvy can be taken as representative. In Chapter 3, Section 3.5, we saw various reasons why apoptosis can be viewed as paradigmatic. In addition to what we said there, we will here discuss two general reasons why we think CM can be generalised.

There are two general reasons to be optimistic that CM can indeed offer a general characterisation of mechanism that captures a concept of practice that is in use in biology and elsewhere. First, our account depends not on very specific features of the particular cases examined but on features that are commonly found in many cases of mechanisms described in biology. That mechanisms correspond to causal pathways, that these pathways are described using the theoretical language of a particular field, that there can be more or less detailed descriptions of causal pathways and that some pathways serve a function (or functions) are very minimal requirements for something to count as a mechanism; for example, it can be easily seen that they apply to all kinds of biochemical (and other) pathways (e.g., signalling pathways, metabolic pathways). This is a main reason why CM can be seen as topic-neutral.

The second reason is that we take the main philosophical contrast to be between CM and metaphysically inflated accounts of mechanism. But if

CM is sufficient to characterise a mechanism such as apoptosis, it is doubtful whether other case studies will make a metaphysically inflated account necessary: if that were the case, to make philosophical sense of the mechanism of apoptosis we would similarly need such a metaphysically inflated account. The reason is that the case of apoptosis exemplifies all the characteristics commonly taken to require a metaphysically inflated characterisation of mechanism: for example, the introduction of both entities and activities to capture both the various macromolecules and their interactions. Hence, if CM is sufficient to understand the present case, it is very doubtful that other cases will require abandoning CM. While we do not want to claim that the prevalent philosophical accounts of mechanisms are ill-motivated, we do claim that they inflate the concept of mechanism *as this is used in science*.

We take it, then, that CM can be generalised to other uses of 'mechanism' within life sciences, but also more broadly: when scientists talk about 'mechanisms' they do so in the context of searching for a process, that is, a *causal pathway*, the identification of which would explain how a particular phenomenon is brought about. If the causal pathway is identified, there is little further interest in understanding it according to a certain theory of causation, or to characterise it in terms of entities bearing powers or engaging in activities, or being involved in activities *and* interactions and the like. In the practice of science, the description of the causal pathway in the *language of theory* is enough for the identification of the mechanism.

4.5.2.2   CM Is Topic-Neutral and Diversifiable

Let us turn now to the criteria of topic-neutrality and diversifiability. To see how it is possible to diversify CM so as to adapt it to specific contexts, consider the relation between the concepts of mechanism and function. In the general accounts presented in Chapter 2, only Bechtel and Abrahamsen explicitly refer to the behaviour of a mechanism in terms of its function. The usual position among new mechanists is that mechanisms are responsible not only for functions but for phenomena in general, where a phenomenon need not be a function.

We agree with this view. If we insist on an account of mechanism broad enough to capture all uses of the concept in life sciences and elsewhere, and given that there are scientific fields where the concept of function is not present (e.g., particle physics, solid state physics, astrophysics, cosmology), an account such as CM seems preferable.[15] Of course, there are contexts

---

[15] Although we are not going to examine in detail examples of mechanisms from these fields in this book, we share the conviction of new mechanists that mechanism is a concept common across

(e.g., in molecular biology) where a mechanism is automatically a mechanism for a certain function; for example, apoptosis is not just a mechanism for cell death in the sense that it leads to cell death, but it also has a homeostatic function within the organism, as we have seen. But it is not the case that a mechanism of star formation, for example, has star formation as its function. The point here is that if we want to claim that a mechanism of cell death and a mechanism of star formation are in some sense the same kind of thing (i.e., that they are both mechanisms) – that is, if we want to give a general account of a mechanism as a concept-in-use across various scientific fields, including non-biological fields – CM (which does not incorporate a robust sense of function) seems a very promising candidate.

At the same time, CM can be easily adapted to capture particular uses of 'mechanism' in various contexts where a specific notion of function is presupposed. This was why we introduced CM-P, as a schema for such more specific versions of CM, in Chapter 3. CM-P, where X is a function that the causal pathway performs, is a typical meaning of what a mechanism is in biological contexts, where a causal pathway is often taken to have a certain function. In the case of apoptosis, the cytological and biochemical descriptions describe the causal pathway, and its role in regulating cell populations identified by Kerr et al. constitutes its function. But even in biology, not all mechanisms have a function in this sense. For example, when necrosis is referred to as a 'mechanism', what is meant is just a causal pathway.

As a further example that shows the prevalence of CM (i.e., causal pathway without a function) within life sciences, take the way pathologists talk about the causes and 'mechanisms of diseases'. They distinguish between causal agents (e.g., viruses), which constitute the *aetiology* of a disease, and the *pathogenesis* of a disease, which concerns the 'mechanism' that leads from the causal agent to a disease state (see Lakhani et al. 2009). We take this to be a clear application of CM. So, when pathologists want to find out how a certain disease is brought about, they look for a specific mechanism, that is, a causal pathway that involves various causal links between, for example, a virus and changes in properties of the organism that ultimately lead to the disease. It is clear, then, that in pathology, to identify a mechanism is to identify a specific causal pathway that connects

many scientific fields, and so want our account to be potentially applicable to as many fields as possible.

an initial 'cause' (the causal agent) with a specific result.[16] Other similar examples are the 'mechanisms' of action of drugs; in these cases too there is no function in the robust sense that we saw in the example of apoptosis. Mechanisms of action of drugs and mechanisms in pathology are just causal pathways.

Pathological and pharmacological mechanisms, then, do not have a function as physiological mechanisms do; that is, the effects of pathological and pharmacological mechanisms have not evolved to aid the survival and reproduction of the organism, as the effects of physiological mechanisms have done; but it would be very strange to insist that the notion of mechanism in pathology is very different from the notion of mechanism when applied in contexts when a robust sense of function is assumed (as in the case of apoptosis). According to our account, in pathology and in other biological fields the common core of the notion of mechanism is captured by CM. To that extent, we take CM to be topic-neutral, in the sense that it can be applied to all biological fields, including pathology. But in non-pathological contexts CM-P, where X is a function the causal pathway performs, may better capture the notion implicit in biological practice. To say this is not to abandon CM as a general and common characterisation of a shared notion; it is, rather, to show how a core concept can be diversified to be adapted to specific contexts.

In sum, all mechanisms, even mechanisms such as apoptosis that serve a specific function, are just causal pathways. This means, of course, that all causal pathways are mechanisms. We take this as the core notion of 'mechanism' present in many scientific fields (not only in biology). A mechanism qua causal pathway need not have a function. But, second, in biology we can find a richer notion of 'mechanism', where a mechanism is a causal pathway that has a function within the organism: both apoptosis and necrosis are causal pathways that lead to cell death and so are mechanisms in the sense of CM; but apoptosis does, and necrosis does not, have a function within the organism (i.e., apoptosis regulates cell populations). This is why apoptosis is considered to be biologically interesting or significant by biologists and is taken to be (in contrast to necrosis) a specifically *biological* mechanism. But although there are contexts where a richer notion of mechanism is presupposed (i.e., CM-P), this does not

---

[16] While in pathology and elsewhere there is a distinction between a 'causal agent' and a 'mechanism', strictly speaking when we have a mechanism the cause should be taken not as a single event or an entity but as a whole sequence of them which lead to the effect. However, we could keep the notion of a 'causal agent' to refer to an event or an entity that initiates a causal pathway, and keep the term 'mechanism' for the causal pathway itself.

correspond to a difference in the world: the worldly reference of all kinds of mechanisms are just causal pathways. These considerations, then, provide strong support for the claim that CM identifies a notion of mechanism that is practice-based, common across fields, topic-neutral and diversifiable.[17]

In a sense, then, we agree with the 'received view' among new mechanists about the relationship between mechanisms and functions: mechanisms in biology and elsewhere need not have functions. But new mechanists also typically insist that there is a sense in which a mechanism is *for* a phenomenon, where a mechanism is not just a causal pathway and where the phenomenon is not a function. As we will see in the next chapter, we don't agree with this as we think such a thin reading of mechanism-for is misleading and not particularly helpful. In the next chapter we will introduce a distinction to classify the various conceptions of mechanisms found in the literature, the distinction between *mechanisms-of* and *mechanisms-for*, and will argue that it is best to restrict the category of 'mechanisms-for' to the 'functional' sense of mechanism.

---

[17] The strategy we have followed in this section, where we take a minimal or core concept of mechanism (CM) and then diversify it (CM-P), shares similarities with Glennan's (2017, chapter 5) discussion about types of mechanisms (see also Glennan & Illari 2017, chapter 7). Glennan argues that there are different types of phenomena – some functions, some not – and these suggest different extensions of Minimal Mechanism.

Mechanisms, Causation and Laws

5.1 Preliminaries

This chapter will focus on the concept of mechanism as an ontological category and will examine the relation between mechanisms, causation, laws and counterfactuals. It will place current conceptions of mechanism within the broader context of mechanistic accounts of causation, notably those of Wesley Salmon and John Mackie.

A central distinction that we develop in the chapter is the one between two different conceptions of 'mechanism': mechanisms-of and mechanisms-for (introduced in Chapter 1). *Mechanisms-of* underlie or constitute a causal process; *mechanisms-for* are systems or processes that function so as to produce a certain behaviour. This distinction is used to revisit the main notions of mechanism found in the literature. Examples of mechanisms-of are the various mechanistic accounts of causation. Examples of mechanisms-for are the more recent accounts of the nature of mechanisms, which are based on the ways this concept features in scientific explanation and more generally in scientific practice. According to some new mechanists, a mechanism should fulfil both of these roles simultaneously.

5.2 Mechanisms and Difference-Making

Causal relations are explanatory. If C causes E, then C explains the occurrence of E. Mechanisms are widely taken to be both what makes a relation causal and what makes causes explanatory. So, typically, if one explains the occurrence of event E by citing its cause C – that is, if one asserts that C brings about E or that E occurs because of C – one is expected to cite the mechanism that links the cause and the effect: it is in virtue of the intervening mechanism that C causes E and hence that C causally explains E. On this account of causation, it is not enough to show that E depends on C, where dependence should be taken to be robust, for example, a difference-making relation. Unless there is a mechanism, there is no causation. Difference-making is taken to be enough for prediction and control but not enough for explanation (see Williamson 2011). But, given all this, what is exactly the relationship between mechanisms and causation? Even if we cannot have a causal relation between C and E unless there is a mechanism linking them, something must be said about how exactly a mechanism brings about E, and about how the causal relations within the mechanism itself are to be understood.

There are two ways to view within-mechanism causal relations: either in terms of difference-making or in terms of production. On the former option, C causes E if and only if C makes a difference to E, where the difference-making is typically seen as counterfactual dependence, namely, if C hadn't happened, then E wouldn't have occurred.[1] On the latter option, C causes E if and only if C produces E. 'Production' is a term of art, of course, with heavily causal connotations. The typical way to account for 'production' is by means of mechanism. So, C produces E if and only if there is a mechanism that links C and E. Now, on the production view, mechanisms do not seem to depend on difference-making relations. As according to CM *difference-making* is crucial in understanding mechanism as a concept-in-use, we have to defend our claim that difference-making is conceptually prior to mechanism.

As we will see, some mechanists tend to refrain from using counterfactuals to account for within-mechanism causal relations. For others, counterfactuals are needed in order to ground the laws that characterise the interactions between the components of a mechanism; but counterfactuals may in turn be grounded in lower-level mechanisms (Glennan 2002). Yet other mechanists try to dispense with both counterfactuals and laws, in favour of activities (Machamer et al. 2000). Hence, understanding the relation between mechanisms and counterfactuals requires also clarifying the relation between mechanisms and laws. Laws will thus be central in the argument of this chapter. The key question then, for our purposes, is: Can

---

[1] As is well known, both views face problems and counterexamples. For instance, the production account cannot accommodate causation by absences. The lack of water caused the plant to die, but there is no mechanism linking the absence of water with death. The difference-making account cannot accommodate cases of overdetermination and pre-emption. For instance, suppose that two causes act independently of each other to produce an effect. There is certainly causation, but no difference-making since the effect would be produced even in the absence of each one of the causes (cf. Williamson 2011).

there be an account of mechanism that does not ineliminably rely on some non-mechanistic account of counterfactual dependence? To be exact, we want to investigate whether a mechanistic theory of causation ultimately relies on a counterfactual theory (and hence, whether it turns out to be a version of the dependence approach) or whether it constitutes a genuine version of the production approach (either because it altogether dispenses with the need to rely on counterfactuals or because it grounds counterfactuals in mechanisms). To be clear on these issues presupposes, as we will argue below, a more careful analysis of the notion of mechanism. Here is the central line of argument we want to investigate: since a mechanism is composed of *interacting* components, the notion of a mechanism should include a characterisation of these interactions, but if (1) these interactions are understood in terms of difference-making relations and (2) these difference-making relations are not in turn grounded in mechanisms, then there is a fundamental asymmetry between mechanistic causation and causation as difference-making, because to offer an adequate account of the former presupposes an account of the latter.

As we will see, not all philosophers who stress the importance of mechanisms for thinking about science are after an account of causation. For some of them, mechanisms are important in understanding scientific explanation and theorising, but it is not the case that causation *itself* is mechanistic (see, e.g., Craver 2007a, 86). Yet even if these philosophers do not have to provide a full-blown account of the ontology of mechanisms, they have to explain the modal force of mechanisms; hence, the issue of the relation between mechanisms and counterfactuals is crucial. So, we can formulate our central question in a more comprehensive way, which will be relevant even for these latter accounts, as follows: given that a mechanism consists of components that interact in some manner, and thus cause changes to one another, does an account of these *interactions* require a commitment to counterfactuals? As we shall see, ultimately, the issue turns not on the need or not to posit relations of counterfactual dependence but on what the suitable truth-makers for counterfactuals are.

5.3   Early Mechanistic Views

5.3.1   Mackie on Causation and Mechanisms

J. L. Mackie's (1974) work on causation is the recent common source of both approaches under discussion. In particular, talk of 'mechanisms' in relation to causation goes back to him.

Mackie explicitly appealed to counterfactuals in his definition of the meaning of singular causal statements. He argued that a causal statement of the form '*c* caused *e*' should be understood as meaning '*c* was necessary in the circumstances for *e*', where *c* and *e* are distinct event-tokens. Necessity-in-the-circumstances, he added, should be understood as follows: if *c* hadn't happened, then *e* wouldn't have happened.

Mackie's counterfactuals are not, strictly speaking, true or false: they do not describe, or fail to describe, 'a fully objective reality' (1974, xi). Instead, they can be reasonable or unreasonable assertions, depending on the inductive evidence that supports them (pp. 229–30). For instance, in assessing the counterfactual 'If this match had been struck, it would have lit', the evidence plays a *double role*. It *first* establishes inductively a generalisation. But *then*, 'it continues to operate separately in making it reasonable to assert the counterfactual conditionals which look like an extension of the law into merely possible worlds' (p. 203). So, for Mackie, it is general propositions (via the evidence there is for them) that carry the weight of counterfactual assertions. If, in the actual world, there is strong evidence for the general proposition 'All *F*s are *G*s', we can reasonably assert that 'if *x* had been an *F* it would have been a *G*' based on the evidence that supports the general proposition. Mackie was no realist about possible worlds. He did not think that they were as real as the actual. Hence, his talk of possible worlds was a mere *façon de parler* (p. 199).

These evidence-based counterfactuals *cannot* ground a fully objective distinction between causal sequences of events and non-causal ones. This created a tension in Mackie's overall project. For although he explicitly aimed to identify an *intrinsic* feature of a causal sequence of events that makes the sequence causal, his dependence on evidence-based counterfactuals jeopardised this attempt: whether a sequence of events will be deemed causal will depend, on his view, on an *extrinsic* feature, namely, on whether there is *evidence* to support the relevant counterfactual conditional. It is for this reason that Mackie went on to try to uncover an *intrinsic* feature of causation, in terms of a *mechanism* that connects the cause and the effect.

As Hume famously noted, the alleged necessary tie between cause and effect is not observable. Mackie thought, not unreasonably, that we might still *hypothesise* that there is such a tie, and then try to form an intelligible theory about what it might consist in. His hypothesis is that the tie consists in a 'causal mechanism', that is, 'some continuous process connecting the antecedent in an observed ... regularity with the consequent' (p. 82).

Where Humeans, generally, refrain from accepting anything other than spatiotemporal contiguity between cause and effect, Mackie thinks that mechanisms might well constitute 'the long-searched-for link between individual cause and effect which a pure regularity theory fails, or refuses, to find' (pp. 228–9).

Mackie's own view was that this mechanism consists in the qualitative or structural continuity, or *persistence*, exhibited by certain processes, which can be deemed causal. There needn't be some general feature (or structure) that persists in every causal process. What persists will depend on the details of the actual 'laws of working' that exist in nature. For instance, it can be 'the total energy' of a system, or the 'number of particles' or 'the mass and energy' of a system (pp. 217–18). But insofar as something persists in a certain process, this feature can be what connects together the several stages of this process and renders it causal.

So, early versions of both current views about causation can be found in Mackie's work. In fact, it turns out that the mechanistic view was more central in Mackie's overall approach, since it promised to offer a more objective account ˏof causation and to avoid the notorious context-sensitivity of counterfactual assertions. Yet Mackie's attempt to spell out mechanisms in terms of persistence was deeply problematic.[2]

After Mackie, the counterfactual and the mechanistic approaches parted their ways. They were separately pursued and developed by other able philosophers. The locus of the standard counterfactual theories of causation is the work of the late David Lewis (1986b). Unlike Mackie, Lewis (1973) put forward a *quasi-objectivist* theory of counterfactuals, based on possible-worlds semantics. We will come back to Lewis's theory in Chapter 7 (see also Psillos 2002, chapter 3). Here we will only make the following point, which is relevant to what follows. Lewis's theory renders causation an *extrinsic* relation between events, since it analyses causation in terms of counterfactual dependence among events and it analyses counterfactuals in terms of relations of similarity among possible worlds. In fact, there is a rather important reason why counterfactual theories *cannot* offer an intrinsic characterisation of causation. If causation amounts to counterfactual dependence among events, then the truth of the claim that c causes e will depend on the absence of causal overdeterminers, since if the effect e is causally overdetermined, it won't be counterfactually dependent on any of its causes. But the presence or absence of overdeterminers is certainly *not* an intrinsic feature of a causal sequence.

[2] See Psillos (2002, 108–10) for a discussion of the central problems.

5.3.2   The Salmon-Dowe Theory

The locus of the standard mechanistic theories of causation is the work of the late Wesley Salmon. Mackie's preferred account of a causal mechanism in terms of qualitative or structural continuity, or *persistence*, exhibited by certain processes faced significant problems that led Salmon (1984) to argue for an account of causal mechanism that is based on the notion of structure-transference (see Psillos 2002, chapter 4, for a detailed account of Salmon's views). Unlike Mackie, Salmon (1984) tried to characterise directly when a process is *causal*, thereby finding the mechanism that links cause and effect. So, he took processes rather than events to be the basic entities in a theory of physical causation.

Salmon kept the basic idea that '[c]ausal processes, causal interactions, and causal laws provide the mechanisms by which the world works; to understand *why* certain things happen, we need to see *how* they are produced by these mechanisms' (p. 132). But he claimed that the distinguishing characteristic of a causal process (and hence of a mechanism) is that it is capable of transmitting its own structure or modifications of its own structure. Generalising Hans Reichenbach's (1956) idea that causal processes are those that are capable of transmitting a mark, Salmon noted that any process, be it causal or not, exhibits 'a certain structure'. A causal process is then a process capable of *transmitting* its own structure. But, Salmon added, 'if a process – a causal process – is transmitting its own structure, then it will be capable of transmitting certain modifications in the structure' (p. 144).

As many critics noted, however, the very idea of structure-transference (a.k.a. mark-transmission) cannot differentiate causal processes from non-causal ones, since *any* process whatever can be such that *some* modification of *some* feature of it gets transmitted after a single local interaction. A typical example was the shadow of a car with a dent – this is a 'dented' shadow, and the mark is transmitted with the shadow for as long as the shadow is there. In response to this Salmon strengthened his account of mark-transmission by requiring that in order for a process P to be causal, it is necessary that 'the process P would have continued to manifest the characteristic Q if the specific marking interaction had not occurred' (p. 148). It should be clear though that this kind of modification takes us back to persistence! In effect, the idea is that a process is causal if (1) a mark made on it (a modification of some feature) gets transmitted after the point of interaction, and (2) in the absence of this interaction, the relevant feature would have *persisted*, where the required persistence is counterfactual.

Salmon's original promise was for a theory of causation that does *not* involve counterfactuals. The promise, however, was not to be fulfilled. Central to Salmon's theory was the *ability* of a process to transmit a mark. But the ability is a capacity or a disposition, and it is essential for Salmon that it is so. For he wants to insist that a process is causal, even if it is *not* actually marked (p. 147). So, a process is causal if it *could* be marked. Counterfactuals loom large! The message is clear: Salmon's original mechanistic approach cannot do away with counterfactuals. In fact, Salmon's appeal to counterfactuals has led some philosophers (e.g., Kitcher 1989) to argue that, in the end, Salmon has offered a *variant* of the counterfactual approach to causation. Salmon has always been very sceptical about the objective character of counterfactual assertions. So, as he said, it was 'with great philosophical regret' (1997, 18) that he took counterfactuals on board in his account of causation.

Salmon did modify this view further by adopting Phil Dowe's (2000) conserved quantity theory, according to which 'it is the possession of a conserved quantity, rather than the ability to transmit a mark, that makes a process a causal process' (p. 89). On what has come to be known as the Salmon-Dowe theory, a *causal process* is a world line of an object that possesses a conserved quantity. And a *causal interaction* is an intersection of world lines that involves exchange of a conserved quantity.

Dowe fixes the characteristic that renders a process causal and, consequently, the characteristic that renders something a mechanism. A conserved quantity is 'any quantity that is governed by a conservation law' (2000, 91), for example, mass-energy, linear momentum, charge and the like. Dowe's Conserved Quantity theory aims to free the mechanistic view of causation from counterfactuals; but it is far from clear whether this theory can indeed avoid them (for a discussion of this issue, see Psillos 2002, 125–7). However, apart from this problem, the main practical concern is that this account of mechanism is too narrow. For even if *physical* causation – and hence physical mechanism – was a matter of the possession of a conserved quantity, it's hard to see how this account of mechanism can even start shedding any light on causal processes in domains outside physics (biological, geological, medical, social). These will have to be understood either in a reductive way or in non-mechanistic (Salmon-Dowe) terms.

5.3.3   Harré on Generative Mechanisms

A rather liberating conception of causal mechanism was offered by Harré in the early 1970s. Harré connected the traditional idea of power-based causation with the traditional idea that causation involves a mechanism. What he called 'generative mechanism' can be put thus:

generative mechanism = powers + mechanisms

As he put it: 'The generative view sees materials and individual things as having causal powers which can be evoked in suitable circumstances' (1972, 121). And he added: 'The causal powers of a thing or material are related to what causal mechanisms it contains. These determine how it will react to stimuli' (p. 137). For example, an explosion is caused both by the detonation and the power of the explosive material, which it has in virtue of its chemical nature.

On this view of causation, the ascription of a power to a particular has the following form:

X has the power to A = if X is subject to stimuli or conditions of an appropriate kind, then X will do A, *in virtue of its intrinsic nature*.   .

But this is not a simple conditional analysis of powers, since as Harré stressed, power-ascriptions involve two *analysantia*:

a *specific conditional* (which says what X will or can do under certain circumstances and in the presence of a certain stimulus) and an *unspecific categorical* claim about the nature of X.

The claim about the nature of X is unspecific, because the exact specification of the nature or constitution of X in virtue of which it has the power to A is left open. (Discovering this is supposed to be a matter of empirical investigation.)

It is a fair complaint that, as stated above, the ascription of powers is explanatorily incomplete unless something specific is (or can be) said about the *nature* of the particular that has the power. Otherwise, power-ascription merely states what needs to be explained, namely, that causes produce their effects. This is where mechanisms come in. Specifying the generative mechanism is cashing the promissory note. As Harré put it: 'Giving a mechanism ... is ... partly to describe the nature and constitution of the things involved which makes clear to us what mechanisms have been brought into operation' (1970, 124). So, the key idea in this mechanistic view of causation is this: causes produce their effects because they have the power to do so, where this power is grounded in the mechanism that connects the cause and the effect, and the mechanism is grounded in the nature of the thing that does the causing.

This is a broad and liberal conception of causal mechanism. Generative mechanisms are taken to be the bearers of causal connections. It is in virtue of them that the causes are supposed to produce the effects. But there is no specific description of a mechanism (let alone one that is couched in terms of physical quantities). A generative mechanism is virtually *any* relatively stable arrangement of entities such that, by engaging in certain interactions, a function is performed or an effect is brought about. As Harré explained, he did not 'intend anything specifically mechanical by the word "mechanisms". Clockwork is a mechanism, Faraday's strained space is a mechanism, electron quantum jumps [are]. . . a mechanism, and so on' (p. 36).

Though this was not quite perceived and acknowledged when Harré was putting forward this conception, this liberal conception of mechanism pointed to a shift from thinking of mechanisms exclusively as the vehicle of causation (what we call *mechanisms-of*) to thinking of mechanisms as whatever implements a certain behaviour or performs a certain function (what we call *mechanisms-for*).

5.4 Mechanisms-for versus Mechanisms-of

Given the views on mechanisms discussed above, as well as the views of new mechanists examined in Chapters 2 and 4, we can say that two traditions have tried to reclaim the notion of mechanism in the philosophical literature of the twentieth century. In exploring the relation between mechanisms and laws/counterfactuals, it is important to distinguish between these two very different senses of 'mechanism'.

The first sense of mechanism is the one typically found in the context of New Mechanism. Central to New Mechanism is the recognition of the fact that scientists try to identify and understand the mechanisms that are responsible for certain phenomena or underlie certain functions, for example, the mechanisms underlying reproduction, gene-transmission, chemical bonding, face-recognition and so on. Mechanisms are always understood as mechanisms *for certain behaviours* (see Glennan 1996; 2002). This has been known in the literature as 'Glennan's law' and is taken to imply that mechanisms are individuated in terms of what they do. For example, there are mechanisms *for* DNA replication or *for* mitosis. What the mechanism does, what the mechanism is a mechanism *for*, determines the boundaries of the mechanism and the identification of its components and operations. Such mechanisms, which we call *mechanisms-for* (i.e., mechanisms *for* certain behaviours/functions), are the mechanisms that, according to new mechanists, figure in explanations in life sciences and elsewhere, and are what many scientists aim to discover. We find mechanisms-for, among others, in MDC (2000), Bechtel & Abrahamsen (2005), Craver (2007a) and Bechtel (2008). This sense of mechanisms is, arguably, the dominant one in various philosophical studies of mechanisms and their role in the various sciences.

The second sense of mechanism is typically found in the context of mechanistic theories of *causation*. These theories, as we have seen in the previous section, aim to characterise the causal link between two events (to fathom Hume's 'secret connexion') in terms of a 'mechanism'. Such theories stem from a general dissatisfaction with standard philosophical views of causation, which fail to explain or take account of the mechanisms by which certain causes bring about certain effects. For this second sense, what the mechanism *does* is not important; what is important is that it is actually there underlying or constituting a certain kind of process. More precisely, what makes a process *causal* is the presence of a mechanism that mediates between cause and effect (or whose parts or moments are the 'cause' and the 'effect'). We call such mechanisms *mechanisms-of*, since they concern the mechanisms of causation. Mechanisms-of are the mechanisms discussed in, for example, theories that view causation as mark transmission (Salmon 1984), persistence, transference or possession of a conserved quantity (Mackie 1974; Salmon 1997; Dowe 2000).[3]

Mechanisms-for are conceived as systems (Glennan 1996; 2002; Bechtel & Abrahamsen 2005) or process-like (Machamer et al. 2000; Craver 2007a; Glennan 2017), performing a function or more generally a behaviour or phenomenon. This is a main difference with mechanisms-of, which are typically conceived as kinds of processes, rather than systems, and need not perform a function. As we have already noted in Chapter 4, the connection between mechanisms and functions is typically not made explicit in general accounts of mechanisms-for (but see Bechtel & Abrahamsen 2005; Garson 2013). New mechanists such as Glennan and

---

[3] Both phrases 'mechanism for X' and 'mechanism of X' are standard locutions in science. Although scientific usage is not entirely consistent on this, there is a tendency to use the phrase 'mechanism *for* X' in biological contexts when X constitutes a certain function and 'mechanism *of* X' more liberally, and also in contexts outside biology, where it seems to mean just a causal process that produces X, without necessarily X being a function. So, there is some justification from scientific practice for using this terminology, as we too take mechanisms-for to be the mechanisms responsible for certain functions. It has been suggested to us that 'productive mechanisms' may be a more accurate term instead of 'mechanisms-of', but we think the term 'mechanism-of' contrasts nicely with 'mechanism-for'. Moreover, although we take this sense of mechanism from mechanistic theories of causation, we use it in a more broad sense. So, 'productive' mechanisms in the technical sense of production are not the only kind of mechanism-of. Causal Mechanism, according to which causation is prior to mechanism and is to be understood in terms of difference-making, is also for us a kind of mechanism-of.

Illari & Williamson who use locutions like 'mechanism for a behaviour' and 'mechanism for a phenomenon' do this precisely in order to have a notion of mechanism-for that is responsible for an X, without the X being necessarily a function.

In contrast to what is perhaps the received view among new mechanists, we do not agree with a non-functional reading of 'mechanisms-for'. If 'mechanism-for' becomes so broad that any causal process can be thought of as a mechanism-for, then we lose the important difference between the two senses of mechanism identified here. But perhaps some new mechanists would insist that 'mechanism-for' should really be understood as a very thin notion; so, whenever we have a causal process and we choose some end point of the process, then we can say that the process is a mechanism *for* that end-state. For, example, if a light beam hits a mirror and gets reflected so as to pass through some particular location, we can say that this process is a mechanism that is responsible for the presence of light at that location[4]. This is partly a terminological issue. The more substantial issue here is that there is a clear distinction between what we call mechanisms-of and mechanisms-for in the functional sense, which is blurred by a very thin reading of 'mechanism-for'.

In order to be conceptually clear, then, we reject the thin reading of 'mechanism-for' and understand all mechanisms-for as mechanisms for certain functions. We think that the thin reading is not really a helpful use of the phrase 'mechanism-for'. For us light reflected by a mirror, for example, is a case of a mechanism-of (i.e., a causal pathway) that is not also a 'mechanism-for'.

The minimal mechanism of Glennan and Illari (2018b), for example, which is defined as a mechanism *for* a phenomenon, can be understood as either a mechanism-of or a mechanism-for, depending on the kind of mechanism to which the definition of minimal mechanism is applied. So, if we apply the definition to the case of the mechanism of apoptosis, which has a particular regulatory function, then we have a mechanism-for. But if we apply the definition to a case like necrosis, where typically there is no function that is performed but what we have is a process that produces a certain end-state, then we have a mechanism-of. In the latter case, even if we characterise necrosis (or any causal process) as a mechanism for a phenomenon, what we really mean is that the mechanism has this

---

[4] Note also here that for someone like Glennan who wants to defend a mechanistic theory of causation, the notion of 'mechanism-for' has to be interpreted as broadly as possible.

phenomenon as its typical effect. What we have in mind in such a case is a mechanism-of.

One natural question may arise at this point. Can a mechanism be *both* what we called a mechanism-for and what we called a mechanism-of? That is, can it be the case that a mechanism *both* underlies or constitutes a causal process *and* is a mechanism for a specific behaviour? Though Harré adopted this view, this position acquired new strength in the early 1990s when Stuart Glennan developed his own mechanistic theory of causation. For him, mechanisms are both what underlie or constitute causal connections between events and thus provide the missing link between cause and effect (mechanisms-of) and at the same time complex systems or processes that are responsible for certain behaviours (mechanisms-for) and are thus individuated in terms of them.

But, as already noted, mechanisms-of are *not* necessarily mechanisms-for. Conceptually, this is obvious if we think of a mechanism as a causal process with various characteristics (such as those discussed above – e.g., they possess a conserved quantity or some kind of persisting structure). There is no reason to think that this kind of mechanism (e.g., the process by means of which the sum of kinetic and potential energy is conserved in some interaction) is a mechanism *for* any particular behaviour. Or consider the reflection of light by a mirror. Although this is clearly a causal process, and so a mechanism-of, there is no sense in which this causal process is *for* something. And of course, if we think of a mechanism as a complex system such that the interactions of its parts bring about a specific behaviour, there is no ipso facto reason to adopt a mechanistic account of causation. In light of this (and if we for the moment restrict the notion of mechanism-of just to mechanisms that feature in mechanistic theories of causation) we arrive at a tripartite categorisation of 'mechanistic' accounts present in the literature (or at *three* different notions of what a mechanism is): mechanisms can be mechanisms-for, mechanisms-of or both.[5]

Where is Causal Mechanism located in terms of this categorisation? For us, the distinction between mechanisms-of and mechanisms-for brings out the contrast between biological mechanisms such as apoptosis that are mechanisms *for* a certain function, and 'mechanisms' in the sense of causes of phenomena. The core sense of mechanism identified by Causal Mechanism says that mechanisms are causal pathways; that is, they are mechanisms-of. But in biological and other functional contexts, the typical notion of mechanism is not CM, but what we called CM-P. So,

---

[5] See Levy (2013) and Andersen (2014a; 2014b) for alternative categorisations.

specifically *biological* mechanisms are mechanisms-for. To return to our main example: both apoptosis and necrosis are mechanisms that lead to cell death, and, hence, they are both mechanisms *of* cell death, but only apoptosis has a function and is thus a mechanism *for* cell death. So, in the case of CM, the mechanisms for/of distinction serves to illustrate the fact that some causal pathways have a function and a functional role to play while other causal pathways do not. The distinction, then, is important for driving home the point that there are mechanisms everywhere (where there are causal pathways) even if there is no function they perform. For example, we can talk about the mechanism of light refraction or the mechanism of beta-decay. So clearly, for us, every mechanism-for is a mechanism-of, but not conversely.[6]

With this map of the conceptual landscape of the philosophical literature on mechanisms in mind, our task now is to examine each case in turn and investigate the relations between each sense of 'mechanism' and laws/counterfactuals. We will briefly look at the Salmon-Dowe theory as the main version of an account of mechanisms-of that are not at the same time mechanisms-for, and then examine in more detail the other two categories, which are the most relevant for our purposes since they represent the main views of what a mechanism is among new mechanists. In Section 5.6 we will discuss mechanisms-for that are not at the same time mechanisms-of. In Chapter 6 we will examine Glennan's theory, in which mechanisms are both mechanisms-of and mechanisms-for.

As noted already, the best known cases of mechanisms-of are those discussed by Salmon and Dowe. Though these accounts of causation are

---

[6] In some cases where scientists refer to 'mechanisms', it may not be clear whether it is the notion of mechanism-of or the notion of mechanism-for (or both) that they have in mind. For example, such uses as 'mechanism of chemical reaction', 'mechanism of speciation' and 'mechanism of action (of a drug)' may in principle be construed in various ways; to insist on a widening of the concept of mechanism-for based on scientific practice (without further argument) seems too quick. On our account, these uses of mechanism are to be viewed as mechanisms-of, that is, as various kinds of causal pathways. Note also what evolutionary biologist Mark Ridley (2004, 35, note 2) writes about the term 'isolating mechanism' in evolutionary biology: 'What is here called an "isolating barrier" has until recently (following Dobzhansky (1970)) usually been called an "isolating mechanism". Some biologists have criticized the word "mechanism" because it might imply that the character that causes isolation evolved in order to prevent interbreeding – that the isolating mechanism is an adaptation to prevent interbreeding.... The use of the term "isolating barrier" is becoming common now, and I follow this usage. However, the older expression could be defended. In biology, a mechanism of X is not always something that evolved to cause X. Compare, for instance, "population regulation mechanism", "mechanism of mutation", "mechanism of speciation", and "mechanism of extinction". Isolating mechanism could mean only a mechanism that isolates, not a mechanism that evolved in order to isolate.' Ridley's two senses of mechanisms clearly correspond to the distinction between mechanisms-for and mechanisms-of.

presented as being compatible with singular causation, it should be quite clear that they rely on counterfactuals. We noted already that in Salmon's account counterfactuals loom large. In fact, counterfactuals play a *double role* in his theory. On the one hand, they secure that a process is causal by making it the case that the process possesses not just an actual uniformity of structure, but also a counterfactual one. On the other hand, they secure the conditions under which an interaction (the marking of a process) is causal: if the marking would have occurred even in the absence of the supposed interaction between two processes, then the interaction is not causal.

On Dowe's account, the very idea of the possession of a conserved quantity for a process to be causal implies that both laws and counterfactuals are in the vicinity. Conserved quantities are individuated by reference to conservation laws, and it is hard to think of a process being causal without the conserved quantity that makes it causal being governed by a conservation law. Counterfactuals are also necessary for Dowe's account of causation, not just because laws imply counterfactuals but also because an appeal to counterfactuals is necessary for claiming that the process is causal. That is, it seems that without counterfactuals there is no way to ground the difference between objects to which conserved quantities may be applied and objects to which they may not. Consider, for example, a single particle with zero momentum versus a shadow with zero quantity of charge; the particle, but not the shadow, is a causal process precisely because it could enter into interactions, which could make its momentum non-zero (see Psillos 2002, 126).

5.5 Mechanisms-for and Mechanisms-of

Let us now turn to an account such as Glennan's, that is, to an account that takes mechanisms to be both mechanisms-for *and* mechanisms-of. There are two parts in Glennan's definition of mechanisms. First, a mechanism consists of components that interact – in this, it is similar to Salmon's account of a mechanism-of as causal process. However, for (early) Glennan the mechanism itself is a system with a stable arrangement of components.[7] In contrast to the view of mechanisms-of as processes (which can in principle be singular causal chains of events), such

---

[7] See Glennan (1996), though in more recent work (Glennan 2010) he drops the stability requirement for some kinds of mechanisms; see Section 5.6.

mechanisms are 'types of systems that exhibit regular and repeatable behaviour' (Glennan 2010, 259).

How should we understand the interactions among the components of such mechanisms? Should they be understood in terms of counterfactuals or not? To answer this question, let us briefly review various possible options.

The first general case we will consider is interactions with laws. Interactions can be governed by laws, where laws are understood in some robust metaphysical sense. For example, according to Fred Dretske (1977), Michael Tooley (1977) and David Armstrong (1983), laws are necessitating relations between universals. So, if there is a necessitating relation between universals A and B, there will be a law between them, and as a result of this law when A is instantiated, so will be B. Suppose we transfer that to the components of a mechanism: when component X instantiates A at some time t_1, some other component Y will instantiate universal B, perhaps at a later time. Or take the rival view (but metaphysically robust too) that laws are embodied in relations between powers. If this is the preferred account of laws, interactions will be understood in terms of powers. Powers are properties possessed by components of a mechanism and produce specific manifestations under specific stimuli. Whereas for Dretske, Tooley and Armstrong the interactions within the mechanism are grounded in the external relation of nomic necessitation, in the powers view, interactions are grounded in the internal relations between the powers of the components of the mechanism. Alternatively, interactions between the components of the mechanism may be viewed as being governed by metaphysically thin laws, for example, by (Humean) regularities. Here, component A can be said to interact with component B in virtue of the fact that this interaction is an instance of a regularity.

If, for the time being, we bracket laws, can we understand the interactions among the components of the mechanism differently? Perhaps counterfactuals can be of direct help here. So, Glennan (2002), following Woodward (2000; 2002; 2003a), understands interactions in terms of change-relating generalisations that are invariant under interventions. Such generalisations are change-relating in the sense that they relate changes in component A to changes in component B. They involve counterfactual situations, in that they concern what would have happened to component B regarding the value of quantity Y possessed by it, if the value of quantity X possessed by component A had changed. These generalisations are invariant under interventions, in that they are about relations between variables that remain invariant under (actual or counterfactual) interventions. These change-relating generalisations, then, are grounded in counterfactuals (called interventionist counterfactuals by Woodward).

But if we are to understand interactions between components in terms of counterfactuals, the next question is: What grounds these counterfactuals? In particular, in virtue of what are interventionist counterfactuals true? The answers here are well known (see, e.g., Psillos 2007). Counterfactuals can be grounded in laws or not. If they are grounded in laws, following what we said previously, these laws can be either metaphysically robust laws of the sort adopted either by Armstrong or by power-based accounts of lawhood, or thin Humean regularities, instances of which are particular token-interactions between components. If the counterfactuals are not grounded in laws, then it's likely that there are counterfactuals 'all the way down', that is, that there are primitive modal facts in the world (see Lange 2009).

In any of these accounts of law-governed within-mechanism interactions, counterfactuals have a central place: either by directly accounting for interactions (as in Woodward's theory), by being part of an account of the nomological dependences that ground the interactions, or as a primitive modal signature of the world.[8] So, if laws regulate the interactions between the components of the mechanism, we cannot do away with counterfactuals in grounding within-mechanism interactions. In Chapter 6 we will consider another option: counterfactuals may be grounded in mechanisms. But for now, let us turn to mechanisms-for.

5.6 Mechanisms-for

So far we have argued that mechanisms-of (mechanisms considered as underlying or constituting causal processes) require laws, and thus difference-making relations must be included in the notion of a mechanism-of. But what about mechanisms-for that view mechanisms as systems or organised entities responsible for certain behaviours? What is the relation between mechanisms-for and laws/counterfactuals? Recall that

---

[8] In Lange's (2009) theory of laws it is a counterfactual notion of stability that determines which facts are lawful and which are accidental. In other theories of lawhood, counterfactuals come 'for free', so to speak, as they must be part of any metaphysically robust theory of laws (such as that of Dretske, Tooley and Armstrong): any such theory must show why laws support counterfactuals. It is not obvious how exactly counterfactuals are part of a regularity view of laws. But note that this is a problem (if at all) for the regularity theorist, and not for the view that interactions have to be understood in terms of laws/counterfactuals. For an attempt to reconcile regularity theory with counterfactuals, see Psillos (2014).

in a mechanism-for the interactions of its parts bring about a certain behaviour or function. A mechanism-for need not commit us to a mechanistic (e.g., à la Salmon-Dowe) account of the causal interactions between its parts.

Here is an argument why mechanisms-for, at least prima facie, have to incorporate laws and/or counterfactuals: a mechanism-for involves components that interact with one another; but laws and/or counterfactuals are needed to account for these interactions; hence, mechanisms-for need to incorporate laws and/or counterfactuals. However, Jim Bogen (2005) has taken the existence of mechanisms that function *irregularly* as an argument against the view that laws and regular behaviour have to characterise the function of mechanisms. In this section we will deal with this argument from irregular mechanisms (in Chapter 6 we will examine the claim that within-mechanism interactions should be viewed in terms of activities and not in terms of laws or difference-making relations).

The first point that we want to stress is that irregular and unrepeatable mechanisms are not as ubiquitous as some philosophers want us to believe. So, consider a claim made by Bert Leuridan (2010). He thinks that mechanisms as complex systems ontologically depend on stable regularities, since there can be no such mechanisms (1) without macrolevel regularities (i.e., the behaviour produced by the mechanism) and (2) without microlevel ones (i.e., the behaviours of the mechanism's parts). Kaiser and Craver (2013, 132) have replied to this that Leuridan's first claim is 'clearly false' since '[o]ne-off mechanisms are mechanisms *without* a macrolevel regularity', where 'one-off mechanisms' are the causal processes discussed by Salmon and others (*mechanisms-of* in our terminology). Moreover, they point to examples where scientists seem to be interested in exactly this kind of mechanism, that is, when they try to explain how a *particular* event occurred (e.g., a particular speciation event).

This kind of reply conflates the two different senses of mechanism we have tried to disentangle. It is not the case that 'singular, unrepeated causal chains . . . are a special, limiting case of [complex system] mechanisms, not something altogether different', as Craver and Kaiser (2013, 131–2) insist. For it is not at all clear that such mechanisms-of are at the same time mechanisms-for, that is, mechanisms *for* a certain behaviour. Similarly, we remain unpersuaded by Glennan's (2010) claim that the mechanisms that produced various historical outcomes are mechanisms-for (he calls them 'ephemeral mechanisms'). In any case, it would be very implausible to insist that any arbitrary causal chain is for a certain behaviour (which is identified with the outcome of the causal chain or, alternatively, with the (higher-level) event constituted by the causal chain). For instance, the reflection of a light ray on a surface is a clear case of a mechanism-of (since it constitutes a causal process), but it is not at all clear that it is a mechanism for a certain behaviour. As we've argued, a thin notion of mechanism-for that can be applied to any causal pathway seems to us misleading and unhelpful.

So, it is not enough to point to singular causal chains in order to argue that there can be irregular mechanisms or one-off mechanisms (mechanisms that function only once) (see also DesAutels 2011; Andersen 2012). Still, one may wonder: Can there be mechanisms-for *without* a corresponding macro-level regularity? Thus, the issue that must be clarified is: What are the conditions for being a mechanism *for* a behaviour? Is it merely to have a function (which is the mechanism's behaviour), or should we, in addition, require that the behaviour be regular?

To make the argument stronger, let us here take function in a wide sense, that is, as not requiring that for something to have a function it has to be the product of conscious design or the result of natural selection (as we have in general assumed in this book). In other words, we are going to take a function in the sense of Robert Cummins (1975), for whom functions are certain kinds of dispositions (see Craver 2001 for such an approach to the functions of mechanisms). In particular, what it is for a mechanism M to have a function F is to have a disposition to F, which contributes to a disposition of a larger system that contains M. Such functions need not be restricted to living systems or artefacts.[9] Yet not anything whatsoever can be ascribed a Cummins-function. In particular, unrepeated causal chains of events, which might well be Salmon's and Dowe's mechanisms-of, need not have a function. We can follow Cummins and say that talk of functions only makes sense when we can apply what Cummins calls the analytical strategy, that is, to explain the disposition of a containing system in terms of the contributions made by the simpler dispositions of its parts.

There can be systems with Cummins functions that exhibit the corresponding behaviour only once; there are many biological functions, the realisation of which requires that the biological entity that has the function cease to exist. An example is the mechanism for apoptosis. Here, the relevant mechanism has a Cummins function; that is, it causally

---

[9] However, note that for some (e.g., Kitcher 1993) we cannot ascribe even Cummins functions to entities that are not products of (either conscious or non-conscious) design, that is, that are not either artifacts or living systems.

contributes to the death of the cell. However, this is a function that, when successfully carried out, can occur only once. But even in such cases, the behaviour of a particular mechanism of apoptosis is a token of *a type of behaviour* that occurs all the time in an organism.

Can there be genuinely *irregular* mechanisms-for, that is, mechanisms-for without a corresponding macrolevel regularity? Bogen (2005) has offered the case of the mechanism of neurotransmitter release as an example of a mechanism-for that behaves irregularly. As this mechanism more often than not fails to carry out its function, there exists no corresponding macro-level regularity; but moreover, and more importantly, Bogen thinks that within-mechanism interactions must themselves be irregular, and thus we must abandon the regularity account of causation in favour of activities.

We do not think that this last conclusion follows from Bogen's example. To see why this is the case, it is useful to distinguish between three cases of what we may call 'irregular' mechanisms-for. The irregularity of mechanisms-for may be only contingent (irregular$_1$), stochastic (irregular$_2$) or (let us assume) more radical (irregular$_3$).

Irregular$_1$ mechanisms-for are mechanisms that could function regularly, but they in fact do not. A defective machine that only functions once in a while is a case in point. Such a machine (1) is a mechanism for a behaviour and (2) functions irregularly. However, it is certainly not the case that a successful operation of the machine is not subject to laws (e.g., laws of electromagnetism, gravity or friction). Nor is it the case that defective machines falsify the regularity account of causation.

Irregular$_2$ mechanisms are like irregular$_1$ mechanisms in that they operate in accordance with laws, but in this case the laws are probabilistic. So, the existence of such mechanisms does not show that within-mechanism interactions are not law-governed, or even that the regularity account of causation is false – regularities can be stochastic. What if the operation of a mechanism is completely chancy, for example, because it involves the radioactive decay of a single atom? Even if we do not have a law here (perhaps because the relevant law concerns a population of atoms rather than a single one), it is not at all clear to us that such a chancy 'mechanism' could be an example of a mechanism-for.[10]

Finally, we can imagine an irregular$_3$ mechanism-for; such a sui generis mechanism operates only once, and its unrepeatability is supposed not to be a contingent matter, but due to the fact that the interactions among its

---

[10] For a notion of 'stochastic' mechanism, see DesAutels (2011) and Andersen (2012).

components cannot *in principle* be repeated. We are not sure that the notion of an irregular₃ mechanism-for actually makes sense. But this is the only kind of example we can imagine, where the irregularity or unrepeatability of a mechanism would be a reason to think that its operation is not law-governed. If that's where the friends of genuinely irregularly operating mechanisms can pin their hopes for a non-law-governed account of mechanism, then so be it!

CHAPTER 6

Against Activities

6.1 Preliminaries

This chapter examines a main feature that for several philosophers is, together with entities, an indispensable component of mechanism as an ontological category, that is, activities. According to several new mechanists, activities constitute a novel ontological category, which is required for an account of the productivity of mechanism, and hence has a central place in a mechanistic world view. This chapter offers a critique of the arguments in favour of activities offered in the recent literature. It thus provides a critique against a commonly accepted metaphysics of mechanisms. The main aim of the chapter is to criticise the popular idea that productivity of mechanisms requires a commitment to activities qua a sui generis ontic category.

6.2 Mechanisms and Counterfactuals

Glennan (1996) suggested that *mechanisms themselves* ground counterfactuals. For him, although interactions are understood in terms of interventionist counterfactuals, these counterfactuals are in turn grounded in (lower-level) mechanisms. In this section we first present Glennan's mechanistic theory of causation in more detail (Section 6.2.1) and then focus on his solution to the problem of counterfactuals (Section 6.2.2) and argue that this solution gives rise to what we call the *asymmetry* problem (Section 6.2.3).

6.2.1 Glennan's Early Mechanistic Theory of Causation

Glennan (1996; 2002) took mechanisms to be complex systems (or objects) which bring about a certain activity or are responsible for a certain behaviour. A thermostat might be a stock example of a mechanism in

Glennan's sense. A conventional thermostat works like an on-off switch. A bimetallic coil tips a small mercury-filled glass bottle. The bimetallic coil is made from two different metal strips that have been sandwiched together and then rolled into a coil. As the temperature changes, the two metals expand differently and the coil winds or unwinds. As it does, it tips the glass bottle and the mercury rolls from one end of the bottle to the other. When the mercury falls to one end, it allows an electric current to flow between two wires and the furnace turns on. When the mercury falls to the other end of the bottle, the current stops flowing and the furnace turns off. According to Glennan (2002, S344):

> (M) A mechanism for a behavior is a complex system that produces that behavior by the interaction of a number of parts, where the interactions between parts can be characterized by direct, invariant, change-relating generalizations.

Mechanisms, he adds, 'are not mechanisms *simpliciter*, but mechanisms *for* behaviors'. For the very same complex system may issue in two different behaviours (e.g., the heart is a mechanism that pumps blood and makes noise). What the mechanism *does* determines its boundaries, its division into parts and the relevant modes of interaction among these parts. Broadly understood, a mechanism consists of some parts (its building blocks) and a certain *organisation* of these parts, which determines how the parts interact with each other to produce a certain output. The parts of the mechanism should be stable and robust; that is, their properties must remain stable, in the absence of interventions. The organisation should also be stable; that is, the system as a whole should have stable dispositions, which produce the behaviour of the mechanism. Thanks to the organisation of the parts, a mechanism is more than the sum of its parts: each of the parts contributes to the overall behaviour of the mechanism more than it would have achieved if it acted on its own. Mechanisms can be contained within larger mechanisms.

There are two major attractions of Glennan's mechanistic theory. The first is that it is descriptively more adequate than the mechanistic approach of Salmon and Dowe. Both of them characterise interactions in terms of the exchange of conserved quantities. To be sure, they do aim at a mechanistic theory of *physical* causation. Still, this account is too narrow to describe cases of causation among higher-level entities. Consider, Glennan says, 'a social mechanism whereby information is disseminated through a phone-calling chain' (p. S346). It is surely otiose and uninformative to try to describe this mechanism in terms of exchange of conserved quantities. Glennan does not deny that the interactions involved in telephone calls supervene on basic physical interactions. But he is surely right in saying that we would miss something if we tried to *explain* them in those terms. We would lose the fact that higher-level interactions form higher-level kinds. So, Glennan's mechanistic view is broad enough to account for mechanisms at levels higher than physics. The explanatory autonomy of higher-level mechanisms is, we think, a lesson that we should take to heart.

The other attraction of Glennan's mechanistic theory relates to his demand that understanding causal claims requires knowing what their underlying mechanisms are (1996, 66). In fact, Glennan wants to make a stronger point, namely, that 'a relation between two events (other than fundamental physical events) is causal when and only when these events are connected in the appropriate way by a mechanism' (p. 56). Although the weaker claim is very plausible, we don't think the stronger claim is warranted. Let us examine the relationship between mechanisms and counterfactuals in Glennan's theory to see the difficulties that arise from this approach to causation.

6.2.2   Mechanisms as Truth-makers of Counterfactuals

Glennan took his mechanistic approach to offer a rather robust solution to the problem of counterfactuals. He took laws that are mechanically explicable (in the sense that there is a mechanism that underpins them) to show in 'an unproblematic way' how 'to understand the counterfactuals which they sustain' (p. 63). The key idea is that the presence of the mechanism (e.g., the thermostat) explains why a certain counterfactual holds (e.g., if the temperature had risen, the furnace would have turned off). Similarly, the breakdown of a mechanism would explain why certain counterfactuals fail to hold. In his 2002 paper (see (M) above), Glennan characterised the interaction of the parts of the mechanism in terms of Woodward's invariant, change-relating generalisations, that is, generalisations that remain invariant under actual and *counterfactual* interventions.

It seems then that there is a tension between Glennan's views in 1996 and in 2002. According to the earlier view, mechanisms explain via mechanical laws when certain counterfactuals hold. According to the later view, certain interventionist counterfactuals explain (or ground) the laws that govern the interaction of the parts of the mechanism. Consider the thermostat: it is a mechanical law (ultimately, the law that metals expand when heated) which explains why it is the case that had the temperature been higher, the switch would have closed. But why is this a *law*? Because, had we intervened to change one magnitude (e.g., the temperature), the law (that metals expand when heated) wouldn't change and the other magnitude (e.g., the length of the metal strips in the bimetallic coil) would have changed. The tension is obvious: mechanical laws support counterfactuals and counterfactuals render mechanical laws *laws*. Though we are not sure we are faced here with a vicious circle, we are also not sure *where* it can be broken so that the described relation between mechanism and interventionist counterfactuals can get going.

A central and stable feature of Glennan's views is a distinction between the fundamental laws of physics and what he calls mechanically explicable laws. He notes, quite plausibly, that the fundamental laws of physics are *not* mechanically explicable and claims that 'all laws are either mechanically explicable or fundamental, *tertium non datur*' (1996, 61). A mechanically explicable law is a law which is underpinned by a mechanism or, as Glennan says, which 'is explained by the behaviour of some mechanism' (p. 62). He takes it that mechanically explicable laws characterise all the special sciences and 'much of physics itself' (p. 50).

Here is then the resulting picture: interactions among components of a mechanism are governed by laws, which are understood in terms of interventionist counterfactuals; these laws are 'mechanically explicable' – that is, there are other mechanisms that ground them – but these (lower-level) mechanisms themselves contain parts, the interactions among which are understood in terms of counterfactuals, and which are in turn grounded in yet other mechanisms, until we finally reach a level where we run out of mechanisms to explain the laws that govern the interactions among components, and thus to ground the relevant counterfactuals. At this fundamental level, interactions among components are *directly grounded in counterfactuals*. But notwithstanding these not mechanically explicable laws, Glennan insists that at all other levels mechanisms can ground interactions. So, even if we need to introduce counterfactuals to account for interactions, mechanisms seem to have priority over counterfactuals, and thus the account is supposed not to be a version of a difference-making theory of causation, but a genuinely mechanical account.

6.2.3   The Asymmetry Problem

If fundamental laws are *not* mechanically explicable, and if they too support counterfactuals (as they do, we suppose), it is not necessary for the truth of a counterfactual that there is a mechanical explanation of it. So, the presence of a mechanically explicable law (and hence of a mechanism) is not a necessary condition for the truth of a counterfactual conditional. Glennan agrees on this; still, he thinks it is a *sufficient* condition. Even if he is right, his theory is incomplete: if some counterfactuals are true even though a mechanism is absent, then there is more to the link between laws and counterfactuals than Glennan's theory admits. Suppose Glennan is right in taking mechanisms to underpin non-fundamental laws. He also subscribes to some kind of supervenience thesis: the non-fundamental laws supervene on the fundamental laws (see 1996, 62 and 66; 2002, S346 and S352). So on Glennan's view, non-fundamental laws are underpinned by mechanisms *and* supervene on fundamental laws, which are not underpinned by mechanisms.

Here is the problem, then. What is the relation between the mechanisms that realise the non-fundamental laws and the more fundamental laws on which the non-fundamental laws supervene? Glennan does not explain. To be sure, he (1996, 66) asserts: 'Although the mechanism responsible for connecting two events may supervene upon other lower-level mechanisms, and ultimately on mechanically inexplicable laws of physics, it is not these laws which make the causal claim true; rather it is the structure of the higher level mechanism and the properties of its parts.'

But this is hardly an explanation of what is going on. One plausible thought is that the fundamental laws govern the interactions of the parts of the mechanism, which realises the non-fundamental law. If this is so, then it would be odd to say that the mechanism that explains, say, Ohm's law is ultimately determined (supervenience *is* a kind of determination) by the fundamental laws that govern the interaction of fundamental particles but that these fundamental laws are *not* (part of) the truth-makers of Ohm's law. Once identified, the mechanism might well have explanatory and epistemic autonomy. But if supervenience holds, the mechanism does not have metaphysical autonomy. We will call this the asymmetry problem.

Here is the problem in more detail: given the existence of not mechanically explicable laws, it is not clear how mechanisms can ground counterfactuals *at any level*. That is, given that the mechanisms at the lowest level depend on counterfactuals, the mechanisms at a level exactly above the fundamental must be equally dependent (albeit *derivatively*) on the fundamental counterfactuals, and so on for every higher level. In other words, to ground counterfactuals at any level, we need the whole lower hierarchy of mechanisms *and* counterfactuals, and since we ultimately arrive at a level where there are either only counterfactuals or only laws (or both), it seems that there is a fundamental asymmetry between mechanisms and laws/counterfactuals. The only way to block the asymmetry would be to argue that we do not need the whole hierarchy in order to ground the counterfactuals at higher levels. Even if this were to be granted for purposes of explanation – that is, even if explanation in terms of mechanisms at level *n* does not require *citing* lower-level mechanisms – metaphysically, the whole hierarchy constitutes the grounds for the mechanism.[1]

We think, then, that the presence of a mechanism is *part* of a sufficient condition for the truth of certain counterfactuals; the fully sufficient condition includes some facts about the fundamental laws that, ultimately, govern the behaviour of the mechanism. This, of course, is entirely consistent with the thought that in most practical situations when it comes to asserting the truth of a certain counterfactual, it is enough to cite the mechanism. The rest of the sufficient condition is not thereby rendered metaphysically redundant, but only explanatorily so.

In sum, given the asymmetry problem and our discussion in Chapter 5, we conclude that if laws are admitted in our notion of mechanism, a reliance on counterfactuals is inevitable. But can we perhaps avoid counterfactuals if we account for within-mechanisms interactions in some other way?

6.3   Activities and Singular Causation

In Chapter 5 we reviewed various options to understand interactions of components of mechanisms, where these interactions are viewed as law-governed. The question now is: Can we have interactions without some notion of law in the background either in terms of regularities or in some more metaphysically robust sense? If yes, then this could be a way to have mechanisms-for without the need to put laws and counterfactuals in the picture.

For some mechanists, the interactions of components have to be understood in terms of *activities*. Activities are a new ontological category that, together with entities, are said to be needed for an adequate ontological account of mechanisms (Machamer et al. 2000; Machamer 2004). Activities are meant to embody the causally productive relations between components. Causation in terms of activities is viewed as a type of singular

---

[1] See Glennan (2011) for an attempt to respond to this argument, and Casini (2016) for a detailed criticism; see also Campaner (2006).

causality, where the causal relation is a local matter; that is, it concerns what happens between the two events that are causally connected, and not what happens at other places and at other times in the universe (as is the case for the regularity theorist). Activities, thus, have been taken to obviate the need for laws.

We disagree. In fact, we are about to argue against the popular activities-based conception of mechanism. We shall start with a criticism of a key argument in favour of activities, that is, that one is led to such a notion if one accepts that causation is *singular* and thus not law-governed.

Does singular causation imply that there are no laws? It would be too quick to infer from singular causal claims that laws are not part of causation. By singular causation we may simply mean that there exist genuine singular causal connections, that is, causal connections between particular event-tokens. But this is not enough to prove that there are no laws in the background. For it is consistent with the existence of singular causal sequences that there are laws under which the causal sequences fall. To use a quick example, on Armstrong's account of laws, singular causation is ipso facto nomological causation since the nomic necessitating relation that relates two universals relates the instances of the two universals too (Armstrong 1997). Interestingly, the same is true if we take singular causation to be grounded in the powers possessed by objects; powers are again *wholly* present in the complex event that constitutes the singular causal sequence. And though there is no nomic relation that relates the two powers, the regular instantiation of the two powers implies the presence of a regularity. So, what both these cases show is that even singular causation can be nomological, that is, subsumed under laws.[2]

Thus, singular causation does not, on its own, constitute an argument in favour of viewing interactions among components of mechanisms as not being law-governed or, more generally, as not depending on difference-making relations. So, friends of activities need to (1) give more reasons to justify the introduction of this new ontological category and (2) explain why activities qua producers of change are themselves counterfactual-free. Although it's conceivable that singular causation just amounts to the local activation of powers which in turn ground activities, powers being

---

[2] There is debate among friends of powers whether such a powers-ontology yields an account of laws in terms of powers (Bird 2007) or a lawless ontology (Mumford 2004). But this need not concern us here.

universals, it's upon the friends of powers to show that we can understand this co-instantiation without also assuming that there is a law present.[3]

We thus reject the widespread claim according to which if causation is singular, this means that mechanisms that embody singular causal relations are not grounded on laws. As we showed, causation can be singular *and* nomological at the same time. Let us now see what other reasons new mechanists have given in favour of an activities-based conception of mechanisms.

6.4   Against Activities I: The MDC Account

As we have already seen, Machamer, Darden and Craver claim: 'Mechanisms are entities and activities organised such that they are productive of regular changes from start or set-up to finish or termination conditions' (2000, 3). On the face of it, the MDC characterisation of a mechanism is fairly similar to Glennan's. On closer inspection, there is a central difference. MDC introduce the concept of *activity* as a means to account for the interaction between the parts of the mechanism and its overall causal efficacy. The MDC approach is exciting, especially when it comes to the detailed description and classification of how mechanisms are taken to operate in neurobiology. But for the purposes of this chapter, we will examine only the notion of activity. This notion is central to MDC's mechanistic view of causation since, as they say, 'activities are types of causes' (p. 6) and 'activities are needed to specify the term "cause"' (p. 8).

As we see it, their view is that an adequate understanding of the concept of mechanism requires an *ontological* shift: we need to accept the existence of activities on top of the usual commitments to entities, properties and processes. This unparsimonious move is recommended on the basis of their claim that mechanisms are 'active': 'they do things' (p. 5). They think that unless activities are accepted as ontological bedfellows of entities, properties and processes, mechanisms will be *passive*: things might be done *via* them, but not *because of* them. They also claim that appeals to causal

---

[3] See Waskan (2011) for a mechanist account of the contents of causal claims that is not based on counterfactuals and Woodward's (2011) answer that causation as difference-making is fundamental in understanding mechanisms; Menzies (2012) provides an illuminating account of mechanisms in terms of the interventionist approach to causation within a structural equations framework; last, Glennan (2017, chapters 5 and 6) offers a detailed treatment of mechanistic causation as a productive account of causation not reducible to difference-making relations. We will examine Glennan's account in detail in Section 6.7.

laws, or to invariant generalisations, fail to capture the productivity of a mechanism, which 'requires the productive nature of activities' (p. 4).

MDC's 'dualism', as they put it, requires that there is a fine distinction between entities (with their properties) and activities. But is there? As is usual in philosophy, we are first given some examples. So, cases such as bonding, diffusion, depolarisation, attraction and repulsion are cases of *activity*. But what do all these share in common in virtue of which they are *activities*? What we are told is that 'activities are the producers of change' (p. 3). But production is itself an activity. So, we are not given an illuminating account of that which some things share in common, in virtue of which they are activities.

MDC say the following of the relation between entities and activities: 'Entities and a specific subset of their properties determine the activities in which they are able to engage. Conversely, activities determine what types of entities (and what properties of those entities) are capable for being the basis of such acts.... Entities and activities are correlatives. They are interdependent' (p. 6). It follows that entities and activities are ontically on a par: they determine each other. They say this more explicitly when they claim that '[t]here are no activities without entities, and entities do not do anything without activities' (p. 8).

We think the supposed ontic parity between entities and activities is wrong-headed. First, it's conceivable that there are entities without activities. Indeed, there may be entities capable of engaging in certain activities, but the prevailing circumstances or the laws of nature may be such that they *fail* to engage in these activities. (If what matters is the ability of an entity to engage in an activity and not the actual occurrence of this activity, then it is clear that MDC have to rely on counterfactuals to illuminate the link between entities and activities.) Second, we cannot see how activities can determine what *types* of entities can engage in them. There may well be an open-ended list of types of objects that can engage in some activity, and they may share very little, if anything, in common. Take the activity of *playing*. It's hard to say that it determines what kinds of entities (and what properties) are involved in this activity. Admittedly, this is a case of a highly generic activity and it might be problematic precisely because of this. There are cases of more specific activities, where the activity is performed by certain *types* of objects. It then might *seem* that the activity does determine what types of object can engage in it. An example of such a specific activity might be the activity of *pushing*. It seems that this activity determines that the objects involved in it must have certain properties, for example, rigidity, bulk and so on. But we think appearances are deceptive.

Epistemically, we might first classify a certain type of activity and then identify what kinds of objects engage in it. But from this it does not follow that this is the order of ontic dependence too. On the contrary, objects can engage in certain activities *because* they have certain properties, and not the other way around.

Consider the activity of chemical bonding. Does this activity determine that entities that engage in it must have a certain electronic structure? Not really. Chemical bonding could not exist without some entities having the right electronic structure. So, not only are the latter presupposed ontically for the activity, but they also fully *determine* this activity: the activity of bonding *consists* in the fact that certain entities with certain electronic structure behave in a certain way when they are in proximity. The dependence of the activity on the properties of entities becomes clear when the activity *fails* to take place. Consider the case where chemical bonding does not take place, for example, the case of noble gases. There, you have the entities without the activity of chemical bonding precisely because the entities and their properties determine that a certain activity *cannot* take place. The situation is exactly symmetrical when the activity *does* take place.

The conclusion we draw is that activities cannot be ontically on a par with entities. But one may wonder: Why should MDC want to hypostatise activities? Why isn't it enough to talk in terms of entities and their properties? MDC may be right in protesting against process-theorists that entities are indispensable in understanding mechanisms; they may rightly claim that the programme of reducing entities to processes is 'problematic at best'. But they also want to argue against 'substantivalists', that is, those who 'confine their attention to entities and properties, believing that it is possible to reduce talk of activities to talk of properties and their transitions' (p. 4). Against them, MDC claim that entities and their properties are not enough for the characterisation of mechanisms: activities are also required. Now, the substantivalists that MDC have in mind take the properties of the entities to be dispositional; they equate them with *capacities* or *active powers*. This is a quite powerful ontology. The friends of active powers would surely protest that given that active powers are granted to entities, talk of activities as *distinct* from these powers is redundant.[4]

---

[4] Consider how Harré (2001, 96) understands an active power: 'a native tendency or inherent capacity to act in certain ways in the appropriate circumstances'. Activities come for free if Harré is right. Note that Harré too favours a mechanistic account of causation.

MDC offer two arguments for activities on top of capacities. We think they are both problematic. The first argument is this:[5] '[I]n order to identify a capacity of an entity, one must first identify the activities in which that entity engages' (p. 4). Even if right, this is irrelevant. It only raises an epistemic point: we cannot know what capacities an entity has unless we first know what it *does*. From this, it does not follow that activities are ontically on a par with capacities. Nor does it follow that it is not the capacities of an entity that determine what activities it engages in. Quite the contrary. To use their own example, it is *because* aspirin has the capacity to relieve headaches (a capacity which we take it to be grounded in its chemical composition) that aspirin engages in this activity, that is, headache- relieving. If capacities are granted, then activities supervene on them. And this remains so, even if, from an epistemic point of view, we need to attend to the (observed) activities in order to conjecture about the capacities.

The second argument that MDC offer is this: '[S]tate transitions have to be more completely described in terms of the activities of the entities and how those activities produce changes that constitute the next change' (p. 5). Here the emphasis is on the *production*. As they explain, activities add the 'productivity' by which changes in properties (state-transitions) are effected. But isn't this question-begging? Many would just deny that there is anything like a productive continuity in state transitions. All there is, they would argue, is just regular succession (or some kind of dependence). In any case, the friends of capacities would argue that there is productive continuity in state-transitions, but that this is grounded in the natures of the entities engaged in state transitions. If water has the capacity to dissolve salt, and if this capacity is grounded in the natures of water and salt, then all that is needed for the dissolution of salt in water (i.e., the activity) is that the circumstances are right and the two substances are brought into contact.

We have a final, but central, objection to MDC: they cannot avoid counterfactuals. Counterfactuals may enter at two places. The first is the activities themselves. Activities, such as bonding, repelling, breaking or dissolving, are supposed to embody causal connections. But one may argue that causal connections are distinguished, at least in part, from non-causal ones by means of counterfactuals. If 'x broke y' is meant to capture the claim that 'x *caused* y to brake', then 'x broke y' must issue in a counterfactual of the form 'if x hadn't struck y, then y wouldn't have broken'. So

<hr />

[5] The essence of this argument is repeated in Machamer (2004).

talk about activities is, in a sense, disguised talk about counterfactuals. The second entry point for counterfactuals is the characterisation of interactions within the mechanism. We have already seen (in Section 6.2) Glennan insisting that this interaction should be captured in terms of the invariance of the relationships among the parts of the mechanism under actual and counterfactual interventions. MDC are not quite clear on what the interaction within the mechanism consists in. Note that it wouldn't help to try to explain the interaction between two parts of a mechanism (say, parts A and B) by positing an intermediate part C. For then we would have to explain the interaction between parts A and C by positing another intermediate part D and so on (ad infinitum?).

We take this to be a crucial problem of the mechanistic approach to causation. In a sense, this approach fills in the 'chain' that connects the cause and the effect with intermediate loops. But there is still no account of how the loops interact. Here, it might well be the case that the most general and informative thing that can be said about these interactions is that there are relations of counterfactual dependence among the parts of the mechanism. Even if we posited activities, as MDC do, we would still need counterfactuals to make sense of them, as we have just seen. In any case, if we are right, there is more to causation than mechanisms.

6.5   Against Activities II: Glennan's Approach

Let us now turn to Glennan's approach to activities, which we will characterise as a top-down approach, partly in order to contrast it with Illari and Williamson's bottom-up approach that we will examine next. Note that, by a top-down approach we do not mean an approach that starts with a preferred ontology and tries to understand science on the basis of it. No new mechanist uses such an approach. Rather, all start with science, but in trying to understand scientific practice new mechanists use metaphysical notions in various degrees and differ in their interest to develop a systematic metaphysical picture. So, for example, some new mechanists may be suspicious of metaphysics and more interested to examine how mechanisms function within science.[6] Others, like Glennan, are interested in developing a systematic mechanical metaphysics. The top-down/bottom-up distinction applies to new mechanists who fall in this latter camp and illuminates, it seems to us, a difference in methodology.

[6] Bechtel is such an example of a new mechanist (although see Craver & Bechtel 2007).

We take our distinction between top-down and bottom-up approaches not to be extremely sharp, but a matter of degree or emphasis. Still, a difference of degree is a difference: some new mechanists focus more on developing a systematic and coherent metaphysical picture than others. So, this new mechanical strategy is not top-down as was, for instance, in the case of someone like Descartes (and his *Principia*). But note that even in the case of Descartes, where the top-down approach is prominent, there is also a focus on methodological issues; so, the resulting Cartesian image, where metaphysics has primacy, can be seen as a reconstruction of a network of philosophical practices that puts the emphasis on metaphysics and not on the methodology, just as in the case of some new mechanists. We think that this similarity justifies calling such an approach 'top-down', even if it is fully naturalistic.

As we have already seen, Glennan has recently put forward what he calls Minimal Mechanism: 'a mechanism for a phenomenon consists of entities (or parts) whose activities and interactions are organised in such a way that they are responsible for the phenomenon' (Glennan 2017, 13). Though minimal, this account is 'an expansive conception of what a mechanism is' (p. 106), mostly because it involves commitment to activities as a novel ontological category. 'Activities', Glennan claims, 'cannot naturally be reduced to properties of or relations between entities' (p. 50).

Here then are some characteristics of activities, according to Glennan. Activities are concrete: 'they are fully determinate particulars located somewhere in space and time; they are part of the causal structure of the world' (p. 20). Activities are the ontic correlate of verbs. They include anything from walking to pushing to bonding (chemically or romantically) to infecting. Given this, activities 'are a kind of process – essentially involving change through time' (p. 20). Some activities are non-relational (unary activities) since they involve just one entity, for example, a solitary walk. But some activities involve interactions: they are non-unary activities, namely, activities that implicate more than one entity (p. 21).

Most activities, Glennan says, 'just are mechanistic processes', that is, spatiotemporally extended processes that 'bring about changes in the entities involved in them' (p. 29). What, then, is a *mechanistic* process? According to Glennan, 'To call a process mechanistic is to emphasise how the outcome of that process depends upon the timing and organisation of the activities and interactions of the entities that make up the process' (p. 26).

Now, it appears that there is a rather tight circle here. A process is mechanistic when the entities that make it up engage in *activities*. But if activities just are *mechanistic processes*, then a process is mechanistic when the entities that make it up engage in mechanistic processes. Not much illumination is achieved. Perhaps, however, Glennan's point is that activities and processes are so tightly linked that they cannot be understood independently of each other. Yet there seems to be a difference: activities (are meant to) imply action. To describe something as an activity is to imply that something acts or that an action takes place. A process need not involve action. It can be seen as a (temporal or causal) sequence of events. In fact, it might be straightforward to just equate the mechanism with the process, namely, the causal pathway that brings about an effect. In the sciences all kinds of processes are characterised as mechanistic irrespective of whether they are 'active' or not.

Let us illustrate this point by a brief discussion of the case of active versus passive membrane transport, which are the two mechanisms of transporting molecules across the cell membrane. The transportation of the molecules takes place across a semi-permeable phospholipid bilayer and is determined by it. Some molecules (small monosaccharides, lipids, oxygen, carbon dioxide) pass freely the membrane through a concentration gradient, whereas other molecules (ions, large proteins) pass the membrane against the concentration gradient and use cellular energy. The main difference between active and passive transport is precisely that in active transport the molecules are pumped using ATP energy, whereas in passive transport the molecules pass through the gradient by diffusion or osmosis. These different mechanisms play different roles. Active transport is required for the entrance of large, insoluble molecules into the cell, whereas passive transport allows the maintenance of homeostasis between the cytosol and extracellular fluid. But they are both causal processes or pathways, even though only one of them is 'active'.

Glennan (2017, 32) takes it that 'the most important feature of activities' is that most or all activities are mechanism-dependent. This, he thinks, suggests that 'the productive character of activities comes from the productive relations between intermediates in the process, and that the causal powers of interactors derive from the productive relations between the parts of those interactors'. But this is not particularly illuminating. Apart from the fact that production is itself an activity, to explain the productive character of activities by reference to the productive activity of intermediaries or of the constituent parts of the mechanism just pushes the issue of the productivity of an activity A to the productivity of the constituent activities $A_1, \ldots, A_n$ of the mechanism that realises A. Far from explaining how activities are productive, it merely assumes it. Now,

Glennan takes an extra step. He takes it that some producings are explained 'in terms of other producings, not in terms of some non-causal features such as regularity, or counterfactual dependence' (p. 33). In the context in which we are supposed to try to understand what distinguishes activities from non-activities, this kind of argument is simply question-begging.

If what makes entities engage in activities are their properties and relations to other entities, in what sense are activities things distinct from them? In what sense are activities 'a novel ontological category'? Here, we find Glennan's argument perplexing. His chief point is that thinking of activities as fixed by the properties and relations of things 'reduces doing to having; it takes the activity out of activities' (p. 50). The language of relations 'is a static language' (p. 50). But activities, we are told, are 'dynamic' (p. 51).

Let us set aside this figurative distinction between doing and having. After all, it is by virtue of having mass that bodies gravitationally attract each other, according to Newton's theory of gravity. More generally, it is by virtue of having properties that things stand in relations to each other, some of which are 'static', for example, being taller than, while others are 'dynamic', for example, being attracted by. To see why activities do not add something novel to ontology, let us stress that for Glennan activities are fully concrete particulars: 'Any particular activity in the world will be fully concrete, though our representations of that activity may be more or less abstract' (pp. 95–6). Now, if activities are always particular, and if they are always specific, like pushings, pullings, bondings, infectings, dissolvings, diffusings, pumpings and so on, there is no need to think of them as comprising a novel ontic category. For each fully concrete activity, there will be some account in terms of entities, their properties and relations. A pushing is an event (or a process) that consists in an object changing its position (over time) due to the impact by another body. Indeed, the very event itself *consists* in a change of the properties of a thing (or of its relations to other things). Similarly, for other concrete activities: there will always be some description of the event or the process involved by reference to the changes of the properties of a thing (that engages in the 'activity') or of the relations with other things.

Take the case of a mechanism such as the formation of a chemical bond. Chemical bonding refers to the attraction between atoms. It allows the formation of substances with more than one atomic component and is the result of the electromagnetic force between opposing charges. Atoms are involved in the formation of chemical bonds in virtue of their valence electrons. There are mainly two types of chemical bonds: ionic and covalent. Ionic bonds are formed between two oppositely charged ions by the complete transfer of electrons. The covalent bond is formed by the equal sharing of electrons between two bonded atoms. These atoms have equal contribution to the formation of the covalent bond. On the basis of the polarity of a covalent bond, it can be classified as a polar or non-polar covalent bond. Electronegativity is the property of an atom by virtue of which it can attract the shared electrons in a covalent bond. In non-polar covalent bonds, the atoms have similar electronegativity. Differences in electronegativity yield bond polarity. In *describing* this mechanism, there was no need to think of particular activities as anything other than events (sharing of electrons) or processes (transfer of valence electrons) that are fixed by the properties of atoms (their valence electrons; electronegativity) and the relations they stand to each other (similar or different electronegativity).

Glennan, however, takes it that 'processes are collections of entities acting and interacting through time' (p. 57). Elsewhere (p. 83), he notes that a mechanism is a 'sequence of events (which will typically be entities acting and interacting)'. If we were to follow Bishop Berkeley's advice to '*think* with the *learned* and *speak with the vulgar*', we could grant this talk in terms of activities, without hypostatising activities over and above the properties and relations by virtue of which entities 'act and interact'. We conclude that 'activity' is an abstraction without ontological correlate.

When he talks about entities, Glennan takes it that a general characteristic of entities is this: 'The causal powers or capacities of entities are what allow them to engage in activities and thereby produce change' (p. 33). What produces the change? It seems Glennan's dualism requires that there are causal powers *and* activities and that the former enable the entities that possess them to engage in activities, thereby producing changes (to other entities). It's as if the activities exist out there ready to be engaged with by entities having suitable causal powers. Glennan is adamant: 'activities are not properties or relations; they are things that an entity or entities do over some period of time' (p. 96).

But this cannot be right. The activities cannot exist independently of the entities and their properties (whether we conceive them as powers or not), as Glennan himself admits. What activities an entity can 'engage with' depends on the properties of this entity. Water can dissolve salt but not gold, to offer a trivial example. The 'activities' an entity can engage in are none other than those that result from the kind of entity it is. If you assume powers, as Glennan does, then the activities of an entity are fixed by the manifestation of its powers (given suitable circumstances). Given a power ontology, the powers are the producers of change; the activities are merely the manifestation of powers.

As Glennan admits: 'The central difference between activities and powers is that activities are actual doings, while powers express capacities or dispositions not yet manifested' (p. 32). As just noted, assuming particulars with powers, activities are the manifestation/exercising of these powers. When a cube of salt is put in water, it dissolves. The dissolving is the manifestation (assuming a power-ontology) of the active power of water to dissolve (water-soluble) materials and the passive power of the salt to get dissolved. The dissolving takes time (and hence it is a process); but it is not acting in any sense; it does not produce any changes in the salt; it *consists* in the changes in the salt. The 'scraping of the skin off the carrot' (Glennan's example) *is* the removal of the skin of the carrot (at least on this particular occasion) and hence it does not cause (or produce) the removal. Activities do not produce anything; they *are* the productions (of effects).

6.6   Against Activities III: Illari and Williamson's Approach

While Glennan's motivation for activities comes from the metaphysics of mechanisms, other philosophers vouch for activities on the grounds that science requires them. The general motivation appears to be that science must constrain metaphysics. Not only is it the case that what there is has to be compatible with what science describes, but also the best route to the fundamental structure of the world should be the descriptions that science offers. Thus, proponents of activities have argued that if we take seriously the descriptions offered in such fields as molecular biology or neurobiology, we find that activities are central in these descriptions (Machamer et al. 2000; Illari & Williamson 2013). Illari and Williamson, in particular, think that '[t]here is a good argument from the successful practice of the biological sciences for the appeal to activities in the characterisation of a mechanism' (Illari & Williamson 2013, 71).

Illari and Williamson (2011) offer a bottom-up argument in favour of what they call an 'active metaphysics' for the workings of mechanisms, by which they mean a metaphysics in terms of capacities (cf. Cartwright 1989), of powers (cf. Gillett 2006) or of activities (cf. Machamer et al. 2000). They contrast active metaphysics with 'passive' metaphysics, which characterises the working of mechanisms in terms of laws or counterfactuals. In this section we examine this kind of bottom-up argument, which we are going to call the 'local argument'.

Although we are here treating the local argument as an argument in favour of activities, Illari and Williamson take the argument to be more general, as it does not differentiate between activities-based and power-based views. In fact, Illari and Williamson (2013) offer reasons to prefer an ontology based on entities and activities over an ontology based on entities and capacities, a main reason being that an ontology of activities is more parsimonious. But since these arguments are largely metaphysical, and we are here focusing on bottom-up arguments, we are going to examine the local argument in its general form.

Illari and Williamson argue that biological practice and, in particular, the fact that mechanisms are taken to be explanatory, constrains the ontology of mechanisms. More specifically, they think that a metaphysics of mechanisms that views within-mechanism interactions in terms of laws or counterfactuals is 'in tension with the actual practice of mechanistic explanation in the sciences, which examines only local regions of spacetime in constructing mechanistic explanations'. So, passive approaches do not 'allow mechanisms to be real and local ... [O]nly active approaches give a local characterisation of a mechanism' (Illari & Williamson 2013, 835). They think, then, that the local argument establishes that a characterisation of mechanism has to be given in terms of an active metaphysics and not in terms of 'counterfactual notions grounded in laws or other possible worlds' (p. 838).

The local argument can be reconstructed as follows:

The practice of mechanistic explanation requires that mechanisms be local (1).

This in turn implies that a characterisation of mechanism has to be local (2).

But only a metaphysics of powers or activities is a local metaphysics (3).

So, a local characterisation of mechanism requires a metaphysics based on powers or activities (4). (pp. 834–8)

In response to this argument for an 'active' metaphysics of mechanisms, it seems to us that 'local' cannot have the same meaning in premises (1) and (2), on the one hand, and in premise (3), on the other: we can have local mechanisms without a local metaphysics. There are three points to note here.

First, it is certainly true that mechanisms are local to the phenomena they produce. In this context, 'local' means that mechanistic explanation involves the localisation of the parts into which the mechanism is decomposed, the operations of which produce the phenomenon for which the mechanism is responsible. Indeed, as Bechtel and Robert Richardson (2010) have argued, localisation is a central strategy in constructing a mechanistic explanation: scientists decompose the phenomenon under study into component operations, and 'localise them within the parts of the mechanism' (p. xxx). But then, localisation of parts can fully capture the sense in which mechanisms are 'local', without entailing a 'local' metaphysics, which is supposed to underlie a characterisation of the interactions among components, and not only the components themselves. Even if we accept a metaphysics of laws, within-mechanism interactions are interactions between 'local' components.

Second, it is not at all easy to account for within-mechanism interactions in terms of a 'local' metaphysics. Energy transformations in biological systems obey the laws of thermodynamics. But it is very difficult to reconcile a power ontology with what it seems to be a global principle, such as the law of conservation of energy. This is something that friends of powers themselves have recognised (cf. Ellis 2001). So, contra Illari and Williamson, a focus on practice seems in fact to imply the opposite conclusion: global principles like the laws of thermodynamics are needed for accounting for within-mechanism interactions (e.g., as studied by bioenergetics; see Nelson et al. (2008, 489); but only a metaphysics in terms of laws seems to offer an adequate account of such global principles; so, a metaphysics of laws is required for a characterisation of the metaphysics of mechanisms. Again, the point here is that 'local' decompositions of mechanistic parts must be kept distinct from 'global' or 'local' ways to characterise interactions.

Third, there is a historical point to be made against the argument that mechanistic explanation is not compatible with a metaphysics of laws. This combination ('local' mechanisms that produce phenomena plus laws of nature) was a dominant view in seventeenth-century mechanical philosophy, as we saw in Chapter 1. Contemporary mechanistic explanations, of course, are very different from their seventeenth-century counterparts, which in many cases just involved parts of matter in motion. But the general pattern of explanation is similar: in giving a mechanistic explanation, one shows how the particular properties of the parts, their organisation and their interactions (which can be captured in terms of the laws that govern them) produce the phenomena.

In view of the previous points, premise (3) above can only be accepted if the meaning of 'local' is disambiguated. An option here is to say that mechanisms have to be local, in the sense that within-mechanism interactions have to be grounded in facts in the vicinity of the mechanism. So, one can think of causation as a local matter, that is, as a relation between the two events that are causally connected, and not as a global matter, that is, as involving a regularity. But note that so-called singular causation is compatible with a metaphysics of laws. One can view causation as a relation between 'local' events, but at the same time adopt an ontology of laws, where laws could be, for example, necessitating relations between universals, or Humean regularities, that is, 'global' facts about the universe (recall Chapter 5).

Note that Illari and Williamson themselves seem to recognise that in understanding scientific practice one need not talk about metaphysics, for they say: 'Understanding the metaphysics of mechanisms on this level is now a philosophical problem with no immediate bearing on scientific method, of course' (Illari & Williamson 2011, 834). But they add: 'It does, however, bear on our understanding of science' (p. 834). While we agree with the first sentence, we believe (and we shall argue below) that an understanding of mechanisms as causal pathways underpinned by difference-making relations is all one needs in order to understand scientific practice. We conclude, then, that there is no reason coming from scientific practice for accepting a power-based or an activities-based account of mechanism.

6.7 Against Glennan on Causation as Production

As we saw in Chapter 4, according to CM difference-making relations are enough to understand mechanisms and hence mechanistic causation and explanations. But those philosophers that view causation in terms of production contest this point. Glennan (2017) is one of the defenders of this view. According to him, mechanisms, qua productive, are the truth-makers of causal claims:

> (MC) A statement of the form 'Event c causes event e' will be true just in case there exists a mechanism by which c contributes to the production of e. (p. 156)

Actually, there are as many causal relations as there are activities. As he puts it: 'There is on this [New Mechanist] view no one thing which is interacting or causing, and when we characterise something as a cause, we are not attributing to it a particular role in a particular relation, but only saying that there is some productive mechanism, consisting of a variety of concrete activities and interactions among entities' (p. 148). This pluralist view leads him to the radical conclusion that '[t]here is ... no such thing as THE ontology or THE epistemology of THE causal relation, but only more localised accounts connected with the particular kinds of producing' (p. 33).

MC tallies with Glennan's singularism about causation. All causings are singular and in fact fully distinct from each other. Singularism is committed to the view that causation is internal (intrinsic, as Glennan puts it) to its relata. Glennan shares this intuition. He says: 'Productive causal relationships are singular and intrinsic. They involve continuity from cause to effect by means of causal processes' (p. 154).

But is causation a relation, after all? And if yes, what are the *relata*? 'Events' is the answer that springs to mind. Glennan agrees but takes events to involve activities: 'Events are particulars – happenings with definite locations and durations in space and time. They involve specific individuals engaging in particular activities and interactions' (p. 149). Or as he put it elsewhere: 'an event is just one or more entities engaging in an activity or interaction' (p. 177).

We have already argued in the previous sections that activity is far from being a sui generis ontic category. Besides, there is the received account of events as property-exemplifications: events are exemplifications of properties (or relations) by an object (or set of objects) at a time (or a period of time). As Glennan admits: 'If exemplifying a property were the same as engaging in an activity, then the two views would coincide.' However, he takes it that 'there are important differences between exemplifying properties and engaging in activities' (p. 177).

The chief difference between property-exemplification and engaging in activities is, Glennan says, that 'properties are paradigmatically synchronic states of an entity that belong to that entity for some time'. Unlike activities, properties 'do not involve change'. Events, Glennan argues, 'involve changes'. It is indeed true that events involve change. The collision of the *Titanic* with the iceberg took time and during it, both the *Titanic* and the iceberg suffered changes in their properties, which resulted in another event, namely the sinking of the *Titanic*. It is true that to account for this we have to introduce relations: the collision is between the *Titanic* and the iceberg. But relations, we are told, are not 'activity-like'. Glennan insists that 'only events (which involve activities) can be causally productive'. Properties, he says, 'cannot produce anything' (p. 178).

When all is said and done, the key question is: Is causation production? Or is it difference-making? Glennan is clear: 'While I grant that production and relevance are two different concepts of cause, I will argue that production is fundamental' (p. 156).

Descriptively, Glennan distinguishes between three kinds of productive relations:

- Constitutive production: An event produces changes in the entities that are engaging in the activities and interactions that constitute the event.
- Precipitating production: An event contributes to the production of a different event by bringing about changes to its entities that precipitate a new event.
- Chained production: An event contributes to the production of another event via a chain of precipitatively productive events. (p. 179)

All this is fine, but what is the chief argument for causation being *production*?

It seems to be this: 'Mechanisms provide the ontological grounding that allows causes to make a difference' (p. 165). Glennan's problem with the claim that a mechanism is itself a network of relations of difference-making between events is that on the difference-making account 'the causal claim depends upon the truth of a counterfactual, whereas on the mechanist account the truth depends upon the existence of an actual mechanism' (p. 167). Furthermore, it is claimed that the truth of the counterfactual requires contrasting an actual situation – where the cause occurs – and a non-actual but possible situation in which the cause does not occur.

Does the production account avoid counterfactuals? Glennan acknowledges that causation as production relies on some notion of relevance but takes this to require actual difference-makers. He takes it that actual difference-makers are 'features of the actual entities and their activities upon which outcome depends' (p. 203).

What is an *actual* difference-maker? A factor such that had it not happened, the effect would not have followed. But (1) in an actual concrete sequence of events which brought about an effect x, all events were necessary in the circumstances; all were difference-makers. If any of them were absent, the effect, in its full concrete individuality, would not follow. A different effect would have followed. But (2) what makes true the counterfactual that 'had x not actually happened, y would not have followed'? To 'delete' x from the actual sequence is to envisage a counterfactual sequence (i.e., a distinct sequence of events) without x. It is then to compare two sequences: the actual and the counterfactual. This requires thinking in terms of counterfactual difference-making. What makes the counterfactual true is not the actual sequence of events but the fact, if it is a fact, that xs are followed by ys, which is a causal law.

Take the example of a ball striking a window while a canary nearby sings. The actual causal situation – the mechanism in all its particularity – includes the process of the acoustic waves of the canary's singing striking the window (say, for convenience, at the moment when the ball strikes the window) as well as the kinetic energy of the ball (which was a red cricket ball) and so on. Despite the fact that the acoustic waves are part of the actual concrete mechanism and clearly contributed to the actual breaking (no matter how little), we would not say that it was the singing that caused the window-breaking. It clearly didn't make a substantial contribution to the breaking. Had it not been there, the window would still have shattered. How can *this* counterfactual be made true by the actual situation? In the actual situation, the singing was a difference-maker since it was part of the mechanism that made the difference. To show that it did not make a difference (better put, that it made a difference without a difference) we have to compare the actual situation in which the singing took place and a non-actual but possible situation in which the singing did not happen. Whatever makes this counterfactual true, it is not the actual situation, in and of itself.

In sum: not only does production not avoid counterfactuals (if actual difference-makers are to be shown that did not make a difference), but it seems that the very idea of production requires difference-making relations if the producer of change is nothing more specific than everything that happened before the effect took place.

6.8   Activities and the Language of Science

In this last section we want to discuss some more general points concerning the methodology of new mechanists that extract metaphysical conclusions from scientific practice. New Mechanists, beginning with MDC, typically stress the 'descriptive adequacy' of characterising mechanisms in terms of organised entities and activities; activity talk, in particular, is commonly taken to be descriptively superior over more standard philosophical talk in terms of the metaphysics of substances, properties, relations and laws. So, a general reason for preferring a metaphysics of activities, which we take to be the underlying rationale of new mechanists that subscribe to activities, is that only such a metaphysics can make sense of the kind of talk one finds in science.

This move, however, gives rise to a crucial and general question that can be posed to all attempts to extract substantive metaphysical conclusions from the kind of talk one finds in a particular discourse, be it everyday language or scientific discourse: Is language a good guide to ontology? More specifically, should we be committed to activities (just) because scientists use verbs and gerunds to describe how mechanisms work? There are well-known arguments, which we tend to accept, that it is not straightforward to read off one's ontology from how language is used. For example, it is not a straightforward manner to decide what is the correct quantum ontology, or whether we should accept the category of relations as a fundamental ontological category distinct from entities and (monadic) properties based simply on certain linguistic considerations. Frank Ramsey's (1925) well-known scepticism concerning the universal/particular distinction and whether we can read it off our language as well as Bertrand Russell's (1905) theory of definite descriptions illustrate some of the difficulties involved here. We have a similar approach concerning mechanical ontology: we think that there are no good reasons to read off a mechanical ontology of activities from the way that molecular biologists, for example, talk.

This thesis, that scientific discourse does not, on its own, favour a particular metaphysical account, may seem to be at odds with the kind of argumentation we used against Glennan's account of activities in Section 6.5. We argued there that activities are just manifestations of powers. One could wonder, then, whether by saying this we are committed to a metaphysics of properties and powers, which, if scientific discourse has no specific metaphysical implications, may seem equally questionable as a scientific ontology. Do we perhaps think that when scientists talk about activities they are in reality talking about the manifestations of powers and so are committed to the existence of powers?

We think that when scientists talk about activities they don't thereby take it that they are reducible to properties or powers; but we think it's also correct to say that scientists do not have a view (i.e., a systematic philosophical conception) of the possible relations between activities and more traditional metaphysical categories. Hence, in line with the general point made above about the relationship between language and ontology, scientists' talk of activities (such as bonding, transmitting, interrupting) does not imply that there really exist activities in the world as a fundamental ontological category. Of course, this is also true regarding other kind of ontologies; in thinking about mechanisms methodologically, and if we just focus on biologists' way of talking, we are not committed to, for example, a power-based ontology or a Humean one (i.e., categorical properties plus laws). For us this is exactly as it should be when it comes to the methodological role of mechanisms.[7]

But although we take both an activities-based metaphysics and other kinds of metaphysics (e.g., a metaphysics based on powers) to be not derivable from scientific practice, we also think that some metaphysical accounts may be conceptually incoherent or more inflated than others. In particular, we take an activities-based metaphysics to be such an inflated account (we also think that it is conceptually incoherent). We have argued in this chapter that according to standard mechanistic approaches, activities, if they are posited as a fundamental category, are taken as distinct and irreducible elements of reality. We do not take this to mean that activities can exist independently from entities and properties. We certainly agree with new mechanists that activities are not capable of existing without one or more entities engaging in them – the situation is similar to relations: relations require relata. Yet that's not the end of the story. Though activities depend on entities in that entities are required for activities, for some new mechanists, activities are irreducible to their entities and properties; they are a genuine ontological add-on. Hence an activities-based ontology is certainly more inflated ontology compared with an ontology of entities and properties.

To clarify, take again the example of relations. We may think of the God-metaphor and ask (as some metaphysicians have asked): Suppose that in the Creation God created all entities and their properties. Did he then rest? Or did he have to add relations? If he rested, then relations can be taken as internal (i.e., given entities with their properties, we thereby have all relations). If he didn't, then (at least some) relations are external; that is, they are something over and above entities and properties. These two broad ontological pictures have to be assessed on the basis of general metaphysical considerations, as well as on the basis of their compatibility with current science. The same is true for mechanical ontologies postulating activities: to take activities as fundamental is to buy into a more inflated metaphysics, where activities enter the world on top of entities and properties (just as in the case of relations). For example, using the God metaphor, the friends of activities would say that, when God created the

---

[7] There is a possible exception here. It is plausible that our commitment to causation as, ultimately, a difference-making relation of (counterfactual) dependence requires a prior commitment to laws as an indispensable part of the truth-makers of counterfactuals (see Chapter 7). But note that even at this high level of generality, there needn't be a commitment to a specific account of lawhood.

chemical elements and their properties (e.g., valence), he didn't thereby fix all facts about chemical compounds; instead, he had also to introduce the activity of 'chemical bonding' as a distinct ontological item.

The question for us, then, is: Does such an ontological picture of added-on activities make sense on general philosophical grounds? There may be reasons why some ontological positions may be better than others, reasons unrelated (or at least not directly related) to the way scientists talk; in this chapter, we have examined some of those reasons as far as activities and production-based mechanical ontologies are concerned. In particular, we think that activities-based ontologies are both inflated compared with more sparse ontologies and not conceptually coherent. Ontologies that include properties, powers and relations, but lack activities, while coherent conceptually, are in turn inflated compared with a Humean ontology. Last, note that all this is not to say that our own methodological position is committed to an ontology of properties (say, qua universals or classes of resembling tropes or what have you), instead of to an ontology of activities. What we argue is that there exist good independent reasons to be sceptical of an activities ontology.

CHAPTER 7

Whither Counterfactuals?

7.1 Preliminaries

There have been two traditions concerning how the 'link' between cause and effect is best understood (Hall 2004; Psillos 2004). According to the first tradition, which goes back to Aristotle, there is a *productive relation* between cause and effect: the cause produces, generates or brings about the effect. This productive relation between cause and effect has been typically understood in terms of powers, which in some sense ground the bringing-about of the effect by the cause. According to the second tradition, which goes back to Hume, the link is some kind of robust relation of dependence between what are taken to be distinct events. On this account, the chief characteristic of causes is that they are *difference-makers*: the occurrence of the cause makes a difference to the occurrence of the effect.[1]

We have already discussed in detail the currently most popular version of the production approach that cashes out the link between cause and effect by reference to mechanisms. When it comes to difference-making, there have been various ways to understand this notion of dependence. It may be nomological dependence (cause and effect fall under a law), counterfactual dependence (if the cause hadn't happened, the effect wouldn't have happened) or probabilistic dependence (the cause raises the probability of the effect). But, arguably, the core notion of difference-making is counterfactual, that is, based on contrary-to-fact hypotheticals. That is, a causal claim of the form 'A caused B' would be

---

[1] Though traditionally causation has been taken to be a single, unitary concept, there has been a tendency, as of late, to question this assumption. The case for there being *two* concepts of causation has been made, quite forcefully, by Ned Hall in his 2004 paper, where he distinguishes between causation as *dependence* and causation as *production*. Hall takes dependence to be *counterfactual* dependence, while he takes the concept of production (*c* produces *e*) as primitive.

understood as implying: if A hadn't happened, B wouldn't have happened either. It is in this sense that A *actually* makes a difference for B.[2]

A currently popular version of the dependence approach is Woodward's *interventionist counterfactual* account, which takes the relationship among some variables X and Y to be causal if, were an intervention to change the value of X appropriately, the relationship between X and Y would remain invariant *and* the value of Y would change.

We shall argue that the interventionist approach, despite its undeniable attractions, faces a couple of important problems. The first relates to what fixes the truth-conditions of counterfactuals, while the second has to do with the role of laws of nature in grounding counterfactuals. Hence, in the end we will favour a Lewis-style account. In any case, we will conclude with a role mechanistic information can play within a difference-making account of causation.

7.2 Counterfactuals: A Primer

7.2.1 The Logic

Subjunctive conditionals or counterfactual conditionals are probably as old as language itself since they give speakers the means to talk about what would or might happen or have happened if certain things were to happen or had happened. In ordinary language, they have the form:

If x were (not) the case, then y would (not) be the case.

or

If x had (not) been the case, then y would (not) have been the case.

Subjunctive conditionals leave open the possibility of the realisation of whatever is expressed in the antecedent, for example, if John were to come to the party, Mary would not go. Counterfactual (or 'contrary-to-fact') conditionals are such that the antecedent is false; the state of affairs expressed in it has not actually obtained, for example, if John had gone to the party, Mary would not have gone. (Here it is an implicit assumption that the actual course of events is that John did *not* go to the party.) Both

---

[2] On a nomological account of causal dependence (i.e., B depends on A if there is a law that connects the two), counterfactuals are required to account for the modal strength of laws (for more on this, see Psillos 2002). So, even if it were to be admitted that the alternative notions of dependence are distinct, counterfactuals play a key role in all versions of the dependence approach to causation.

kinds of conditional contrast to indicative conditionals of the form: *if x is the case, then y is the case*. Though there are differences between them, we will not be detained by them, and concentrate on counterfactual conditionals. (From now on, we will follow customary usage and use $\square\!\!\rightarrow$ to express the counterfactual 'if . . ., then . . .'.)

Counterfactuals fail a number of principles that indicative conditionals satisfy. Most importantly, they are non-monotonic; that is, they fail the principle of strengthening of the antecedent:

$$X\square\!\!\rightarrow Y \text{ does not entail } X\&Z\square\!\!\rightarrow Y.$$

Example: If John had been poisoned, he would have died. This does not entail: if John had been poisoned and taken an antidote, he would have died.

Transitivity:

$$X\square\!\!\rightarrow Y \text{ and } Y\square\!\!\rightarrow Z \text{ does not entail } X\square\!\!\rightarrow Z.$$

Example: If John had gone to the market, he would have taken the bus; if John had taken the bus, then he would have gone to his office. These two do not entail: if John had gone to the market, then he would have gone to his office.

Contraposition:

$$X\square\!\!\rightarrow Y \text{ does not entail } \text{not-}Y\square\!\!\rightarrow \text{not-}X.$$

Example: If John had lived in a euro-zone country, he would have used euros. This is not equivalent to: if John had not used euros, he would not have lived in a euro-zone country.

If we assume that classical semantics apply to indicative conditionals (the indicative conditional is true iff either the antecedent is false or the consequent is true), trying to apply classical semantics to counterfactuals leads to their trivialisation: given the actual falsity of the antecedent of a counterfactual, both the counterfactual with the actual consequent and the counterfactual with the negation of the actual consequent end up being true.

Example: given that the vase was not struck with a hammer, both of the following two conditionals (treated as material conditionals) are true:

If this vase had been struck with a hammer, it would have broken.

and

If this vase had been struck with the hammer, it would not have broken.

The failure of the three principles and this unwanted consequence is a *reductio* of the view that classical semantics apply to counterfactuals. But then, what is the right semantics for counterfactuals? What are the truth-conditions of a counterfactual conditional? Or, at least, what are their assertibility conditions? This problem came under sharp focus in the 1940s, when philosophers started to realise that the concept of counterfactual conditionals is instrumental for the explication/understanding of a number of other philosophical concepts. As Nelson Goodman put it in one of the first papers to deal with this issue: 'if we lack the means of interpreting counterfactual conditionals, we can hardly claim to have any adequate philosophy of science' (1947, 113).

Note that in assessing a counterfactual assertion X$\square\!\!\rightarrow$ Y, we should replace, as it were, the actual non-occurrence of X with the supposition that X has occurred. But given that the laws of nature and the actual course of events led to non-X, in supposing the actual occurrence of X we need to make counterfactual suppositions concerning either the laws or the actual course of events such that X actually occurred. In particular, we have to assume either that some laws were broken (so that X did happen after all) and/or that some actual particular matters of fact did not occur. Hence, in specifying the semantics of counterfactuals, we have to take into account considerations concerning the laws of nature and other particular matters of fact prior to the conditions specified in the antecedent of the counterfactuals.

There are two major views concerning the semantics of counterfactuals, the first being introduced by Goodman himself, while the second was developed by Robert Stalnaker and David Lewis (but introduced by William Todd in 1964). Let us examine them in turn.

7.2.2 *The Semantics*

7.2.2.1 *The Metalinguistic or 'Support' View*
On the first major view, known as 'support view' or 'metalinguistic view', a counterfactual conditional X$\square\!\!\rightarrow$ Y is an elliptic or telescoped argument (or a linguistic construction *about* an argument) such that the antecedent X (taken in its indicative form) together with suitable auxiliary premises entails the consequent (taken in its indicative form). Hence, X$\square\!\!\rightarrow$ Y should not be taken to be a statement at all; its assertoric content is captured by the following argument-type:

$$X \& S \& L \text{ (materially) imply Y,}$$

where L are statements capturing laws of nature and S are singular statements capturing background or collateral conditions which should be 'cotenable' with the antecedent X and express necessary conditions for the consequent to follow.

Example: If this match had been struck (X), it would have lit (Y). For a struck match to light, it is necessary that the match is well made, that it is dry, that there is oxygen and so on. But even these conditions (collectively designated by S) are not sufficient for the lighting of the match; various laws are required (collectively designated by L). Hence, in asserting the counterfactual 'if this match had been struck, it would have lit', we are committed to the truth of the various statements that describe the laws and the relevant background conditions.

The first general problem with this view concerns the characterisation of the relevance relation when it comes to the background/collateral conditions. It cannot be too permissive. If we allowed all true statements to be relevant to the argument, the falsity of the antecedent X (which is *actually* false) would be relevant too; but then the counterfactual would be trivially true. It cannot be too restrictive either. The consequent of the counterfactual is false too. Hence not-Y is the case. It's not hard to see that given $(X \& S \& L \to Y)$ and not-X and L, it follows (by obvious steps) that $X \to$ not-S. (Assuming, for simplicity, that S is 'the match is dry', the conclusion would be: if the match is struck, it will not be dry!) The point, then, is that only those background/collateral conditions which are 'cotenable' with the antecedent should be admitted. But which are they? Those conditions S which are such that if X had been true, S would have been true too. This is a counterfactual assertion, and Goodman thought that this kind of circularity impairs the metalinguistic analysis of counterfactuals.

The second general problem with this view concerns the characterisation of laws, which are indispensable for the *connection* between the antecedent and the consequent of a counterfactual. The key thought here is that some generalisations, though true, are unable to 'support' counterfactuals because they are accidental. *Example*: Compare the following two counterfactuals:

(A)   If x had been a golden sphere, its diameter would not have been more than one mile long.
(B)   If x had been a plutonium sphere, its diameter would not have been more than one mile long.

(A) is false, while (B) is true – we rightly suppose. And this is because there is a law of nature backing up (B), while the generalisation related to (A) is merely accidental (intuitively: if we have had enough gold, we could build a sphere of it with the required diameter; not so with plutonium). So the general statements L that are part of the premises of the argument whose telescopic form is X□→ Y must express *laws of nature* and not merely accidentally true generalisations. But how exactly are we to distinguish between laws and accidents? If we felt that laws are those generalisations that support counterfactuals while accidents are those that do not (see the example above), then we would move in a(nother) circle. So there is need to look for ways to distinguish between laws and accidents that do not rely (in the first instance, at least) on their modal force (at least when expressed in their support of counterfactuals). When Goodman brought this problem to the attention of philosophers, the prevailing view of laws was that they are simply regularities (cf. Chisholm 1946); hence the distinction between laws and accidents (which are regularities too) was taken to be mostly an 'honorific' distinction which is captured by the different epistemic attitudes we have towards them. For instance, laws are those regularities that are projected to the future or are conformable by their instances and so on. Wilfrid Sellars (1958, 268), however, pointed out that even if laws are taken to be regularities, they are those regularities that are characterised by 'neck-sticking-out-ness', where this characteristic is captured in the subjunctive mood: 'If this were an A-situation, it would be accompanied by a B-situation.' The counterfactual content of a law, then, is seen as a 'contextual implication' of a law-statement.

This idea of 'contextual implication' is captured by the supposition view of counterfactuals which is akin to (though interestingly different from) the metalinguistic view, as this was developed by John Mackie (1973). As we already saw in Section 5.3.1, according to this view, to assert something like X□→ Y is to assert Y *within the scope of the supposition that X*. In other words, we suppose X and then we envisage various possibilities and consequences. This account brings to light the contextuality of counterfactual conditionals, which is not resolvable without some degree of arbitrariness: X did not happen; supposing that X did happen, what else do we have to assume or suppose? What features of the background (including laws and particular matters of fact) should we retain or change? There is no uniquely determined answer to this question, though contextual matters (including a fuller specification of the antecedent of the conditional) might (and as a rule do) help us. *Example*: Is the counterfactual: 'If I had let go of this stone, it would have fallen to the ground' true or false (or assertible/not-assertible)? It depends on the context! There are certain conversational contexts, in which it would be false to assert it, for example, if this were a precious stone and the owner was very careful with it, so if the stone were to be let go, it would have been caught in mid-air. *Another example*: Consider the following pair of counterfactuals: 'If Julius

Caesar had been in charge of UN Forces during the Korean War, then he would have used nuclear weapons' and 'If Julius Caesar had been in charge of UN Forces during the Korean War, then he would have used catapults.' Only contextual assumptions can tell us which one, if any, and in what context is true (or assertible).

The supposition view takes it that counterfactuals are *not* truths about possible words but ways to express an attitude towards a possible state of affairs made within the scope of a supposition. Suppose that the sole ground for believing the law L (e.g., All Fs are G) is an enumeration of *all* actual instances (Fai and Gai) of L. Then adding the supposition X, namely, that a *further* a is F, removes the ground for accepting L. We can no longer draw the conclusion that this further a is G. Hence we cannot assert the counterfactual X□→ Y. More generally, if the reasons for accepting L survive placing L within the scope of *the supposition that there are further instances of the law's subject term*, then we can say that the law supports the relevant counterfactual conditional. The required reasons are ordinary inductive reasons, namely, good inductive evidence for the law. Good inductive evidence, in other words, is evidence for the 'neck-sticking-out-ness' of the law.

An interesting related thought comes from Julius Weinberg (1951), who claimed the following. A counterfactual X□→ Y is best seen not as the indicative statement (the statement of a generalisation) X → Y plus some further antecedent conditions (including that X did not actually happen) but rather as asserting something about the evidence there is for X → Y, namely, that there is evidence for, and no evidence against, the generalisation: for all X (X → Y). Hence, the additional strength a counterfactual is supposed to have over the corresponding generalisation is captured by the evidence there is for the generalisation.

7.2.2.2   The Possible-Worlds View

Taking literally the view that counterfactuals are used in contemplating *possibilities*, the second major view of the semantics of counterfactuals appeals to possible worlds. In first suggesting this view, Todd (1964, 107) noted that when we allow for the possibility that the antecedent of a counterfactual be true, we are 'hypothetically substituting a different world for the actual one'. On this view, the core meaning of a counterfactual X□→ Y is (roughly): In the possible (but not actual) world where X, Y too.

A possible world is a way the world might be or might have been. For instance, it is possible that gold is not yellow, that planets describe circular orbits, that birds do not fly or that beer doesn't need yeast to brew. But are there really possible worlds? There are three views here. The first is that talk of possible worlds is a mere *façon de parler*, though useful when it comes to assessing counterfactuals. (We take it that an extension of this view is that possible worlds are useful *fictions*.) The second is 'extreme realism', according to which the way the world *actually* is, is one among the many ways the world could be; hence, the actual world is one among the many possible worlds, the latter being no less real than the actual. The chief advocate of this view was David Lewis (1973). The third view is 'abstract realism', according to which possible worlds are maximally consistent sets of propositions: total ways things might be. A 'possible world' then is fit to represent a concrete reality, but only one possible world actually represents anything, namely, the actual world (see Bennett 2003).

Stalnaker (1968) developed the core meaning of counterfactuals as follows:

> Consider a possible world W in which X is true, but is otherwise similar to the actual world @. X□→ Y is true iff Y is true in W.

The similarity relation among worlds (a selection function, as Stalnaker put it) is an ordering of possible worlds with respect to their resemblance to the actual world.

Calling an X-world a possible world in which X holds, counterfactuals might be taken to be strict conditionals of the form:

> X□→ Y is true in a world W iff Y is true in all X-worlds such that ___.

where the blank is filled by a general condition that X-worlds should satisfy. Hence, whatever goes into the blank places a restriction on the admissible (or accessible) possible worlds. This idea would model counterfactuals along the lines of strict conditionals of the form:

> It is physically necessary that ___

or

> It is logically necessary that ___

where the first restriction is to all worlds with the same laws as the actual, while the second 'restriction' would be to all possible worlds *simpliciter*.

But this analysis cannot be correct. There is no set of possible worlds W such that X → Y throughout W (this is another way to state the fact that counterfactuals are non-monotonic). So Lewis (1973) suggested that counterfactuals X□→ Y are *variably strict conditionals*: each of them is a strict conditional, i.e., every X-world *of a certain sort* is a Y-world; but the relevant set of worlds varies with different conditionals.

Like Stalnaker, Lewis takes it that worlds are ordered in terms of similarity, or closeness to the actual world. According to this *primitive* notion of 'comparative overall similarity': 'we may say that one world is closer to actuality than another if the first resembles our actual world more than the second does, taking account of all the respects of similarity and difference and balancing them off against one another' (1986b, 163).

But unlike Stalnaker, Lewis took it that in assessing the counterfactual X$\square\!\!\rightarrow$ Y it does not make good sense to talk about *the* closest-to-actual possible X-world. It's not just that there might be more than one closest-to-the-actual possible worlds. It is mainly that there might not be even one rightly deemed *the* closest (even in a limiting sense). Hence, on Lewis's view:

> X$\square\!\!\rightarrow$ Y is true at a world W iff some (accessible) X-world in which Y holds is closer to W than any X-worlds which Y does not hold.

For instance, take the counterfactual that if this pen had been left unsupported (X), it would have fallen to the floor (Y). Neither X nor Y is true of the actual world. The pen was never removed from the table, and it didn't fall to the floor. Take all X-worlds. The counterfactual X$\square\!\!\rightarrow$ Y is true (in @) iff the X-worlds in which Y is true (i.e., the pen is left unsupported and falls to the floor) are closer to @ than any of the X-worlds in which Y is false (i.e., the pen is left unsupported but does not fall to the ground, e.g., it stays still in mid-air). As Lewis (1986b, 164) put it: 'a counterfactual . . . is true iff it takes less of a departure from actuality to make the consequent true along with the antecedent than it does to make the antecedent true without the consequent'.

The key idea behind the possible-world semantics is that in specifying the truth-conditions of a counterfactual conditional we should imagine a state of affairs in which X obtains and which is such that *all else is pretty much as they actually were*. But as noted already, this is not quite possible. In the possible world in which X did happen, many other things (including the laws) were different from the actual world @ in which X did not occur. Can we find comfort in the notion of comparative similarity? Now, though 'comparative overall similarity' is not strictly defined, a lot can be said of it. Notably, it imposes a weak ordering on the set of possible worlds which are accessible from @, namely, the relation of comparative similarity is connected and transitive. (It also imposes a centring assumption: @ is closer to itself than any other world is to it.) More importantly, however, similarity is clearly not one-dimensional, but rather the resultant of many compo-nent similarities. Lewis (1986a, 47–8) ranks possible worlds according to the following dimensions of similarity (put in order of importance).

- Avoid big, widespread violations of the laws of nature of the actual world (very important).
- Maximise the spatiotemporal perfect match of particular matters of fact.
- Avoid small, localised violations of the laws of nature of the actual world.
- Secure approximate similarity of particular matters of fact (not at all important).

So, a world W_1 which has the same laws of nature as the actual world @ is closer to @ than a world W_2 which has different laws. But insofar as there is exact similarity of particular facts in large spatiotemporal regions between @ and a world W_3, Lewis allows that W_3 is close to @ even if some of the laws that hold in @ are violated in W_3.

All this implies that there is quite a lot of vagueness in the notion of overall comparative similarity, which accounts for the fact that counter-factuals themselves are vague, at least in the sense that it is a contextual matter what to keep fixed and what to change when we assert a counter-factual conditional. A more serious worry relates to the issue of the motivation behind the foregoing ranking of dimensions of similarity among worlds. It has been observed by many that Lewis's initial theory yielded the wrong truth-values for a type of counterfactual conditional which can be schematised thus:

$$X \square \!\!\rightarrow \text{BIG DIFFERENCE.}$$

For instance:

(C) If the president had pressed the button, a nuclear war would have ensued.

Intuitively, (C) is true. But on Lewis's initial account, it would be false. For a possible world W_1 in which the president did press the button and a nuclear war did erupt is more distant from (because more dissimilar to) the actual world than a world W_2 in which the president did press the button but, somehow, a nuclear war did *not* follow. Addressing this worry, Lewis noted that intuitive judgements of the truth and falsity of counterfactuals are prior to the similarity relation that is required for the semantics of counterfactuals; hence the similarity relation should be such that it tallies with the right intuitive judgements concerning counterfactuals. The similarity ranking above is meant to solve this problem. To see how the foregoing counterfactual is indeed true, Lewis invites us to consider the following. Take a world W_1 in which nothing extraordinary happened between the president's pressing the button and the activation of the nuclear missiles. In W_1 the nuclear war did erupt. Take, now, a world W_2 in which the president did press the button but the nuclear war did *not* follow. For this to happen, many miracles would need to take place (or, to put it in a different way, a really *big* miracle would have to occur). For all the many and tiny traces of the button pushing would have to be wiped out. Hence, appearances to the contrary, W_2 would be more distant from (because more dissimilar to) actuality @ than W_1. The *big* violation of laws of nature in W_2 is outweighed by the maximisation of the perfect spatio-temporal match of particular matters of fact between W_1 and @. So, with the help of the refined criteria of similarity among possible worlds, the president-counterfactual comes out true. Still, one may follow Horwich (1987, 171–2) in wondering how psychologically plausible Lewis's theory becomes: the similarity criteria are so tailored that the right counterfactuals come out true, but they have little to do with our pre-theoretical under-standing of judgements of similarity.

As noted already in relation to the 'support' view, there are two hurdles that an adequate theory of counterfactuals has to jump. The first relates to cotenability. Lewis solves this problem by taking it that some conditions S are cotenable with X (the antecedent of the counterfactual $X \square\!\!\!\rightarrow Y$) iff some X-world is closer to the actual world than any not-S world. The second hurdle relates to the distinction between laws and accidents. Here, the possible-world approach is on safe ground, though the ground can support any decent theory of counterfactuals. David Lewis (1973) revamped a long tradition that goes back to John Stuart Mill, via Frank Ramsey, according to which the regularities that constitute the laws of nature are those that are expressed by the axioms and theorems of an ideal deductive system of our knowledge of the world and, in particular, of a deductive system that strikes the *best* balance between simplicity and strength. Simplicity is required because it disallows extraneous elements from the system of laws. Strength is required because the deductive system should be as informative as possible about the laws that hold in the world. Whatever regularity is not part of this *best system* is merely accidental. The gist of this approach is that no regularity, taken in isolation, can be deemed a law of nature. The regularities that constitute laws of nature are deter-mined in a kind of holistic fashion by being parts of a structure. An advantage of this approach is that it can sustain, in a non-circular way, the view that laws can support counterfactuals. For it identifies laws *independently* of their ability to support counterfactuals.

A key objection to the possible-world approach to counterfactuals is that counterfactual conditionals are not purely objective; an irremediably subjective element enters into the judgement of similarity (and, arguably, into the distinction between laws and accidents). Not only are the truth-conditions of counterfactuals 'a highly volatile matter' as Lewis (1973, 92) himself noted, but also what counterfactuals are true turns out to depend on various partly non-objective judgements concerning similarity weights and conversational contexts. This objection, however, might not be as fatal as it first seems precisely because counterfactual conditionals should be taken not to be pointers to necessary connections, powers and the like, but (in either of the two theories we have examined) summaries of attitudes we have towards statements that are supposed to express a *connection* between a hypothetical antecedent and a consequent. To exploit an idea of Sellars's, the core idea behind counterfactual reasoning is to assert that there are not good inductive reasons to affirm simultaneously a generalisation and the physical possibility of an exception to it.

7.3 Counterfactual Manipulation and Causation

In a series of papers and a book, Woodward (1997; 2000; 2003a; 2003b) developed a new account of the semantics of counterfactuals which he put to the service of a counterfactual account of causation. His account is *counterfactual* in the following sense: what matters is what *would* happen to a relationship, *were* interventions to be carried out. A relationship among some variables X and Y is causal if, were one to intervene to change the value of X appropriately, the relationship between X and Y wouldn't change *and* the value of Y would change. To use a stock example, the force exerted on a spring *causes* a change of its length, because if an intervention changed the force exerted on the spring, the length of the spring would change too (but the relationship between the two magnitudes – expressed by Hooke's law – would remain invariant, within a certain range of interventions).

Woodward (1997; 2000; 2003a) has analysed further the central notions of invariance and intervention. The gist of his characterisation of an *intervention* is this. A change of the value of X counts as an intervention I if it has the following characteristics:

1. The change of the value of X is entirely due to the intervention I.
2. The intervention changes the value of Y, if at all, only through changing the value of X.

The first characteristic makes sure that the change of X does not have causes other than the intervention I, while the second makes sure that the change of Y does not have causes other than the change of X (and its possible effects).[3] These characteristics are meant to ensure that Y-changes are exclusively due to X-changes, which, in turn, are exclusively due to the intervention I. As Woodward notes, there is a close link between intervention and manipulation. Yet his account makes no special reference to human beings and their (manipulative) activities. Insofar as a process has the right characteristics, it counts as an intervention. So, interventions can occur 'naturally', even if they can be highlighted by reference to 'an idealised experimental manipulation' (2000, 199).

Woodward links the notion of intervention with the notion of *invariance*. A certain relation (or a generalisation) is invariant, Woodward says, 'if it would continue to hold – would remain stable or unchanged – as various other conditions change' (p. 205). What really matters for the characterisation of invariance is that the generalisation remains stable under a set of actual and counterfactual *interventions*. So Woodward (p. 235) notes:

> the notion of invariance is obviously a modal or counterfactual notion [since it has to do] with whether a relationship would remain stable if, perhaps contrary to actual fact, certain changes or interventions were to occur.

Counterfactuals have been reprimanded on the ground that they are context-dependent and vague. Take, for instance, the following counterfactual: 'If the smoker had not smoked so heavily, they would have lived a few years more.' What is it for it to be true? Any attempt to say whether it is true, were it to be possible at all, would require specifying what else should be held fixed. For instance, other aspects of the smoker's health should be held fixed, assuming that other factors (e.g., a weak heart) wouldn't cause a premature death anyway. But what things to hold fixed is not, necessarily, an objective matter.

7.3.1   Interventions

To address this problem, Woodward devises a new theory of the semantics of counterfactuals. The key idea is that only counterfactuals which are

---

[3] There is a *third* characteristic too, namely, that the intervention I is not correlated with other causes of Y besides X.

related to *interventions* can be of help when it comes to assessing their test- or assertibility conditions. An intervention gives rise to an 'active counter-factual', that is, to a counterfactual whose antecedent is made true by (hypothetical) interventions. Woodward (2003b, 3) very explicitly char-acterises the appropriate counterfactuals in terms of *experiments*: they 'are understood as claims about what would happen if a certain sort of experiment were to be performed'.

Consider a case he discusses (pp. 4–5). Take Ohm's law (that the voltage E of a current is equal to the product of its intensity I times the resistance R of the wire) and consider the following two counterfactuals:

(1)   If the resistance were set to $R = r$ at time t, and the voltage were set to $E = e$ at t, then the intensity I would be $i = e/r$ at t.

(2)   If the resistance were set to $R = r$ at time t, and the voltage were set to $E = e$ at time t, then the intensity I would be $i* \neq e/r$ at t.

There is nothing mysterious here, says Woodward, 'as long as we can describe how to test them' (p. 6). We can perform the experiments at a future time $t*$ in order to see whether (1) or (2) is true. If, on the other hand, we are interested in what *would* have happened had we performed the experiment in a past time t, Woodward invites us to rely on the 'very good evidence' we have 'that the behaviour of the circuit is stable over time' (p. 5). Given this evidence, we can assume, in effect, that the *actual* performance of the experiment at a future time $t*$ is as good for the assessment of (1) and (2) as a *hypothetical* performance of the experiment at the past time t.

For Woodward, the truth-conditions of counterfactual statements (and their truth-values) are not specified by means of an abstract metaphysical theory, for example, by means of abstract relations of similarity among possible worlds. He calls his own approach 'pragmatic'. That's how he puts it:

> For it to be legitimate to use counterfactuals for these goals [understanding causal claims and problems of causal inference], I think that it is enough that (a) they be useful in solving problems, clarifying concepts, and facil-itating inference, that (b) we be able to explain how the kinds of counter-factual claims we are using can be tested or how empirical evidence can be brought to bear on them, and (c) we have some system for representing counterfactual claims that allows us to reason with them and draw infer-ences in a way that is precise, truth-preserving and so on. (p. 4)

At the same time, Woodward's view is meant to be realist and objec-tivist. He is quite clear that counterfactual conditionals have non-trivial truth-values independently of the actual and hypothetical experiments by virtue of which it can be assessed whether they are true or false. He says:

> On the face of things, doing the experiment corresponding to the antecedent of (1) and (2) doesn't *make* (1) and (2) have the truth values they do. Instead the experiments look like ways of *finding out* what the truth values of (1) and (2) were all along. On this view of the matter, (1) and (2) have non-trivial truth values – one is true and the other false – even if we don't do the experiments of realizing their antecedents. Of course, we may not *know* which of (1) and (2) is true and which false if we don't do these experiments and don't have evidence from some other source, but this does not mean that (1) and (2) both have the same truth-value. (p. 5)

So though 'pragmatic', Woodward's theory is also objectivist. But it is minimally so. As he notes, his view

> requires only that there be facts of the matter, independent of facts about human abilities and psychology, about which counterfactual claims about the outcome of hypothetical experiments are true or false and about whether a correlation between C and E reflects a causal relationship between C and E or not. Beyond this, it commits us to no particular metaphysical picture of the 'truth-makers' for causal claims. (2003a, 121–2)

7.3.2   Truth-Conditions

There are a few delicate issues here to be reckoned with. We will restrict ourselves to the following: *What are the truth-conditions of counterfactual assertions?* Woodward doesn't take all counterfactuals to be meaningful and truth-valuable. As we have seen (see also 2003a, 122), he takes only a subclass of them, the active counterfactuals, to be such. However, he does not want to say that the truth-conditions of active counterfactuals are fully specified by (are reduced to) actual and hypothetical experiments. If he said this, he could no longer say that active counterfactuals have determinate truth-conditions independently of the (actual and hypothetical) experiments that can test them. In other words, Woodward wants to distinguish between the truth-conditions of counterfactuals and their evidence- (or test) conditions, which are captured by certain actual and hypothetical experiments. The problem that arises is the following. Though we are given a relatively detailed account of the evidence-conditions of counterfactuals, we are not given anything remotely like this for their *truth-conditions*. What, in other words, is it that makes a certain counterfactual conditional true?

A thought here might be that there is no need to say anything more about the truth-conditions of counterfactuals other than offering a Tarski-style meta-linguistic account of them of the form

(T) 'If x had been the case, then y would have been the case' is true iff if x had been the case, then y would have been the case.

This move is possible indeed but not terribly informative. We don't know when to assert (or hold true) the right hand-side. And the question is precisely this: When is it right to assert (or hold true) the right-hand side? Suppose we were to tell a story in terms of actual and hypothetical experiments that realise the antecedent of the right-hand side of (T). The obvious problem with this move is that the truth-conditions of the counterfactual conditional would be specified in terms of its evidence-conditions, which is exactly what Woodward wants to block. Besides, if we just stayed with (T) above, without any further explication of its right-hand side, *any* counterfactual assertion (and not just the active counterfactuals) would end up meaningful and truth-valuable. Here again, Woodward's project would be undermined. Woodward is adamant: 'Just as non counterfactual claims (e.g., about the past, the future, or unobservables) about which we have no evidence can nonetheless possess non-trivial truth-values, so also for counterfactuals' (2003b, 5). This is fine. But in the case of claims about the past or about unobservables, there are well-known stories to be told as to what the difference is between truth- and evidence-conditions. When it comes to Woodward's counterfactuals, we are *not* told such a story.

In light of the above, there are two options available. The first is to *collapse* the truth-conditions of counterfactuals to their evidence-conditions. One can see the prima facie attraction of this move. Since evidence-conditions are specified in terms of actual and hypothetical experiments, the right sort of counterfactuals (the active counterfactuals) and only those end up being meaningful and truth-valuable. But there is an important drawback. Recall counterfactual assertion (1) above. On the option presently considered, what makes (1) true is that its evidence-conditions obtain. Under this option, counterfactual conditionals lose, so to speak, their counterfactuality. (1) becomes a shorthand for a future prediction and/or the evidence that supports the relevant law. If t is a *future* time, (1) gives way to an actual conditional (a prediction). If t is a past time, then, given that there is good evidence for Ohm's law, all that (1) asserts under the present option is that there has been good evidence for the law.

7.3.3   Laws to the Rescue?

In any case, Woodward seems keen to keep evidence- and truth-conditions apart. Then (and this is the *second* option available), some informative story should be told as to what the truth-conditions of counterfactual conditionals *are* and *how* they are connected with their evidence-conditions (i.e., with actual and hypothetical experiments). There may be a number of stories to be told here.[4] The one we favour ties the truth-conditions of counterfactual assertions to *laws of nature*. It is then easy to see how the evidence-conditions (i.e., actual and hypothetical experiments) are connected with the truth-conditions of a counterfactual: actual and hypothetical experiments are symptoms for the presence of a law. There is a hurdle to be jumped, however. It is notorious that many attempts to distinguish between genuine laws of nature and accidentally true general-isations rely on the claim that laws do, while accidents do not, support counterfactuals. So counterfactuals are called for to distinguish laws from accidents. If at the same time laws are called for to tell when a counter-factual is true, we go around in circles. Fortunately, there is the Mill-Ramsey-Lewis view of laws (see Psillos 2002, chapter 5). Laws are those regularities which are members of a coherent system of regularities, in particular, a system which can be represented as an ideal deductive axiom-atic system striking a good balance between *simplicity* and *strength*. On this view, laws are identified independently of their ability to support counter-factuals. Hence, they can be used to specify the conditions under which a counterfactual is true.[5]

Let us consider here one relevant thought that is central to Woodward's approach. He takes laws to be relations that remain invariant under

---

[4] One might try to keep truth- and evidence-conditions apart by saying that counterfactual assertions have excess content over their evidence-conditions in the way in which statements about the past have excess content over their (present) evidence-conditions. Take the view (roughly Dummett's) that statements about the past are meaningful and true insofar as they are verifiable (i.e., their truth can be known). This view may legitimately distinguish between the *content* of a statement about the past and the present or future evidence there is for it. Plausibly, this excess content of a past statement may be cast in terms of counterfactuals: a meaningful past statement *p* implies counterfactuals of the form 'if *x* were present at time *t*, *x* would verify that *p*'. This move presupposes that there are meaningful and true counterfactual assertions. But note that a similar story *cannot* be told about counterfactual conditionals. If we were to treat their supposed excess content in the way we just treated the excess content of past statements, we would be involved in an obvious regress: we would need counterfactuals to account for the excess content of counterfactuals.

[5] Obviously, the same holds for the Armstrong-Dretske-Tooley view of laws (see Psillos 2002, chapter 6). If one takes laws as necessitating relations among properties, then one can explain why laws support counterfactuals and, at the same time, identify laws *independently* of this support.

(a range of) actual and counterfactual interventions. If this is so, when checking whether a generalisation or any relationship among magnitudes or variables is invariant we need to subject it to some variations/changes/interventions. What changes will it be subjected to? The obvious answer is: those that are permitted or are permissible by the prevailing laws of nature. Suppose that we test Ohm's law. Suppose also that one of the interventions envisaged was to see whether it would remain invariant if the measurement of the intensity of the current was made on a spaceship, which moved faster than light. This, of course, cannot be done, because it is a *law* that nothing travels faster than light. So, some *laws* must be in place before, based on considerations of invariance, it is established that some generalisation is invariant under some interventions. Hence, Woodward's notion of 'invariance under interventions' (2000, 206) cannot offer an adequate analysis of lawhood, since laws are required to determine what interventions are possible.

Couldn't Woodward say that even basic laws – those that determine what interventions and changes are possible – express just relations of invariance? Take, once more, the law that nothing travels faster than light. Can the fact that it is a law be the result of subjecting it to interventions and changes? Hardly. For it itself establishes the *limits* of possible interventions and control.[6] We do not doubt that it may well be the case that genuine laws express relations of invariance. But this is not the issue. For the manifestation of invariance might well be the *symptom* of a law, without being constitutive of it.[7]

Before we move on we want to address a possible objection. It might be that Woodward aims only to provide a *criterion* of meaningfulness for counterfactual conditionals without also specifying their truth-conditions. This would seem in order with his 'pragmatic' account of counterfactuals, since it would offer a criterion of meaningfulness and a description of the 'evidence conditions' of counterfactuals, which are presumed to be enough to understand causation. In response to this, we would not deny that Woodward has indeed offered a sufficient condition of meaningfulness. Saying that counterfactuals are meaningful if they can be interpreted as

---

[6] Woodward (2000, 206–7) too agrees that this law cannot be accounted for in terms of invariance.
[7] We take to heart Marc Lange's (2000) important diagnosis: either *all* laws, taken as a whole, form an invariant-under-interventions set, or, strictly speaking, no law, taken in isolation, is invariant-under-interventions. This does not yet tell us what laws *are*. But it does tell us what marks them off from intuitively accidental generalisations.

claims about actual and hypothetical experiments is fine (and a step forward in the relevant debate). But can this also be taken as a necessary condition? Can we say that *only* those counterfactuals are meaningful which can be seen as claims for actual and hypothetical experiments? If we did say this, we would rule out as meaningless a number of counterfactuals that philosophers have played with over the years. Consider the following pair of counterfactuals: 'If Julius Caesar had been in charge of UN forces during the Korean War, then he would have used nuclear weapons' and 'If Julius Caesar had been in charge of UN forces during the Korean War, then he would have used catapults.' It is hard to see how we could possibly tell which of them, if any, is true. And yet it's even harder to think of them as *meaningless*.

Take one of Lewis's examples, that had he walked on water, he would not have been wet. We don't think it is meaningless. One may well wonder what the point of offering such counterfactuals might be. But whatever it is, they are understood and, perhaps, are true. Perhaps, as Woodward (2003a, 151) says, the antecedents of such counterfactuals are 'unmanipulable for conceptual reasons'. But if they are understood (and if they are true), this would be enough of an argument *against* the view that manipulability offers a necessary condition for meaningfulness.

It turns out, however, that there are more sensible counterfactuals that fail Woodward's criterion. Some of them are discussed by Woodward himself (2003a, 127–33). Consider the true causal claim: Changes in the position of the moon with respect to the earth and corresponding changes in the gravitational attraction exerted by the moon on the earth's surface cause changes in the motion of the tides. As Woodward adamantly admits, this claim cannot be said to be true on the basis of interventionist (experimental) counterfactuals, simply because realising the antecedent of the relevant counterfactual is physically impossible. His response to this is an alternative way of assessing counterfactuals. This is that counterfactuals can be meaningful if there is some 'basis for assessing the truth of counterfactual claims concerning what would happen if various interventions were to occur'. Then, he adds, 'it doesn't matter that it may not be physically possible for those interventions to occur' (p. 130). And he sums it up by saying that 'an intervention on X with respect to Y will be "possible" as long as it is logically or conceptually possible for a process meeting the conditions for an intervention on X with respect to Y to occur' (p. 132). Our worry then is this. We now have a much more liberal criterion of meaningfulness at play, and it is not clear, to say the least, which counterfactuals end up meaningless by applying it.

7.4   Causal Inference and Counterfactuals

In the last decades, there has been an increasing interest in causal inference among statisticians and social scientists, and counterfactuals have loomed large in some key attempts to model it. Prominent among them is Rubin's model, which has been advanced by Donald Rubin (1978) and Paul Holland (1986).[8] This model focuses on the discovery of the effects of causes. Suppose, to use a simple example, we want to find out whether taking an aspirin makes a difference to a *specific* subject's relief from headache. We would like to give a certain subject u an aspirin in order to see what happens to the headache episode – let's call the result Y. But we would also like, at the same time, to withhold giving aspirin to the very same subject u, in order to see what happens to the headache episode – let's call this result Y'. The difference, if any, between Y and Y' would naturally be considered the actual causal effect of aspirin-taking on the headache episode of subject u. But this kind of experiment is impossible: the experimenter cannot both give and *not* give an aspirin to the *same* subject u at the *same* time. Rubin's and Holland's main idea is that an appeal to counterfactuals allows us to make an inference about the causal effect.

Let's consider a population U of individuals, or units, $u \in U$. In a typical experiment, the experimenter applies one treatment, say, i, out of a set of possible treatments, T, to each unit u and observes the resulting responses Y. The experimental units are chosen and separated into two groups (the experimental group and the control group) by randomisation. To simplify matters, let the treatment set T consist of two possible actions (treatment t, and control c). For instance, t may be taking the aspirin and c may be taking a placebo. Let, also, Y consist of two possible responses, for example, headache relief Y_t, and headache persistence Y_c. Though it is crucial that each unit is potentially exposable to any one of the treatments, to each unit u just one treatment is *actually* given, that is, either t or c. Similarly, for each unit u, there is just one response that is actually observed, that is, either $Y_t(u) = Y(t, u)$ or $Y_c(u) = Y(c, u)$. Rubin's model defines the two responses in counterfactual terms. That is, $Y(t, u)$ is the value of the response that would be observed if the unit u were exposed to treatment t and $Y(c, u)$ is the value that would be observed *on the same unit u* if it were exposed to c. A key assumption of Rubin's model is that both

---

[8] See also Holland (1988), Stone (1993), Cox and Wermuth (2001), Maldonado and Greenland (2002) and Kluve (2004).

values $Y(t, u)$ are $Y(c, u)$ are well defined and determined. In particular, it is assumed that even if subject u is actually given treatment t and has response $Y(t, u)$, there is still a fact of the matter about what the subject's u response would have been, had they been given treatment c. The task is to figure out the so-called *individual causal effect*, that is, the difference

$$(1) \quad \tau(u) = Y(t, u) - Y(c, u)$$

which measures the effect of treatment t on u, relative to treatment c.

In each particular experiment, either $Y(t, u)$ or $Y(c, u)$ (but not both) ceases to be counterfactual. Yet, given that one of $Y(t, u)$ and $Y(c, u)$ becomes observable, the other *has to* be unobservable. Holland has called a situation such as this 'the *fundamental problem of causal inference*'. As he (1986, 947) put it: 'It is impossible to *observe* the value of $Y(t, u)$ and $Y(c, u)$ on the same unit and, therefore, it is impossible to observe the effect of t on u.' Does it follow that figuring out (1) above is impossible?

Suppose that we give treatment t to u and we observe $Y(t, u)$. The question, then, is how could we possibly figure out the value of $Y(c, u)$? Recall that $Y(c, u)$ is a counterfactual: the response that would be observed if the unit u were exposed to treatment c (given that it was in fact exposed to treatment t and the observed value was $Y(t, u)$). The important insight of Rubin's model is that when certain assumptions are in place, there are ways to assess counterfactuals such as the above. The following is how we may proceed.

Given that unit u got treatment t, we may try treatment c to a different unit u', which is very much like u, except that it was given treatment c instead. That is, instead of assessing the counterfactual conditional $Y(c, u)$, which is impossible, we assess the factual conditional $Y(c, u')$ – the response of unit u' if the subject is given treatment c – and claim that this tells *indirectly* what the value of $Y(c, u)$ is. If this move is to be plausible at all, we need an assumption of *unit homogeneity*. We need to assume that u and u' are so similar that the actual response of u' to treatment c is the same as the response that unit u *would* have to treatment c. Under this assumption, we take it that $Y(t, u) = Y(t, u')$ and $Y(c, u) = Y(c, u')$. Then, the individual causal effect can be calculated, since (1) becomes thus:

$$(2) \quad \tau(u) = Y(t, u) - Y(c, u) = Y(t, u) - Y(c, u').$$

This is all fine and we are prepared to say that, modulo the uniformity assumption, it does tell us something about the individual causal effect. But something strange has happened. Expression (1) involves essentially a counterfactual conditional $(Y(c, u))$; (2) does not. Expression (2) is indeed measurable, but the counterfactuals are gone. Instead, (2) has two factual conditionals, one for unit u who received treatment t and another for unit u' who received treatment c. In a sense, we are still asking: What would have happened to u, had we given it treatment c? But it also seems that we have now *reduced* this question to two different ones: (a) What *does* happen to u', if we give it treatment c? and (b) Assuming unit homogeneity, $Y(c, u')$ and $Y(t, u)$, what *is* the causal effect of t on u? These questions involve no counterfactuals. The content of the counterfactual conditional $Y(c, u)$ seems exhausted by the joined content of the factual conditional $Y(c, u')$ and the unit homogeneity assumption. In other words, the unit homogeneity assumption renders the counterfactual conditional $Y(c, u)$ not so much a claim about the *specific* unit u but rather a claim about *any* of the homogeneous units. It is because of this fact that the counterfactual becomes testable.

There is another way we might proceed in our attempt to calculate $\tau(u)$. This time, instead of giving treatment t to unit u and treatment c to (uniform) unit u', we give treatment c to unit u at time t_1 and treatment t to the *very same unit* u at a later time t_2. As Holland (1986, 948) notes, this move requires another assumption, namely, *temporal stability*. This, he says, 'asserts the constancy of response over time'. It also requires an assumption of 'causal transience', since it implies that 'the effect of the cause c and the measurement process that results in $Y(c, u)$ is transient and does not change u enough to affect $Y(t, u)$ measured later' (p. 948). So, if Alice's taking a placebo at time t_1 changes some properties of Alice enough to affect her response to taking an aspirin at a later time t_2, the causal effect of taking aspirin on Alice's headache episode ceases to be calculable. Under these assumptions, we take it that $Y(t_{t1}, u) = Y(t_{t2}, u)$ and $Y(c_{t1}, u) = Y(c_{t2}, u)$. If this is so, then the individual causal effect can be calculated, since (3) becomes thus:

(3) $\tau(u) = Y(t, u) - Y(c, u) = Y(t_{t2}u) - Y(c_{t1}, u)$.

The points made about (2) can be repeated about (3) too. Expression (3) has no counterfactuals and it seems that the content of (1) – which does involve the counterfactual $Y(c, u)$ – *reduces* to the joined content of two factual conditionals $(Y(t_{t2}, u)$ and $Y(c_{t1}, u))$ together with the two further assumptions of causal transience and temporal stability.

We are willing to allow that we may be wrong here. That is, it might be the case that counterfactuals such as the ones we have been discussing *do* have excess content over the joint content of the relevant factual conditionals and the relevant assumptions. Still, what matters is that counterfactual conditionals can be assessed in terms of truth and falsity only when certain assumptions are in place. Those assumptions might fail. If, however, there are reasons to believe they do not, then causal inference seems quite safe. This is really an important achievement of Rubin's model. But we shouldn't lose sight of the fact that these assumptions are characteristics of *stable causal or nomological structures*.[9] Consider *unit homogeneity*. For it to hold, it must be the case that two units u and u' are alike in all causally relevant respects other than treatment status. If this is so, we can substitute u for u' and vice versa. This simply means that there is a causal law connecting the treatment and its characteristic effect which holds for all homogeneous units and hence is independent of the actual unit chosen (or could have been chosen) to test it. In effect, this holds for temporal stability too, since the latter is the temporal version of unit homogeneity. It does indeed make sense to wonder what the value of the voltage in a resistor would have been if the intensity of the current was I instead of the actual I_0 precisely because Ohm's law provides a stable nomological structure to address this counterfactual. But suppose we wanted to check the counterfactual that had the election taken place at an earlier time, the government would have been re-elected. Here it is obvious that temporal stability cannot be assumed because there is no stable nomological structure to back it up. Law-backed counterfactuals can indeed be assessed precisely because the laws make sure that the required assumptions are in place.[10]

In light of the above, it might not be surprising that according to Judea Pearl (2000, 428), who is one of the champions of the counterfactual approach, 'the word "counterfactual" is a misnomer'. In the case of individual causal effects, Pearl notes, we are interested in finding out things such as this:

> Q_{JI}: The probability that John's headache would have stayed had he not taken aspirin, given that he did in fact take aspirin and the headache has gone.

It does not matter for present purposes that Pearl formulates the issue in terms of probabilities. What matters is that Q_{JI} is a counterfactual claim of which Pearl stresses:

---

[9] Our favourite way to spell out this notion is given by Herbert Simon and Nicholas Rescher (1966). In fact, in showing how a stable structure can make some counterfactuals true, they blend the causal and the nomological in a fine way.

[10] It goes without saying that this causal or nomological structure should be characterised independently of counterfactuals, but as Simon and Rescher (1966) show, this can be done.

Counterfactual claims are merely conversational shorthand for scientific predictions. Hence Q_{II} stands for the probability that a person will benefit from taking aspirin in the next headache episode, given that aspirin proved effective for that person in the past.... Therefore, Q_{II} is testable in sequential experiments where subjects' reactions to aspirin are monitored repeatedly over time. (p. 249)

Nothing said so far is meant to belittle causal inference. Whether or not we view it as involving an ineliminably *counterfactual* element, we can certainly draw safe causal conclusions when the relevant assumptions are fulfilled. Actually, both the advocates of the counterfactual approach (e.g., Holland 1986; Cox & Wermuth 2001) and their opponents (e.g., Dawid 2000) agree that we can get valuable information about the so-called *average causal effect*. This is the average causal effect on the whole population, that is, the difference between the expected value of responses to treatment t and the expected value of responses to treatment c. Indeed, randomised controlled experiments are important precisely because they let us know about average causal effects.[11] However, to get from the average causal effect in a population to the *individual causal effect* on a specific unit u, we need the further assumption of 'constant effect' (Holland 1986, 948) or 'unit-treatment additivity' (Cox 1986, 963). According to this, the effect of treatment t on each and every unit u is the same.[12] Whether this holds or not is a largely empirical matter.[13]

7.5 Using a Black Box versus Looking into it

Given the difficulties of the interventionist semantics for counterfactuals, it seems prudent to assume the simpler and more comprehensive Lewis-style semantics. In any event, the differences between the various counterfactual accounts of difference-making are less important than the common difference from the mechanistic account. We have already stressed towards the end of Section 6.4 that there is a fundamental asymmetry between dependence and mechanistic approaches: the counterfactual approach (a fortiori

---

[11] For some interesting (but manageable) complication, see Kluve (2004, 86–7).

[12] In fact, the constant effect assumption is a consequence of unit homogeneity (cf. Holland 1986, 949). For some criticism of this assumption, see Cox and Wermuth (2001, 68).

[13] The counterfactual approach to causal inference has been severely criticised by Philip Dawid (2000) and has been vigorously defended by others (see the discussion that follows Dawid's article). Dawid has a number of important complaints. But the thrust of his critique is that the counterfactual approach relies on untestable metaphysical assumptions and, in particular, on a hopeless attempt to calculate the value of an unobservable quantity. Dawid's reaction, though invariably interesting, may be too positivistic.

the *dependence* approach) is more *basic* than the mechanistic (a fortiori the *productive*) one in that a proper account of mechanisms depends on counterfactuals, while counterfactuals need not be supported (or depend on) mechanisms.

The argument is worth repeating because of its centrality. The mechanistic approaches fail to account for the interactions among the parts of the mechanism unless they assume difference-making relations under actual and counterfactual interventions. MDC are not quite clear on what the interaction within the mechanism consists in. The mechanistic approach to causation fills in the 'chain' that connects the cause and the effect with intermediate loops. But there is still no account of how the loops interact. So, the interaction between parts *A* and *B* of mechanism M is accounted for by positing an intermediate part *C*. But then, there is need to account for the interactions between *A* and *C*, and *C* and *B*. Then we posit other intermediate parts *D* and *E* and so on. It might well be the case that the most general and informative thing that can be said about these interactions is that there are relations of counterfactual dependence among the parts of the mechanism.

Imagine a perfectly randomised experiment in which *t* (for treatment) produces higher response than *c* (for control). Has a causal connection been established? If we treat the randomised experiment as a black box, then insofar as it is a *good* experiment, we have established a causal connection. But what is inside the *black box*? Some might think that without a specification of the mechanism by which the higher response *t* was effected, the causal connection has *not* been established.[14]

This is a delicate issue. As noted in the end of the last section, establishing the causal status of each part of a mechanism would require finding out (or estimating) its causal effect. And the best way to do this is by non-mechanistic means, and in particular by means of counterfactual dependence. So, there seems to be a genuine asymmetry here. The causal effect can be found out, at least in favourable circumstances, *without* understanding the causal mechanisms, if any, involved; but the causal mechanisms, even if they are present, cannot be understood without the notion of the causal effect, that is, without some notion of (counterfactual) dependence.

[14] Notably, this is the view of D. R. Cox (1992, 297). He claims that this was also R. A. Fisher's view. When asked, at a conference, for his view on the step from association to causation, Fisher is reported to have responded: make your theories elaborate (cf. Cox 1992, 292). It's also the thrust of the Russo-Williamson Thesis discussed in Section 4.3.2.

But there are at least three things that show how mechanistic considerations can help the counterfactual approach to causal inference. First, mechanistic considerations can help testing the stability assumptions (unit homogeneity, temporal stability) that are necessary for the counterfactual inference. We take this to be fairly obvious, so we won't elaborate on it further.

Second, mechanistic considerations can help deal with the endogeneity problem. Briefly put, the problem of endogeneity is this. It might happen that the values taken by the so-called explanatory (or causal) variable are consequences, rather than causes, of the values of the dependent variable. In a perfectly controlled experiment this cannot happen because the variables that are manipulated are the explanatory variables. But in cases where the research is qualitative or where an experiment is not possible at all, the counterfactual approach might well fail to solve the endogeneity problem. Consider one of the classic problems of the early twentieth-century social science: Max Weber's claim that a certain type of economic behaviour – the capitalist spirit – was induced by the Protestant ethic. Many social scientists have argued that this claim falls foul of the endogeneity problem. Opponents of Weber's thesis claimed that the order of dependence goes in the other direction: Europeans who already have had an interest in breaking free of the pre-capitalist mode of production might have broken free of the Catholic Church precisely for that purpose. That is, it was the economic interests of certain groups that caused the Protestant ethic and not conversely. In cases such as this, a controlled experiment is out of the question. Besides, the assessment of intuitively relevant counterfactuals will be, to say the least, precarious. But an understanding of the mechanisms at play can well help resolve the endogeneity problem. These mechanisms, we presume, include a more detailed description of the explosion of the capitalist economic activity in the sixteenth century and of the economic behaviour of certain groups, for example, in Venice and Florence or in England and Holland, which predate the emergence of Protestantism.

The third way in which mechanistic considerations can help the counterfactual approach concerns the possible confounders. In a perfectly randomised trial, the problem of confounding variables does not arise. The experimental method itself makes it very unlikely that the explanatory variable is correlated with possible confounders. But in qualitative research, or even when matching techniques are used, it is possible that the explanatory variable is correlated with a confounding variable. Take, for instance, the dependent variable to be participation in demonstrations and the explanatory variable to be the age of the participants. It might well be that a confounding variable (e.g., radicalness of beliefs) is correlated with the explanatory variable and has an influence on the dependent variable. In cases such as this, knowledge of mechanisms can help identify possible confounders and control for them. Conversely, knowledge of mechanisms can explain why the experimenter need not control for some variables (e.g., the colour of the eyes of those who participate in demonstrations).

Mechanisms cannot be the surrogate of a careful experiment. If we think of an experiment as a black box, then counterfactuals have a key role to play. After all, when certain assumptions hold, they can establish a causal relation. But without some knowledge of the mechanism inside the black box, we won't have *full* understanding of the causal relation. Nor can we solve, at least as effectively, some methodological problems of causal inference.

Using the black box carefully does establish a causal link. *Looking into* the box does offer extra understanding, even if it does *not*, in and of itself, establish the causal relation.

PART III

Beyond New Mechanism

CHAPTER 8

Constitution versus Causation

8.1 Preliminaries

New mechanists typically distinguish between causal and constitutive mechanisms, and between causal and constitutive mechanistic explanations.[1] Whereas causal mechanisms are taken to cause phenomena, constitutive mechanisms are taken to constitute them. Constitution, according to Craver's (2007a) widely adopted account, or 'constitutive relevance' as he calls it, is viewed as a non-causal dependency relation between the phenomenon and the components of the mechanism.

To illustrate what it is for a mechanism to constitute a phenomenon, Craver introduced his well-known diagrams (which have been called 'Craver diagrams'; see Figure 8.1). At the lower part of the diagram, we have the mechanism's component entities and activities that are taken to constitute the phenomenon, which is depicted as a dark oval above the mechanism. While Craver does not offer a theory of what constitution amounts to, he takes the phenomenon to supervene on the mechanism as a whole (i.e., the oval in the lower part that includes all the components of the mechanism). Craver describes the components of the mechanism as '$X_1\varphi$-ing', '$X_2\varphi$-ing' and so on, where the term '$X \varphi$-ing' refers to an entity X engaging in an activity φ. He describes the phenomenon as $S\psi$-ing, where the S is an entity that engages in the activity ψ and where $S\psi$-ing is the phenomenon that is taken to be constituted by the mechanism. To illustrate, when a neuron generates an action potential, S is the neuron, $S\psi$-ing the neuron generating an action potential, and the Xs that φ are the various components of the mechanism that is responsible for the generation of the action potential (in what follows we will adopt Craver's terminology to refer to the components of the mechanism and to the phenomenon).

---

[1] This distinction goes back to Salmon (1984).

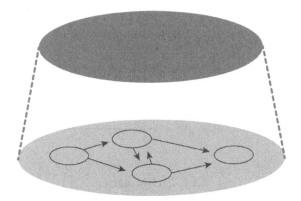

Figure 8.1   A Craver diagram (based on figure 1.1 from Craver 2007a, 7).

Importantly, not all parts of the neuron are components of the mechanism for the action potential. In general, in the case of an S that ψs, the S will typically have several parts that are not components of the mechanism responsible for S's ψ-ing. So, not all parts of S are also components of the mechanism responsible for S's ψ-ing. The components are said to be the parts that are 'constitutively relevant' to S's ψ-ing. As Craver puts it, 'the very idea of a mechanism presupposes the idea of constitutive relevance ... This difference between mechanisms and machines turns at least in part on the fact that all of the constituents of a mechanism are relevant to what it does (they are components), while only some of the parts of machines are relevant to what they do (namely, those that are components in one or more of its mechanisms)' (2007b, 6).

8.2   Craver on Constitutive Relevance

Craver gives two conditions for constitutive relevance. According to the mutual manipulability condition (1) there has to be some change to X's φ-ing that changes S's ψ-ing, and (2) there has to be some change to S's ψ-ing that changes X's φ-ing. The parthood condition says that the Xs are parts of S.[2] Craver uses Woodward's notion of ideal intervention to give an

---

[2] For Craver, then, constitutive relevance does not amount to a supervenience or identity claim. It is a relationship between the phenomenon (S's ψ-ing) and one of the components of the mechanism responsible for it. S's ψ-ing does not supervene on X's φ-ing; rather, it supervenes on the organised activities of all of the components in the mechanism. Note also that Craver says that the mutual manipulability account provides a sufficient condition for something to be a component of a mechanism; but he also says that the failure of mutual manipulability is a sufficient condition for

account of mutual manipulability. As we saw in Chapter 7, an ideal intervention I on X with respect to Y changes the value of Y only via the change in X. Craver defines an analogous notion of ideal intervention for constitutive mechanisms. The main idea is that 'an *ideal* intervention I on φ with respect to ψ is a change in the value of φ that changes ψ, if at all, *only via* the change in φ', and similarly for an ideal intervention on ψ with respect to φ (2007a, 154). So, for φ to be constitutively relevant to ψ, '[t]here should be some ideal intervention on φ under which ψ changes, and there should be some ideal intervention on ψ under which φ changes' (p. 154).

The adequacy of Craver's account is a contested issue in the mechanistic literature. A main problem for this account is whether it is indeed possible to apply Woodward's notion of ideal intervention to the case of constitutive relevance. It is not possible to intervene on one of the components of S's ψ-ing without also intervening on S's ψ-ing, and vice versa. As Michael Baumgartner and Lorenzo Casini (2017) note, an intervention I on ψ with respect to φ must be viewed as the common cause of the changes in both ψ and φ, and so is not an ideal intervention on ψ with respect to φ, in contrast to what Craver assumes. Baumgartner and Casini (2017, 220) conclude that 'the types of interventions required by [mutual manipulability] cannot possibly exist for any mechanistic system. [Mutual manipulability] is hence unsatisfiable, which means that constitutive relations as defined by [mutual manipulability] are inexistent, which again entails that friends of mechanistic explanations who rely on [mutual manipulability] chase a chimera.'[3]

If Craver's account of constitutive relevance is problematic, what options are open for the mechanist? Craver holds both that constitutive relevance is needed in order to understand (constitutive) mechanisms and that this can be done in terms of interventionism. A first option, then, is to retain both claims, but to try to find an alternative way to reconcile

something to fail to be a component (2007a, 159). As Baumgartner and Casini note (2017), this amounts to a necessary and sufficient condition for something to be a component of a mechanism.
[3] Several other criticisms have been made to Craver's account. Leuridan, for example, has argued that mutual manipulability fails to provide a sufficient account for constitutive relevance, since 'plenty of cases of bidirectional causation would unintentionally fit the mutual manipulability account of constitutive relevance' (2012, 400). Franklin-Hall (2016) has argued that although Craver's account can show whether a part of a whole is relevant to some behaviour of the whole, it does not solve what she calls the 'carving problem', that is, how to provide appropriate decompositions of mechanisms. Another concern is that Craver's account seems to offer a way to decide whether a part is a component of a mechanism, but does not tell us what constitutive relevance really is (Couch 2011). This is also Glennan's main complaint (2017, 44).

interventionism with mechanistic constitution.[4] A second option is to retain the first claim, but deny that interventionism can be used to give an account of constitutive relevance and attempt to find an alternative account.[5] There exists a third, more radical, option, that is, to deny that constitutive relevance, qua a non-causal dependency relation, is in fact needed in order to understand what a mechanism is. This (unpopular) option leads to a deflationary view of 'constitutive' explanation, as being a certain kind of causal explanation, and is the view that will be defended in this chapter.

8.3   Are There Constitutive Mechanisms?

Craver first introduced the 'Craver diagram' representation of a (constitutive) mechanism in *Explaining the Brain* (2007a). He begins with the example of the mechanism of neurotransmitter release by a neuron. Consider Craver's description of this mechanism:

> The mechanism begins, we can say, when an action potential depolarizes the axon terminal and so opens voltage-sensitive calcium (Ca^{2+}) channels in the neuronal membrane. Intracellular Ca^{2+} concentrations rise, causing more Ca^{2+} to bind to Ca^{2+}/Calmodulin dependent kinase. The latter phosphorylates synapsin, which frees the transmitter containing vesicle from the cytoskeleton. At this point, Rab_3A and Rab_3C target the freed vesicle to release sites in the membrane. Then v-SNARES (such as VAMP), which are incorporated into the vesicle membrane, bind to t-SNARES (such as syntaxin and SNAP-25), which are incorporated into the axon terminal membrane, thereby bringing the vesicle and the membrane next to one another. Finally, local influx of Ca^{2+} at the active zone in the terminal leads this SNARE complex, either acting alone or in concert with other proteins, to open a fusion pore that spans the membrane to the synaptic cleft. (pp. 4–5)

Craver goes on to offer a general characterisation of a mechanism by saying that the mechanism of neurotransmitter release 'is a set of entities and

---

[4] See, for example, the accounts given in Romero (2015) and Harinen (2018) – we will discuss Harinen's account in Section 8.5. One possibility is to adopt a modified form of interventionism on which it is not required, when intervening on some variable, to hold other variables fixed, if these are related by, for example, mereological or supervenience relations to the variable on which the intervention is applied (see Woodward 2015). Adopting such a framework makes interventions on variables related by constitutive relations possible, but it seems to lead to a problem of empirical underdetermination when we try to infer constitutive relations (see Baumgartner & Casini 2017).

[5] See, for example, the 'regularity' accounts of constitutive relevance offered in Harbecke (2010) and Couch (2011); see also Gillett (2013) for an account based on the dimensioned view of realisation.

activities organized such that they exhibit the phenomenon to be explained' (p. 5). The word 'exhibit' here is meant to capture the specific relation between the organised entities and activities, on the one hand, and the phenomenon, on the other; the organised entities and activities, that is, the mechanism, do not *produce* the phenomenon, since they are not its antecedent causes, but they *exhibit* it. Craver then offers his well-known diagram for the abstract representation of a mechanism and notes that this account leads to a type of explanation different from 'etiological causal explanation' which says 'how a phenomenon is produced by its causes', namely, to 'constitutive (or componential) causal-mechanical explanation', which is 'the explanation of a phenomenon ... by the organization of component entities and activities' (p. 8).

According to us, the mechanism of neurotransmitter release is an example of a mechanism as a causal pathway; similar to the case of the signalling pathways of apoptosis, we have a sequence of causal steps that causally link the depolarisation of the axon terminal to the opening of the fusion pore and the release of neurotransmitters.[6] Such a case, then, as well as the cases of biological pathways we examined in Chapter 3, cytological cases like apoptosis or mitosis, pathological mechanisms, cases like the mechanism of DNA replication and protein synthesis often discussed by mechanists and the several other examples we have seen in the book, are best viewed, we think, as causal sequences that produce an effect, and not in terms of a 'Craver diagram'.

Given the prevalence of mechanisms as causal pathways it is perhaps strange that constitutive mechanisms have become central among new mechanists. It is important to note that mechanists recognise that some mechanisms are best viewed as causal sequences. Although this is perhaps not very explicit in Craver (2007a), new mechanists commonly distinguish between two types of mechanisms. For example, Craver and Darden (2013, 65–6) distinguish between 'productive mechanisms' and 'underlying mechanisms', where in each case there is a different kind of relationship between the mechanism and the phenomenon. In the case of productive mechanisms 'one typically starts with some understanding of the end product and seeks the components that are assembled and the processes by which they are assembled and the activities that transform them on the way to the final stage'. In the case of underlying mechanisms, in contrast, 'one typically breaks a system as a whole into component parts

---

[6] Craver seems to agree, as he later (2007a, 22) notes that this is an etiological explanation, where the explanans 'is the mechanism linking the influx of Ca^{2+} into the axon terminal'.

that one takes to be working components in a mechanism, and one shows how they are organized together, spatially, temporally, and actively such that they give rise to the phenomenon as a whole'. Productive mechanisms are similar to mechanisms as causal pathways (although of course we would not characterise causal pathways in terms of productivity or in terms of the distinction between entities and activities, as we think that such talk unjustifiably metaphysically inflates the notion of mechanism).

Similarly, Craver and Tabery (2015) note that some mechanisms can be conceived 'as a causal sequence terminating in some end product: as when a virus produces symptoms via a disease mechanism or an enzyme phosphorylates a substrate'. But not all mechanisms are to be viewed in such terms. They claim that for 'many physiological mechanisms . . . it is more appropriate to say that the mechanism *underlies* the phenomenon'. Examples are the mechanism of the action potential and the mechanism of working memory. The phenomenon here is not the production of something, but 'a capacity or behavior of the mechanism as a whole'.

New mechanists, then, can accept that at least some mechanisms in biology are better viewed as causal pathways. The question, then, is not whether mechanisms as causal pathways exist (or whether mechanism talk in life sciences is to be understood in terms of causal pathways), but whether we have to accept a second type of mechanism, that is, constitutive mechanisms. It is not our claim that the notion of a constitutive mechanism is conceptually problematic (although we think that a clear account of what constitutive relevance amounts to is still lacking). For us, the main question is: Is the notion of a constitutive mechanism important in illuminating the notion of mechanism as a concept-in-use central in practice? Given the prima facie adequacy of Causal Mechanism to capture this concept, what would be a convincing argument to accept that at least some instances of the notion are to be understood in terms of constitutive mechanisms? Ideally, we need a more general reason than simply to point out that a limited number of cases are better viewed in such terms.

Is, then, a 'constitutive' sense of mechanism present in scientific practice? A point to be noted here is that this is not an easy question to answer if we don't know what constitution or constitutive relevance is. For example, Craver notes that mechanistic constitution is not material constitution, which is a notion about which we may have various intuitions. And as we have seen, Craver's account of it is not without problems. But then we face the following difficulty: If we do not have a clear account of what a constitutive mechanism is, how exactly can we look for it within scientific practice? A solution could be to examine what scientists themselves say about mechanistic constitution or to look at some paradigmatic cases of mechanistic constitution that are regarded as such by scientists in order to try to analyse this notion. One might think that the situation here is similar to what we do when we want to analyse the notion of causation as it functions within scientific practice: for instance, we look at the structure of experiments performed to reveal causal relations. The difference-making account of causation, for example, is grounded in this way in scientific practice, as we saw in Chapter 4 with the case of scurvy. But there is an asymmetry here: in contrast to the case of causation or mechanism in the sense of CM, it is not easy to argue that there is a notion of mechanistic constitution present in scientific practice.[7]

Be that as it may, we can use Craver's account of what a constitutive mechanism is, even if it is not entirely clear what mechanistic constitution amounts to, to try to see whether something like this can be found in scientific practice. We can identify two strategies to provide such a general reason in order to argue that constitutive mechanisms are important. Both strategies attempt to show that if our account of mechanism as a concept-in-use is confined to viewing a mechanism as a causal pathway, we fail to capture a central aspect of biological practice; that is, our account fails to be descriptively adequate.

The first strategy is simply to say that when we look at scientific practice, we find many instances of constitutive mechanistic explanations; to understand how exactly these explanations function we need to view a mechanism in 'constitutive' terms, that is, as constituting (or underlying) the phenomenon to be explained. The second strategy is to say that there are certain kinds of experiments in biology that can be understood only by adopting a 'constitutive' account of mechanisms. If these strategies succeed, the account offered by CM leads to an impoverished notion of mechanism as a concept-in-use since it fails to capture some salient features of scientific practice.

---

[7] This emphasis on scientific practice when offering an account of mechanisms is a central feature of Baetu's (2019) analysis. In relation to Craver's constitutive relevance account, he says: 'Instead of first committing to a noncausal interpretation and then attempting the impossible task of demonstrating that evidence for causation can somehow be used to demonstrate something else than causation, one can simply let go of ready-made metaphysical intuitions borrowed from physics and the philosophy of mind and consider the relationship between mechanisms and phenomena from the strictly experimental standpoint' (p. 33). He thinks that this 'strictly experimental standpoint' leads to what he calls the 'causal mediation account' concerning the relationship between mechanisms and phenomena. Baetu's analysis, with its emphasis on methodology, shares many similarities with CM.

An example of the first strategy can be found in Marie Kaiser and Beate Krickel's (2017) analysis of constitutive mechanisms. According to them,

> one can etiologically explain protein synthesis by describing how a certain sequence of causes leads to the synthesis of a protein, or one can constitutively explain protein synthesis by referring to the components of a cell and describing how they act and interact such that the cell synthesizes proteins. On a closer inspection, however, it turns out that what we are explaining is not the same phenomenon, but two different phenomena: the etiological MEx [mechanistic explanation] explains the end-result (there being a protein) and the constitutive MEx explains the process of protein synthesis (we want to know what happens at every step of protein synthesis). (p. 8)

The claim here is that a constitutive mechanistic explanation explains something that an etiological mechanistic explanation fails to explain. If this is correct, then we have a general reason why constitutive mechanisms are central: they explain an aspect of the phenomenon that cannot be explained otherwise. On this view, constitutive mechanisms seem to be as important and ubiquitous as mechanisms qua causal sequences: when we have a causal sequence mechanism, we have a constitutive one, and vice versa.

But this does not sound very plausible. Consider again protein synthesis; what more is there to be explained, if we know everything that is to be known about the causal sequence that leads to the production of the protein? And what more is there to be explained if we know all the causal steps in a metabolic or signal transduction pathway? A mechanistic explanation in our sense, that is, a description of the causal pathway of protein synthesis, is sufficient to explain 'the process of protein synthesis' or 'what happens at every step' of the mechanism.

The second strategy is exemplified by Craver's analysis of what he calls 'interlevel' experiments. As he puts it, '[t]he norms of constitutive relevance are implicit in the experimental strategies that neuroscientists use to test claims about componency and in the rules by which neuroscientists evaluate applications of those strategies' (2007a, 144). Interlevel experiments for Craver can be bottom-up or top-down. In bottom-up experiments one intervenes to change the components of the mechanism and detects how the phenomenon (Sψ-ing) changes. In top-down experiments the intervention is applied to the phenomenon and one then detects how the behaviour of the components of the mechanism change. Craver uses the bottom-up/top-down distinction and the distinction between an inhibitory and an excitatory intervention to classify the most common kinds of experiments in neuroscience, i.e., interference, stimulation and activation experiments (pp. 144–52). Interference experiments are bottom-up and inhibitory (e.g., lesion experiments), stimulation experiments bottom-up and excitatory (e.g., stimulating the brain to produce movements in specific muscles) and activation experiments top-down and excitatory (e.g., engaging the subject in some task so that a cognitive system gets activated and monitoring what happens in the brain as in PET and fMRI studies). Craver reviews various inferential challenges that these experiments face; the combined use of these experiments, that is, performing both a bottom-up and a top-down experiment, is a way to overcome these challenges. For example, in an interference experiment a lesion may lead to a change in the phenomenon, but this does not necessarily mean that a particular brain region is a component in the mechanism responsible for the phenomenon; an alternative hypothesis is that the disrupted brain region leads to a change in some other part of the brain, and this other part is a component in the mechanism.

These sort of challenges and the ways they are overcome are used by Craver 'as data points in building a descriptively and normatively adequate account of constitutive relevance' (p. 147) and provide a main motivation for his mutual manipulability account. His strategy, then, is to use inter-level experiments to offer an adequate account of what it is for X to be a component of a mechanism. Craver seems to take it for granted that interlevel experiments are to be viewed in terms of 'Craver diagrams' (p. 146); he thus applies a 'constitutive' framework to analyse the experiments and asks what they suggest about the norms of constitutive relevance. But one may use these experiments to justify the 'constitutive' framework itself and the salience of constitutive mechanisms in scientific practice. The argument here, then, is that only 'constitutive' mechanisms can make sense of interlevel experiments; the view of mechanisms as causal pathways cannot capture this important feature of scientific practice. The task for us, then, is to show that Causal Mechanism can lead to an alternative account of interlevel experiments. This alternative account will involve a different view of levels than Craver's; we will introduce the main idea of this alternative account in this chapter, but we will develop it more fully in Chapter 9.

8.4 Against Constitutive Mechanisms

We have seen that new mechanists think that in certain cases it is more appropriate to view mechanisms in constitutive terms. But when exactly is it more appropriate to do so? In a constitutive mechanism the organised entities and activities explain the behaviour of some entity of which they are parts, that is, what Craver refers to as 'Sψ-ing' in his diagram. Our question thus becomes: When is it appropriate to talk about an S that ψs? A natural suggestion is that we apply such language when there are natural boundaries around a mechanism, that is, when the mechanism is part of a biological object or structure. For example, the metabolic pathway of glycolysis occurs inside the cell, so we can take the cell as the S in this case.[8] A problem here is that many mechanisms are not confined within natural boundaries. For example, in signal transduction pathways the signal comes from outside the cell (as in the case of the extrinsic pathway of apoptosis), and so, although a component of the pathway, it is not part of the cell. Kaiser and Krickel (2017, 29) mention the example of the mechanism for muscle contraction which also includes an external signal as one of its components, that is, 'neurotransmitters that bind to receptor molecules outside of the muscle fibre'; another example is from David Kaplan (2012, 552), the gecko adhesion mechanism, which is 'spatially distributed to include external components spanning the boundary between the gecko and its environment'. In pathological mechanisms, too, the causal agent of a disease can be an environmental factor, for example, radiation, extreme temperature, toxic substances or viruses.

Craver is aware of this point; as he says, 'mechanisms frequently transgress compartmental boundaries' (2007a, 141). His main example, the mechanism of the action potential, is a case in point. This mechanism 'relies crucially on the fact that some components of the mechanism are inside the membrane and some are outside. The membrane allows the intracellular and extracellular concentrations of ions to be different, allows a diffusion gradient to be set up, and allows for a separation of charge' (p. 141). Also, cognitive mechanisms 'draw upon resources outside of the brain and outside of the body to such an extent that it may not be fruitful to see the skin, or the surface of the CNS, as a useful boundary' (p. 141).

---

[8] Of course, if a mechanism such as glycolysis occurs inside a cell, there will be several other biological structures of which it is a part, for example, the tissue in which the cell is a part, the organ in which the tissue is a part, and finally the organism. However, it is more natural in this case to regard the cell as the S, the behaviour of which the mechanism explains. In some cases, however, it may not be clear which biological structure should be regarded as the S; for example, what about a mechanism that occurs inside a cell organelle? DNA replication in eukaryotes occurs in the nucleus; is the nucleus or the cell the S in this case? Moreover, it may seem strange to attribute some behaviour to the cell itself in virtue of the fact that DNA is replicated inside its nucleus, or some protein is synthesised, or glycolysis occurs. It is, we think, more appropriate to view all these processes as pathways that occur inside a cell, without taking the cell as an acting entity and viewing the pathways in terms of a constitutive mechanism that involves the behaviour of the cell as a whole.

We have mentioned earlier that for Craver the components of a mech-anism have to be parts of the entity S. In view of the previous examples, however, this is problematic; in all these examples, some components of the mechanism are not parts of the entity S.[9] Kaiser and Krickel (2017) suggest weakening the parthood condition, that is, to require most, but not all, of a mechanism's components to be parts of S. In this case, we will still have a constitutive mechanism, that is, an S that contains part of a mechanism that explains some behaviour of S.[10]

We draw a different lesson from the existence of components of a mechanism that are external to the entity S. Let us ask the following question: When is it appropriate to apply constitutive terms to a specific example of a mechanism? In other words, when is it appropriate to view a particular mechanism in terms of a Craver diagram? We think that the existence of components that are external to a biological structure that is a candidate for being the entity S undermines treating this biological struc-ture as the entity that ψs; hence, we should either look for some other entity or treat the mechanism not as a constitutive one but as a causal sequence. It is not clear, however, which other biological structure can be the entity S; even if the whole organism is taken as the S (which is implausible for a mechanism such as a signal transduction pathway), we have seen that there are mechanisms the components of which are external to the organism. If, alternatively, the S is taken as the mechanism as a whole, as Craver seems to suggest, this leads to problems (as we mention in note 9).

The problem of external components, then, is not a reason to reject or weaken the parthood condition of constitutive relevance but a reason to view the mechanism in terms of a causal sequence and not in 'constitutive' terms. We take typical and paradigmatic biological mechanisms to be causal pathways, where these pathways are not necessarily confined within specific biological objects or structures, but can transgress natural bound-aries. In many cases of pathways, that is, there will not exist an appropriate

---

[9] Although we take S to be an entity such as a neuron that contains the components of a mechanism that explains some behaviour of S, Craver also says that S refers to 'the mechanism as a whole' (2007a, 7). We agree with Kaiser and Krickel (2017, 26) that this latter claim is problematic; as they say, '[i]t is not the mechanism of muscle contraction that is contracting. Nor is it the action potential mechanism that fires, or the spatial memory mechanism that navigates through the Morris Water Maze. Rather, the relevant objects are individuals that in most cases are larger objects or systems that contain one or often more mechanisms.'

[10] Note that the option to reject entirely the parthood condition of constitutive relevance leads to problems, as cases of causal feedback mechanisms would count as examples of constitutive relevance (cf. Leuridan 2012).

biological object S, where every component of the pathway is part of
S. Such mechanisms are better viewed in etiological, rather than constitutive, terms.[11]

Let us now consider a mechanism, for example, protein synthesis, all the
components of which are parts of a biological object, that is, a cell. Would
we be justified to regard the cell as the S that ψs? A problem here is that the
mechanism of protein synthesis, as well as several other mechanisms, can
exist outside the cell. In fact, this is how the molecular details of these
mechanisms were discovered. This process involves what are known as cell-
free systems; to develop a cell-free system cells must be disrupted and
fractionated. The mechanism of protein synthesis was first studied by
using a cell homogenate that translated RNA molecules, producing proteins. As Alberts et al. (2014, 451) explain, '[f]ractionation of this homogenate, step by step, produced in turn the ribosomes, tRNAs, and various
enzymes that together constitute the protein-synthetic machinery'.
Individual purified components 'could be added or withheld separately
to define its exact role in the overall process'. As the authors stress,

> [a] major goal for cell biologists is the reconstitution of every biological
> process in a purified cell-free system. Only in this way can we define all of
> the components needed for the process and control their concentrations,
> which is required to work out their precise mechanism of action … [A]
> great deal of what we know today about the molecular biology of the cell
> has been discovered by studies in such cell-free systems. They have been
> used, for example, to decipher the molecular details of DNA replication and
> DNA transcription, RNA splicing, protein translation, muscle contraction,
> and particle transport along microtubules. (p. 451)

The fact that a mechanism such as protein synthesis can operate outside
a cell means that the components of the mechanism (the Xφ-ings) can
exist, and the mechanism can function, without the entity that they
constitute (the Sψ-ing). But is this possible? The organised entities and
activities that are constitutively relevant to the phenomenon P (the mechanism for P) are taken to constitute the phenomenon (cf. Kaiser & Krickel
2017, 9). But it seems strange to have a mechanism that, if present in a
cell, constitutes phenomenon P, but if outside the cell, although it operates

---

[11] But is there a fact of the matter about where exactly the starting point of a mechanism is? If it is
more or less arbitrary what we take as the starting point of a causal sequence, why not just focus on
the part of a causal sequence that is part of an entity S? That way we may avoid the problem of
external components. Our answer is that what is arbitrary is to insist that all the components of a
mechanism have to be parts of the entity S; as we have seen, paradigmatic mechanisms transgress
natural boundaries.

normally, does not constitute anything or perhaps constitutes something different.[12] Moreover, if we take S's ψ-ing to supervene on the mechanism for S's ψ-ing in the sense that 'there can be no difference in S's ψ-ing without a difference in the mechanism for S's ψ-ing' (Craver 2007a, 153, n. 33), then in the case of cell-free protein synthesis we have a difference in S's ψ-ing (trivially, since we have no cell that ψs – and we cannot have just the ψ-ing without the entity that engages in it) but no difference in the underlying mechanism. The conclusion, then, is that the entity S in such a case cannot be the cell; it is best to view the mechanism as a causal pathway, without also having an S that ψs.[13]

We take the arguments in this section to show that it is not appropriate or useful to view typical and paradigmatic biological mechanisms in constitutive terms. Paradigmatic mechanisms such as the mechanism of protein synthesis and metabolic pathways are not constitutive mechanisms to be represented by Craver diagrams, and many of them transgress natural boundaries. Biological mechanisms are better viewed, then, as pathways and analysed in terms of causation (and not constitution).

8.5   Causal Mechanism and Constitutive Relevance

We will close this chapter by arguing that some of the elements of Craver's analysis still hold if we adopt the view that mechanisms are causal pathways. Consider again the distinction between etiological causal-mechanical explanation and constitutive or componential causal-mechanical explanation. This distinction can be retained, even if we reject Craver's account of constitutive mechanisms. In the case of an etiological explanation we focus on the antecedent causes of a phenomenon. In componential explanation we start with a previously established causal relationship between X and

---

[12] In the case of material constitution, if A materially constitutes B, it is possible for A to exist without constituting B; but as Craver stresses, mechanistic constitution is not material constitution, so we have no reason to expect that an analogous claim holds in the case of mechanistic constitution.

[13] Another option is to take as the entity S not the cell, but the mereological sum of all the components of the mechanism. For example, in the case of the protein synthesis the S could be the mereological sum of the DNA and RNA molecules, the amino acids, various enzymes and so on. But this seems implausible: a mereological sum of all the components of a biological pathway is not a 'natural' biological object, and it seems very strange to attribute a behaviour to that object, which the mechanism is supposed to explain. In general, it is not plausible to view a mechanism as a whole as an object or as an acting entity. Mechanisms, to use a distinction from metaphysics that Kaiser and Krickel apply in their analysis, are better viewed as occurrents, that is, things that occur like football matches and weddings, and not as continuants, that is, things like stones and tables. But if mechanisms are occurrents, it again seems strange to think that they constitute the ψ-ing of an entity S, where S is presumably a continuant, for example, a cell.

Y and focus on the intermediate steps, that is, the components, of the causal pathway that leads from X to Y.

As we have noted, this is how a mechanism is viewed in statistics; that is, it concerns a causal structure that links a cause with an effect variable. In our example from Chapter 4, vitamin C is the mediating variable in the causal relationship between citrus fruits and scurvy. Pathological mechanisms, in general, link a causal agent such as a virus to the development of a disease. Similarly, the signalling pathways of apoptosis link an initial trigger such as a toxic agent to the cytological process of apoptosis described by Kerr et al. (1972). Identifying the components of a causal pathway that links an initial cause to some effect can then be usefully distinguished from etiological causal explanation that identifies antecedent causes, although in both cases what we are identifying are causal relationships between variables or components of a pathway. Mechanism talk is more appropriate if what we are interested in is finding the mediator or, more generally, 'extending' a causal pathway by identifying intermediate causal steps. This is because by a 'mechanism' biologists typically mean the way a cause produces the effect. In such contexts, causal explanation can be viewed as 'componential' and not a mere matter of identifying some antecedent cause or establishing a causal relationship between two variables.

However, several other claims that Craver makes will have to be rejected. In particular, we do not accept a non-causal relation of constitutive relevance; components of pathways are only *causally* relevant to the outcome of the mechanism. We can perhaps say things like 'the components of the signalling pathway of apoptosis constitute the pathway', but this should be viewed in a deflationary way and not as referring to some synchronic and symmetric relationship between relata that are not wholly distinct (as Craver characterises the relationship of constitutive relevance). In fact, we do not think that it is very useful (and it makes much sense, at least in the case of the typical and paradigmatic cases of biological mechanisms we have been considering) to talk about 'the behaviour of a mechanism as a whole'. This kind of talk can lead to a metaphysical inflation of the notion of mechanism, where we think of the mechanism itself as a kind of entity that engages in some activity (an S that ψs) and then ask about the nature of the relation between the components of the mechanism and S's ψ-ing. We have argued in this chapter that this general picture is not appropriate when we look at typical biological mechanisms.

Attempts to view the relation of constitutive relevance in causal terms lead to accounts that are not very far from the picture we want to defend (see especially Menzies 2012 and Harinen 2018). The main idea here is that S's ψ-ing can be viewed as an input-output relationship where the mechanism is causally between the input and output.[14] Totte Harinen, in particular, represents the phenomenon that the mechanism exhibits 'as a causal relation holding between two variables, ψ_{in} and ψ_{out}, corresponding to the input and output conditions characteristic of the relevant regularity' (2018, 14). On this view, in assessing mutual manipulability we have to take three variables into account: the higher-level variables ψ_{in} and ψ_{out}, which are the input and output variables of the mechanism, and a lower-level variable φ_i that is causally between ψ_{in} and ψ_{out}. In a top-down intervention we intervene on the input variable ψ_{in} and observe whether the lower-level variable φ_i changes. In bottom-up experiments we intervene on the lower-level variable φ_i and observe whether the output variable ψ_{out} changes. In claiming that constitutive relevance can be viewed as causal relevance, Harinen accepts interlevel causation, which is denied by several philosophers who think that there can be no interlevel causal relations between a mechanism (or Sψ-ing) and its components (cf. Craver & Bechtel 2007; Romero 2015; Baumgartner & Gebharter 2016).[15]

We agree with Harinen that mechanisms can include interlevel causation, but we do not view 'levels' in terms of Craver's levels of mechanisms; levels for us are levels of composition (or organisation), where levels and mechanisms are distinct notions. In Chapter 9 we will develop this view in some detail and claim that it can offer an alternative account of interlevel experiments. Our view, then, is that causal pathways can contain components from various levels of organisation, and representations of mechanisms can contain both higher-level and lower-level variables. So, we can have pathways like those suggested by Harinen ($\psi_{in} \rightarrow \varphi_i \rightarrow \psi_{out}$),

[14] Menzies (2012) uses a structural equations framework to provide an analysis of the causal structure of a mechanism 'in terms of the composition of functional dependences, or, in material mode, in terms of the programmed exercise of modular capacities' (p. 804). Although he rejects Craver's account of constitutive relevance, he thinks of his own account as retaining a notion of constitutive relevance distinct from causal relevance. For him 'any variable that lies on a pathway between the input variable and the output variable of the capacity to be explained counts as part of the mechanism underlying the capacity' (p. 801). So, intervening variables on a pathway 'are constitutively relevant to the mechanism by virtue of being parts of the pathway that is the mechanism'. But these variables 'are also causally relevant to the input and output variables'. Menzies thus takes the two relevance relations to be different. Whereas talk of systems having capacities that are explained by the mechanism may be appropriate for psychology and cognitive science, we are sceptical whether this is a useful interpretation of the biological examples we have been considering in this book.

[15] Harinen's account uses the modified interventionist framework developed in Woodward (2015) and so does not have the problem that Craver's original account had in relation to the feasibility of ideal interventions.

where the input and the output variables are at a higher level of organisation, whereas the other components are at lower levels; in the case of scurvy, for example, the mediator (vitamin C) is at a lower level of organisation than the input and output variables (a diet of citrus fruits and scurvy). But it is also possible to have other kinds of pathways, with various combinations of levels (in the next chapter we will review various such examples).

So-called constitutive explanation is for us a version of causal explanation, and there is no need to posit non-causal relations such as constitution, constitutive relevance or supervenience in order to understand what a mechanism is in biology. To do so is, as in the case of general characterisations of mechanisms that refer to entities and activities, to inflate the notion of mechanism as concept-in-use. Importantly, as we will see, such a view of mechanisms leads to a different understanding of the notion of levels and the relationship between levels and mechanisms from the dominant view among new mechanists, which is the account of levels of mechanisms developed by Craver.

CHAPTER 9

Multilevel Mechanistic Explanation

9.1 Preliminaries

There are various notions of 'levels' in science and philosophy (see Craver 2007a); we take it, however, that a main use of the notion of 'levels', especially in the life sciences, is to refer to levels of organisation or composition, where entities at one level are parts of or compose entities at a higher level (cf. Eronen & Brooks 2018). Consider, for example, the biological hierarchy: smaller molecules, for example, nucleotides, compose bigger ones, for example, DNA; DNA proteins and other molecules are parts of cells; cells compose tissues; tissues compose organs; organs are parts of organisms, which in turn are parts of ecosystems. This is a traditional view of levels in science (cf. Oppenheim & Putnam 1958). On this view, it is a more or less objective matter on what level an entity is, as level-membership is grounded in mereological or compositional relations that are taken to be objective features of the world.[1]

We can now introduce another notion of 'levels', what we might call levels of scope. Levels of scope concern the sense in which we say that physics has a wider scope than biology. Where levels of composition concern entities composing other entities, levels of scope concern the laws of nature governing entities at various levels. Thus, laws of physics have a much wider scope than laws of biology, for example (or whatever plays the role of laws in biology), since laws of physics govern entities at all levels of the compositional hierarchy, whereas laws of biology govern specifically biological entities. Laws of special sciences, then, have a narrower scope; whereas physical laws that range over all entities have the widest scope possible. In other words, physical laws range over entities at many (and perhaps all) levels of composition, but biological 'laws' range

---

[1] We take it to be an open question whether there exists a fundamental level or, alternatively, whether there is an infinite hierarchy of levels, where every entity is composed of more basic entities.

209

over entities at particular levels of composition. Levels of composition give us one kind of hierarchy; levels of scope give us another. That is, we can arrange scientific disciplines according to their levels of scope, that is, the generality of their laws (or of whatever plays the role of laws).[2]

We take this picture of levels of composition or organisation to be a picture that emerges from our best science. Scientists have revealed a world of entities of diverse sizes and complexity, ranging from fundamental particles, at one end of the spectrum, to superclusters of galaxies, at the other end, in terms of size, and from fundamental particles to brains and organisms, in terms of complexity. Thus, we take levels of composition and levels of scope, irrespective of the specific topology of the structure to which they give rise, to be an uncontroversial picture.

9.2   Levels of Composition in Biology

Levels of composition are, moreover, a typical way to understand talk of levels in science, and in particular in biology. In life sciences, one often talks about 'levels of organisation' or 'levels of complexity', with each level being characterised by its distinctive sets of entities and rules of operation. We have, for example, the genetic level, where one focuses on genes, how they are expressed and so on, and the anatomical level, a higher level that involves parts of organisms such as the brain and the spinal cord. Although the behaviour of the entities at some level depends on what happens at lower levels, biologists typically think that each level is characterised by its own principles not reducible to the levels below.

By way of illustration, consider the following quotation from a well-known textbook on developmental biology by Scott Gilbert:

> The properties of a system at any given level of organization cannot be totally explained by those of levels 'below' it. Thus, temperature is not a property of an atom, but a property that 'emerges' from an aggregate of atoms. Similarly, voltage potential is a property of a biological membrane

---

[2] These hierarchies need not give rise to a simple linear structure. In Oppenheim and Putnam's account, we have indeed a single hierarchy of levels of composition, with every level of composition corresponding to a scientific discipline for which it constitutes its subject matter. But the hierarchies can be part of a much more complex structure, such as the one given in Wimsatt (1976). In such more complex structures, we still have mereological and compositional relations with different entities being governed by different kinds of laws. So, even in such structures, which give a much more realistic picture than Oppenheim and Putnam's hierarchy, it is still possible to have levels of composition and levels of scope. Note also that even if we agree with new mechanists that the life sciences are structured around mechanisms rather than laws, it does not follow that there are no laws in biology; see, for example, Waters (1998) and Mitchell (2000) for accounts of biological laws.

but not of any of its components. Higher-level properties result from lower-level activities, but they must be understood in the context of the whole. (2010, 618)

Gilbert goes on to generalise this picture:

> Parts are organized into wholes, and these wholes are often components of larger wholes. Moreover, at each biological level there are appropriate rules, and one cannot necessarily 'reduce' all the properties of body tissues to atomic phenomena.... When you have an entity as complex as the cell, the fact that quarks have certain spins is irrelevant. This is not to say, however, that each level is independent of those 'below' it. To the contrary, laws at one level may be almost deterministically dependent on those at lower levels; but they may also be dependent on levels 'above'. (p. 620; emphasis added)

Gilbert (p. 620) quotes Joseph Needham (1943), who wrote:

> The deadlock [between mechanism and vitalism] is overcome when it is realized that every level of organization has its own regularities and principles, not reducible to those appropriate to lower levels of organization, nor applicable to higher levels, but at the same time in no way inscrutable or immune from scientific analysis and comprehension.

We see here, first, the notion of level developed above: a membrane is at a higher level of organisation than the molecules that compose it. Second, entities at different levels can have properties that none of its constituent parts has (e.g., temperature, voltage potential). Third, these higher-level properties are governed by a specific set of laws or rules and are taken to be in some sense irreducible to the levels below it. We will come back to this last point about the irreducibility of levels and to how it can be made more precise in Section 9.6. For now, the main point is to establish that the notion of levels of composition (or organisation) is an important notion in biology. Moreover, this fact can be used to argue that the world itself has a level-structure: levels of composition or organisation reflect how the world is organised. We take, then, the main argument in favour of the existence of levels of composition to be an a posteriori one: namely, this is how the world according to our best science is structured.

An important question here is whether this picture leads to some form of anti-reductionism or whether levels of composition are compatible with a reductionistic picture. Despite Gilbert's and Needham's apparent rejection of reductionism, we do not think that to accept that levels of organisation are real automatically leads to an acceptance of anti-reductionism or some form of ontological emergence. One can then be a realist about levels but remain agnostic about the ultimate truth of reductionism. Thus, Hilary Putnam and Paul Oppenheim wrote about ontological levels but at the same time adopted the micro-reduction of each whole at some ontological level to its constituents. More generally, the discovery that entities are typically composed of other entities does not of itself entail an account about the best way to understand the ontological nature of levels.

What then does the thesis of the reality of levels of composition amount to? What is the ontology that a realist about levels commits to, if not to some form of anti-reductionism? Our answer is that, even if the realist about levels can remain agnostic about the issue of reductionism versus anti-reductionism, the thesis is strong enough to rule out various ontological accounts. For example, vitalism – the thesis that some wholes have vital powers that do not in some sense depend on the constituents of the whole – and dualism are ruled out. In addition, the realist about levels takes it that wholes really exist and thus rejects an eliminativist stance about higher-level entities. Hence, although minimal as an ontological position, realism about levels has enough content to rule out some ontological accounts. At the same time, it leaves open the issue of exactly how properties of wholes depend on properties of their parts.[3]

9.3   Craver on Levels of Mechanisms

We take it to be a very plausible view, to say the least, that levels of composition constitute a typical sense of levels within life sciences. According to a very influential account by Craver (2007a), levels of composition in biology, and in particular in neuroscience, are to be viewed in terms of levels of mechanisms. Craver (2007a) has developed an account of multilevel explanation in neuroscience that is based on his account of what a mechanism is in biology. Below, after briefly presenting the main features of this account, we argue for an alternative conception of multilevel mechanistic explanation.

In explaining how we are to understand levels of mechanisms, Craver and Bechtel (2007, 548) write:

> In levels of mechanisms, an item X is at a lower level than an item S if and only if X is a component in the mechanism for some activity ψ of S. X is a

---

[3] See also Gillett (2016), who argues that the existence of compositional relations between wholes and their constituents leaves open the issue of reductionism versus emergence. Gillett views mechanistic explanation as a kind of compositional explanation, which is a view that we reject.

component in a mechanism if and only if it is one of the entities or activities organized such that S ψs.

Consider, for example, a dividing cell. In this case, the chromosomes and other entities that together compose the mechanism responsible for cell division are at a lower level than the cell. The activities of the components of the mechanism are in turn phenomena that can be accounted for in terms of lower-level components, and so on. We have thus a hierarchy of mechanisms that grounds a level-relation among all the components of mechanisms in a given hierarchy. These are what Craver calls 'levels of mechanisms'.

On this picture, therefore, the concepts of mechanism and levels are interrelated, in the sense that the same relation both underlies the level relation and is used to understand what a (constitutive) mechanism is. In contrast, on the picture sketched earlier, levels of composition are not viewed in terms of mechanistic levels. On the account we want to defend, in particular, levels and mechanisms are distinct notions, in the sense that there is no relation that underlies them both. This means that we do not think that a relation of constitutive relevance is needed in order to understand what a mechanism is as a concept-in-use; causation in terms of difference-making is enough.

We do not of course deny that there exist relations of composition. What we deny is that in order to understand what a mechanism is in the life sciences, we need to give an account of what this composition relation consists in. What we want to claim is the following: the typical sense of mechanism in the life sciences is Causal Mechanism, and the typical sense of levels is levels of composition. Thus, in order to understand what levels and what mechanisms are in the life sciences, we should keep these notions apart.

To further see the difference between levels of composition in our sense and levels of mechanisms, consider the following strange consequence of Craver and Bechtel's account: on that view a protein is not necessarily at a lower level than a cell. If entity X is not a component in a mechanism for some activity ψ of S, it is not at a lower level than S. But a particular protein may not be a component in the mechanism for some activity of a particular cell. Also, this account does not provide an answer about whether two adjacent cells, which are not components of the same mechanism, are at the same level or not. We find both of these consequences counterintuitive.[4] When biologists talk about the cellular level, for example, what they mean is the level of organisation that concerns cells and their activities, which is above the genetic level and below the anatomical

---

[4] Craver is aware of these consequences of his view; see Craver (2007a, 192–3).

Figure 9.1   A causal pathway with a single level of composition.

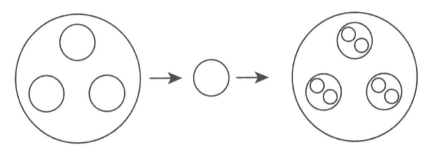

Figure 9.2   A causal pathway with multiple levels of composition.

level, for instance. In this picture, all cells, even if they are not part of the same mechanisms or wholes (because, e.g., some cells compose the muscle tissue, and other cells compose the neural tissue), are nevertheless at the same level.

We agree with new mechanists that many explanations in life sciences are multilevel. But since for us mechanistic constitution cannot be used to ground a hierarchy of levels of composition, we need now to explain how it is possible to have explanations spanning multiple levels if we accept Causal Mechanism.

9.4   Levels of Composition and Multilevel Mechanistic Explanation

How can we understand multilevel mechanistic explanation on our view? The key idea is the following: causal pathways may contain components from one level of composition only, but they may also contain components from multiple levels of composition (see Figures 9.1 and 9.2). In saying this we do not mean that causal relata in pathways are objects; components of mechanisms qua causal pathways and the entities that stand in compositional relations are different kinds of 'things'. Thus, components in a pathway are things that can stand in causal relations, for example, events or whatever one takes to be the relata of causal relations, and so typically not entities. When a pathway is represented using variables, we talk about causal relations among variables that are taken to correspond to properties of various entities. In contrast, it is typically entities that stand in compositional relations.[5] However, the causally relevant properties in a pathway are possessed by entities that can be at various compositional levels (e.g., membrane potential is a property of the membrane, which is composed of phospholipids and other molecules), and so we will say that in this case a pathway contains entities from various levels of composition.

We will say that whatever causally contributes to the phenomenon at hand is *part of the same pathway*, and in this sense it can be viewed as being at the same explanatory level, irrespective of its level of composition. To be at the same explanatory level is to be part of the same causal pathway. Explanatory levels are to be contrasted with compositional levels. Compositional levels are formed out of compositional or mereological relations. Explanations that describe causal pathways that contain entities from many compositional levels can then be viewed as multilevel explanations. Of course, the notion of 'level' in such multilevel explanations is the compositional notion; explanatorily, as we said, components of a pathway are at the same 'level'. But since, as we saw earlier, the language of levels in biology commonly refers to levels of composition or organisation, it makes sense to emphasise *compositional* levels and call these explanations '*multilevel*'.

A crucial point here is that the issue of how exactly to understand composition– that is, what it is for a set of entities to compose a whole, and what ontological view best accounts for compositional relations – does not matter for explaining how a phenomenon is brought about. This means that one can have an account of mechanisms as causal pathways and of multilevel explanation in our sense without being committed to reductionism or anti-reductionism. The reason is that causation in Causal Mechanism is viewed in terms of difference-making: *something is a component in a pathway if it makes a difference to the effect produced by the pathway*. Now, a whole can be viewed in reductionist terms, as not having any causal powers over and above the causal powers of its organised component parts; alternatively, it can be viewed in anti-reductionist terms as possessing causal powers over and above the causal powers of its organised components parts.[6] Either way, the whole (or the whole's properties) can act as a difference-maker and can thus be a component

[5] This is not accepted by all. Gillett (2016), for example, thinks that not only entities, but properties, powers and processes can stand in compositional relations too. In Craver's picture 'acting entities', that is, a mechanism's components, can stand in compositional relations to a higher-level acting entity, that is, the phenomenon for which the constitutive mechanism is responsible.

[6] This is just one way to view the difference between ontological reductionism and anti-reductionism (cf. Gillett 2016). It does not matter, for our argument, how exactly one understands the

in a causal pathway. We can thus understand its causal role in a pathway without being committed to a specific view about the ontology of composition. As in the case of the metaphysics of causation, this is another issue where a proponent of Causal Mechanism can remain agnostic.[7]

Let us now illustrate the idea that we can have causal pathways containing entities from several levels of composition by considering various examples. This will also establish that we find many multilevel mechanistic explanations in the life sciences in the sense recommended by CM.

9.5　Examples of Multilevel Mechanisms

We start with an example we have examined in detail, that is, the mechanism of apoptosis. We have seen in Chapter 3 that Kerr et al. described this mechanism at the cytological level. As they stressed, at the time it was not known which factors trigger apoptosis and what kind of cellular mechanisms are active before the morphological changes associated with apoptosis can be observed. The biochemical descriptions of the signalling pathways of apoptosis are exactly these cellular mechanisms that act as the trigger for the process described by Kerr et al.

What is the relationship between the pathway described at the biochemical level and the pathway described in cytological terms by Kerr et al.? The answer is that the biochemical pathway is a cause of the morphological changes. We could then combine the extrinsic pathway, for example, and the cytological mechanism as follows:

> Fas ligand binds to Fas receptor → adaptor protein binds to Fas receptor → procaspase-8 or 10 binds to adaptor protein → formation of DISC → activation of caspases-8 or 10 → caspases-8 or 10 activate effector caspases → destruction of proteins → condensation of nucleus and cytoplasm → budding → formation of apoptotic bodies → apoptotic bodies are phagocytosed.

disagreement between ontological reductionists and their opponents. The important point is that a whole can act as a difference-maker irrespective of the correct ontological account of composition.

[7] Agnosticism about the nature of composition is an important difference between Causal Mechanism and an account such as Craver's. In Craver's account, as we saw, the notions of a 'constitutive' mechanism and a multilevel mechanistic explanation both depend on the notion of constitutive relevance. Thus, an account of constitutive relevance is required. For us, such an account is not required and we can remain agnostic. One can here say that mutual manipulability is not an account about the nature of constitutive relevance, but is to be viewed in epistemic terms; but then the account seems incomplete. See also Glennan (2021, 11441), who argues that since 'objects are counted among the components of mechanisms ... an account of corporeal composition is required to properly elucidate mechanistic constitution'.

Here we have a causal pathway containing entities from various levels of composition. Part of the pathway is described in biochemical terms and thus refers to entities that belong to the biochemical level; and part of the pathway is described in cytological terms and thus refers to higher-level structures like the nucleus and the apoptotic bodies. This, then, is a case of a multilevel mechanism.

Another example of a multilevel mechanism is the pathway of visual transduction. When light enters the eye, it is focused by a lens on the retina. Light then activates rhodopsin, a receptor protein located in rod cells, which then activates G-proteins. G-proteins, in turn, activate PDE6 (a protein complex) which catalyses the conversion of cyclic GMP to GMP. This leads to the closing of sodium channels, the cell hyperpolarises and the voltage-gated calcium channels close. As a result, glutamate release drops, which leads to the depolarisation of on-centre bipolar cells. This activates ganglion cells that activate the optic nerve, which results in the receiving of the signal by the brain (cf. Tortora & Derrickson 2012). Here, we have a pathway that contains entities from various levels of organisation. It includes photons, the lens, various proteins, protein complexes, channels and cell membranes (since hyperpolarisation is a property of the cell membrane).

Our account of the visual transduction pathway can be contrasted with how Craver and Bechtel (2007) view this example, that is, in terms of levels of mechanisms, distinguishing between processes at higher and at lower levels. The transduction of light into 'a pattern of neural activities in the optic nerve' (p. 549) is the process at the highest level. This process is then decomposed in lower-level entities and their activities. The changes in rods and cones are components in a lower-level process, and the components that explain rod activation are at an even lower level. As they put it, '[e]ach new decomposition of a mechanism into its component parts reveals another lower-level mechanism until the mechanism bottoms out in items for which mechanistic decomposition is no longer possible' (p. 549). We think that our construal of the pathway is simpler, in that it shows how a process (i.e., visual transduction) is decomposed into sub-processes, without adopting the constitutive account of mechanism (and levels of mechanisms), which leads to positing non-causal constitutive relations between 'acting entities' (in Craver's sense) at different levels. For us, what decomposition amounts to is finding intermediate causal steps between an initial cause (e.g., the entering of light into the eye) and an effect (e.g., the activation of the optic nerve).

In Chapter 4 we examined the case of scurvy and we saw that scientists discovered the following causal pathway:

Citrus fruits → vitamin C → scurvy.

This is a very simplified description of the causal pathway, but the important point for our purposes is that this is again a multilevel causal pathway, as it contains entities from several levels of organisation. The outcome of the pathway in particular, that is, the development of scurvy, is something that concerns the whole organism, as the symptoms of scurvy include weakness, feeling tired, sore arms and legs, which are properties of the organism as a whole. Similarly, the dietary habits of the organism, in particular the presence or not of citrus fruits in the diet, is again something that has to do with the whole organism. In contrast, the mediator, that is, presence or not of vitamin C, concerns a lower level of organisation. A more complete description of the pathway will have to explain how lack of vitamin C disrupts various biosynthetic pathways such as the synthesis of collagen, dopamine, epinephrine and carnitine; these pathways concern the biochemical level of organisation. It will also have to mention how lack of vitamin C affects various tissues, such as skin, gums and bones, which concern a higher level of organisation. Thus, the pathway that leads to scurvy contains entities from several levels of organisation, as it describes entities at the biochemical level, at the level of tissues and at the level of the whole organism.

Like the pathway of scurvy, many pathological mechanisms include entities from various levels of organisation. A description of a pathological mechanism may mention entities at the levels of genes, biochemical pathways, cells, tissues, behaviour of organs and properties of the whole organism. Causes of a disease, in particular, include environmental factors like radiation and temperature extremes. Let us look at an example that exhibits many of these features: the mechanism of development of type 2 diabetes.

Diabetes is a syndrome characterised by hyperglycaemia (high blood sugar) due to deficiency of insulin (Bugianesi et al. 2005; Gardner & Shoback 2017). Insulin is one of the main hormones that regulate glucose homeostasis by stimulating glucose uptake. Main symptoms of type 2 diabetes are increased thirst and hunger, frequent urination, weight loss and feeling tired. Development of type 2 diabetes is due to insulin resistance (which means that cells do not respond normally to insulin) and due to a defect in pancreatic β-cells. Pancreatic β-cells release insulin as a response to increased levels in glucose in the blood. Normally, β-cells can compensate for insulin resistance by increasing insulin production, but defective β-cells cannot. This failure of β-cells is mainly due to underlying genetic factors. So, in type 2 diabetes glucose homeostasis is impaired. As diabetes progresses, various mechanisms may cause the function of β-cells to decline further, resulting in worsening hyperglycaemia.

Type 2 diabetes has a 'natural history' that progresses from an early stage with insulin resistance but no symptoms to a stage with mild hyperglycaemia and finally to a stage where pharmacological intervention is required. This natural history is roughly as follows. The development of insulin resistance is influenced by both genetic and environmental factors such as obesity and physical inactivity. There is an initial period before the development of type 2 diabetes, where insulin resistance leads to hyperinsulinemia, which means that the pancreatic β-cell can produce high levels of insulin and overcome insulin resistance, so that normal glucose homeostasis is maintained. But eventually defective β-cells in combination with insulin resistance cannot maintain glucose homeostasis; at this point, type 2 diabetes is diagnosed and can lead to microvascular and cardiovascular complications (e.g., diabetic kidney disease, retinopathy, coronary artery disease, stroke).

The pathophysiology of the disease includes three main defects, which are typically described at the level of organs and tissues and involve factors such as levels of glucose and insulin. These defects concern the activities of the pancreas, the liver, and the muscle and adipose tissues. The first defect concerns the pancreatic β-cells that, as we have seen, cannot produce enough insulin. The second defect is that due to the decrease in insulin and insulin resistance, glucose uptake at muscle and adipose tissues is decreased. The third defect is an increase in glucose production in the liver, which is normally regulated by insulin; due to the decrease in insulin, hepatic glucose overproduction can no longer be restrained. All three of these defects result in hyperglycaemia.

Many of the molecular mechanisms involved in type 2 diabetes are known. For example, the action of insulin in stimulating glucose uptake can briefly be described as follows: insulin binds to its receptor and activates a pathway that eventually allows glucose to enter the cell. The binding of insulin to its receptor located in the cell membrane causes tyrosine phosphorylation of insulin receptor substrate (IRS) proteins, which eventually causes glucose transporter 4 to be translocated to the cell membrane, where it allows glucose to enter the cell. Various defects in this pathway can lead to insulin resistance. In the muscle cell, for example, the cause of insulin resistance is serine rather than tyrosine phosphorylation of IRS proteins. In such a case, glucose is prevented from entering the cell.

The mechanism of development of type 2 diabetes, then, involves many organs and functions, many interrelated physiological and molecular mechanisms, is described at various levels of organisation and is characterised by a natural history.

Evolutionary mechanisms provide more examples of mechanisms that involve entities from various levels of composition. Evolutionary mechanisms such as natural selection have presented problems for new mechanists (as we will see in Chapter 10), since it is not evident what exactly counts as entities, activities and organisation in these mechanisms. Mechanism talk, however, is very common in evolutionary biology, so the question arises in what sense processes like natural selection and genetic drift are mechanisms. We think that CM provides an easy answer; the main reason, we think, to say that these processes are mechanisms is that they identify the causal steps of how evolutionary change comes about (we will come back to this point in Chapter 10). The important thing to note here is that descriptions of processes of natural selection can include genes and their frequencies in populations, traits of organisms, properties of the environment as well as properties of developmental systems, such as robustness, phenotypic plasticity and modularity, that is, entities from several levels of organisation.

As a last case, let us also briefly discuss a main example that Craver (2007a) examines: the multilevel mechanism of spatial memory. Craver takes this mechanism as having four main levels. The level of spatial memory is at the highest level; then, we have the level of spatial map formation, the cellular-electrophysiological level (which includes the mechanism of long-term potentiation) and, last, the molecular level that includes the molecular mechanisms (e.g., the function of NMDA receptors that underlie the chemical and electrical properties of nerve cells). Craver's general argument is that these four levels of spatial memory are more appropriately viewed in terms of levels of mechanism, and not in terms of other notions of levels, such as levels of size or of spatial containment. Craver, then, takes the mechanism of spatial memory to 'include NMDA receptors as components in LTP mechanisms, LTP as a component in a hippocampal spatial map mechanism, and spatial map formation as a component in a spatial memory mechanism' (p. 266), and he thinks experiments provide powerful evidence that 'the phenomena at each of these levels – NMDA receptor function, LTP, spatial map formation, and spatial memory – is constitutively relevant to the next' (p. 265).

For us, the relevance relations among all these components are causal relations: the function of NMDA receptors is causally relevant to the operation of LTP mechanisms, which are parts of the spatial map mechanism. When we go up a level, we consider a more extended causal network. The mechanism of spatial memory, the 'highest level' in Craver's account, can be viewed as the most extended causal network, which includes NDMA receptors as components. When we have such a network we can zoom in all the way to the molecular level, that is, to a size scale where we can identify the molecular details of the mechanism, or zoom out to the scale of neurons or brain regions.

Moreover, in the extended causal network we have entities from several levels of composition. This is important for making sense of interlevel experiments. For example, Craver mentions a bottom-up interlevel experiment where researchers knock out a gene that encodes a subunit of the NMDA receptor, so that knockout mice do not have functional NMDA receptors, and test their performance in the Morris water maze; the result is that knockout mice perform far worse than controls. Change in the performance of the knockout mice, which we take as the causal outcome of the bottom-up change, is a change in the behaviour of an entity at a higher level of organisation than the NMDA receptors, that is, an organism as a whole. Multilevel mechanisms containing components from various levels of composition, then, can account for interlevel experiments.[8]

9.6 Mechanisms and Interlevel Causation

We take the examples of mechanisms discussed in Section 9.5 to illustrate a main point of this chapter, that is, that mechanisms that include components from several levels of composition are very common. Scientists, then, commonly make causal claims that involve interlevel causes, which, however, are often viewed with suspicion by philosophers. In this last section we will argue that interlevel causation is unproblematic, contrasting again our account with Craver and Bechtel's analysis.

A difficulty of accepting interlevel causation is that things related by mereological relations are not distinct in a way that would enable them to stand in causal relations. This consideration is not a problem for the view that there exist multilevel causal pathways in our sense, since in several of the cases we have examined the components of the pathway do not stand

---

[8] For a critical examination of Craver's example of spatial memory, see also Eronen (2015). Eronen adopts what he calls a 'deflationary approach' to levels; he suggests that we should prefer more well-defined concepts, especially when we think about causation, such as scale and composition.

in mereological or compositional relations to each other. For example, the apoptosome is composed of several proteins, and is thus at a higher level of composition than, for example, an Apaf-1 protein (which is one of its components). But there is no difficulty in saying that the apoptosome is a component in a causal pathway that also includes entities at a lower level of composition as components. Such causal claims are unproblematic, as are claims that smaller entities can cause changes in bigger ones, and vice versa (however, in cases such as the example of type 2 diabetes, components of molecular pathways are spatiotemporally contained within the individual that develops type 2 diabetes; we will discuss this case below).

However, why posit an interlevel causal relation at all? Consider a causal pathway such as the case of scurvy. Here, the cause is at a higher level, since it concerns dietary habits, while the presence or absence of vitamin C concerns a lower level of organisation. But then one can argue as follows. When we have a case of such interlevel causal claims, what really happens is that the lower-level constituents of the putative higher-level cause do the real causal work, and thus we do not need to posit an interlevel causal relation. According to Craver and Bechtel, for example, there is no inter-level causation. They view cases that seem to involve interlevel causal relations either as cases of constitutive relevance, if the putative causal relation is between a component and the mechanism as a whole, or in terms of what they call 'mechanistically mediated effects'. These are 'hybrids of constitutive and causal relations in a mechanism, where the constitutive relations are interlevel, and the causal relations are exclusively intralevel' (2007, 547).

Take, for example, the claim that infection with a virus led to the death of the general; here, according to Craver and Bechtel, we have a causal claim that concerns how the virus interferes with various mechanisms in the organism, ultimately producing 'the physiological conditions that constitute the general's death' (p. 557). Similarly, in cases of putative top-down causes, the top-down cause is constituted by some mechanism, which then produces an outcome. This is why, then, cases of putative interlevel causation are described as 'hybrids', since 'the putative interlevel claim is analyzed into a causal claim coupled with one or more constituency claims' (p. 561).[9]

---

[9] Bechtel (2017) argues that the account in Craver and Bechtel (2007) in effect renders higher levels epiphenomenal, as it 'suggests a highly reductionistic picture of levels according to which causal relations that were supposed to be between entities at higher levels of organization dissolve into causal interactions at the lowest level considered' (p. 262).

We do not think that we need to adopt a hybrid picture such as the one suggested by Craver and Bechtel in order to make sense of interlevel causal relations in multilevel causal pathways. First, as we have argued in Chapter 8, we reject the constitutive account of mechanism that is presupposed in Craver and Bechtel's view. We think that mechanism as a concept-in-use in the life sciences is captured by CM and so it does not incorporate any compositional relations, just causation as difference-making. Second, we think that interlevel causal claims in science should be taken at face value and that we should try to develop a philosophical account that is adequate to capture these claims, rather than dismissing a literal construal of interlevel causal claims in part because of philosophical intuitions such as that wholes cannot causally influence their parts and vice versa. To return to the example of the mechanism of development of type 2 diabetes, to say that a defect in β-cells and insulin resistance cause type 2 diabetes or that type 2 diabetes can result in cardiovascular complications is to make causal claims (thus taking scientific talk of interlevel causation literally) and not to say something about the molecular mechanisms that underlie the natural history of type 2 diabetes and constitute the higher levels of mechanisms or organisation involved in this case. Third, to adopt interlevel causal claims where the causal relata involve properties of entities that are themselves (i.e., the entities) related as part and whole is not to say that parts cause wholes or vice versa in a synchronic, and thus problematic, manner. It is important here to take into account the temporal dimension, where, for example, defects in the insulin pathway and β-cells over time lead to changes that concern higher levels of organisation, irrespective of the fact that insulin pathways and β-cells are spatiotemporally contained in the organism.

We thus take interlevel causal relations to be conceptually coherent; moreover, we take it that interlevel causal claims are very often to be preferred to causal claims that involve only lower levels of organisation – for example, trying to couch all mechanistic causal explanations in molecular terms. This is because, as Gilbert put it, each level of organisation is in some sense irreducible to those below it. This can be made more precise by using Woodward's notion of conditional independence that we think captures part of what Gilbert means. Woodward (2020) defines conditional independence as follows. Suppose that we have a set of variables L that are causally related to E, but that we can use higher-level variables U that 'correspond to a coarsening of the L variables' and can be used to 'summarize the impact of the L variables on E'. This means that conditional on the values of U, further details about L won't matter – U 'screen off L from E. Variables L in this case 'are *independent* of E, *conditional* on U' (p. 428). As Woodward puts it, 'on this view of the matter, claims of downward causation (and claims of interlevel causation more generally) can be thought of as claims about the irrelevance of certain kinds of information conditional on other sorts of information – we can legitimately make claims of interlevel causation when such conditional irrelevance relations are present' (p. 444). For example, when we say, in the case of the Hodgkin-Huxley model, that the membrane potential V is a cause of the ionic currents and channel conductances, 'any further information about how that potential is realized in the electromagnetic forces associated with individual atoms and molecules does not matter' (p. 444) for the causal impact that V has on the other variables. Woodward's account then shows when we have reason to appeal to interlevel causation.[10]

In sum, then, causal pathways can be multilevel in the sense that they can contain entities from various levels of organisation. Interlevel causal relations are not problematic and, as we have seen, are ubiquitous in the life sciences. This account of multilevel mechanistic explanation is simpler than Craver's account, which relies on the notion of levels of mechanisms, as well as Craver and Bechtel's hybrid picture of interlevel causation, as we do not use any constitutive relations to give an account of what a mechanism is; the notion of a mechanism and the notion of levels (of composition) are for us distinct notions. Last, this account of multilevel mechanistic explanation can remain agnostic regarding the issue of how exactly composition is to be understood.

---

[10] Apart from conditional independence, another reason to appeal to interlevel causation, as Woodward notes, is that explanation in terms of lower-level causes is often computationally and epistemically intractable. It is important to note also that Woodward frames his analysis in terms of what he calls an 'interactionist' notion of levels that he takes to be conceptually distinct from the notion of levels of composition or size.

CHAPTER 10

Methodological Mechanism

10.1 Preliminaries

In the previous chapters we have defended what we have called *Causal Mechanism*, that is, the view that mechanisms (especially in the life sciences) are causal pathways that are described in theoretical language, where the pathway is underpinned by networks of difference-making relations. We have also characterised CM as metaphysically agnostic. This is an especially important feature of CM, since it differentiates it from what we have been calling inflationary accounts of mechanism, which we take all major accounts to be. In Chapters 8 and 9 we also argued that non-causal constitutive relations are not required to understand what a mechanism is in biological practice. CM is best seen in the context of a thesis that we call, following Woodger (1929) and Brandon (1984), *Methodological Mechanism* (MM). It will be the main aim of this chapter to develop and argue for MM as a general framework for understanding the search for mechanisms.

10.2 Methodological Mechanism: Historical Predecessors

10.2.1 Woodger on 'Methodological Mechanism'

Woodger (1929) distinguished between two ways in which a certain notion can be employed: a metaphysical or ontological way and a methodological one. The latter is when a notion is used for the purposes of description 'independently of its metaphysical interpretation'. In this case, Woodger says, the notion 'is employed methodologically, that is, simply for the purpose of investigation' (p. 31). The advantage of this use is that the notion can be used in a certain practice and cast light on it *independently* of whatever difficulties (and controversies) are raised by the intricate metaphysical debates concerning what its worldly reference is really like. In his discussion of mechanism, Woodger says that the notion of mechanism can be employed in precisely this methodological way, independently of how it is metaphysically interpreted.[1]

As a justification for this view, Woodger cites embryologist Gavin de Beer, who understands the mechanistic viewpoint in biology as follows:

> [The mechanistic point of view] in no way commits one to the 'materialistic' idea of life. Neither does it mean that life is nothing but physics and chemistry. What this point of view does stand for is that whatever the processes of life may be, they work in an orderly way, producing similar effects under similar conditions. *Steering between 'materialism' and 'vitalism', this conception has become known as mechanistic.* (quoted in Woodger 1929, 258; emphasis added)

For de Beer, then, the mechanistic viewpoint is not an ontological thesis, but concerns the nature of the object of study that must be presupposed for (mechanistic) science to be possible. In contrast to materialism and vitalism, it affirms no specific ontology for biology. Here is Woodger's own reading of this:

> It seems clear from this that all that is here meant by mechanism is the belief in the 'law of causation' or the 'uniformity of nature'. This is commonly regarded as a necessary methodological postulate of natural science.... Psychologists for example, speak of the 'mechanism of a neurosis' referring thereby I suppose to the ordered 'structure' of psychical processes. They assume that there is some such orderly structure and call it a mechanism without implying anything further about the ontological nature of those processes. *Thus the term may be used for the methodological postulate that there is some sort of order, and then it may be applied to that order itself.* (emphasis added)

He explains further:

> [For] de Beer, the term [mechanism] was used for a methodological postulate which asserts that the object of study is in some way orderly, and we saw how this term may then be extended to that which is thus assumed to be orderly but without any further metaphysical assumptions about its 'nature'. (pp. 259–60)

A 'mechanism' as a concept-in-use, then, according to Woodger's construal of de Beer's view, can be taken to mean an ordered causal

---

[1] Woodger (1929, chapter 5) distinguishes between four different senses of mechanism. These are (1) mechanism in the sense of classical mechanics, (2) mechanism as an explanation that uses only concepts from physics and chemistry, (3) mechanism in the sense of an analogy to a machine and (4) mechanism as a kind of a methodological postulate.

structure that scientists discover and that they describe in theoretical terms, without any further specification as to the fundamental nature of this order. The mechanistic point of view amounts to the methodological thesis that the aim of science is to discover such ordered causal structures. In the account developed in this book, instead of de Beer's ordered structures, we have causal pathways responsible for the phenomena. We can retain the idea of an ordered causal structure (except in indeterministic cases), adding that what makes a pathway *causal* are the difference-making relations among the components of the pathway.

According to Woodger, we need to 'ask the methodological mechanist what he has to say in support of his contention that the mechanical explanation is the only one which is admissible in science' (p. 231). Hence, Methodological Mechanism is a view about mechanistic explanation and its admissibility, and not about the blueprint of the universe. It's not about the metaphysics of mechanism, but about the use of the concept of mechanism in science and in particular about the importance of identifying causal pathways. In adopting this view, Woodger noted that mechanism is a 'methodological postulate' which as such 'makes no assertions about the nature of the processes studied, but merely asserts that they take place according to law, or "work in an orderly way"' (p. 258).

Taking a cue from Woodger's Methodological Mechanism, we want to claim that commitment to mechanism in science is adopting a *methodological postulate* which licenses looking for the causal pathways for the phenomena of interest. Hence, MM licenses adopting Causal Mechanism. CM, as we saw, allows for a rich understanding of the use of this concept in biology (and other sciences) without getting embroiled in a debate about what things in the world mechanisms *really* are and what kind of metaphysical categories their (theory-described) components fall into. Viewing mechanism as a methodological thesis allows that the sought-after identification of the causal pathway by which a specific result is produced is fully captured in the language of the specific theory, using deeply theory-laden concepts. It forfeits any further need, for the purposes of understanding how mechanisms explain, to offer a general metaphysical account of how the theory-described entities and processes – the *causal pathway* – fall into neat metaphysical categories.

It bears stressing that the key feature of MM is that it is non-committal about fundamental ontology. It adopts the postulate that scientists should always try to identify the way that a particular phenomenon is produced, but it says very little about how causation itself is to be understood: it asserts only what is required for making sense of the practice of looking for mechanisms. Hence, MM is philosophically neutral. But this does not mean that MM is scientifically neutral. Insofar as it is adopted it licenses mechanistic (and only mechanistic) explanations of the phenomena or the behaviours to be explained.

Though MM does not commit us to a specific view about how causation is to be understood from a metaphysical point of view (e.g., it need not commit itself to the view that interaction is the transmission of conserved quantities, etc.), MM can still clarify the close relations between causation, explanation and (the identification of) mechanisms: at least when there is no genuine indeterminism, whatever happens has a prior cause and identifying the way the cause brings about the effect is identifying the causal pathway by means of which the cause operates.

MM, we will argue below, illuminates practice in a way that ontologically inflated accounts of mechanism do not. It accounts for the centrality of mechanisms in scientific discovery and explanation, since according to it, discovering mechanisms (i.e., causal pathways) is the central task of science. At the same time, however, it refrains from imposing on scientific practice ontic constraints that are not licensed by it. According to MM, the mechanistic view need not be taken as something stronger than a certain methodological commitment to a kind of explanation.[2]

10.2.2   Brandon on 'Mechanism'

Brandon (1984) also argues for a methodological position he calls mechanism. He claims that mechanism, in this sense, should not be identified with reductionism, and that '[b]iological methodology is thoroughly mechanistic' (p. 348). He takes this position to have been described by Marjorie Grene (1971), when she writes:

> [L]et us look for a mechanism which might underlie the phenomena we hope to understand, seeking wherever we may relevant sources from which

---

[2] Consider also the following quotation by the zoologist Lancelot Hogben (as given by Brandon 1984), who views mechanism as a primarily epistemological view: '[I]n any discussion between the two [mechanist and holist or vitalist], the combatants are generally at cross purposes. The mechanist is primarily concerned with an epistemological issue. His critic has always an ontological axe to grind. The mechanist is concerned with how to proceed to a construction which will represent as much about the universe as human beings with their limited range of receptor organs can agree to accept. The vitalist or holist has an incorrigible urge to get behind the limitations of our receptor organs and discover what the universe is really like' (Hogben 1930, 100).

to derive, first, an analogue of a possible mechanism, and then, if we are shrewd and lucky and experience bears us out, maybe a description of the mechanism itself. (pp. 63–4)[3]

For Grene and Brandon, then, the central task of science is to search for mechanisms that produce the phenomena. This is the methodological position that Brandon calls 'mechanism'.[4] This leads to a mechanistic explanation that 'tells us how in fact those phenomena are produced' (Grene 1971, 64). But what is here meant by a mechanism? Here is Brandon again:

> Here I cannot be precise. Sometimes old-fashion spring-wound clocks and watches are called mechanical devices, in contrast modern battery-powered digital watches are called electronic devices. Clearly, I cannot use 'mechanism' in such a narrow sense. Mechanisms may consist of springs and gears, they may consist of computer chips and electrical pulses, they may consist of small peripheral populations and geographic isolating barriers. I cannot delimit all possible mechanisms because it is the business of science to discover the mechanisms of nature. At best I could list the sorts of mechanisms science, or more specifically, biology has discovered.... To model a process is to offer a more or less plausible hypothesis concerning the mechanism underlying the process. *Thus any process capable of being modelled is a mechanistic process.* (1984, 346, emphasis added)

Brandon suggests that the opposition between mechanism and various non-reductionistic ontologies, such as vitalism and holism, rests on the mistake that 'mechanistic methodology has been seen as implying (or somehow supporting) a reductionistic ontology' (p. 347). Brandon thinks, following Grene, that mechanism supports a multilevel ontology (Grene calls this 'level-pluralism').

Brandon (1990, 185) returns to the question 'What is a mechanism?' He says:

> A causal/mechanical explanation is one that explains the phenomenon of interest in terms of the mechanisms that produced the phenomenon. What is a mechanism?... [T]his question has no general metaphysical answer, because the business of science is the discovery of mechanisms; so we cannot delimit in any a priori manner the mechanisms of nature.... The best we can do is to give an open-ended answer: *a mechanism is any describable causal process.* (emphasis added)

---

[3] Grene follows Harré, saying that 'the central task of science in Harrean terms is the imaginative construction of theoretical models which suggest ways in which particular sets of phenomena may be produced'. For Harré's views on mechanism, see Chapter 5.

[4] Grene does not use 'mechanism' as a label for this methodological position.

We think that Brandon's position here can be generalised as follows: concepts such as *mechanism* that are central in scientific practice should be viewed as methodological postulates rather than as presupposing robust metaphysical commitments. But methodological postulates should be 'open-ended'; otherwise, they would unnecessarily limit research. This, we think, does not render MM a trivial thesis (we will return to the triviality objection in Section 10.6). Far from being a trivial commitment, MM is flexible enough to foster searching for mechanisms, whatever the ontic signature of the world might be.

Consider again Woodger's question: What has the methodological mechanist to say 'in support of his contention that the mechanical explanation is the only one which is admissible in science'? Although we do not think that mechanistic explanation is the only admissible form of explanation in biology and elsewhere, we agree with Woodger that this question should be answered.[5] We agree with Grene and Brandon, as well with the more recent mechanists, that the search for mechanisms is an important aim in science in general, and in life sciences in particular. For many new mechanists, however, this methodological norm is understood in terms of inflationary metaphysics. In what follows, we will argue that such accounts weaken the normative force of the main mechanistic methodological norm. We will argue that for the methodological norm to have its full force, mechanism should be understood along the lines of CM. We take this argument to constitute an answer to Woodger's question, in the sense that the argument will illuminate why the main mechanistic methodological norm has the place it has in current science. So, this argument will be a further reason to accept CM.

10.3   General Characterisations of Mechanism

10.3.1   Descriptive Adequacy

We take it that any general concept of mechanism adequate to the aims of many new mechanists has to satisfy at least two important adequacy conditions. First, the concept we seek to clarify must be central in scientific practice; second, it should be common across scientific fields. The first

---

[5] Here is Brandon's answer to this: 'I've argued that biological methodology is thoroughly mechanistic, but why should it be? As discussed in the first section of this paper, Grene has argued that mechanistic explanations give, or attempt to give, one an understanding of how the phenomena to be explained are really produced. I have nothing more to add to that discussion except to baldly state that that is what science, or more specifically, biology, ought to be up to. Thus, on this view not only is biological methodology mechanistic, it ought to be' (1984, 350).

condition stems from New Mechanism's central aim to give an account of science that is as close as possible to actual scientific practice. It says that the general concept has to feature in scientific practice and be in conformity with how scientists themselves use the concept, that is, to be a *concept-in-use*, as we like to call it. The second condition is justified given New Mechanism's aspiration to be a framework to be applied not only to a particular area within science, but to life sciences in general, as well as to many other scientific disciplines, from the social sciences to physics. The hypothesis that there exists such a concept common across diverse fields is explicitly endorsed by many new mechanists (see Illari & Williamson 2012; Glennan 2017).

To further appreciate the importance of this commonality condition, consider the trend in recent mechanistic literature to characterise in general terms various specific instances of mechanisms found in particular scientific fields (see, e.g., Glennan & Illari 2018a, chapters in part 4). Such more local accounts of mechanism cannot remove the need for searching for a characterisation of the general concept, even if there are more specific uses of this concept in particular cases. Clearly, one can always raise the question: *in virtue of what* are all those different species of mechanisms members of the same genus? What makes them all mechanisms? But then, a general characterisation of mechanism is still needed; such a characterisation would in that case be an explication of the general concept of which the various more particular kinds of mechanisms are instances.[6]

10.3.2 Normative Adequacy

General characterisations of mechanism have also to conform to a kind of normative adequacy condition. If we take mechanism to be a concept that is really present in actual scientific practice, then it has to have an influence in directing or regulating practice.[7] This regulating role can best be seen in

---

[6] Of course, a more radical pluralist stance is also available: perhaps there is no overarching concept of mechanism, but various distinct notions that have to be distinguished. Such a view, however, would undermine both the main working hypothesis of many new mechanists that there exists a general concept and, more importantly, the significance of the methodological and ontological theses of New Mechanism (see Sections 10.3 and 10.4).

[7] Our approach shares similarities with what Woodward (2015) calls a 'functional' project, which he contrasts with a metaphysical project, among others. He explains what a functional project is in the case of causation as follows: 'by a functional approach to causation, I have in mind an approach that takes as its point of departure the idea that causal information and reasoning are sometimes useful or functional in the sense of serving various goals and purposes that we have. It then proceeds by trying to understand and evaluate various forms of causal cognition in terms of how well they conduce to the achievement of these purposes' (pp. 693–4). In contrast to the metaphysical and other projects, the functional project has a 'normative or methodological dimension' (p. 694). Our examination of

the context of what we can call the *Methodological Tenet*, which can be formulated as follows:

(MT) Scientists, in investigating the phenomena, should search for mechanisms responsible for them.

MT has to be viewed in the context of the condition of descriptive adequacy mentioned earlier: since mechanism is taken to be a central concept-in-use active in diverse scientific fields, MT must be pervasive in science. MT is best seen as a methodological norm that guides scientific practice; it says what a main aim of science should be and how scientists should proceed in investigating the phenomena.

There are two points that should be stressed concerning MT. The first is that, as formulated, MT does not specify what a mechanism is. But without unpacking the meaning of 'mechanism', it is not clear what MT amounts to; unless we do this, its informational content remains unclear and thus MT is unhelpful as a guide for research. The following (incomplete) version of MT, then, needs to be fleshed out, by inserting a specific characterisation of mechanism:

(MT*) Scientists, in investigating the phenomena, should search for mechanisms responsible for them, where a mechanism is ⟨...⟩.

Second, MT being a thesis about methodology, one can ask whether there exists any general argument to the effect that searching for mechanisms must be a central aim of science (this is a version of Woodger's question). The more normative force we take MT to have, the more urgent this request is.

These two points are interrelated: the completion of MT* by providing a characterisation of mechanism must not be such as to weaken the normative force of MT. In that sense, MT constrains potential candidates for a general characterisation of mechanism. This, then, is a second kind of adequacy condition to any general account of mechanism, over and above its descriptive adequacy.

10.4 Inflationary New Mechanism

Most new mechanists describe mechanisms as things with a specific ontic signature; they put emphasis on the metaphysics of mechanisms.[8] Some,

---

mechanism as a concept-in-use can be viewed as a functional project, since we are interested in the role and usefulness of this concept within scientific practice.

[8] Glennan (2017) is a representative recent example; see also Krickel (2018).

admittedly, focus primarily on epistemology and methodology rather than ontology, analysing how scientists discover mechanisms and construct mechanistic explanations (Bechtel 2006; 2008; Craver 2007a; Craver and Darden 2013). However, some version of this ontological viewpoint is implied by all dominant formulations of the concept of mechanism qua concept-in-use: as we saw in Chapters 1 and 4, the general characterisations offered are formulated in terms that are more or less ontological. New mechanists thus share the view that there is a route that leads from the philosophical elucidation of practice to substantial conclusions about the blueprint of the world.

New Mechanism, then, combines two attitudes towards mechanism as a concept-in-use. On the one hand, the concept of mechanism is taken to play a central role in practice, in discovery and in explanation. On the other hand, it is taken to tell us something about the structure of the world: mechanisms are what science discovers, and they are taken to be the building blocks of reality. Let us call this second thesis the *Ontological Tenet*. Analogously to MT* we then have:

(OT*) The world consists of mechanisms, where a mechanism is ⟨...⟩.

We will now argue that putting these two tenets together leads to a tension which is detrimental to New Mechanism.

10.4.1 A Central Dilemma for New Mechanism

Here is how the dilemma arises. On the one hand, New Mechanism tries to flesh out MT* and OT* by means of the general characterisation of mechanism that it abstracted from scientific practice. Since OT is offered as a substantial ontological thesis about the world, the general characterisation must have sufficient content for OT to be able to function as such. Moreover, as the underlying motivation here is that metaphysical conclusions must be directly derived from practice, there is no independent source to provide content to flesh out OT*: all such content must be provided by the general characterisation of mechanism that is grounded in scientific practice.

On the other hand, the general characterisation of mechanism must be normatively adequate. So, it must be able to guide scientific practice in a way that the usefulness of MT as a methodological norm is maximised. But then, MT should avoid being overly specific, so that instead of regulating scientific practice it ends up constraining it. Now, if mechanism as a concept-in-use were to be formulated in very specific ontological terms so as to accurately describe the building blocks of a mechanistic world and thus to provide OT* with ontological oomph, the normative force of MT would be very weak; thus, the general characterisation would fail to be normatively adequate, which is an essential adequacy condition for any general account of mechanism.

Here, then, is how we can describe the unstable nature of the combination of MT and OT in general terms: the more content the general characterisation of mechanism has so that OT can be a robust ontological thesis about the world, the more this weakens MT; and the more defensible MT is as a central methodological maxim that regulates scientific practice, the more this weakens OT. The challenge, then, is to find a general characterisation of mechanism as a concept-in-use that can simultaneously satisfy both OT and MT.

10.4.2   The Central Dilemma in More Detail

Let us assume that one wants to provide that content to OT* so that it becomes a robust thesis about the world, by taking a mechanism to be a causal process that only involves material particles in motion, as seventeenth-century corpuscularians had hypothesised – for the sake of example, let us take these particles to be characterised only in terms of a typical seventeenth-century list of mechanical properties, for example, extension, shape, size, impenetrability and motion. Surely, as an ontological tenet, this is very informative: everything that exists in the world consists in matter describable in terms of a specific list of 'mechanical' properties. But, as our argument about Newton in Chapter 1 showed, such a construal of mechanism did in fact put limits on the science of the seventeenth century. One can view the criticisms against Newton from such figures as Leibniz to arise from a commitment to the ontological constraints that should regulate scientific explanation. Newton's project can similarly be seen as involving a liberation of scientific explanation from such ontological constraints. Note here that this liberation has been fully embraced by new mechanists, who have adopted a much more liberal construal of mechanism than their seventeenth-century predecessors.

This point can be generalised: if mechanism were to be explicated in strong reductionist terms, this would surely lead to a substantial OT but to a very weak MT. Why suppose, for example, that the only legitimate mechanistic explanations in biology are explanations in terms of what happens at the molecular level? As we have seen in Chapter 9, a causal pathway can involve entities from various levels of composition; such higher-level components, then, feature in legitimate mechanistic explanations and are causally relevant and even indispensable.

But does mechanism instead support anti-reductionism? Recall that methodological mechanists are not committed to a version of materialism, which is an ontological thesis, and similarly are not committed to some form of vitalism or holism, which are also typically viewed as ontological theses. Grene and Brandon, of course, as well as most new mechanists, view mechanism as an anti-reductionist position. We agree with new mechanists that mechanism leads to explanatory anti-reductionism. But we do not think that, ontologically, the mechanist needs to cling to a firm view on this matter. This is because mechanism as a concept-in-use does not involve any commitments about whether a whole is something over and above the properties of its parts and their organisation.[9] To view properties of wholes as being causally relevant and capable of featuring in causal pathways is not a reason to opt for ontological anti-reductionism. Indeed if, as Brandon insists, mechanism should not be opposed to ontological reductionism, similarly mechanism should not be viewed as ontological anti-reductionism; for then, it would indeed be correct to oppose it to ontological reductionism too. The proper attitude for a methodological mechanist, then, should be suspension of judgement regarding this issue.

Actually, we want to claim something stronger: mechanism as concept-in-use does not (and should not) involve a commitment to any kind of ontological theses that do not seem to have a function within scientific practice.

10.4.2.1   Inflationary Accounts Have Ontological Excess Content

Consider again the Newtonian move against New Mechanism that we described in Chapter 1. We argued that as old mechanists like Leibniz put ontic constraints to scientific explanation, new mechanists too put ontic constraints to current mechanistic explanation. They do this by requiring that a mechanistic explanation conform to a description of mechanism given in metaphysical terms. Glennan expresses a typical view among new mechanists when he says that a mechanistic explanation shows 'how the organized activities and interactions of some set of entities cause and

---

[9] It does not matter here how exactly ontic reductionism or anti-reductionism is construed; the point is that a methodological mechanist need not have a view about this ontological issue (see Gillett (2016) for a recent discussion about how the ontic versions of reductionism and anti-reductionism should be understood).

constitute the phenomenon to be explained' (2017, 223) and that it 'always involves characterizing the activities and interactions of a mechanism's parts' (p. 223). But to say all this is to subject mechanistic explanation to ontic constraints not warranted by scientific practice.

New mechanists motivate the inflationary accounts by arguing that they are descriptively adequate to capture specific cases of mechanisms in neurobiology and molecular biology. But certainly CM, which takes mechanisms to be theoretically described *pathways*, is descriptively adequate too (more on this in Section 10.5). Moreover, we take CM to be a very clear account of what a mechanism is, which is fully grounded in scientific practice. By contrast, the inflationary accounts give rise to a series of questions about how activities relate to entities, the metaphysics of causation or the nature of the constitution relation that are heavily debated among new mechanists. As the answers to these questions are important in order to understand the content of the inflationary accounts we think it is fair to say that the inflationary accounts are not as close to practice as CM. So, descriptive adequacy cannot be the reason why inflationary accounts are to be preferred.

10.4.2.2 Inflationary Accounts Can Constrain Practice

We in fact think that mechanism as a concept-in-use shouldn't be inflationary, the reason being that any ontological commitments that go further than those of CM would weaken the methodological tenet of mechanism. Consider again Old Mechanism; if mechanism were to be viewed in very strong reductionist terms, this would greatly weaken the regulative force of mechanism as a concept-in-use. But mechanism as a methodological tenet is taken by new mechanists to have a central place within scientific practice. Any very strong ontological commitments would count against this central place, as they would weaken MT.

Consider, for example, the MDC account or Minimal Mechanism: What possible reason could there be for insisting that all mechanistic explanations should be in terms of organised entities and activities? Even if this requirement is much more minimal and plausible than insisting that explanations should be couched in mechanical or physicochemical terms, it nevertheless puts a constraint on practice. There is, for example, much discussion among new mechanists about whether evolutionary mechanisms such as natural selection can be captured by the dominant general characterisations of mechanism. So, Robert Skipper and Roberta Millstein (2005) have argued that the mechanism of natural selection cannot be captured by the MDC and Glennan's earlier complex system account.

According to Skipper and Millstein (2005), natural selection lacks the decomposability and organisation that (early) Glennan and MDC view as requirements for something to be a mechanism. For example, they note that it is not clear whether the environment or perhaps some parts of the environment (and how many exactly?) should be viewed as parts of the mechanism of natural selection. Concerning the MDC account, they note that 'the activities of organisms do not have any particular temporal order', 'any particular rate' or a particular duration. In general, due to the variation that exists within populations they think that 'it is unlikely that natural selection has the degree of organization required by either MDC or Glennan' (p. 338). Prima facie at least, it seems that these points can be raised against Minimal Mechanism too. If natural selection and other population-level mechanisms are not mechanisms in the sense of Minimal Mechanism, then insisting that all mechanistic explanations should conform to Minimal Mechanism would be methodologically misleading.[10]

Another interesting case is the case of developmental mechanisms. In developmental mechanisms the constituents and organisation of the mechanism change in the course of the operation of the mechanism; moreover, developmental mechanisms involve constituents, such as morphogenetic fields, that are diffuse entities, unlike the discrete ones in typical examples of mechanisms given by new mechanists (see McManus 2012). It has yet to be shown that such cases can be captured by Minimal Mechanism or some other inflationary account. In addition, we take it that a lesson of our discussion of constitution in Chapter 8 is that insisting that (some) typical mechanisms in biology are to be understood in terms of Craver's account

---

[10] See Illari and Williamson (2011) for criticisms of Skipper and Millstein's argument. DesAutels (2018) thinks that the mechanism of natural selection can be captured by Minimal Mechanism. Skipper and Millstein view the mechanism of natural selection 'as a chain of temporal steps or stages' (2005, 329) that are causally connected, which is exactly how CM views mechanisms. Interestingly, as Newton's achievement can be seen as introducing a more liberal notion of mechanism, so Charles Darwin introduced a new kind of evolutionary explanation, that is, variational explanation as opposed to transformational explanations of evolution, as, for example, in the case of Lamarck's theory (the distinction is due to Richard Lewontin; see Sober (1984) for discussion). Relatedly, Ernst Mayr has argued that Darwin introduced a new way of thinking into biology, which he called 'population thinking' (1959). As in the case of Newton, Darwin's innovation was met with suspicion by some of his contemporary naturalists and philosophers. Arguably, Darwin's achievement can be interpreted as an introduction of a wholly new type of mechanism to explain evolutionary phenomena. (Darwin never uses the phrase 'mechanism of natural selection'; he usually calls it an 'action', a 'principle', and a 'process' (1859/1964). But such talk – especially talk of the 'process' of selection – can easily be interpreted in terms of CM.)

of constitutive mechanisms puts similar constraints on mechanistic explanation.

The point here is not that there is no way for new mechanists to make the modifications required to capture some of these cases, by clarifying, for example, how processes involving populations and interactions between organisms and environment are to be handled, or what exactly counts as a part. Rather, the point is that a characterisation of mechanism that is not flexible enough to easily accommodate new cases is not useful methodologically. Since we cannot know in advance what mechanisms there are in the world, our general characterisation has to be as open-ended as possible, in order not to constrain scientific practice.

10.4.2.3   Scientific Practice and Inflationary Accounts

Proponents of inflationary accounts might object that talk about entities and activities, for example, cannot function as a constraint on practice in the same way as Old Mechanism constrained seventeenth-century explanations of gravity. To say that mechanisms involve activities, one could argue, should be viewed more as a philosophical gloss on practice, rather than as the elucidation of a concept inherent in practice itself. More generally, new mechanists might object that the purpose of the general characterisation of mechanism is not to guide practice; rather, it is just an abstraction from typical and paradigmatic cases of mechanism and its main purpose is to clarify what a mechanism in general is.

Our answer is as follows: we are mainly interested in identifying a concept-in-use that has a regulatory role within practice. For us, then, the examination of particular cases to derive a general concept is a method to identify the concept-in-use.[11] Since, then, the aim is to understand a concept that functions within practice, the general characterisation cannot just be viewed as a philosophical gloss that is not necessarily relevant for practice, but should identify elements present in practice. And even if we were to agree that talk about activities in no way constrains scientific investigation, the problem of ontological excess content would remain: if talk about activities does not somehow constrain the form of a mechanistic explanation, this would most probably be because it is not really an actual element of practice, but ontological excess content. In view of the sufficiency of CM as an account of practice, this excess content can be omitted from the general characterisation.

---

[11] Note also that cases such as evolutionary and developmental mechanisms are problematic for new mechanists even if they have just the aim of clarifying what a mechanism is.

As we see it, this points to a dissimilarity between current inflationary accounts of mechanism and accounts such as Old Mechanism (or other ontologically inflated views of mechanism). For Boyle and Leibniz, mechanism was a concept of practice but was also viewed in ontological terms. However, in putting constraints on legitimate mechanistic explanations, the ontological content had a role within practice. For new mechanists who adopt inflationary accounts the situation seems different: on the one hand, they do not want to constrain practice; on the other, they give inflationary characterisations of a concept-in-use. But one cannot do both: if practice is not to be constrained, then the ontological content of mechanism should not have an important role within practice; but if it has no such role, it should not be viewed as an element of the concept-in-use.

10.5 Causal Mechanism as a Way Out of the Dilemma

We have seen that new mechanists use a general characterisation of mechanism to flesh out both OT* and MT* and that this leads to a dilemma, because these two theses pull in opposite directions. If we just abandon, or suitably modify, one of these two theses, the dilemma might be resolved. But which one? Since the starting point is to find a general characterisation of mechanism as a concept-in-use, it is clear that MT has priority. After all, one of the primary aims of New Mechanism is to give an account of the role of mechanism in scientific practice. So, the obvious solution is to abandon OT or to weaken it so as to be compatible with a robust version of MT. We can then add another kind of adequacy condition to the general characterisation of mechanism, apart from descriptive and normative adequacy: the general characterisation has to be as minimally ontologically committed as possible. Only such an account would be suitable for making MT as strong as possible. Let us call this *ontological adequacy*.

Causal Mechanism is an account that, as we will now argue, succeeds in being descriptively, normatively and ontologically adequate.

10.5.1   Descriptive Adequacy of CM

Recall that

(CM) A mechanism is a causal pathway, described in theoretical language.

We have already said a lot in the previous chapters about the descriptive adequacy of CM. CM captures a typical use of 'mechanism' in life sciences, which, as we have argued in Chapter 4, identifies a notion of mechanism that is (1) practice-based, (2) common across fields, (3) topic-neutral and (4) diversifiable; it can thus serve as a general characterisation of mechanism in the life sciences.

What CM purports to do is to find the *common denominator* of all uses of the term 'mechanism' in scientific contexts. So, CM is fully compatible with the possibility that in different fields, 'mechanism' can have more specific meanings. Still, CM gives the reason why we can regard all these more specific cases as being instances of a common general concept-in-use. We have then, as we saw in Chapter 3, the following schematic form for more specific kinds of mechanisms:

(P-CM) A mechanism is a causal pathway + X, where X is some external feature of the causal pathway.

The schematic form captures the requirement that it is scientific *practice* itself that will identify what further criteria a causal pathway must fulfil in order to count as a proper mechanism of a particular scientific field. So, an apparent pluralism in scientific practice in how the concept of mechanism is used does not count against CM as a general characterisation that captures a common notion underlying the more specific uses.

10.5.2   Ontological Adequacy of CM

But what exactly *is* a causal pathway? Can we offer CM without fleshing it out in terms, for example, of entities and interactions? But if we do this, we need also to further explain what the entities and interactions are supposed to be, and how they relate to each other. This is where the requirement that the causal pathway should be described in theoretical language comes in. There is simply no more informative and comprehensive way to describe a causal pathway than by reference to the relevant theoretical language. When one of us (S.P.) asked his doctor about the mechanism of Parkinson's, the answer was a description of a pathway. Parkinson's is an incurable progressive neurodegenerative disorder which, because of the depletion of dopamine in the brain, leads to severe motor and movement coordination malfunction and other effects (e.g., dysarthria, bradykinesia and others). As the disease progresses, patients eventually experience severe disability and sometimes dementia in later stages. Well, roughly put, the doctor said, when the supply of dopamine, a key neurotransmitter of motion-related signals, from the substantia nigra (a network of basal ganglia) to the striatum (the part of the brain with neurons that control and coordinate movement) is cut off (for reasons still not entirely clear), there follows a progressive loss of motor functions. Understanding even this rough sketch requires immersion into a theoretical language. It adds nothing by way of understanding to describe this in the language of the metaphysical theory of entities and activities. This sketchy account is a lot more informative than the even sketchier, because abstract and general, account of the new mechanist.

According to CM, the general characterisation is open-ended, in the sense that it avoids any commitment to a specific way to describe a mechanism. The description of the mechanism should in every case be specified in terms of the theoretical language of the appropriate scientific field (or fields). In other words, we should let practice itself decide what are the appropriate theoretical descriptions of a mechanism.

This point can be put as follows: the question 'what is a mechanism?' can be answered, first, by pointing to specific instances of mechanisms in the sciences. If it is asked what all these instances have in common, CM offers a general answer: they are all theoretically describable causal pathways. This is sufficient in order to answer the initial question; any more robust answer would amount to an ontologically inflated characterisation that would unjustifiably constrain scientific practice, a main aim of which is precisely the search for mechanisms.

However, this may still seem unsatisfactory; for, even if by adopting CM we seem not to be committed to any specific account about ontology or the metaphysics of causation, we are still claiming that the pathway is a *causal* one. But then, even if we abstain from saying anything more about what metaphysically grounds causal claims, should we not be committed to a general theory about causation that would distinguish causal pathways from mere sequences of events that are not causally related? This worry can be answered as follows: what is important for the identification of causal pathways in scientific practice is the identification of difference-making relations between the components of the pathway. The identification of the extrinsic pathway of apoptosis, for instance, required the specification of a series of steps, described in molecular terms, that form a causal pathway in the sense that each step makes a difference to what happens next: the initial signal (e.g., the binding of the Fas ligand of T-lymphocytes to the Fas receptor) leads to the activation of the FADD domain of the death receptor, which leads to the recruitment of an adaptor protein, which in turn causes procaspase-8 or 10 to bind to the adaptor protein, which leads to the formation of active caspases 8 and 10, which causes the activation of caspase 3 and the caspase cascade that causes apoptotic cell death.

Let us repeat a key point. While CM is agnostic regarding the metaphysics of causation and abstains from viewing mechanisms in terms of ontological categories, it does not follow that mechanisms qua causal pathways are not parts of the furniture of the world. While a thesis about ontology, this thesis is best viewed as part of a general realist stance concerning science, rather than as part of a new mechanical ontology that science has discovered.

10.5.3   Normative Adequacy of CM

We have identified MT as the main component of New Mechanism and offered CM as a general characterisation of mechanism. Two questions arise now. First, let us grant that explanations in terms of mechanisms are commonly offered in science. Should we say anything more than this, and in particular should we view the search for mechanisms in (strong) normative terms? Second, even if we accept MT, is this a reason to accept CM? MT and CM (or any general characterisation of mechanism) are logically independent theses: MT says nothing about what a mechanism is, just that science should try to discover them. So, acceptance of MT does not necessitate acceptance of CM. Why, then, not accept both MT and any of the existing general accounts of mechanisms? As should be evident by now, MT is best defended when understood in terms of CM. So, on the one hand, if we already accept MT, CM should be the preferred choice for a characterisation of mechanism. On the other hand, acceptance of CM allows us to understand how MT (even when interpreted in strong normative terms) is plausible in the first place.

The reason that explanations in terms of mechanisms are important is that discovering the causal pathways that in fact produce the phenomena that we investigate is one of the main aims in science. And to do that, what is required is to describe causal pathways in terms of the theoretical language of the particular scientific field (or fields) that studies the phenomenon of interest. MT* then becomes:

(MT) Scientists, in investigating the phenomena, should search for mechanisms responsible for them, where a mechanism is *a causal pathway described in theoretical language.*

10.6   The Triviality Problem

How informative, really, is MT understood in terms of CM? MT may seem trivial and even vacuous due precisely to the refusal of methodological mechanists to commit themselves to some robust version of OT. The underlying assumption behind this triviality problem, as we will call it, is that in order to have normative force, MT should exclude alternative methodological viewpoints. But if it remains ontologically non-committal, it seems that MT is too minimal to do this. The problem, then, is that by adopting CM as a characterisation of mechanism as concept-in-use, we in effect trivialise MT.

However, MT is far from a trivial thesis. It has both a negative and a positive role in guiding scientific practice. Historically, at least, there certainly have been alternative methodological viewpoints to mechanism. A central motivation for the introduction of the corpuscularian philosophy in the seventeenth century was precisely the presence of such an alternative methodology, namely, the explanation of phenomena in terms of substantial forms, Aristotelian powers and other unintelligible – from the point of view of old mechanists – metaphysical entities.[12] In early twentieth-century biology, vitalism also constituted an alternative non-mechanistic methodological viewpoint (see Allen 2005). Even if seventeenth-century Aristotelianism and early twentieth-century vitalism are usually viewed as ontological theses, they are also theses about biological methodology and the forms that biological explanation should take.

What is common among non-mechanistic viewpoints is their recourse to what seem for the point of view of mechanists as unintelligible notions (e.g., sui generis powers, substantial forms and entelechies) that incorporate some kind of teleology.[13] If, for example, an Aristotelian explains how X produces Y by saying that X has the power to produce Y without saying anything about the way (i.e., the causal pathway) that Y is in fact produced, or if a vitalist explains the course of development by postulating an entelechy, then, mechanists think, no explanation has really been given. What is missing is the means (i.e., the causal pathway) by which these powers and entelechies do their causal work.[14]

Note also that for MT to be informative it is not important that there actually exists a rival scientific tradition, as was the case in the seventeenth

---

[12] See our discussion of Boyle in Chapter 1.

[13] As Brandon (1984) stresses, there is a kind of teleology that is compatible with mechanism, namely, the teleology involved in evolutionary explanations of adaptations, where an adaptation is explained in terms of its effects on the organism that possesses it. Since such explanations involve the mechanism of natural selection, the teleology in question is not opposed to mechanistic methodology.

[14] Except, of course, in cases like fundamental physics, where there can be no mechanism that mediates between cause and effect. Note also that commitment to this explanatory norm is not to deny the existence of powers; powers may be a way to ground causation in a mechanism, but what explains a phenomenon is the mechanism itself.

century. Even in the absence of this, in regulating scientific practice, part of the role of MT is to block certain kinds of explanations among biologists or to guard against implicit assumptions. In the following quotation by the cell biologist Richard Lockshin we see this negative role of MT in action:

> a cell ... *neither plans its future nor considers its relationship to the organism.* In the *mechanistic view of cell biology,* biochemical and biophysical changes within the cytoplasm beget adjustments that activate autophagy, apoptosis, or other responses. (Lockshin 2016, 14; emphasis added)

We take the point that Lockshin makes here to be that the mechanistic viewpoint is a non-teleological viewpoint, in the sense that it does not allow explaining biological phenomena by attributing folk psychological properties to biological entities such as cells. Instead, what must be explained is how changes in the cell give rise to subsequent events; and to do this, the causal pathways that produce the phenomena must be identified. So, even if minimal as a methodological norm, the consequences of MT for scientific practice can be quite drastic.

Apart from its negative role in blocking certain kinds of explanation, there is also a positive role that MT has. Mechanisms offered in science are often incomplete, in the sense that various details of the causal pathway can be unknown. For example, when Kerr et al. (1972) first proposed the existence of apoptosis as a mechanism of cell death, the biochemical mechanisms responsible for the apoptotic process were completely unknown. In Chapter 3 (Section 3.5.2) we mentioned a vertical and a horizontal dimension that are involved in making theoretical descriptions more detailed. Take again the cytological description of apoptosis. The horizontal dimension involves identifying more details at the cytological level of organisation, whereas the vertical dimension involves identifying processes at lower levels of organisation. In instructing scientists to always identify causal pathways, MT directs them to fill in the missing details in incomplete descriptions of mechanisms.[15] To the extent that existing descriptions of mechanisms qua causal pathways can always be made more

---

[15] As we saw in the case of apoptosis, incomplete knowledge of the details of a causal pathway does not prevent scientists from identifying it and describing it as a new 'mechanism'. The importance of filling in the missing details in incomplete descriptions of mechanisms is a point that new mechanists have emphasised; see, for example, Darden's (2002) distinction between mechanism schemas and mechanism sketches (on schemas and sketches, see also Machamer et al. (2000); on various strategies for discovering mechanisms, see Craver and Darden (2013)).

detailed, this positive role of MT is pervasive in science. Note also that to take scientific practice seriously is not only to account for concepts such as mechanism that are central in practice but also to account for central methodological norms such as MT. By showing that MT is a compelling view, CM allows us to appreciate why searching for mechanisms is widespread in science in a way that ontologically inflated accounts cannot. Pervasiveness should not be confused with triviality. The fact that the mechanistic methodological viewpoint, in the sense we have been explicating it, is so widespread as to perhaps seem trivial cannot lead to a criticism of MT as a thesis that adequately describes what scientists are doing: the descriptive adequacy of a philosophical account of a piece of scientific methodology should certainly count in favour of the philosophical account in question. Last, if the accusation of triviality concerns the ontologically minimal account of mechanism that MT incorporates, then let us note once again that mechanism in science cannot mean anything more robust than CM if it is going to be a concept useful in practice: far from being a trivial thesis, MT advocates searching for mechanisms irrespective of what the fundamental ontological structure of the world really is.[16]

In sum: if we accept, together with new mechanists, that a common notion of mechanism is present across scientific fields, then CM is the best candidate for a general characterisation of this notion. CM captures the unifying character of the concept of mechanism; it is an account that remains close to practice, without incorporating elements that do not seem to be needed by practicing scientists (such as a commitment to powers or activities); at the same time, it is diversifiable, as it can easily be adapted to account for how the concept functions in specific scientific fields; in addition, in showing why MT is central in science, CM succeeds in being

---

[16] A possible worry here about CM is that it (and a fortiori MT) seems almost vacuous, since everything is (or has) a mechanism: Are there things-in-the-world that are not mechanisms? The worry, then, is that 'mechanism' becomes a concept devoid of real empirical content. Note, by way of reply, that this kind of worry can be effective, if at all, against 'thicker' accounts of mechanism too. It is not clear, for instance, what does not count as a mechanism on Glennan's Minimal Mechanism account – though on Glennan's earlier views there are non-mechanisms (only) at the level of fundamental physics. Be that as it may, our answer would simply be: something is not a mechanism in the CM sense if it is not a causal pathway. More importantly, however, the objection has a bite against MT only if MT is taken to be a metaphysical thesis, which it is not. As such, the proper contrast (as noted above) is not what-in-the-world-is-not-a-mechanism versus what-in-the-world-is-a-mechanism, but rather: Are there alternative methodological standpoints that explain non-mechanistically?

non-trivial and thus illuminating and informative both as a general account of practice as well as a concept central to scientific practice itself. Acceptance of CM resolves the dilemma faced by New Mechanism, since to accept CM means to abandon a robust version of the ontological tenet. New Mechanism as a framework to think about science is then best viewed as built around a primarily methodological thesis; New Mechanism is Methodological Mechanism.

Finale

The central question of this book has been: How should we characterise the concept of mechanism as a concept-in-use of scientific/biological practice? We have argued that the most appropriate general characterisation is what we have called Causal Mechanism. In this finale we examine to what extent CM is a properly 'mechanistic' thesis, that is, to what extent it can be seen as a descendant of the original notion of mechanism developed in seventeenth century. In so doing, we examine possible extensions of the seventeenth-century notion of mechanism and discuss whether they can be used to characterise mechanism as a concept-in-use.

Our strategy will be to examine how we can extend the original notion of mechanism by abstracting from the details of physical theory, while at the same time retaining enough content so that the resulting notions can be seen as more general notions of the same family of concepts. Following this strategy, we identify two central and independent conditions that a biological explanation has to satisfy in order to count as mechanistic, both of which were central in Old Mechanism: the condition of intelligibility (which says that a mechanistic explanation should not resort to unintelligible principles such as vital powers) and the condition of the priority of the parts over the whole. In order to clarify this second condition, we develop the notion of mechanistic reduction, which is motivated by the historical discussion of Chapter 2. We argue that mechanistic reduction entails causal modularity and thus that the failure of causal modularity is an indication that the parts have to be seen as in some sense dependent on the whole.

We use our two conditions to distinguish between two notions of mechanism: a more narrow one that incorporates both the intelligibility condition and the condition of the priority of the parts and a broader one that incorporates only the intelligibility condition and is thus a weakened form of mechanism. We claim that an account of mechanism as a concept-in-use requires the weakened notion, which when viewed in terms of CM has nevertheless enough content so that it can be seen as a descendant of the original concept.

Two Conditions for Mechanistic Explanation

As we have seen in Chapter 1, a central consideration in the context of Old Mechanism is what we can call the intelligibility argument. Here is how Leibniz puts it in his essay 'Against Barbaric Physics':

> That physics which explains everything in the nature of body through number, measure, weight, or size, shape and motion, which teaches that nothing is moved naturally except through contact and motion, and so teaches that, in physics, everything happens mechanically, that is, intelligibly, this physics seems excessively clear and easy. (1989, 312)

Here, Leibniz equates the mechanical with the intelligible. Indeed this argument, that is, that only an explanation in terms of the mechanical affections of parts of matter is intelligible, whereas an explanation in terms of scholastic powers, substantial forms and real qualities is obscure ('barbaric' according to Leibniz), is central in many thinkers of the seventeenth century. Descartes and Boyle, for example, both argue that only a mechanical explanation is intelligible. In Old Mechanism, then, explanations of phenomena in terms of sui generis powers of things are non-mechanical.

The condition of intelligibility is not just a negative thesis; that is, it does not just say what a mechanistic explanation should not do. It has some positive content too. This positive content can be seen very clearly in Boyle's writings examined in Chapter 1; a mechanistic explanation, for Boyle, has to state how exactly the cause operates to bring about the effect. In other words, in giving a mechanistic explanation we have to specify the causal steps (or the causal pathway) leading from an initial cause to an effect.

We can retain this condition on mechanistic explanation, which we will call the *condition of intelligibility*, even if we abandon the specific details of seventeenth-century mechanistic physics. So, we can expand the concept of mechanism to allow for chemical and other kinds of interactions among the components of a pathway. We thereby satisfy the spirit of Old Mechanism as long as in doing so we do not reintroduce the sui generis powers of scholastic Aristotelianism.[1]

---

[1] What is important here is that biological phenomena should not be explained in terms of illegitimate forms of teleology, which was a central aspect of scholastic powers; see our discussion in Chapter 10. The use of teleology and sui generis vital forces to contrast mechanical with non-mechanical

As we saw in Chapter 2, mechanistic explanation has historically been associated with a second idea, that is, that the properties and behaviours of the parts are in some sense independent of the whole, and so the properties and behaviour of the whole can be explained in terms of its parts. We have here a second condition for mechanistic explanation, that is, the priority of the parts over the whole. This form of explanation was central in Old Mechanism: for thinkers such as Boyle and Descartes, the capacity of a clock to tell time and the capacity of fire to burn wood are both explained in terms of the motions of the parts of matter that make up the clock or fire. This condition too need not be viewed just in the context of seventeenth-century physics.

The rejection of these conditions leads to non-mechanistic explanations. The rejection of the first condition, that is, the positing of sui generis vital powers and final causality, leads to vitalism. Such a position, for example, was adopted by Hans Driesch at the end of the nineteenth century, who posited what he called 'entelechies' that somehow regulate the operation and development of organisms. The rejection of the priority of the parts over the whole leads to holism (or 'organicism'). Twentieth-century holists argued that biological systems, despite being, in contrast to the vitalistic doctrine, 'mere' physicochemical systems, form a unified whole and thus are not 'mechanisms'. According to holists, higher biological levels of organisation are indispensable in giving biological explanations of phenomena such as reproduction, purposeful reaction to stimuli and complex self-regulation. As Allen puts it, '[i]t is in their appreciation of the concept that each level of organization in a complex system has its own special properties, and that these must by studied by techniques appropriate for that level, that holists differ in one significant way from Mechanists' (2005, 168).

We have thus identified two conditions that a biological explanation has to satisfy in order to count as mechanistic. These two conditions are logically independent. For example, holists adopted the first condition but not the second. Although vitalism as developed by Driesch rejects both conditions, it is also possible to think that the parts have priority over the whole, which can be explained in terms of them, but to accept that the parts have vital powers.[2]

---

explanation is justified historically; see our discussion of Kant's views and Broad's Substantial Vitalism in Chapter 2. Note also that current versions of neo-Aristotelianism are not at odds with the intelligibility condition, since they do not postulate sui generis vital powers.

[2] Bertoloni Meli gives the example of the eighteenth-century anatomist Xavier Bichat, who explained the activity of organs in terms of their component parts, but took it that those parts had vital

Given the above, we can ask: Do contemporary notions of mechanism satisfy these conditions? The first condition seems the most easy to satisfy: no biologist nowadays thinks that we need to introduce sui generis vital powers to understand the behaviour of biological systems. When we turn to the second condition, however, things are more difficult. Can a whole always be explained in terms of the properties of parts that are taken not to depend in some sense on the whole, or are there cases where the whole has some degree of autonomy with respect to the parts?

Causal Modularity

We can answer this question by considering the notion of causal modularity. According to some philosophers, for an explanation to be mechanistic, or for a system to be regarded as a mechanism, it should have a modular structure: it should be in principle possible to change a particular causal relationship within the system (e.g., by removing one of the components of the system) without changing other causal relationships in the system (see Woodward 2002; Menzies 2012). This kind of modularity as 'independent disruptability' (Hausman & Woodward 1999) makes possible the decomposition of the whole effect of a causal system into the independent causal contributions of its constituents. Modularity is thought to capture the sense in which the behaviour of the components of the mechanism are independent of the behaviour of the mechanism as a whole. The idea is that only if this independence obtains can we explain the whole in terms of the parts. Modularity has thus been viewed as a necessary condition for mechanistic explanation.

Woodward, in particular, takes mechanistic explanation to be a species of causal explanation. He formulates modularity within his interventionist framework of causation, which we discussed in Chapter 7. A mechanism is

properties not found in non-living matter. Bertoloni Meli uses the example of Bichat to criticise Bechtel's claim that 'Bichat was pursuing a program of mechanistic explanation' (Bechtel 2006, 45). He stresses that in fact Bichat 'actively opposed the mechanistic program because he deemed it erroneous'. He notes that 'if the defining feature of a mechanism is that it operates "in virtue of its component parts" [as Bechtel argues], Bechtel should argue that Aristotle and Galen too, despite their teleology, in crucial respects were "pursuing a program of mechanistic explanation" because they "attempted to explicate the behavior" of bodies in terms of the organs "out of which they were constructed"' (Bertoloni Meli 2019, 6). Although Bertoloni Meli makes this point in the context of arguing that an approach such as Bechtel's 'may be adequate for systematic concerns and analyses of the role of mechanism in biology, more sophisticated tools are needed for a meaningful historical analysis' (p. 6), we take the point to be that an adequate characterisation of mechanism and mechanistic explanation has to incorporate both the condition of intelligibility and the condition of the priority of parts.

a set of components, which are characterised by variables that stand in causal dependence relations to each other. Woodward takes a representation of these variables standing in causal relations to be modular 'to the extent that each of the individual G_i [generalisations that describe the causal relationships among the components of the system] remain at least somewhat stable under interventions that change the other G_i' (Woodward 2013, 51). Woodward thinks that the requirement that causal relationships within the mechanism have to be modular captures the sense in which a mechanistic explanation explains the behaviour of the whole in terms of the intrinsic behaviour of the parts. He cites biologists George von Dassow and Ed Munro, who claim that:

> Mechanism, per se, is an explanatory mode in which we describe what are the parts, how they behave intrinsically, and how those intrinsic behaviors of parts are coupled to each other to produce the behavior of the whole. This common sense definition of mechanism implies an inherently hierarchical decomposition; having identified a part with its own intrinsic behavior, that part may in turn be treated as a whole to be explained. (von Dassow & Munro 1999, 309)

According to Woodward, in a case of a modular system:

> each such subset of causally related components continues to be governed by the same set of causal relationships, independently of what may be happening with components outside of that subset, so that the behaviour is (in this respect) 'intrinsic' to that subset of components ... modularity seems to me to capture at least part of what is involved in their notion of 'intrinsicness'. (2013, 51)

Note here exactly what modularity requires. It does not require that, were we to intervene to disrupt a causal relationship between a pair of variables, the mechanism would produce the same result as before the intervention; in the case of such a disruption, the production of the outcome of the mechanism would be disrupted too. The claim, rather, is that the disruption of a causal dependence, for example, by removing a component, would not disrupt what other causal dependencies there are in the system.

The important point for our discussion is that mechanistic reduction implies modularity. That is, if modularity captures the sense of 'intrinsicness', as Woodward claims, which is the central feature of mechanistic reduction as explained above, then, if we have a system such as a mechanical clock where organisation just concerns the spatial relations among independently existing components with intrinsic properties, such a system is modular. But then, if a system fails to be modular, this means that mechanistic reduction fails too.

Many biological systems do not satisfy the condition of modularity as independent disruptability (see Mitchell 2008).[3] In such 'robust' systems, a change in some causal interactions within the system may lead to a restructuring of the system. So, failure of modularity shows that the nature of biological components is not entirely independent of the whole in which they are located. Rather, some of their properties are dependent on their being components of a certain whole. As Woodward puts it, in non-modular systems 'how some of the components behave depends in a global, "extrinsic" way on what is going on in other components' (2013, 54).

Moreover, parts of biological systems can depend on the whole for their existence. For example, the cell regulates various parameters that are necessary for the operation of cellular mechanisms. Thus, the existence of these mechanisms depends upon the existence of a properly functioning cell. In general, organisation in biological systems does not just concern a network of spatial relations, as in the case of a mechanical clock. In the case of a cell, organisation is not imposed on the mechanism from 'outside', so to speak, but is itself in part the result of the operation of the mechanism.[4]

Causal Mechanism Is Mechanism Enough

Some of the notions we have examined in Chapters 1 and 2 satisfy both conditions of intelligibility and priority of parts, that is, Cartesian mechanism, mechanical mechanism (i.e., the more liberal post-Newtonian notion), Ewing's quasi-mechanical mechanism and Broad's Pure Mechanism and Biological Mechanism. Quasi-mechanical and Biological Mechanism are here the more general notions, that is, the ones that do not make any specific assumptions about the underlying physics or chemistry. But can such concepts capture the mechanism talk that is ubiquitous in life sciences?

---

[3] More generally, according to Woodward, 'many biological systems require explanations that are relatively non-mechanical or depart from expectations one associates with the behaviour of machines' (2013, 39), for example, dynamical systems explanations.

[4] Biological systems such as cells and organisms are what have been called 'autonomous' systems. An autonomous system is 'a far-from-equilibrium system that constitutes and maintains itself establishing an organisational identity of its own, a functionally integrated (homeostatic and active) unit based on a set of endergonic-exergonic couplings between internal self-constructing processes, as well as with other processes of interaction with its environment' (Ruiz-Mirazo et al. 2004).

Mechanism as concept-in-use seems to be a very liberal notion, applied to all sorts of systems irrespective of their causal organisation. For example, Lakhani et al. (2009, 4) write about the 'strongly biomedical concept of disease' they adopt, that it is 'a mechanistic model that regards the body as a machine with repairable or replaceable parts. It looks for specific underlying biological causes and places a high emphasis on the scientific evidence-base for untangling cause and effect in both the disease and its treatment, because this is important for patient care and prognosis' (p. 4). But although they adopt a mechanistic model, 'it is a complex model with multiple parts that interconnect. A change in one area is likely to affect another. Thus maintaining homeostasis is not a simple single feedback loop and it is perfectly acceptable that a new equilibrium is achieved under a new set of circumstances, a new baseline; you do not have to return to the original state' (p. 4). On this account, to adopt a 'mechanistic model' concerning the body is just to say that the body contains various interacting parts that underlie bodily functions and sometimes result in diseases, such that it is possible to identify cause and effect relationships and to intervene to treat parts that malfunction.[5]

A way to capture the more liberal notion of mechanism is to reject the condition of the priority of the parts. So we are left only with the intelligibility condition. Would that be enough for mechanism? One might be tempted to add the claim that though modularity (or priority of the parts) doesn't hold, something near it does, that an activity of a whole is explained in terms of the organised entities and activities that constitute it. However, this leads to a form of mechanism where a whole is explained in terms of its parts, but the parts may not be independent of the whole. The main problem here is that we lose the clear contrast with what a non-mechanistic explanation would be. Broad's Substantial Vitalism does certainly count as a non-mechanistic position on this view, but what about Emergent Vitalism, or the holist and organicist positions of the early twentieth century? If these are to be counted as mechanistic too, why use the term 'mechanism' at all? There is here the danger that the concept of mechanism may become vacuous. In addition, it becomes difficult to view this weakened notion as a descendant of the original notion of mechanism.

Our alternative is to view mechanism as combining intelligibility with Causal Mechanism. As we view mechanism as a concept-in-use of

[5] For a detailed examination and criticism of the claim that the cell can be seen as a machine, see Nicholson (2019).

biological practice, the aim of this practice is not to explain a whole in terms of its components but to trace the causal steps, that is, to identify the causal pathway, leading from an initial cause to an effect.[6] The centrality of the concept in practice is an indication that the notion is not vacuous, but has an important methodological role (as we have argued in Chapter 10). Adopting CM fully embraces the processual character of biological mechanisms; moreover, it does not blur the distinction between mechanism and holist or emergentist positions; last, it can easily be viewed as a descendant of the original seventeenth-century concept of mechanism. Causal Mechanism is Mechanism Enough!

[6] Nicholson's (2011) analysis of the concept of mechanism in biology shares some similarities with our approach. Nicholson distinguishes between what he calls 'machine mechanism', which concerns 'the internal workings of a machine-like structure', and 'causal mechanism', which concerns 'the causal explanation of a particular phenomenon'. He claims that new mechanists like Craver, Darden and Bechtel conflate these two senses of mechanism and 'inappropriately endow causal mechanisms with the ontic status of machine mechanisms, and this invariably results in problematic accounts of the role played by mechanism-talk in scientific practice' (p. 152).

References

Alberts, B., Johnson, A., Lewis, J., Raff, M., Roberts, K. & Walters, P. (2014). *Molecular Biology of the Cell*, 6th ed. New York: Garland Science.

Allen, G. E. (2005). Mechanism, vitalism and organicism in late nineteenth and twentieth-century biology: the importance of historical context. *Studies in History and Philosophy of Biological and Biomedical Sciences*, 36, 261–83.

Andersen, H. (2012). The case for regularity in mechanistic causal explanation. *Synthese*, 189, 415–32.

(2014a). A field guide to mechanisms: part I. *Philosophy Compass*, 9, 274–83.

(2014b). A field guide to mechanisms: part II. *Philosophy Compass*, 9, 284–93.

Armstrong, D. M. (1983).*What Is a Law of Nature?* Cambridge: Cambridge University Press.

(1997). Singular causation and laws of nature. In J. Earman and J. Norton, eds., *The Cosmos of Science*. Pittsburgh, PA: University of Pittsburgh Press, pp. 498–511.

Baetu, T. M. (2019). *Mechanisms in Molecular Biology*, Cambridge Elements. Cambridge: Cambridge University Press.

Baron, R. M. & Kenny, D. A. (1986). The moderator-mediator variable distinction in social psychological research: conceptual, strategic, and statistical considerations. *Journal of Personality and Social Psychology*, 51, 1173–82.

Bartholomew, M. (2002). James Lind's treatise of the scurvy (1753). *Postgraduate Medical Journal*, 78, 695–6.

Baumgartner, M. & Casini, L. (2017). An abductive theory of constitution. *Philosophy of Science*, 84, 214–33.

Baumgartner, M. & Gebharter, A. (2016). Constitutive relevance, mutual manipulability, and fat-handedness. *The British Journal for the Philosophy of Science*, 67, 731–56.

Bechtel, W. (2006). *Discovering Cell Mechanisms: The Creation of Modern Cell Biology*. Cambridge: Cambridge University Press.

(2008). *Mental Mechanisms: Philosophical Perspectives on Cognitive Neuroscience*. New York: Routledge.

(2017). Explicating top-down causation using networks and dynamics. *Philosophy of Science*, 84, 253–74.

Bechtel, W. & Abrahamsen, A. (2005). Explanation: a mechanistic alternative. *Studies in History and Philosophy of Biological and Biomedical Sciences*, 36, 421–41.

Bechtel, W. & Richardson, R. C. (2010) [1993]. *Discovering Complexity: Decomposition and Localization as Strategies in Scientific Research*, 2nd ed. Cambridge, MA: MIT Press/Bradford Books.

Beiser, F. (2005). *Hegel*. New York: Routledge.

Bennett, J. (2003). *A Philosophical Guide to Conditionals*. Oxford: Oxford University Press.

Bertoloni Meli, D. (2019). *Mechanism: A Visual, Lexical, and Conceptual History*. Pittsburgh, PA: University of Pittsburgh Press.

Bird, A. (2007). *Nature's Metaphysics: Laws and Properties*. Oxford: Oxford University Press.

Boas, M. (1952). The establishment of the mechanical philosophy. *Osiris*, 10, 412–541.

Bogen, J. (2005). Regularities and causality; generalizations and causal explanations. *Studies in History and Philosophy of Biological and Biomedical Sciences*, 36, 397–420.

Boniolo, G. & Campaner, R. (2018). Molecular pathways and the contextual explanation of molecular functions. *Biology and Philosophy*, 33, 24. https://doi.org/10.1007/s10539-018-9634-2.

Bork, A. (1967). Maxwell and the electromagnetic wave equation. *American Journal of Physics*, 35, 83–9.

Boyle, R. (1991). *Selected Philosophical Papers of Robert Boyle*, ed. M. A. Stewart. Indianapolis, IN: Hackett Publishing Company.

Brandon, R. N. (1984). Grene on mechanism and reductionism: more than just a side issue. In P. Asquith and P. Kitcher, eds., *PSA: Proceedings of the Biennial Meeting of the Philosophy of Science Association*, vol. 2. East Lansing, MI: Philosophy of Science Association, pp. 345–53.

(1990). *Adaptation and Environment*. Princeton, NJ: Princeton University Press.

Breitenbach, A. (2006). Mechanical explanation of nature and its limits in Kant's *Critique of Judgement*. *Studies in History and Philosophy of Biological and Biomedical Sciences*, 37, 694–711.

Broad, C. D. (1925). *Mind and Its Place in Nature*. London: Routledge and Kegan Paul.

Brown, S. R. (2003). *Scurvy: How a Surgeon, a Mariner, and a Gentleman Solved the Greatest Medical Mystery of the Age of Sail*. Chichester: Summersdale Publishers.

Bugianesi, E., McCullough, A. & Marchesini, G. (2005). Insulin resistance: a metabolic pathway to chronic liver disease. *Hepatology*, 42, 987–1000.

Cairrão F. & Domingos, P. M. (2010). Apoptosis: molecular mechanisms. In *Encyclopedia of Life Sciences*. Chichester: John Wiley & Sons.

Campaner, R. (2006). Mechanisms and counterfactuals: a different glimpse of the (secret?) connexion. *Philosophica*, 77, 15–44.

Carpenter, K. J. (2012). The discovery of vitamin C. *Annals of Nutrition and Metabolism*, 61, 259–64.

Cartwright, N. D. (1989). *Nature's Capacities and Their Measurement*. Oxford: Clarendon Press.

Casini, L. (2016). Can interventions rescue Glennan's mechanistic account of causality? *British Journal for the Philosophy of Science*, 67, 1155–83.

Chisholm, R. (1946). The contrary-to-fact conditionals. *Mind*, 55, 289–307.

Clarke, B., Gillies, D., Illari, P., Russo, F. & Williamson, J. (2014). Mechanisms and the evidence hierarchy. *Topoi*, 33, 339–60.

Clarke, P. G. H. & Clarke, S. (1996). Nineteenth century research on naturally occurring cell death and related phenomena. *Anatomy and Embryology*, 193, 81–99.

Couch, M. B. (2011). Mechanisms and constitutive relevance. *Synthese*, 183, 375–88.

Cox, D. R. (1986). Comment. *Journal of the American Statistical Association*, 81, 963–4.

(1992). Causality: some statistical aspects. *Journal of the Royal Statistical Society Series A*, 155, 291–301.

Cox, D. R & Wermuth, N. (2001). Some statistical aspects of causality. *European Sociological Review*, 17, 65–74.

Craver, C. F. (2001). Role functions, mechanisms and hierarchy. *Philosophy of Science*, 68, 31–55.

(2007a). *Explaining the Brain: Mechanisms and the Mosaic Unity of Neuroscience*. Oxford: Oxford University Press.

(2007b). Constitutive explanatory relevance. *Journal of Philosophical Research*, 32, 3–20.

(2013). Functions and mechanisms: a perspectivalist view. In P. Huneman, ed., *Functions: Selection and Mechanisms*. Dordrecht: Springer, pp. 133–58.

Craver, C. F. & Bechtel, W. (2007). Top-down causation without top-down causes. *Biology and Philosophy*, 22, 547–63.

Craver, C. F. & Darden, L. (2005). Introduction. *Studies in History and Philosophy of Biological and Biomedical Sciences*, 36, 233–44.

(2013). *In Search of Mechanisms: Discoveries across the Life Sciences*. Chicago: University of Chicago Press.

Craver C. & Tabery, J. (2015). Mechanisms in science. In E. N. Zalta, ed., *The Stanford Encyclopedia of Philosophy* (Summer 2019 edition), https://plato.stanford.edu/archives/sum2019/entries/science-mechanisms/.

Cummins, R. (1975). Functional analysis. *Journal of Philosophy*, 72, 741–64.

Darden, L. (2002). Strategies for discovering mechanisms: schema instantiation, modular subassembly, forward/backward chaining. *Philosophy of Science*, 69, S354–S365.

(2006). *Reasoning in Biological Discoveries*. Cambridge: Cambridge University Press.

Darwin, C. (1859/1964). *Origin of Species*. Cambridge, MA: Harvard University Press.

Dawid, P. (2000). Causal inference without counterfactuals. *Journal of the American Statistical Association*, 95, 407–24.

DesAutels, L. (2011). Against regular and irregular characterizations of mechanisms. *Philosophy of Science*, 78, 914–25.

(2018). Mechanisms in evolutionary biology. In S. Glennan and P. Illari, eds., *The Routledge Handbook of Mechanisms and Mechanical Philosophy*. New York: Routledge, pp. 296–307.

Descartes, R. (1982). *Principles of Philosophy*, trans. V. R. Miller and R. P. Miller. Dordrecht: D. Reidel Publishing Company.

(2004). *René Descartes: The World and Other Writings*, ed. S. Gaukroger. Cambridge: Cambridge University Press.

Dowe, P. (2000). *Physical Causation*. Cambridge: Cambridge University Press.

Dretske, F. I. (1977). Laws of nature. *Philosophy of Science*, 44, 248–68.

Ellis, B. (2001). *Scientific Essentialism*. Cambridge: Cambridge University Press.

Ellis, H. M. & Horvitz, H. R. (1986). Genetic control of programmed cell death in the nematode C. elegans. *Cell*, 44, 817–29.

Eronen, M. I. (2015). Levels of organization: a deflationary account. *Biology and Philosophy*, 30, 39–58.

Eronen, M. I. & Brooks, D. S. (2018). Levels of organization in biology. In E. N. Zalta, ed., *The Stanford Encyclopedia of Philosophy* (Spring 2018 edition), https://plato.stanford.edu/archives/spr2018/entries/levels-org-biology/.

Ewing, A. C. (1969). *Kant's Treatment of Causality*. Hamden, CT: Archon Books.

Fink, S. L. & Cookson, B. T. (2005). Apoptosis, pyroptosis, and necrosis: mechanistic description of dead and dying eukaryotic cells. *Infection and Immunity*, 73, 1907–16.

Franklin-Hall, L. (2016). New mechanistic explanation and the need for explanatory constraints. In K. Aizawa and C. Gillett, eds., *Scientific Composition and Metaphysical Ground*. London: Palgrave Macmillan, pp. 41–74.

Funk, C. (1912). The etiology of the deficiency diseases. *The Journal of State Medicine*, 20, 341–68.

Gardner, D. G. & Shoback, D. (2017). *Greenspan's Basic & Clinical Endocrinology*, 10th ed. New York: McGraw-Hill.

Garson, J. (2013). The functional sense of mechanism. *Philosophy of Science*, 80, 317–33.

(2018). Mechanisms, phenomena, and functions. In S. Glennan and P. Illari, eds., *The Routledge Handbook of Mechanisms and Mechanical Philosophy*. New York: Routledge, pp. 104–15.

Gilbert, S. F. (2010). *Developmental Biology*, 9th ed. Sunderland, MA: Sinauer Associates.

Gillett, C. (2006). The metaphysics of mechanisms and the challenge of the new reductionism. In M. Schouten and H. L. de Jong, eds., *The Matter of the Mind*. Oxford: Blackwell, pp. 76–100.

(2013). Constitution, and multiple constitution, in the sciences: using the neuron to construct a starting framework. *Minds and Machines*, 23, 309–37.

(2016). *Reduction and Emergence in Science and Philosophy*. Cambridge: Cambridge University Press.

Gillies, D. (2011). The Russo-Williamson thesis and the question of whether smoking causes heart disease. In P. Illari, F. Russo and J. Williamson, eds., *Causality in the Sciences*. Oxford: Oxford University Press, pp. 110–25.

(2017). Mechanisms in medicine. *Axiomathes*, 27, 621–34.

(2019). *Causality, Probability and Medicine*. London: Routledge.

Ginsborg, H. (2004). Two kinds of mechanical inexplicability in Kant and Aristotle. *Journal of the History of Philosophy*, 42, 33–65.

Glennan, S. (1992). Mechanisms, models and causation. PhD dissertation, University of Chicago.

(1996). Mechanisms and the nature of causation. *Erkenntnis*, 44, 49–71.

(2002). Rethinking mechanistic explanation. *Philosophy of Science*, 69, S342–S353.

(2010). Ephemeral mechanisms and historical explanation. *Erkenntnis*, 72, 251–66.

(2011). Singular and general causal relations: a mechanist perspective. In P. Illari, F. Russo and J. Williamson, eds., *Causality in the Sciences*. Oxford: Oxford University Press, pp. 789–817.

(2017). *The New Mechanical Philosophy*. Oxford: Oxford University Press.

(2021). Corporeal composition. *Synthese*, 198, 11439–62.

Glennan, S. & Illari, P., eds. (2018a). *The Routledge Handbook of Mechanisms and Mechanical Philosophy*. New York: Routledge.

(2018b). Introduction: mechanisms and mechanical philosophies. In S. Glennan and P. Illari, eds., *The Routledge Handbook of Mechanisms and Mechanical Philosophy*. New York: Routledge, pp. 1–9.

Goodman, N. (1947). The problem of counterfactual conditionals. *Journal of Philosophy*, 44, 113–28.

Grene, M. (1971). Reducibility: another side issue? In M. Grene, ed., *Interpretations of Life and Mind*. New York: Humanities Press, pp. 14–37 (reprinted in R. S. Cohen and M. W. Wartofsky, eds., The Understanding of Nature, Boston Studies in the Philosophy of Science, vol. 23 [Dordrecht: Reidel, 1974], pp. 53–73).

Hall, N. (2004). Two concepts of causation. In J. Collins, N. Hall and L. Paul, eds., *Causation and Counterfactuals*. Cambridge, MA: MIT Press, pp. 225–76.

Harbecke, J. (2010). Mechanistic constitution in neurobiological explanations. *International Studies in the Philosophy of Science*, 24, 267–85.

Harinen, T. (2018). Mutual manipulability and causal inbetweenness. *Synthese*, 195, 35–54.

Harré, R. (1970). *The Principles of Scientific Thinking*. London: Macmillan.

(1972). *The Philosophies of Science: An Introductory Survey*. Oxford: Oxford University Press.

(2001). Active powers and powerful actors. In A. O'Hear, ed., *Philosophy at the New Millennium*. Cambridge: Cambridge University Press, pp. 91–109.

Hausman, D. M. & Woodward, J. (1999). Independence, invariance and the Causal Markov condition. *British Journal for the Philosophy of Science*, 50, 521–83.

Hegel, G. W. F. (1832/1991). *The Encyclopaedia Logic, Part I of the Encyclopaedia of Philosophical Sciences with the Zusätze*, trans. T. F. Geraets, W. A. Suchting and H. S. Harris. Indianapolis: Hackett Publishing Company.

(2002). *Science of Logic*, trans. A. V. Miller. London: Routledge.

Hertz, H. (1894/1955)]. *The Principles of Mechanics Presented in a New Form*, first English trans. 1899, reprinted by Dover. New York: Dover Publications.

Hill, B. (1965). The environment of disease: association or causation? *Proceedings of the Royal Society of Medicine*, 58, 295–300.

Hogben, L. (1930). *The Nature of Living Matter*. London: Kegan Paul, Tranch, Trubner and Co.

Holland, P. (1986). Statistics and causal inference. *Journal of the American Statistical Association*, 81, 945–60.

(1988). Comment: causal mechanism or causal effect: which is best for statistical science? *Statistical Science*, 3, 186–8.

Horwich, P. (1987). *Asymmetries in Time*. Cambridge, MA: MIT Press.

Hughes, R. E. (1983). From ignose to hexuronic acid to vitamin C. *Trends in Biochemical Sciences*, 8, 146–7.

Hunt, B. (1991). *The Maxwellians*. Ithaca, NY: Cornell University Press.

Huygens, C. (1690/1997). *Discourse on the Cause of Gravity*, trans. K. Bailey. Mimeographed.

Illari, P. & Williamson, J. (2010). Function and organization: comparing the mechanisms of protein synthesis and natural selection. *Studies in History and Philosophy of the Biological and Biomedical Sciences*, 41, 279–91.

(2011). Mechanisms are real and local. In P. Illari, F. Russo and J. Williamson, eds., *Causality in the Sciences*. Oxford: Oxford University Press, pp. 818–44.

(2012). What is a mechanism? Thinking about mechanisms across the sciences. *European Journal of Philosophy of Science*, 2, 119–35.

(2013). In defense of activities. *Journal for General Philosophy of Science*, 44, 69–83.

Ioannidis, S. & Psillos, S. (2017). In defense of methodological mechanism: the case of apoptosis. *Axiomathes*, 27, 601–19.

Kaiser, M. I. (2018). The components and boundaries of mechanisms. In S. Glennan and P. Illari, eds., *The Routledge Handbook of Mechanisms and Mechanical Philosophy*. New York: Routledge, pp. 116–30.

Kaiser, M. & Craver, C. F. (2013). Mechanisms and laws: clarifying the debate. In H. K. Chao, S. T. Chen and R. L. Millstein, eds., *Mechanism and Causality in Biology and Economics*. Dordrecht: Springer, pp. 125–45.

Kaiser, M. I. & Krickel, B. (2017). The metaphysics of constitutive mechanistic phenomena. *The British Journal for the Philosophy of Science*, 68, 745–79.

Kanduc, D., Mittelman, A., Serpico, R., Sinigaglia, E., Sinha, A. A., Natale, C., Santacroce, R., Di Corcia, M. G., Lucchese, A., Dini, L., Pani, P., Santacroce, S., Simone, S., Bucci, R. & Farber, E. (2002). Cell death: apoptosis versus necrosis. *International Journal of Oncology*, 21, 165–70.

Kant, I. (1790/2008). *Critique of Judgement*, trans. N. Walker and J. C. Meredith. Oxford: Oxford University Press.

Kaplan, D. M. (2012). How to demarcate the boundaries of cognition. *Biology and Philosophy*, 27, 545–70.

Kerr, J. F. R. (1971). Shrinkage necrosis: a distinct mode of cellular death. *Journal of Pathology*, 105, 13–20.

(2002). History of the events leading to the formulation of the apoptosis concept. *Toxicology*, 181–2, 471–4.

Kerr, J. F. R., Wyllie, A. H. & Currie, A. R. (1972). Apoptosis: a basic biological phenomenon with wide-ranging implications in tissue kinetics. *British Journal of Cancer*, 26, 239–57.

Kitcher, P. (1989). Explanatory unification and causal structure. In P. Kitcher and W. Salmon, eds., *Scientific Explanation*, Minnesota Studies in the Philosophy of Science, vol. 13. Minneapolis: University of Minnesota Press, pp. 410–505.

(1993). Function and design. *Midwest Studies in Philosophy*, 18, 379–97.

Klein, M. J. (1972). Mechanical explanation at the end of the 19th century. *Centaurus*, 17, 58–82.

Kluve, J. (2004). On the role of counterfactuals in inferring causal effects. *Foundations of Science*, 9, 65–101.

Kreines, J. (2004). Hegel's critique of pure mechanism and the philosophical appeal of the Logic project. *European Journal of Philosophy*, 12, 38–74.

Krickel, B. (2018). *The Mechanical World: The Metaphysical Commitments of the New Mechanistic Approach*. Cham: Springer.

Lakhani, S., Dilly, S. & Finlayson, C. (2009). *Basic Pathology: An Introduction to the Mechanisms of Disease*, 4th ed. London: Hodder Arnold.

Lange, M. (2000). *Natural Laws in Scientific Practice*. Oxford: Oxford University Press.

(2009). *Laws and Lawmakers: Science, Metaphysics, and the Laws of Nature*, New York: Oxford University Press.

Larmor, J. (1894). A dynamical theory of the electric and luminiferous medium (part I). *Philosophical Transactions of the Royal Society*, 185, 719–822 (reprinted in J. Larmor, *Mathematical and Physical Papers*, vol. 1, [Cambridge: Cambridge University Press, 1929]).

Leibniz, G. W. (1989). *G. W. Leibniz: Philosophical Essays*, ed. R. Ariew and D. Garber. Indianapolis, IN: Hackett Publishing Company.

Leuridan, B. (2010). Can mechanisms really replace laws of nature? *Philosophy of Science*, 77, 317–40.

(2012). Three problems for the mutual manipulability account of constitutive relevance in mechanisms. *The British Journal for the Philosophy of Science*, 63, 399–427.

Levin S., Bucci, T. J., Cohen, S. M., Fix, A. S., Hardisty, J. F., LeGrand, E. K., Maronpot, R. R. & Trump, B. F. (1999). The nomenclature of cell death: recommendations of an ad hoc committee of the society of toxicologic pathologists. *Toxicologic Pathology*, 27, 484–90.

Levy, A. (2013). Three kinds of New Mechanism. *Biology and Philosophy*, 28, 99–114.

Lewis, D. (1973). *Counterfactuals*. Cambridge, MA: Harvard University Press.

(1986a). *Philosophical Papers*, vol. 2. Oxford: Oxford University Press.

(1986b). Causation. In *Philosophical Papers*, vol. 2. Oxford: Oxford University Press, pp. 159–213.

Lind, J. (1753). *A Treatise of the Scurvy, in Three Parts, Containing an Inquiry into the Nature, Causes, and Cure of That Disease. Together with a Critical and Chronological View of What Has Been Published on the Subject*. Edinburgh: Sands, Murray and Cochran.

Linster, C. L. & Van Schaftingen, E. (2007). Vitamin C biosynthesis, recycling and degradation in mammals. *FEBS Journal*, 274, 1–22.

Lockshin, R. A. (2008). Early work on apoptosis, an interview with Richard Lockshin. *Cell Death and Differentiation*, 15, 1091–5.

(2016). Programmed cell death 50 (and beyond). *Cell Death and Differentiation*, 23, 10–17.

Lockshin, R. A. & Williams, C. M. (1964). Programmed cell death – II. Endocrine potentiation of the breakdown of the intersegmental muscles of silkmoths. *Journal of Insect Physiology*, 10, 643–9.

Lockshin, R. A. & Zakeri, Z. (2001). Programmed cell death and apoptosis: origins of the theory. *Nature Reviews Molecular Cell Biology*, 2, 545–50.

Machamer, P. (2004). Activities and causation: the metaphysics and epistemology of mechanisms. *International Studies in the Philosophy of Science*, 18, 27–39.

Machamer, P., Darden, L. & Craver, C. F. (2000). Thinking about mechanisms. *Philosophy of Science*, 67, 1–25.

Mackie, J. L. (1973). *Truth, Probability and Paradox*. Oxford: Clarendon Press.

(1974). *The Cement of the Universe*. Oxford: Clarendon Press.

Magiorkinis, E., Beloukas, A. & Diamantis, A. (2011). Scurvy: past, present and future. *European Journal of Internal Medicine*, 22, 147–52.

Majno, G. & Joris, I. (1995). Apoptosis, oncosis, and necrosis. An overview of cell death. *American Journal of Pathology*, 146, 3–15.

Maldonado, G. & Greenland, S. (2002). Estimating causal effects. *International Journal of Epidemiology*, 31, 422–9.

Matthews, L. J. & Tabery, J. (2018). Mechanisms and the metaphysics of causation. In S. Glennan and P. Illari, eds., *The Routledge Handbook of Mechanisms and Mechanical Philosophy*. New York: Routledge, pp. 131–43.

Mayr, E. (1959). Typological versus population thinking. In B. J. Meggers, ed., *Evolution and Anthropology: A Centennial Appraisal*. Washington, DC: Anthropological Society of Washington, pp. 409–12.

Maxwell, J. C. (1873). *A Treatise on Electricity and Magnetism*, 3rd ed., vol. 2. Oxford: Clarendon Press.

McLaughlin, P. (1990). *Kant's Critique of Teleology in Biological Explanation*. Lewiston, NY : Edwin Mellon Press.

McManus, F. (2012). Development and mechanistic explanation. *Studies in History and Philosophy of Biological and Biomedical Sciences*, 43, 532–41.

Menzies, P. (2012). The causal structure of mechanisms. *Studies in History and Philosophy of Biological and Biomedical Sciences*, 43, 796–805.

Millikan, R. G. (1984). *Language, Thought, and Other Biological Categories: New Foundations for Realism*. Cambridge, MA: MIT Press.

Mitchell, S. D. (2000). Dimensions of scientific law. *Philosophy of Science*, 67, 242–65.

(2008). Causal knowledge in evolutionary and developmental biology. *Philosophy of Science*, 75, 697–706.

Mumford, S.(2004). *Laws in Nature*. London: Routledge.

Neander, K. (1991). Functions as selected effects: the conceptual analyst's defense. *Philosophy of Science*, 58, 168–84.

Needham, J. (1943). *Time: The Refreshing River*. London: Allen and Unwin.

Nelson, D. L., Lehninger, A. L. & Cox, M. M. (2008). *Lehninger Principles of Biochemistry*, 5th ed. New York: W. H. Freeman.

Newton, I. (2004). *Philosophical Writings*, ed. A. Janiak. Cambridge: Cambridge University Press.

Nicholson, D. J. (2011). The concept of mechanism in biology. *Studies in History and Philosophy of Biological and Biomedical Sciences*, 43, 152–63.

(2019). Is the cell *really* a machine? *Journal of Theoretical Biology*, 477, 108–26.

Oppenheim, P., & Putnam, H. (1958). The unity of science as a working hypothesis. In H. Feigl, M. Scriven and G. Maxwell, eds., *Concepts, Theories, and the Mind-Body Problem*. Minneapolis: University of Minnesota Press, pp. 3–36.

Pearl, J. (2000). Comment. *Journal of the American Statistical Association*, 95, 428–31.

Pearl, J. & Mackenzie, D. (2018). *The Book of Why: The New Science of Cause and Effect*. New York: Basic Books.

Poincaré, H. (1890/1901). *Électricité et optique: la lumière et les théories électromagnétiques*, 2nd ed. Paris: Gathier-Villairs.

(1897). Les idées de Hertz sur la mécanique. *Revue Générale des Sciences*, 8, 734–43 (reprinted in *Ouevres de Henri Poincaré* [Paris: Gauthier-Villars, 1952], vol. 7, 231–50).

(1900). Relations entre la physique expérimentale et de la physique mathématique. *Revue Générale des Sciences*, 11, 1163–5.

(1902/1968). *La science et l'hypothése*. Paris: Flammarion.

Povich, M. & Craver, C. F. (2018). Mechanistic levels, reduction and emergence. In S. Glennan and P. Illari, eds., *The Routledge Handbook of Mechanisms and Mechanical Philosophy*. New York: Routledge, pp. 185–97.

Proskuryakov, S. Y. & Gabai, V. L. (2010). Mechanisms of tumor cell necrosis. *Current Pharmaceutical Design*, 16, 56–68.

Psillos, S. (1995). Poincaré's conception of mechanical explanation. In J.-L. Greffe, G. Heinzmann and K. Lorenz, eds., *Henri Poincaré: Science and Philosophy*. Berlin: Academie Verlag; Paris: Albert Blanchard.

(1999). *Scientific Realism: How Science Tracks Truth*. London: Routledge.
(2002). *Causation and Explanation*. Chesham Acumen; Montreal: McGill-Queens University Press.
(2004). A glimpse of the secret connexion: harmonising mechanisms with counterfactuals. *Perspectives on Science*, 12, 288–319.
(2007). Causal explanation and manipulation. In J. Person and P. Ylikoski, eds., *Rethinking Explanation*, Boston Studies in the Philosophy of Science, vol. 252. Dordrecht: Springer, pp. 97–112.
(2014). Regularities, natural patterns and laws of nature. *Theoria*, 79, 9–27.
Ramsey, F. P. (1925). Universals. *Mind*, 34, 401–17.
Rang, H. P., Ritter, J. M., Flower, R. J. & Henderson, G. (2016). *Rang & Dale's Pharmacology*, 8th ed. London: Elsevier Churchill Livingstone.
Reichenbach, H. (1956). *The Direction of Time*. Berkeley: University of California Press.
Ridley, M. (2004). *Evolution*, 3rd ed. Malden, MA: Blackwell Publishing.
Romero, F. (2015). Why there isn't interlevel causation in mechanisms. *Synthese*, 192, 3731–55.
Ross, L. N. (2021). Causal concepts in biology: how pathways differ from mechanisms and why it matters. *The British Journal for the Philosophy of Science*, 72, 131–58.
Rubin, D. B. (1978). Bayesian inference for causal effects: the role of randomization. *The Annals of Statistics*, 6, 34–58.
Ruiz-Mirazo, K., Peretó, J. & Moreno, A. (2004). A universal definition of life: autonomy and open-ended evolution. *Origins of Life and Evolution of the Biosphere* 34, 323–46.
Russell, B. (1905). On denoting. *Mind*, 14, 479–93.
Russo, F. & Williamson, J. (2007). Interpreting causality in the health sciences. *International Studies in the Philosophy of Science*, 21, 157–70.
Salmon, W. (1984). *Scientific Explanation and the Causal Structure of the World*. Princeton, NJ: Princeton University Press.
(1997). Causality and explanation: a reply to two critiques. *Philosophy of Science*, 64, 461–77.
Saunders, J. W., Jr. (1966). Death in embryonic systems. *Science*, 154, 604–12.
Schiemann, G. (2008). *Hermann von Helmholtz's Mechanism: The Loss of Certainty*. Berlin: Springer.
Schofield, R. E. (1970). *Mechanism and Materialism: British Natural Philosophy in An Age of Reason*. Princeton, NJ: Princeton University Press.
Schultz, J. (2002). *The Discovery of Vitamin C by Albert Szent-Gyögyi*. American Chemical Society National Historic Chemical Landmarks. www.acs.org/content/acs/en/education/whatischemistry/landmarks/szentgyorgyi.html (accessed February 11, 2019).
Sellars, W. (1958). Counterfactuals, dispositions, and the causal modalities. In H. Feigl, M. Scriven and G. Maxwell, eds., *Concepts, Theories, and the Mind-Body Problem*. Minneapolis: University of Minnesota Press, pp. 225–308.

Shiozaki, E. N. & Shi, Y. (2004). Caspases, IAPs and Smac/DIABLO: mechanisms from structural biology. *Trends in Biochemical Sciences*, 39, 486–94.

Simon, H. A. & Rescher, N. (1966). Cause and counterfactual. *Philosophy of Science*, 33, 323–40.

Skipper, R. & Millstein, R. (2005). Thinking about evolutionary mechanisms: natural selection. *Studies in History and Philosophy of Biological and Biomedical Sciences*, 36, 327–47.

Slack, J. M. W. (2005). *Essential Developmental Biology*, 2nd ed. Malden, MA: Blackwell Publishing.

Sloviter, R. S. (2002). Apoptosis: a guide for the perplexed. *Trends in Pharmacological Sciences*, 23, 19–24.

Sober, E. (1984). *The Nature of Selection*. Chicago: University of Chicago Press.

Stalnaker, R. (1968). A theory of conditionals. In N. Rescher, ed., *Studies in Logical Theory*. Oxford: Blackwell, pp. 98–112.

Stone, R. (1993). The assumptions on which causal inferences rest. *Journal of the Royal Statistical Society B*, 55, 455–66.

Thagard, P. (1999). *How Scientists Explain Disease*. Princeton, NJ: Princeton University Press.

Todd, W. (1964). Counterfactual conditionals and the presuppositions of induction. *Philosophy of Science*, 31, 101–10.

Tooley, M. (1977). The nature of laws. *Canadian Journal of Philosophy*, 7, 667–98.

Tortora, G. J. & Derrickson, B. (2012). *Principles of Anatomy & Physiology*, 13th ed. Hoboken, NJ: John Wiley & Sons.

von Dassow, G. & Munro, E. (1999). Modularity in animal development and evolution: elements of a conceptual framework for evo-devo. *Journal of Experimental Zoology*, 285, 307–25.

Walsh, D. M. (2006). Organisms as natural purposes: the contemporary evolutionary perspective. *Studies in History and Philosophy of Biological and Biomedical Sciences*, 37, 771–91.

Waskan, J. (2011). Mechanistic explanation at the limit. *Synthese*, 183, 389–408.

Waters, C. K. (1998). Causal regularities in the biological world of contingent distributions. *Biology and Philosophy*, 13, 5–36.

Weinberg, J. (1951). Contrary-to-fact conditionals. *Journal of Philosophy*, 48, 17–22.

Williamson, J. (2011). Mechanistic theories of causality part II. *Philosophy Compass*, 6, 433–44.

Williamson, J. & Wilde, M. (2016). Evidence and epistemic causality. In W. Wiedermann and A. von Eye, eds., *Statistics and Causality: Methods for Applied Empirical Research*. Hoboken, NJ: Wiley and Sons, pp. 31–41.

Wilson, M. D. (1999). *Ideas and Mechanism: Essays on Early Modern Philosophy*. Princeton, NJ: Princeton University Press.

Wimsatt, W. C. (1976). Reductionism, levels of organization, and the mind–body problem. In G. Globus, I. Savodnik and G. Maxwell, eds., *Consciousness and the Brain*. New York: Plenum Press, pp. 199–267.

Wolpert, L., Beddington, R., Jessell, T., Lawrence, P., Meyerowitz, E. & Smith, J. (2002). *Principles of Development*, 2nd ed. Oxford: Oxford University Press.

Wong, R. S. Y. (2011). Apoptosis in cancer: from pathogenesis to treatment. *Journal of Experimental and Clinical Cancer Research*, 30, 87.

Woodger, J. H. (1929). *Biological Principles: A Critical Study*, London: Routledge & Kegan Paul.

Woodward, J. (1997). Explanation, invariance and intervention. *Philosophy of Science*, 64, S26–S41.

(2000). Explanation and invariance in the special sciences. *British Journal for the Philosophy of Science*, 51, 197–254.

(2002). What is a mechanism? A counterfactual account. *Philosophy of Science*, 69, S366–S377.

(2003a). *Making Things Happen: A Theory of Causal Explanation*. New York: Oxford University Press.

(2003b). Counterfactuals and causal explanation. *International Studies in the Philosophy of Science*, 18, 41–72.

(2011). Mechanisms revisited. *Synthese*, 183, 409–27.

(2013). Mechanistic explanation: its scope and limits. *Aristotelian Society Supplementary Volume*, 87, 39–65.

(2014). A functional account of causation; or, a defense of the legitimacy of causal thinking by reference to the only standard that matters – usefulness (as opposed to metaphysics or agreement with intuitive judgment). *Philosophy of Science*, 81, 691–713.

(2015). Interventionism and causal exclusion. *Philosophy and Phenomenological Research*, 91, 303–47.

(2020). Levels: What are they and what are they good for?. In K. S. Kendler, J. Parnas and P. Zachar, eds., *Levels of Analysis in Psychopathology: Cross Disciplinary Perspectives*. Cambridge: Cambridge University Press, pp. 424–49.

Index

Printed in the United States
by Baker & Taylor Publisher Services

New Materials for
Microphotonics

MATERIALS RESEARCH SOCIETY
SYMPOSIUM PROCEEDINGS VOLUME 817

New Materials for Microphotonics

Symposium held April 13–15, 2004, San Francisco, California, U.S.A.

EDITORS:

Jung H. Shin
Korea Advanced Institute of
Science and Technology (KAIST)
Daejeon, Korea

Mark Brongersma
Stanford University
Stanford, California, U.S.A.

Christoph Buchal
Forschungszentrum
Jülich, Germany

Francesco Priolo
MATIS-INFM and University of Catania
Catania, Italy

Materials Research Society
Warrendale, Pennsylvania

CAMBRIDGE UNIVERSITY PRESS
Cambridge, New York, Melbourne, Madrid, Cape Town,
Singapore, São Paulo, Delhi, Mexico City

Cambridge University Press
32 Avenue of the Americas, New York NY 10013-2473, USA

Published in the United States of America by Cambridge University Press, New York

www.cambridge.org
Information on this title: www.cambridge.org/9781107409187

Materials Research Society
506 Keystone Drive, Warrendale, PA 15086
http://www.mrs.org

© Materials Research Society 2004

This publication has been registered with Copyright Clearance Center, Inc.
For further information please contact the Copyright Clearance Center,
Salem, Massachusetts.

First published 2004
First paperback edition 2012

Single article reprints from this publication are available through
University Microfilms Inc., 300 North Zeeb Road, Ann Arbor, MI 48106

CODEN: MRSPDH

ISBN 978-1-107-40918-7 Paperback

CONTENTS

*Invited Paper

*Invited Paper

NANOCRYSTALS

NEW CONCEPTS AND DEVICES

*Invited Paper

ELECTRO-OPTIC MATERIALS

*Invited Paper

LUMINESCENT MATERIALS

PREFACE

This volume contains papers presented at Symposium L, "New Materials for Microphotonics," held April 13–15 at the 2004 MRS Spring Meeting in San Francisco, California.

The Materials Research Society has an excellent reputation of bringing together scientists from different fields with a common interest in materials synthesis, properties and analysis. In this spirit, we have organized a symposium for all those who are interested or working in optoelectronics, integrated optics, microphotonics and related research areas. We were very pleased with the resonance. The speakers came from very distinct fields, giving their personal perspective of new developments in microphotonics. As can be seen from the table of contents, the talks covered many classes of materials, including semiconductors, insulators, ferroelectrics and polymers. A broad and widely interested audience was present during the talks and especially during the preceding tutorials. The proceedings itself is divided into six sections: Rare-Earth Doped Structures, Photonic Bandgap Crystals, Nanocrystals, New Concepts and Devices, Electro-optic Materials, and Luminescent Materials. However, we hope that the authors and readers will understand that the field of microphotonics, by its multi-disciplinary nature, cannot be simplified by such divisions, and that many manuscripts are relevant to sections other than the ones into which we chose to place them. This publication offers the chance to peruse the different issues of microphotonics, deepen the understanding and identify progress and challenges of the exciting new field of microphotonics.

Jung H. Shin
Mark Brongersma
Christoph Buchal
Francesco Priolo

June 2004

MATERIALS RESEARCH SOCIETY SYMPOSIUM PROCEEDINGS

Volume 782— Micro- and Nanosystems, D. LaVan, M. Mcnie, A. Ayon, M. Madou, S. Prasad, 2004,
ISBN: 1-55899-720-2
Volume 783— Materials, Integration and Packaging Issues for High-Frequency Devices, P. Muralt, Y.S. Cho,
J-P. Maria, M. Klee, C. Hoffmann, C.A. Randall, 2004, ISBN: 1-55899-721-0
Volume 784— Ferroelectric Thin Films XII, S. Hoffmann-Eifert, H. Funakubo, A.I. Kingon, I.P. Koutsaroff,
V. Joshi, 2004, ISBN: 1-55899-722-9
Volume 785— Materials and Devices for Smart Systems, Y. Furuya, E. Quandt, Q. Zhang, K. Inoue,
M. Shahinpoor, 2004, ISBN: 1-55899-723-7
Volume 786— Fundamentals of Novel Oxide/Semiconductor Interfaces, C.R. Abernathy, E. Gusev,
D.G. Schlom, S. Stemmer, 2004, ISBN: 1-55899-724-5
Volume 787— Molecularly Imprinted Materials—2003, P. Kofinas, M.J. Roberts, B. Sellergren, 2004,
ISBN: 1-55899-725-3
Volume 788— Continuous Nanophase and Nanostructured Materials, S. Komarneni, J.C. Parker, J. Watkins,
2004, ISBN: 1-55899-726-1
Volume 789— Quantum Dots, Nanoparticles and Nanowires, P. Guyot-Sionnest, N.J. Halas, H. Mattoussi,
Z.L. Wang, U. Woggon, 2004, ISBN: 1-55899-727-X
Volume 790— Dynamics in Small Confining Systems—2003, J.T. Fourkas, P. Levitz, M. Urbakh, K.J. Wahl,
2004, ISBN: 1-55899-728-8
Volume 791— Mechanical Properties of Nanostructured Materials and Nanocomposites, R. Krishnamoorti,
E. Lavernia, I. Ovid'ko, C.S. Pande, G. Skandan, 2004, ISBN: 1-55899-729-6
Volume 792— Radiation Effects and Ion-Beam Processing of Materials, L. Wang, R. Fromknecht, L.L. Snead,
D.F. Downey, H. Takahashi, 2004, ISBN: 1-55899-730-X
Volume 793— Thermoelectric Materials 2003—Research and Applications, G.S. Nolas, J. Yang, T.P. Hogan,
D.C. Johnson, 2004, ISBN: 1-55899-731-8
Volume 794— Self-Organized Processes in Semiconductor Heteroepitaxy, R.S. Goldman, R. Noetzel,
A.G. Norman, G.B. Stringfellow, 2004, ISBN: 1-55899-732-6
Volume 795— Thin Films—Stresses and Mechanical Properties X, S.G. Corcoran, Y-C. Joo, N.R. Moody,
Z. Suo, 2004, ISBN: 1-55899-733-4
Volume 796— Critical Interfacial Issues in Thin-Film Optoelectronic and Energy Conversion Devices,
D.S. Ginley, S.A. Carter, M. Grätzel, R.W. Birkmire, 2004, ISBN: 1-55899-734-2
Volume 797— Engineered Porosity for Microphotonics and Plasmonics, R. Wehrspohn, F. Garcial-Vidal,
M. Notomi, A. Scherer, 2004, ISBN: 1-55899-735-0
Volume 798— GaN and Related Alloys—2003, H.M. Ng, M. Wraback, K. Hiramatsu, N. Grandjean, 2004,
ISBN: 1-55899-736-9
Volume 799— Progress in Compound Semiconductor Materials III—Electronic and Optoelectronic
Applications, D. Friedman, M.O. Manasreh, I. Buyanova, F.D. Auret, A. Munkholm, 2004,
ISBN: 1-55899-737-7
Volume 800— Synthesis, Characterization and Properties of Energetic/Reactive Nanomaterials,
R.W. Armstrong, N.N. Thadhani, W.H. Wilson, J.J. Gilman, Z. Munir, R.L. Simpson, 2004,
ISBN: 1-55899-738-5
Volume 801— Hydrogen Storage Materials, M. Nazri, G-A. Nazri, R.C. Young, C. Ping, 2004,
ISBN: 1-55899-739-3
Volume 802— Actinides—Basic Science, Applications and Technology, L. Soderholm, J. Joyce, M.F. Nicol,
D. Shuh, J.G. Tobin, 2004, ISBN: 1-55899-740-7
Volume 803— Advanced Data Storage Materials and Characterization Techniques, J. Ahner, L. Hesselink,
J. Levy, 2004, ISBN: 1-55899-741-5
Volume 804— Combinatorial and Artificial Intelligence Methods in Materials Science II, R.A. Potyrailo,
A. Karim, Q. Wang, T. Chikyow, 2004, ISBN: 1-55899-742-3
Volume 805— Quasicrystals 2003—Preparation, Properties and Applications, E. Belin-Ferré, M. Feuerbacher,
Y. Ishii, D. Sordelet, 2004, ISBN: 1-55899-743-1
Volume 806— Amorphous and Nanocrystalline Metals, R. Busch, T. Hufnagel, J. Eckert, A. Inoue,
W. Johnson, A.R. Yavari, 2004, ISBN: 1-55899-744-X

MATERIALS RESEARCH SOCIETY SYMPOSIUM PROCEEDINGS

Volume 807— Scientific Basis for Nuclear Waste Management XXVII, V.M. Oversby, L.O. Werme, 2004, ISBN: 1-55899-752-0

Volume 808— Amorphous and Nanocrystalline Silicon Science and Technology—2004, R. Biswas, G. Ganguly, E. Schiff, R. Carius, M. Kondo, 2004, ISBN: 1-55899-758-X

Volume 809— High-Mobility Group-IV Materials and Devices, M. Caymax, E. Kasper, S. Zaima, K. Rim, P.F.P. Fichtner, 2004, ISBN: 1-55899-759-8

Volume 810— Silicon Front-End Junction Formation—Physics and Technology, P. Pichler, A. Claverie, R. Lindsay, M. Orlowski, W. Windl, 2004, ISBN: 1-55899-760-1

Volume 811— Integration of Advanced Micro- and Nanoelectronic Devices—Critical Issues and Solutions, J. Morais, D. Kumar, M. Houssa, R.K. Singh, D. Landheer, R. Ramesh, R. Wallace, S. Guha, H. Koinuma, 2004, ISBN: 1-55899-761-X

Volume 812— Materials, Technology and Reliability for Advanced Interconnects and Low-k Dielectrics—2004, R. Carter, C. Hau-Riege, G. Kloster, T-M. Lu, S. Schulz, 2004, ISBN: 1-55899-762-8

Volume 813— Hydrogen in Semiconductors, N.H. Nickel, M.D. McCluskey, S. Zhang, 2004, ISBN: 1-55899-763-6

Volume 814— Flexible Electronics 2004—Materials and Device Technology, B.R. Chalamala, B.E. Gnade, N. Fruehauf, J. Jang, 2004, ISBN: 1-55899-764-4

Volume 815— Silicon Carbide 2004—Materials, Processing and Devices, M. Dudley, P. Gouma, P.G. Neudeck, T. Kimoto, S.E. Saddow, 2004, ISBN: 1-55899-765-2

Volume 816— Advances in Chemical-Mechanical Polishing, D. Boning, J.W. Bartha, G. Shinn, I. Vos, A. Philipossian, 2004, ISBN: 1-55899-766-0

Volume 817— New Materials for Microphotonics, J.H. Shin, M. Brongersma, F. Priolo, C. Buchal, 2004, ISBN: 1-55899-767-9

Volume 818— Nanoparticles and Nanowire Building Blocks—Synthesis, Processing, Characterization and Theory, O. Glembocki, C. Hunt, C. Murray, G. Galli, 2004, ISBN: 1-55899-768-7

Volume 819— Interfacial Engineering for Optimized Properties III, C.A. Schuh, M. Kumar, V. Randle, C.B. Carter, 2004, ISBN: 1-55899-769-5

Volume 820— Nanoengineered Assemblies and Advanced Micro/Nanosystems, J.T. Borenstein, P. Grodzinski, L.P. Lee, J. Liu, Z. Wang, D. McIlroy, L. Merhari, J.B. Pendry, D.P. Taylor, 2004, ISBN: 1-55899-770-9

Volume 821— Nanoscale Materials and Modeling—Relations Among Processing, Microstructure and Mechanical Properties, P.M. Anderson, T. Foecke, A. Misra, R.E. Rudd, 2004, ISBN: 1-55899-771-7

Volume 822— Nanostructured Materials in Alternative Energy Devices, E.R. Leite, J-M. Tarascon, Y-M. Chiang, E.M. Kelder, 2004, ISBN: 1-55899-772-5

Volume 823— Biological and Bioinspired Materials and Devices, J. Aizenberg, C. Orme, W.J. Landis, R. Wang, 2004, ISBN: 1-55899-773-3

Volume 824— Scientific Basis for Nuclear Waste Management XXVIII, J.M. Hanchar, S. Stroes-Gascoyne, L. Browning, 2004, ISBN: 1-55899-774-1

Volume 825E—Semiconductor Spintronics, B. Beschoten, S. Datta, J. Kikkawa, J. Nitta, T. Schäpers, 2004, ISBN: 1-55899-753-9

Volume 826E—Proteins as Materials, V.P. Conticello, A. Chilkoti, E. Atkins, D.G. Lynn, 2004, ISBN: 1-55899-754-7

Volume 827E—Educating Tomorrow's Materials Scientists and Engineers, K.C. Chen, M.L. Falk, T.R. Finlayson, W.E. Jones Jr., L.J. Martinez-Miranda, 2004, ISBN: 1-55899-755-5

Prior Materials Research Society Symposium Proceedings available by contacting Materials Research Society

Rare-Earth Doped
Structures

Mat. Res. Soc. Symp. Proc. Vol. 817 © 2004 Materials Research Society

Rare-earth doped Si nanostructures for Microphotonics

D. Pacifici, G. Franzò, F. Iacona,[1] A. Irrera,[1] S. Boninelli, M. Miritello, and F. Priolo
MATIS-INFM and Dipartimento di Fisica e Astronomia, Via S. Sofia 64,
I-95123 Catania, Italy
[1] CNR-IMM, Sezione di Catania, Stradale Primosole 50, I-95121 Catania, Italy

ABSTRACT

In the present paper, we will review our work on rare-earth doped Si nanoclusters. The samples have been obtained by implanting the rare-earth (e.g. Er) in a film containing pre-formed Si nanocrystals. After the implant, samples have been treated at 900°C for 1h. This annealing temperature is not enough to re-crystallize all of the amorphized Si clusters. However, even if the Si nanoclusters are in the amorphous phase, they can still efficiently transfer the energy to nearby rare-earth ions. We developed a model for the Si nanoclusters-Er system, based on an energy level scheme taking into account the coupling between each Si nanocluster and the neighboring Er ions. By fitting the data, we were able to determine a value of 3×10^{-15} cm^3 s^{-1} for the Si nanocluster-Er coupling coefficient. Moreover, a strong cooperative up-conversion mechanism between two excited Er ions and characterized by a coefficient of 7×10^{-17} cm^3 s^{-1}, is shown to be active in the system, demonstrating that more than one Er ion can be excited by the same nanocluster. We show that the overall light emission yield of the Er related luminescence can be enhanced by using higher concentrations of very small nanoaggregates. Eventually, electroluminescent devices based on rare-earth doped Si nanoclusters will be demonstrated.

INTRODUCTION

Among the different approaches developed to overcome the intrinsic low efficiency of silicon as a light emitter, quantum confinement and rare earth doping of silicon have dominated the scientific scenario of silicon-based microphotonics. Recently, Er doping of Si nanocrystals has been recognized as an interesting way of combining the promising features of both previous methods [1-3]. Indeed, it has been demonstrated that Si nanocrystals can act as efficient sensitizers for Er [4-7]. In particular, the nanocrystal, once excited, transfers quasi-resonantly [8] its energy to the nearby Er ion, which then decays emitting a photon at 1.54 μm. The effective excitation cross section for Er in presence of Si nanocrystals is more than two orders of magnitude higher with respect to the resonant absorption of a photon in a silica matrix [4]. The recent determination of net optical gain at 1.54 μm in Er-doped Si nanocluster sensitized waveguides [9] and the demonstration of efficient room temperature electroluminescence from Er-Si nanoclusters devices [10] opened the route towards the future fabrication of electrically driven optical amplifiers based on this system. Several basic issues remain however to be first addressed, namely i) the role of amorphous Si clusters in exciting the Er^{3+} ions, ii) the coupling strength between Si nanoclusters and Er ions and the microscopic details of the interaction, iii) the role and strength of the competitive non radiative processes such as co-operative up-conversion among interacting Er^{3+} ions, iv) the optimization of key parameters ruling the strength of the Si nanoclusters-Er coupling, v) the realization of efficient electrical excitation of the system. In the present work, we will explore all of this important issues. In particular, we will show that amorphous as well as crystalline Si nanoclusters can act as efficient sensitizers for the Er luminescence. Through a comparison of simulated and measured photoluminescence data, both the coupling and the up-conversion coefficients have been determined [11]. A new processing of the material, able to increase the light emission yield of Er in presence of Si aggregates, will also be explored and light emitting devices based on different rare-earths (Er, Tm, Yb) will be demonstrated.

EXPERIMENTAL

Si nanocrystals were produced by 1250°C annealing of a 0.2 μm thick substoichiometric SiO_x film (with 42 at.% Si) grown by plasma-enhanced chemical vapor deposition on a Si substrate. Thermal annealing induces the separation of the Si and SiO_2 phases and hence Si nanocrystals with a mean radius of ~ 2 nm are formed, as evidenced by dark field plan view transmission electron microscopy (TEM). Er ions were then implanted at different energies (in the range between 170-500 keV) and different doses in order to produce an almost constant Er concentration (in the range 2.2×10^{20}-1.2×10^{21}/cm^3) all over the film thickness. The very same Er implants were also performed in SiO_2 layers not containing Si nanocrystals in order to have reference samples. Other implants have been performed on a sample containing Si nanocluster obtained by annealing the SiO_x film at a lower temperature of 800°C for 1h. All of the samples were eventually annealed at 900°C for 1h in order to activate Er, preventing it from clustering.

Photoluminescence (PL) measurements were performed by pumping with the 488 nm line of an Ar laser. The pump power was varied in a wide range, between 1-10^3 mW, and focused over a circular area of ~ 0.3 mm in radius. The laser beam was chopped through an acousto-optic modulator at a frequency of 11 Hz. Electroluminescence (EL) measurements were taken by biasing the device with a square pulse at a frequency of 11 Hz, using a fast Agilent Pulse Generator. The luminescence (PL or EL) signals were then analyzed by a single grating monochromator and detected by a photomultiplier tube for the visible range (0.4-0.9 μm) or by a Ge detector for the infra-red (0.8-1.7 μm spectral region). Spectra were recorded with a lock-in amplifier using the chopping frequency as a reference. All the spectra have been measured at room temperature and corrected for the spectral system response. Time resolved luminescence measurements were performed by first detecting the modulated luminescence signal with a Hamamatsu infrared-extended photomultiplier tube having an almost constant spectral response in the range 0.4-1.7 μm. The signal was hence analyzed with a photon counting multichannel scaler, triggered by the acousto-optic modulator (for PL) or by the pulse generator (for EL). The overall time resolution of our system is of ~ 5 ns.

DISCUSSION

Structural and optical properties

The structural characterization was performed by using a 200 kV energy filtered transmission electron microscope (EFTEM) Jeol Jem 2010F with Gatan Image Filter. This system consists of a conventional TEM coupled with an electron energy loss spectrometer (EELS). Through EFTEM it is possible to create an image of the sample by using only electrons that have lost a specific amount of energy. This is a very suitable method to detect silicon nanograins (both crystalline and amorphous and independently of the crystal orientation) dispersed in a silica matrix, in fact the plasmon loss energy in silicon is 16.7 eV, that is well separated from the plasmon loss energy in silica (23.2 eV).

In figure 1, TEM plan view images of samples annealed at different temperatures are shown. figure 1a reports the EFTEM image of the SiO_x sample annealed at 1250°C for 1 hour obtained by selecting an energy window centered at 16 eV (corresponding to the value of the Si plasmon loss). A large number of Si grains appears; the clusters are well separated and are characterized by a mean radius of 2.2 nm. A dark field TEM image, reported in figure 1b, was obtained from the same sample by selecting a small portion of the diffraction ring of the {111} Si planes. The dark field image confirms that the Si clusters are crystalline; also in this case their mean radius is about 2.2 nm. The great difference between the number of Si clusters detected by the two techniques is due to the fact that EFTEM is able to detect all the Si agglomerates present in the layer, independently of their orientation and/or crystalline structure.

Figure 1c reports an EFTEM plan view image of the sample annealed at 1250° C, implanted with Er and subsequently annealed at 900°C for 1h. The image shows a Si nanoclusters distribution very similar to the one reported in figure 1a, before the Er implantation process. Therefore, the implantation process is not producing a mixing of the Si clusters with the SiO_2

Figure 1. (a) EFTEM image of a SiO_x sample annealed at 1250°C for 1h and (b) corresponding dark field image. (c) EFTEM image of a SiO_x sample annealed at 1250°C for 1h, implanted with Er and subsequently annealed at 900°C for 1h. (d) EFTEM image of the sample pre-annealed at 800 °C for 1h, implanted with Er and finally annealed at 900 °C for 1h.

matrix. However, for this sample it was impossible to obtain a dark field image, since the electron diffraction pattern reveals the absence of a crystalline phase. It is therefore reasonable to conclude that the nanocrystals are amorphized by the Er ion beam, and after implantation the grains preserve their shape but remain amorphous even if a thermal process at 900 °C for 1h is performed. Indeed, it has been demonstrated that in order to completely induce the recrystallization of amorphized Si nanocrystals it is necessary to perform thermal processes at higher temperature than those needed to crystallize bulk amorphous silicon, and for longer times [12]. For comparison, figure 1d reports an EFTEM image of a sample annealed at 800°C prior to

Er implantation. Only extremely small amorphous Si grains are visible in this case. We will discuss in more details the implications of these structural data in the following.

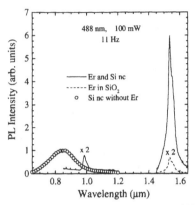

Figure 2. Room temperature photoluminescence (PL) spectra of Si nanocrystals, Er in SiO_2 and Er in presence of Si nc. The Er concentration in the two implanted samples is $6.5 \times 10^{20}/cm^3$.

In figure 2, the room temperature PL spectra of three different samples consisting of Er in presence of Si nanoclusters (continuous lines), Er in SiO_2 (dashed line) and Si nanocrystals before Er implantation (open circles) are shown. All of the spectra have been obtained in the very same conditions, i.e. exciting the systems with the 488 nm Ar-laser line at a pump power of 100 mW and at a chopper frequency of 11 Hz. First of all, it is worth noticing that the efficient luminescence at around 0.8 μm due to Si nanocrystals alone completely disappears when introducing Er in the sample. Indeed, an intense peak at 1.54 μm from Er^{3+} ions appears at the expense of the Si nanocrystal related emission. Moreover, the 1.54 μm PL intensity due to Er ions in presence of Si nanocluster is over one order of magnitude higher with respect to the sample with Er in SiO_2. It has been demonstrated that this intensity increase is due to the sensitizing action made by Si nanoclusters for the Er ion [4]. Since the electronic structures of amorphous and crystalline Si nanoclusters have been shown to be very similar [13], the same arguments valid for nanocrystals can be extended to amorphous Si clusters. Indeed, we can imagine that each Si nanocluster absorbs an incident photon and, once excited, promptly transfers its energy to a nearby Er ion which will return to its ground state emitting a photon at 1.54 μm, corresponding to the transition from the first excited level $^4I_{13/2}$ to the ground state $^4I_{15/2}$ of Er^{3+}. Another important feature in figure 2 is the presence of the 0.98 μm line in the spectrum related to the Er-doped Si nanocluster sample, due to the radiative transition from the second excited state $^4I_{11/2}$ to the ground state of Er^{3+}. This is a clear evidence of the energy transfer between Si nanoclusters and Er ions, since it has been postulated [3] and recently demonstrated [8] that a quasi-resonant energy transfer can occur between the Si nanocrystals emitting at 0.8 and 0.98 μm and the two manifold Er-related levels $^4I_{9/2}$ and $^4I_{11/2}$, respectively. Thus the appearance of the efficient 0.98 μm Er line can again be attributed to the sensitizing action of Si nanoclusters which are strongly coupled with the Er-related levels. Indeed this line is still visible at room temperature even at laser pump powers as low as a few mW, being in that range totally

6

undetectable for the Er-doped silicon dioxide sample. Another effect which could in principle produce a strong emission at 0.98 μm is the up-conversion among two excited Er3+ ions, which we are going to study in details in the following section.

Up-conversion mechanism

Figure 3a reports the typical PL spectra for Er in presence of Si nanoclusters at three different Er concentrations and for a reference sample of Er in SiO$_2$ at high pump power (800 mW, corresponding to a photon flux of about 7×10^{20} cm^{-2} s^{-1}).

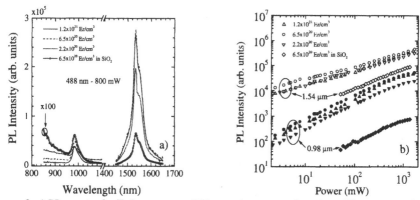

Figure 3. a) PL spectra for Er in presence of Si nanoclusters as a function of Er concentration (lines) and for a reference sample of Er in SiO$_2$ (line+circles). b) Power dependence of the 1.54 and 0.98 mm lines in different samples.

It is possible to observe that the PL intensity at 1.54 μm for Er in presence of Si nanoclusters increases less than linearly with Er concentration, and it slightly decreases for the sample containing the highest Er concentration (1.2×10^{21} Er/cm^3). This phenomenon is partly determined by a strong concentration quenching occurring between an excited Er ion and a nearby ground state ion, leading to a decrease in the mean decay time of level ^{4}I$_{13/2}$ [4]. However, the presence of the 0.98 μm line suggests the occurrence of another strong non radiative decay channel under high pump power conditions for an Er ion in the first excited level, i.e. up-conversion. Indeed, two excited Er ions can interact if their distance is sufficiently small, leading to the de-excitation of one Er ion and the following excitation of the other in higher lying excited levels. It is worth noticing that under the very same excitation conditions, the PL intensity emitted by the reference sample of Er in SiO$_2$ at 0.98 μm is two orders of magnitude lower than the samples containing Si nanoclusters. Indeed, in figure 3b the pump power dependencies of both the 1.54 and 0.98 μm PL intensities, reported for the three samples containing Er in Si nanoclusters and for Er in SiO$_2$, show that the 0.98 μm emission for Er in presence of Si nanocrystals is two orders of magnitudes higher than for the reference sample all over the pump power range. Moreover, the 1.54 μm PL intensity for Er in presence of Si nanoclusters is shown to strongly saturate as a function of power, while the trend for the Er doped SiO_2 sample is linear. In order to investigate the presence of up-conversion, we plotted in figure 4 the ratio of the 0.98 μm PL intensity to the respective 1.54 μm intensity as a function of the 1.54 μm one, for each of the pump powers used and for all the samples. For Er in presence of Si nanocrystals

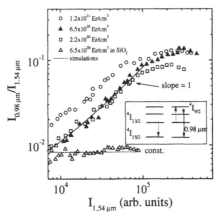

Figure 4. Correlation graph reporting a quadratic dependence of the 0.98 μm PL intensity with respect to the 1.54 μm one for the samples containing Si nanoclusters. The Er doped SiO_2 sample show no correlation between the two lines. Inset: up-conversion scheme.

the ratio of the two intensities increases linearly as a function of the 1.54 μm intensity, meaning indeed that the 0.98 μm emission is proportional to the square of the 1.54 μm line intensity. Since the intensity at a certain wavelength is proportional to the number of emitted photons, which is moreover proportional to the number of excited centers emitting at that wavelength, it is possible to conclude that the number of Er ions excited in the $^4I_{11/2}$ level, which is responsible for the 0.98 μm emission, is proportional to the square of the number of Er ions which are excited in the $^4I_{13/2}$ level. That is to say, in order to have an Er ion excited in the $^4I_{11/2}$ level we need two Er ions excited in the $^4I_{13/2}$ level. This is the result of a typical cooperative up-conversion mechanism, where, as reported in the inset of figure 4, one of two nearby interacting Er ions, both in the first excited state, gives up resonantly and non radiatively its energy to the other, collapsing to the ground state and bringing the other in the $^4I_{9/2}$ level. A fast non radiative de-excitation occurs from this level to the lower lying $^4I_{11/2}$ level and the subsequent radiative recombination from this level produces a 0.98 μm photon. We get a 0.98 μm photon at the expenses of two 1.54 μm photons which can no more be emitted. This is the physical origin of the quadratic power law behavior observed in figure 4. The continuous line is a simulation obtained by using a value of $3x10^{-15}$ cm^3 s^{-1} for the coupling coefficient between a Si nanocluster and an Er ion, and a value of $7x10^{-17}$ cm^3 s^{-1} for the up-conversion coefficient between two Er ions in the first excited level. For comparison, the ratio of the 0.98 μm to the 1.54 μm intensity for an Er doped SiO_2 sample is reported in figure 4 as a function of the 1.54 μm intensity (open triangles). The trend is characterized by a constant slope, meaning indeed that only a linear correlation exists between the two lines. This is due to the fact that we are pumping directly the higher lying Er levels, and from there the two excited levels responsible for the PL at the two wavelengths are populated with constant branching ratio. It is worthwhile to notice that in order to get the same emission intensity at 1.54 μm from Er in SiO₂, we have to pump the system with powers more than one order of magnitude higher with respect to the sample with the same Er concentration but in presence of Si nanoclusters, as can be easily observed in figure 3b. The presence of cooperative up-conversion in our Er-doped Si nanoclusters sample is a clear demonstration of the fact that under cw pumping more than one Er^{3+} ion can actually be excited by a single Si nanocluster.

Effect of up-conversion on the excitation rate of Er

Since we are eventually interested in the emission of 1.54 μm photons, we want to better investigate the real effect of up-conversion in the dynamics of the $^4I_{13/2}$ level population.

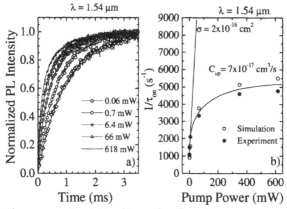

Figure 5. Risetime curves a) and $1/\tau_{on}$ b) as a function of pump power.

Figure 5a shows the normalized luminescence intensities recorded at 1.54 μm, for a sample containing 6.5×10^{20} Er/cm^3, as a function of time after switching on the laser beam at t=0, and for different pump powers. The luminescence signal reaches the saturation value in a time which is shorter the higher the excitation power. We define the typical experimental risetime τ_{on} as the time it takes the PL signal to reach the 63% (i.e. $1-e^{-1}$) of the saturation value. In figure 5b, the reciprocal of the experimental risetimes (solid circles) extracted from figure 5a is plotted as a function of the pump power. The experimental trend is linear up to 1 mW. Indeed, within this low power regime it is possible to demonstrate that the reciprocal of the risetime τ_{on} follows the law:

$$\frac{1}{\tau_{on}} = \sigma\phi + \frac{1}{\tau} \qquad (1)$$

where ϕ is the photon flux, τ the total lifetime of level $^4I_{13/2}$, comprising the radiative and all of the non radiative de-excitation processes, and σ_{eff} is an effective excitation cross section for Er in presence of Si nanoclusters. From a linear fit of the experimental data through equation 1 a value of $\sim 2 \times 10^{-16}$ cm^2 can be estimated for the Er excitation cross section [4]. At higher pump powers, the linear approximation is no more valid, and indeed the trend of the experimental data in figure 5b shows a strong saturation, which can be attributed to the up-conversion mechanism, as indicated by the good agreement of the simulated data (open circles) obtained within the model.

Role of nanocluster density on the Er-luminescence yield

Cooperative up-conversion can be easily avoided by simply reducing the concentration of Er^{3+} ions, hence decreasing the mean Er-Er distance and therefore suppressing the short range dipole-dipole interaction among pairs of Er ions. On the other hand, much less is known as far as the key parameters ruling the Si nanocluster-Er coupling are concerned. For sure the mean distance between Si nanoclusters and Er ions and the mean size of the nanograins play a crucial role. Indeed, even if there's no complete agreement in literature about the exact law describing the energy transfer mechanism between Si nanoclusters and Er ions, i.e. whether it is resonant dipole-dipole interaction [16] or direct electron exchange [17-19], it is experimentally known

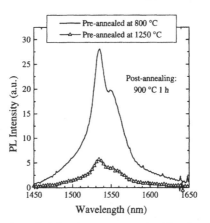

Figure 6. Room temperature photoluminescence spectra taken by exciting with a 10 mW laser beam SiO$_x$ samples pre-annealed at different temperatures (800°C and 1250°C for 1h) and then implanted with Er. The post-implantation thermal process was 900 °C 1h for both the samples.

that the interaction range has to be ≤ 1 nm [19]. This poses a limit on the minimum concentration of sensitizers and emitting centers that have to be inserted in the sample for the interaction to become active. Moreover, it is known that, within a certain size distribution which guarantees the resonance condition for the energy transfer, the smaller is a nanograin, the stronger is the interaction mechanisms, i.e. the energy transfer efficiency [20]. In the following we are going to investigate these important and still open issues.

Figure 6 reports the room temperature photoluminescence spectra taken by exciting with a 10 mW laser beam two Er doped SiO$_x$ samples obtained by first annealing the SiO$_x$ layer at different temperatures (800 and 1250 °C for 1h, respectively), then implantated with Er and eventually annealed both at 900 °C for 1h. The spectra show the typical features of the Er emission with a main peak at 1.54 μm and a shoulder at 1.55 μm. However, it is worth noticing that for a pre-annealing process of 1250 °C, the intensity at 1.54 μm is about a factor of 5 lower with respect to the maximum PL signal observed in the sample pre-annealed at 800°C. The reasons for the observed enhancement have to be searched at first in the different excitation and/or de-excitation processes acting in the samples. It turns out that the decay time for the Er luminescence at 1.54 μm has a value of ~ 2 ms for both samples. This indicates that the difference in the PL signals cannot be attributed to a different contribution of non-radiative decay channels. As far as the excitation mechanism is concerned, we have measured the Er excitation cross section in the two samples finding also in this case comparable values (we found a value of 2.5×10^{-16} cm^2 for the sample pre-annealed at 800 °C and 1.3×10^{-16} cm^2 for the sample pre-annealed at 1250 °C). The factor of 2 between the two excitation cross sections cannot completely account for the observed difference by a factor of 5 in the PL signals.

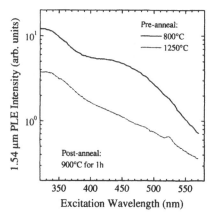

Figure 7. Photoluminescence excitation (PLE) spectra obtained by measuring the PL intensity at 1.54 μm versus the excitation wavelength, for the sample pre-annealed at 800 °C for 1h and for that pre-annealed at 1250 °C for 1h. Both samples were implanted with Er and then annealed at 900 °C for 1 h.

In figure 7 the PL signal at 1.54 μm is reported as a function of the excitation wavelength for the two samples, by using a 150W Xenon lamp coupled to a monochromator as the excitation source. The typical photon flux at 488 nm was 4×10^{16} cm^{-2} s^{-1}. The PL signal at 1.54 μm monotonically decreases with increasing the excitation wavelength, reflecting the behavior of the Si nanoclusters absorption, and no resonances are observed, indicating that Er is excited in both cases through an electron-hole mediated process and not by a direct photon absorption. Therefore in both samples, no matters which is the pre-annealing temperature, there are centers absorbing the energy from the laser beam and transferring it to Er afterwards. However, the PLE spectrum related to the sample pre-annealed at 800°C is higher and it shows a broad absorption band at around 450 nm. This band could be related to very small silicon nanoclusters which are not present in the sample annealed at higher temperature, since they disappear for an Ostwald ripening process. The increase in PL intensity in the 400-500 nm range can therefore be explained by an increased number of sensitizers in the sample annealed at low temperature. Indeed, figure 1d reports an EFTEM plan view image of the sample annealed at 800° C prior to Er implantation (this is the sample for which the luminescence signal at 1.54 μm is higher). It is possible to distinguish some very small Si grains, having a radius of 0.5 nm or lower. A careful analysis of the image reveals that the clusters are not separated very well, but they seem to be connected by a Si network. This effect is due to the fact that the phase separation process is not completed, as a result of the low temperature used for the annealing process. Note that the same kind of structure is not present in as deposited SiO$_x$ films, whose EFTEM images are characterized by the absence of any relevant contrast. Moreover, the diffraction pattern (not shown) provides no evidence for the presence of the crystalline phase.

Electroluminescence

In order to form the devices, a substoichiometric SiO$_x$ (x < 2) film, 70 nm thick, was deposited on top of a low resistivity p-type Si substrate by plasma enhanced chemical vapor deposition. The Si concentration in the film was fixed to 46 at.%. After deposition, different SiO$_x$ films were implanted with different rare-earths (Er, Tm, Yb) to a dose of $7\times10^{14} - 1\times10^{15}$cm^{-2}; the energy was chosen in order to locate the rare-earth profiles in the middle of the dielectric layer. After the implantation step, the samples were annealed at 900 °C for 1 h in N$_2$ atmosphere in order to activate Er and to induce the separation of the Si and SiO$_2$ phases. EFTEM measurements have shown a structure similar to that reported on figure 1d, i.e. small Si aggregates, possibly very small Si nanocrystals or amorphous clusters, are formed in these conditions. The ion-implanted SiO$_x$ films were then used as dielectric layers in metal–oxide–semiconductor devices having an active area of 0.09 mm^2 [22-24].

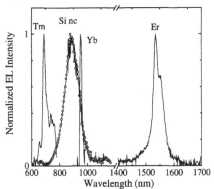

Figure 8. Normalized room temperature electroluminescence spectra of light emitting devices based on Si nanocrystals (nc) or rare-earth doped Si nanoclusters.

As a matter of fact, by changing the characteristics of the emitting centers in the active layer it has been possible to realize electrically pumped light sources operating at different wavelengths, as shown in figure 8. In particular, by using an active layer with only the Si nanocrystals dispersed in the matrix, room temperature light emission centered at around 850 nm has been obtained. On the other hand, by doping the Si rich layer with rare-earth ions (Er, Tm, Yb) it has been possible to obtain an efficient electroluminescent emission at different wavelengths (1.54, 0.78 and 0.98 μm, respectively), with current densities of the order of 10-20 A/cm^2. It should be noted that these devices are very stable and can work continuously for several days without deteriorating. This high stability is probably due to the large amount of closely spaced and very small Si nanoclusters dispersed in the SiO$_2$ matrix that facilitate the current flow through the insulating layer. Moreover, the presence of Si nanoclusters in the matrix allows for an efficient electrical pumping of the rare-earths to occur. Indeed, through time-resolved EL measurements (not shown) the value of the excitation cross section σ for Er in presence of Si nanoclusters under electrical pumping has been measured to be ~1×10^{-14} cm^2 [23]. It is noteworthy that this value is two orders of magnitude higher than the effective excitation cross section of Er ions through Si nanoclusters under optical pumping at 488 nm and it is comparable to the value found for the impact excitation of undoped Si nanoclusters by hot electrons, which is ~4×10^{-14} cm^2 [24], confirming the fundamental role of Si nanoclusters in the excitation mechanism of the rare-earth.

CONCLUSIONS

In conclusion, we have demonstrated that Si nanocrystals are completely amorphized when Er ions are introduced in the film through ion implantation, and that thermal treatments at 900°C for 1h, necessary for the optical activation of Er^{3+}, are not able to produce the recrystallization of the clusters, which therefore remain amorphous. Nevertheless, it has been shown that even in the amorphous phase, Si nanoclusters are still able to efficiently transfer the energy to nearby Er ions, thus acting as sensitizers for the Er luminescence at 1.54 μm. We developed a phenomenological model for the Si nanoclusters-Er interaction, able to quantitatively describe the measured optical properties. By solving the system of rate equations, describing the density populations of interacting Si nanoclusters-Er levels, we were able to determine both the nanoclusters-Er coupling constant and the up-conversion coefficient, through a fit of the experimental data. The presence of cooperative up-conversion definitely demonstrates that each Si nanocluster can actually excite more than one Er ion. On the other hand we have shown that, given a fixed Er concentration, the luminescence yield of the Er-Si nanoclusters system can be enhanced by increasing the density and decreasing the mean size of the sensitizing centers. Eventually, we have demonstrated that efficient room temperature electroluminescence can be obtained from different rare-earths due to the presence of Si nanoclusters, which are responsible for a good current injection and act as efficient sensitizers for all the rare-earths. These results open the route towards the fabrication of electrically-driven optical amplifiers based on the promising Er-doped Si nanoclusters system.

ACKNOWLEDGEMENTS

The authors wish to thank N. Marino, A. Spada, C. Bongiorno, A. Marino, S. Pannitteri, for their expert technical collaboration. This work has been partially supported by the EU IST-SINERGIA project and by MIUR through the project FIRB.

REFERENCES

1. A. J. Kenyon, P. F. Trwoga, M. Federighi, and C. W. Pitt, *J. Phys.: Condens. Matter* **6**, L319 (1994).
2. M. Fujii, M. Yoshida, Y. Kanzawa, S. Hayashi, and K. Yamamoto, *Appl. Phys. Lett.* **71**, 1198 (1997).
3. G. Franzò, V. Vinciguerra, and F. Priolo, *Appl. Phys. A: Mater. Sci. Process.* **69**, 3 (1999).
4. F. Priolo, G. Franzò, D. Pacifici, V. Vinciguerra, F. Iacona, and A. Irrera, *J. Appl. Phys.* **89**, 264 (2001).
5. G. Franzò, D. Pacifici, V. Vinciguerra, F. Iacona, and F. Priolo, *Appl. Phys. Lett.* **76**, 2167 (2000).
6. P. G. Kik, M. L. Brongersma, and A. Polman, *Appl. Phys. Lett.* **76**, 2325 (2000).
7. S. -Y. Seo, and J. H. Shin, *Appl. Phys. Lett.* **78**, 2709 (2001).
8. K. Watanabe, M. Fujii and S. Hayashi, *J. Appl. Phys.* **90**, 4761 (2001).
9. H. -S. Han, S. -Y. Seo, J. H. Shin, and N. Park, *Appl. Phys. Lett.* **81**, 3720 (2002)
10. F. Iacona, D. Pacifici, A. Irrera, M. Miritello, G. Franzò, F. Priolo, D. Sanfilippo, G. Di Stefano, and P. G. Fallica, *Appl. Phys. Lett.* **81**, 3242 (2002)
11. D. Pacifici, G. Franzò, F. Iacona, Luca Dal Negro, and F. Priolo, *Phys. Rev. B* **67**, 245301 (2003).
12. D. Pacifici, E.C. Moreira, G. Franzò, V. Martorino, F. Iacona, and F. Priolo, *Phys. Rev. B* **65**, 144109 (2002).
13. G. Allan, C. Delerue, and M. Lannoo, *Phys. Rev. Lett.* **78**, 3161 (1997).
14. A.Yu. Kobitski, K. S. Zhuravlev, H. P. Wagner, and D. R. T. Zahn, *Phys. Rev. B* **63**, 115423 (2001).
15. M. V. Wölkin, J. Jorne, P. M. Fauchet, G. Allan and C. Delerue, *Phys. Rev. Lett.* **82**, 197 (1999).
16. T. Förster, *Ann. Phys.* (N.Y.) **2**, 55 (1948).
17. D.L. Dexter, *J. Chem. Phys.* **21**, 836 (1953).
18. D. Kovalev, E. Gross, N. Künzner, F. Koch, V. Yu. Timoshenko, M. Fujii, *Phys. Rev. Lett.* **89**, 137401 (2002).
19. J.-H. Jhe, J. H. Shin, K.J. Kim, and D.W. Moon, *Appl. Phys. Lett.* **82**, 4489 (2003).
20. M. Fujii, M. Yoshida, S. Hayashi, and K. Yamamoto, *J. Appl. Phys.* **84**, 4525 (1998).
21. J. H. Shin, M.-J. Kim, S.-Y. Seo, and C. Lee, *Appl. Phys. Lett.* **72**, 1092 (1998).
22. G. Franzò, A. Irrera, E.C. Moreira, M. Miritello, F. Iacona, D. Sanfilippo, G. Di Stefano, P.G. Fallica, and F. Priolo, *Appl. Phys. A* **74**, 1 (2002).
23. F. Iacona, D. Pacifici, A. Irrera, M. Miritello, G. Franzò, F. Priolo, D. Sanfilippo, G. Di Stefano, and P. G. Fallica, *Appl. Phys. Lett.* **81**, 3242 (2002).
24. A. Irrera, D. Pacifici, M. Miritello, G. Franzò, F. Priolo, F. Iacona, D. Sanfilippo, G. Di Stefano, and P. G. Fallica, *Appl. Phys. Lett.* **81**, 1866 (2002).

Mat. Res. Soc. Symp. Proc. Vol. 817 © 2004 Materials Research Society   L1.4

The dot size effect of amorphous silicon quantum dot on 1.54-μm Er luminescence

Nae-Man Park[1]*, Tae-Youb Kim[1], Gun Yong Sung[1], Baek-Hyun Kim[2], Seong-Ju Park[2], Kwan Sik Cho[3], Jung H. Shin[3], Jung-Kun Lee[4], and Michael Nastasi[4]

[1]*Basic Research Laboratory, Electronics and Telecommunications Research Institute, Daejeon 305-350, Korea*
[2]*Department of Materials Science and Engineering, Kwangju Institute of Science and Technology, Kwangju 500-712, Korea*
[3]*Department of Physics, Korea Advanced Institute of Science and Technology, Daejeon 305-701, Korea*
[4]*Materials Science & Technology Division, Los Alamos National Laboratory, NM 87545, USA*
(*Electronic mail: nmpark@etri.re.kr)

ABSTRACT

The role of the size of amorphous silicon quantum dots in the Er luminescence at 1.54 μm was investigated. As the dot size was increased, the more Er ions were located near one dot due to its large surface area and more Er ions interacted with other ones. This Er-Er interaction caused a weak photoluminescence intensity despite the increase in the effective excitation cross section. The critical dot size, needed to take advantage of the positive effect on Er luminescence, is considered to be about 2.0 nm, below which a small dot is very effective in the efficient luminescence of Er.

INTRODUCTION

Er-doped silicon has attracted a great deal of interest because of its promising future in the development of light-emitting diodes and lasers operating at a wavelength of 1.54 μm, which coincides with the absorption minimum of optical fibers.[1, 2] However, the intensity of the photoluminescence (PL) of Er in this matrix is very weak at room temperature. Attention is currently focused on Er-doped Si nanocrystals in SiO_x which holds some promise for efficiently generating light emission, since it has been demonstrated that Si nanocrystals in presence of Er act as efficient sensitizers for Er ions.[3-6] Amorphous Si quantum dots (a-Si QDs) have been fabricated previously and their role as an active layer in visible light-emitting diode demonstrated, which stimulated interest in the control of dot size in a small dimension compared to nanocrystals.[7-9] Theoretical calculation also showed that the radiative recombination rate for a-Si QD is higher by two to three orders of magnitude than that for crystalline Si QD,[10] indicating that better performance can be obtained for a 1.54 μm light source when it is fabricated by using an Er-doped a-Si QD.

In a recent report, the density effect of a very small a-Si cluster on Er PL in Si-rich SiO_2 was investigated, where a high density of a-Si clusters enhanced PL efficiency compared to Si nanocrystals.[11] In this work, we report on the effect of the size of a-Si QDs on Er luminescence.

EXPERIMENTAL DETAILS

50-nm thick silicon nitride films containing a-Si QDs were grown on Si substrates by plasma enhanced chemical vapor deposition with various dot sizes.[8] Er^+ ions were then implanted with an ion dose of 1×10^{21}/cm^3 into the silicon nitride films. The profile of implanted Er ions was monitored by Rutherford backscattering spectroscopy. Finally, the samples were annealed at a temperature of 900 °C for 0.5 h in order to reduce the residual defects left by the implantation process. PL of the Er ions in the annealed films was measured using an Ar laser. The samples were classified into three groups, referred to as large-dot, medium- dot, and small-dot samples in accordance with dot sizes (diameters) of 2.5, 1.8, and 1.4 nm, respectively. The change of dot size after the Er ion implantation and the thermal annealing was confirmed by PL measurement which showed the same peak position before and after each process and, therefore, the ion implantation and the thermal annealing at 900 ℃ for 30 min are considered not to influence on the dot size.

RESULTS AND DISCUSSION

Figure 1 shows the PL spectra as a function of the dot size at room temperature. The spectrum shows the luminescence of a-Si QDs in a visible range and a sharp peak at around 1.54 μm, which is a characteristic transition between $^4I_{13/2}$ and $^4I_{15/2}$ manifolds in Er^{3+} ions. The emission peak position of a-Si QDs was controlled by the dot size due to a quantum size effect.[7] The small-dot sample shows a nearly 4 times higher Er PL intensity than the large-dot sample. Since the luminescence of Er ions is sensitive to energy transfer between Si dots and Er ions, these data show that a small-dot is very effective in Er luminescence and, as a result, enhances the luminescence efficiency.

Figure 2 shows the power dependency of Er PL versus dot size. The PL intensity is nearly proportional to the square root of the pump power. The sublinear power dependence in Er:Si can be rationalized as being due to the saturation of excitation of active Er ions,[12] where

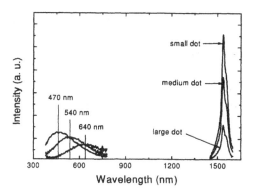

Figure 1. PL spectra as a function of the size of an *a*-Si QD at room temperature. PL spectra of various sized *a*-Si QDs correspond to the dot sizes of 2.5, 1.8, and 1.4 nm, respectively.

the saturation power is dependent on the PL decay time, that is, the high saturation power is indicative of a short decay time. However, the saturation power increases with increasing decay time in our samples when the dot size is decreased. The inset in Fig. 2 summarizes the decay time (τ_{decay}) and the effective excitation cross section (σ) of Er PL for different dot size samples at room temperature. The decrease in τ_{decay} value with increasing dot size indicates that the non-radiative process is likely to be enhanced as the dot size increases. The effective excitation cross section is, on the other hand, increased with increasing dot size.

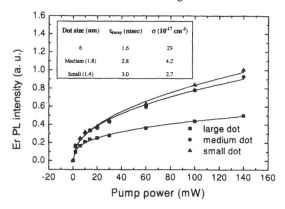

Dot size (nm)	τ_{decay} (msec)	σ (10^{-17} cm^{-3})
6	1.6	29
Medium (1.8)	2.8	4.2
Small (1.4)	3.0	2.7

Figure 2. PL intensity of Er ions as a function of excitation power for various dot-sized samples, measured at 25 K using a 488 nm pump light. The inset shows the decay time and the excitation cross section of Er PL.

Although the Er ion concentration is the same for all samples, σ is larger in a large-dot sample than that in a small-dot sample. A previous study showed that the density of a-Si QD was increased about eightfold when the dot size was decreased from 2.0 to 1.4 nm.[8] Considering the increases in the dot density and the total volume fraction of dots, the enhanced Er PL intensity with decreasing dot size is reasonable because luminescent Er ions excited by dots are increased. Therefore, it would be expected that σ must increase with decreasing the dot size, but it follows an opposite trend, which cannot be explained by the dot density.

The increase in σ value with increasing dot size can be explained by the fact that the absorption coefficient (α) of the film is increased with increasing dot size because σ is proportional to α/n_{Er}. In reality, α is increased about 3 times with increasing dot size from 1.4 to 2.5 nm. However, σ value was increased about 10 times as dot size was increased, which means that n_{Er} should be decreased about 1/3 times with increasing dot size. In this case, the saturation power can be decreased due to the decrease in n_{Er}. The decrease of τ_{decay} can be also understood by considering the increase in the refractive index with increasing dot size.[13] Because Er PL intensity is linearly proportional to n_{Er} in a relation of $I_{PL} \sim n_{Er}/\tau_{rad}$, where τ_{rad} is the radiative lifetime[14] which is correlated to the decay time ($1/\tau_{decay} = 1/\tau_{rad} + 1/\tau_{nr}$ in which τ_{nr} is the nonradiative lifetime), the decrease in Er PL intensity with increasing dot size results in the similar τ_{rad} for all samples. However, the low temperature PL decay measurement showed that τ_{decay} at 25K was almost the same as that at room temperature within 0.2 msec for all samples, which means that τ_{rad} is different to each other. Therefore, another effect must be considered in order to clearly explain the trend of Er PL intensity with varying the dot size in addition to the variation of the absorption coefficient.

Recently, luminescent Er ions were observed to be located nearly at the surface of the nanoclusters.[4] Jhe et al.[15] also observed that the characteristic carrier-Er interaction distance is 0.5 nm, over which the interaction between carriers in an a-Si well and Er ions becomes very weak. This is considered to be reasonable in our case. The surface area of a dot is increased as the dot size is increased. A large surface area of a dot is considered to be very effective in sensitizing Er ions because more Er ions can be located near one dot and interact with the dot. However, an excited Er ion is also able to readily interact with other Er ions in the presence of one dot because the luminescent Er ions are close to each other. This constitutes the negative effect in terms of luminescence efficiency. In Er-doped Si nanocrystals,[16, 17] the effective excitation cross section was increased due to the interaction of close Er pairs with increasing the Er content in the film and the decay time was simultaneously decreased. In our study, the same mechanism may be responsible for the correlation between the effective excitation cross section and PL intensity in various dot sizes although the Er content was the same for all samples. As the dot size increased, the number of optically active Er ions near one a-Si QD increased and the

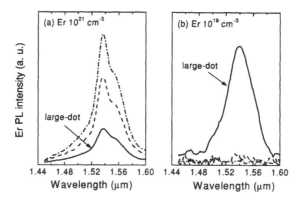

Figure 3. PL spectra for various dot-sized samples with (a) an Er concentration of 10^{21} cm^{-3} and (b) 10^{19} cm^{-3} at room temperature.

close space between Er ions allows for energy exchange between neighboring ions, which is well-known concentration quenching effect. This effect caused the increase in the effective excitation cross section and the decrease in the saturation power with increasing dot size because Er ions are additionally excited due to the resonant excitation by other Er ions. Er-Er interactions, however, can be easily coupled to the nonradiative quenching sites and, as a result, the PL decay time is decreased and the PL intensity becomes weak. This effect was not observed in the medium- and small-dot samples. If this explanation is correct, a large-dot sample must show an efficient PL at a small Er dose, compared to a small-dot sample. In reality, PL was observed only in a large-dot sample implanted with an Er dose of 1×10^{19}/cm^3 as shown in Fig. 3. Therefore, our work suggests that the maximum dot size, needed to take advantage of the positive effect of a-Si QDs on Er luminescence without being affected by quenching phenomena due to Er-Er interactions, is about 2.0 nm.

CONCLUSION

Er PL properties as a function of the size of the a-Si QD were investigated in a silicon nitride film. The effect of dot size on Er PL is associated with an increase in the close Er pair interactions, which are more prevalent in a large-dot sample because a large dot sample has a large surface area and more luminescent Er ions are located near one dot. This indicates that dot size is very important in the efficient luminescence of Er.

ACKNOWLEDGMENTS

This work was supported by the Ministry of Information and Communication in Korea. Authors of Los Alamos National Laboratory thank the technical staff of the Ion Beam Materials Lab for their assistance.

REFERENCES

1. J. Palm, F. Gan, B. Zheng, J. Michel, and L. C. Kimerling, Phys. Rev. B 54, 17603 (1996).
2. A. Polman, J. Appl. Phys. **82**, 1 (1997).
3. M. Fujii, M. Yoshida, S. Hayashi, and K. Yamamoto, J. Appl. Phys. 84, 4525 (1998).
4. S.-Y. Seo and J. H. Shin, Appl. Phys. Lett. **78**, 2709 (2001).
5. M. Schmidt, J. Heitmann, R. Scholz, and M. Zacharias, J. Non-Cryst. Solids **299-302**, 678 (2002).
6. F. Iacona, D. Pacifici, A. Irrera, M. Miritello, G. Franzò, F. Priolo, D. Sanfilippo, G. Di Stefano, and P. G. Fallica, Appl. Phys. Lett. **81**, 3242 (2002).
7. N.-M. Park, C.-J. Choi, T.-Y. Seong, and S.-J. Park, Phys. Rev. Lett. **86**, 1355 (2001).
8. N.-M. Park, S. H. Kim, G. Y. Sung, and S.-J. Park, Chem. Vap. Deposition **8**, 254 (2002).
9. N.-M. Park, T.-S. Kim, and S.-J. Park, Appl. Phys. Lett. **78**, 2575 (2001).
10. K. Nishio, J. Koga, T. Yamaguchi and F. Yonezawa, Phys. Rev. B **67**, 195304 (2003).
11. G. Franzò, S. Boninelli, D. Pacifici, F. Priolo, F. Iacona, and C. Bongiorno, Appl. Phys. Lett. **82**, 3871 (2003).
12. S. Coffa, F. Priolo, G. Franzò, V. Bellani, A. Carnera, and C. Spinella, Phys. Rev. B. **48**, 11782 (1993).
13. H. P. Urbach and G. L. J. A. Rikken, Phys. Rev. A **57**, 3913 (1998).
14. M. S. Bresler, O. B. Gusev, P. E. Pak, E. I. Terukov, and I. N. Yassievich, Phys. Solid State 43, 625 (2001).
15. J.-H. Jhe, J. H. Shin, K. J. Kim, and D. W. Moon, Appl. Phys. Lett. **82**, 4489 (2003).
16. P. G. Kik and A. Polman, J. Appl. Phys. **88**, 1992 (2000).
17. F. Priolo, G. Franzò, D. Pacifici, V. Vinciguerra, F. Iacona, and A. Irrera, J. Appl. Phys. **89**, 264 (2001).

Mat. Res. Soc. Symp. Proc. Vol. 817 © 2004 Materials Research Society　　　　　　L1.6

Rare-earth doped microlasers for microphotonic applications

Lan Yang, Bumki Min and K. J. Vahala
Department of Applied Physics, California Institute of Technology, Pasadena, CA 91125, U.S.A.

ABSTRACT

Sol gels provide a highly flexible technique for preparation of both planar and non-planar oxide thin films. They also enable the incorporation of various dopants into the films. In this work we describe the application of erbium-doped solgel films to surface functionalize optical microresonators. The resulting microlaser devices are especially interesting because their emission band falls in the important 1.5 µm window used for optical fiber communications. Both microsphere and ultra-high-Q microtoroid resonators-on-a-chip were functionalized into lasers and then characterized [1]. The erbium-doped sol-gel films were applied to the resonator surface and subsequently a CO_2 laser was used to induce flow and densification of the sol-gel film on the surface. Optical quality thin films were obtained after the CO_2 laser induced anneal. By varying the doping concentration and thickness of the applied sol-gel layers in microsphere resonators, we can vary the laser dynamics so that both continuous-wave and pulsation operation are possible. Single mode performance with high differential quantum efficiency was also obtained using the ultra-high-Q microtoroid resonator. These chip-based microlasers enable integration with other optical or electronic functions [2-3].

INTRODUCTION

Microcavities formed by surface tension can exhibit quality factors in excess of 1×10^8 and are of interest in nonlinear optics, optical fiber communications, and sensing. Among these applications, Er^{3+}-doped microlasers are especially interesting due to the erbium $4f$ transition $^4I_{13/2} \rightarrow {}^4I_{15/2}$ which falls in the 1.55 µm telecommunication window. The combination of ultra-high quality factor and small mode volume can lead to ultra-low threshold microlasers. In addition, microtoroid lasers are fabricated on a silicon wafer and are thus integrable with other optical or electronic components. In this work, we study and characterize the laser performance of both microsphere and microtoroid lasers achieved by solgel surface functionalization of ultra-high Q microresonators. To gain insight into the microlaser operation and performance, a model was developed and predictions were compared with experimental data.

EXPERIMENTAL DETAILS

Two different microcavities, microspheres and microtoroids on silicon wafer [1], were used as the base resonator structure for surface functionalization. In the case of spheres, the initial silica microsphere was formed by heating the end of a tapered fiber tip with a CO_2 laser as describe by Knight et al [4]. On the other hand, ultra-high Q microtoroids were fabricated upon silicon wafer with 2-µm layer of silica by using a combination of lithography, dry etching and selective reflow process. This process, as described in reference 1, begins by creation of a series of circular silica pads on a base silicon wafer through optical lithography and buffered HF etching. Subsequently, these pads served as etch masks for isotropic etching of silicon using XeF_2. During this process, the silica disks are undercut equally, leaving the silica disks supported by a silicon pillar. Finally, a CO_2 laser was used to selectively reflow the silica disks, during which the disks collapse into toroids due to surface tension.

The solgel starting solution was prepared by hydrolyzing tetraethoxysilane (TEOS) using water in acidic condition (pH~1) with isopropanol as a co-solvent. Er^{3+} was introduced by adding $Er(NO_3)_3$ with an appropriate ratio to obtain the desired Er^{3+} concentration in the solgel film. Then the mixture was stirred vigorously at 70 °C for 3~10 h to form a viscous solution. After aging, the silica microresonators were immersed in the solution at room temperature for several hours. Then, the coated microcavities were heated in an oven at around 160 °C to drive off the solvents inside the solgel films. After this process, the coated microcavities were irradiated using a CO_2 laser for several seconds. The laser intensity was sufficient to induce flow and densification of the solgel layer. In addition, microcracking that was present in the solgel films can be annealed out by the laser reflow process. The exceptionally smooth surface of the coated microcavities endows these structures with their high-quality-factor properties. By repeating the process, we varied the coating thickness.

A tunable single frequency, narrow-linewidth (<300 kHz) external cavity laser in the 1480 nm band was used to pump the Er^{3+}-doped microlasers. A fiber optic taper [5-6]was used to couple both pump light into the resonators as well to extract laser emission. A three-axis translator was used to control the position of the microlasers and a rotator was used to adjust the angle of the microlaser relative to the fiber taper. The microlaser-taper coupling zone was monitored by two CCD cameras in both horizontal and vertical direction. The fiber taper with a diameter of 1~2 µm was made by heating and stretching a single mode fiber with a hydrogen flame. An optical spectrum analyzer (OSA) with resolution of 0.5 nm was used to monitor both the pump and signal spectra.

RESULTS AND DISCUSSION

Whispering-gallery mode (WGM) resonances correspond to light trapped in circulating orbits just within the surface of the cavities. Figure 1(a) shows representative lateral emission distribution for a fundamental WGM in the microsphere-taper coupling zone. The green rings are upconverted photoluminescence of the Er^{3+}. Like all whispering-gallery type microresonators microtoroids feature optical modes that are confined near the resonator periphery. But in contras to the microsphere, this structure has a simpler mode spectrum, which, as discussed below facilitates single-mode laser operation in the microtoroids. Figure 1(b) presents a top view of ar Er^{3+}-doped microtoroid laser coupled by a fiber taper.

Figure 1. (a) Image of the fundamental WGM in the taper-microsphere coupling zone. (b) Image of WGM in the taper-microtoroid coupling zone. The green rings are due to the upconverted transition of Er^{3+}.

A typical laser spectrum of a microtoroid laser is presented in Figure 2(a). Unlike microspheres, single line operation is easily achieved in the microtoroid lasers owing to their simplified (i.e., less densely populated) mode spectrum. Multi-line operation was also possible in these devices, however, single line operation could always be obtained by proper choice of the coupling condition. Figure 2(b) shows the typical input pump power versus output lasing power relation for a toroidal microcavity with major diameter of 60 μm. In this case, the threshold power was around 14 μW and the unidirectional slope efficiency was 1.4 %. During the data collection, the lasing spectrum remained in single mode operation within the resolution of optical spectrum analyzer. When compared with Er^{3+}-doped microsphere laser, a toroidal microcavity laser of comparable Q factor and major diameter can achieve a much lower threshold power theoretically due to its compressed mode volume. The spectral features of these devices also benefit from reduced mode structure enabling single mode lasing.

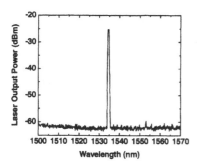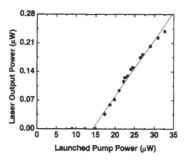

Figure 2: (a). Typical laser spectrum of an Er^{3+}-doped microtoroid laser. (b).Measured laser output power versus the absorbed pump power for a microtoroid laser with principal diameter of 60 μm. The threshold in this case is 14 μW.

Modeling shows that threshold power is optimal at a certain erbium ion concentration, which depends upon the intrinsic Q-factor. In the low concentration limit, the threshold power increases sharply because erbium ions are not able to give sufficient gain required for loss compensation; while in high concentration limit, the threshold increases due to concentration-dependent loss mechanisms such as upconversion and ion-pairing. As noted, the optimized erbium concentration depends mainly on intrinsic Q-factor of the cavity (assuming that the overlap integral between mode field pattern and erbium doping profile is the same). If the intrinsic Q-factor is increased, the optimized erbium concentration is lowered. This trend can be expressed roughly as a constant (erbium concentration)×(intrinsic Q-factor) effect. Figure 3 illustrates theoretical absorbed threshold power as a function of erbium concentration for different intrinsic Q-factors. As can be seen in this graph, sub-microwatt absorbed threshold power is possible in theory for intrinsic Q-factors easily obtained in toroidal microcavities. This theoretical plot is based on the coupled harmonic oscillator model, which includes erbium ions as the active gain material [7]. Because precise control of erbium concentration is possible using the solgel method, further optimization of concentration can result in a sub-microwatt threshold laser [8].

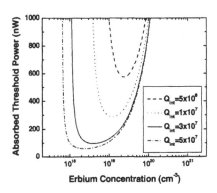

Figure 3: Absorbed threshold power versus erbium concentration for different microcavity quality factors.

CONCLUSIONS

In conclusion, we have demonstrated that we can achieve microlasers by applying rare-earth doped solgel films to the surface of ultra-high-Q microspheres and microtoroids. This technique provides a way to achieve a range of possible gain media in the microresonator system. Furthermore, it is also possible to use the solgel process for creation of the base silica film itself, thereby eliminating the need to surface functionalize a thermal silica layer. This direct method will be described elsewhere [8]. Finally, other applications, such as nonlinear optics, can benefit from the surface-functionalization technique by applying layers of nonlinear media to the surface of the microresonator [9].

ACKNOWLEDGEMENTS

This work was supported by the Defense Advanced Research Project Agency, the National Science Foundation and the Caltech Lee Center.

REFERENCES

1. D.K Armani, T. J. Kippenberg, S. M. Spillane, and K. J. Vahala, Nature, **421**, 925 (2003)
2. L. Yang, D.K Armani, and K. J. Vahala, Appl. Phys. Lett. **83**, 825 (2003)
3. A. Polman, B. Min, J. Kalkman, T. J. Kippenberg, and K. J. Vahala, Appl. Phys. Lett. **7**, 1037 (2004)
4. J.C.Knight, G.Cheung, F.Jaques, and T.A.Birks, Opt. Lett. **22**, 1129 (1997)
5. M. Cai, O. Painter, K. J. Vahala, Phy. Rev. Lett. **85**, 74, (2000)
6. S M Spillane, T J Kippenberg, O J Painter, K .J. Vahala, Phy. Rev. Lett. **91**, 043902 (2003)
7. B. Min, T. J. Kippenberg, L. Yang, K. J. Vahala, J. Kalkman, and A. Polman, submitted to Phys. Rev. A
8. L. Yang, S. Spillane, B. Min, T. Carmon, and K. Vahala, submitted to Appl. Phys. Lett
9. S. Spillane, T. Kippenberg, and K. J. Vahala, Nature **415**, 621 (2002)

Mat. Res. Soc. Symp. Proc. Vol. 817 © 2004 Materials Research Society   L1.7

A High Index Contrast Silicon Oxynitride Materials Platform for Er-doped Microphotonic Amplifiers

Sajan Saini, Jessica G. Sandland, Anat Eshed, Daniel K. Sparacin, Luca Dal Negro, Jurgen Michel and Lionel C. Kimerling
Microphotonics Center, Dept. of Materials Science and Engineering,
Massachusetts Institute of Technology, Cambridge, MA 02139, U.S.A.

ABSTRACT

Er-based optical amplification continues to be the ideal low noise, WDM crosstalk free, broadband candidate for waveguide amplifiers. Design analysis of the applicability of Er-Doped Waveguide Amplifiers (EDWAs) for micron-scale integrated photonics in a planar lightwave circuit concludes: (i) an >80× increase in gain efficiency, and (ii) a >40× increase in device shrink can be realized, for a high index contrast EDWA (with a core-cladding index difference of $\Delta n = 0.1 \leftrightarrow 0.7$), compared to a conventional Er-doped fiber amplifier. The materials challenge now is to establish a robust materials system which meets this high index difference design requirement while simultaneously leveraging the capability of silicon (Si) processing: a host platform for EDWAs must be found which can integrate with Si Microphotonics. Silicon nitride (Si_3N_4), silicon oxide (SiO_2) and a miscible silicon oxynitride alloy (SiON) of the two meet this materials challenge. We present the results of reactive and conventional magnetron sputtering based materials characterization for this high index host system. Room temperature and 4 K photo-luminescence studies for annealed samples show the reduction of non-radiative de-excitation centers while maintaining an amorphous host structure. Atomic force microscopy shows less than 1 nm peak-to-peak roughness in deposited films. Prism coupler measurements show a reliable reproducibility of host index of refraction with waveguide scattering loss <2 dB/cm. We conclude that the SiON host system forms an optimal waveguide core for an SiO_2-clad EDWA. Initial gain measurements show a gain coefficient of approximately 3.9 dB/cm.

INTRODUCTION

Integrated Circuit (IC) chip design relies on lithography as the organizing principle for creation of integrated electronic device elements on a common material substrate (the semiconductor silicon (Si)). More importantly, the use of lithography defines the integrated circuit as a scalable technology, correlating smaller devices and denser IC architectures with lithography wavelength. Moore's Law[1] has mapped the exponential impact of this lithography principle ultimately to materials and processing performance limitations. One such limitation is the 'interconnection bottleneck,' referring to the increase in resistive and capacitive (RC) delay in the global metal interconnect lines[2] running across an IC chip. A proposed solution to the interconnect bottleneck problem is the replacement of global metal interconnects with 'optical interconnects,' a term referring to planar waveguides that carry IC signals as encoded light pulses, free of parasitic RC delay[3].

Planar waveguide technology developed from fiber optic telecommunications has matured in the last ten years into the field of photonics. Micron-scale photonics, or Microphotonics, aims to converge the solutions of the fiber optic link to resolve the interconnection bottleneck problem. The goal of Microphotonics is to reproduce a complete fiber optic link communications architecture as an integrated lithography based scalable technology: a 'planar lightwave circuit.' The most critical device element to the long-distance fiber optic link was the Er-Doped Fiber Amplifier (EDFA), which served to amplify signal power at λ=1.55μm wavelength light[4]. The EDFA is an optical fiber with higher numerical aperture, i.e. a higher index difference Δn between waveguide core and cladding, than conventional fiber. This higher Δn of ~0.01, results in improved power efficiency. The metric for evaluating amplifier power performance is gain efficiency, defined as the amount of decibel (dB) signal gain output, per unit milliwatt (mW) pump power input.

Previously published work[5] summarizes our theoretical investigations into the dependence of gain efficiency on Δn. This work concludes a cumulative performance benefit of $\Delta n^{2.6}$ for amplifier performance, quantified as the Figure of Merit:

$$\text{Figure of Merit} = \frac{\text{Gain Efficiency}}{\text{Footprint}} \tag{1}$$

where Footprint is the amount of two-dimensional planar area occupied by a coiled channel waveguide planar structure.

To experimentally realize these results, we investigate an Er-doped waveguide core material host with a refractive index on the order of Δn~0.1$\leftrightarrow$0.75 higher than an SiO_2 waveguide cladding. We chose SiON and Si_3N_4 as a Si processing compatible dielectric host that yields high Er-doped luminescence for low pump powers.

EXPERIMENTAL PROCEDURE

RF magnetron sputtering was used to deposit 0.3-0.5 μm thick films of Er-doped SiO_2, SiON and Si_3N_4. The SiO_2 and Si_3N_4 films were sputtered using Ar gas with SiO_2 and Si_3N_4 targets, respectively. The SiON films were grown both by reactive sputtering of a Si_3N_4 target using an Ar/O_2 (10%) gas, and by co-sputtering of the SiO_2 and Si_3N_4 targets using Ar gas. In all cases, Er doping was done by co-sputtering a metal Er target at low target powers in order to ensure doping concentrations on the order of 1% atomic fraction. Sputtering was done in a Kurt-Lesker UHV sputter system, with a base pressure of 1-5$\times 10^{-8}$ torr.

Film thickness was measured by profilometry and refractive index was measured by both ellipsometry and prism coupler measurements. Optical loss in the films was measured by a modified prism coupler measurement detecting scattered light decay as a function of distance. Er dopant concentrations were determined by Rutherford Back-Scattering. Room temperature, 4 Kelvin (K) photoluminescence (PL) intensity and lifetime measurements were done in an Oxford Instruments liquid helium cooled cryostat, using a mechanically chopped Ar ion 488 nm laser as the excitation source. Luminescence

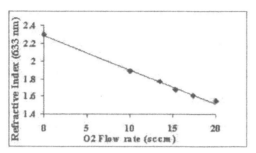

Figure 1. Adjustable control of refractive index for reactively sputtered SiON films, as a function of Ar/O₂ (10% O₂) gas flow rate.

intensity was collected using a Spex spectrometer and liquid nitrogen cooled Hamamatsu photo-multiplier tube.

DISCUSSION

The principal advantage of Ar/O₂ reactive or conventional Ar sputtered SiON and SiO₂/Si₃N₄ films, respectively, is the absence of hydrogen (H) incorporation in the thin films. Plasma Enhanced Chemical Vapor Deposition, a widespread high deposition rate tool, produces H-incorporated SiON films, whose N-H bond creates an absorption band in the 1.55 μm wavelength spectral range (the range of emission from the Er atom).

We have shown reliable reproducibility of refractive index control in reactively sputtered SiON films (see Fig.1), as a function of the Ar/O₂ (10% O₂) gas flow rate. Lastly, the choice of sputtering allows for doping of Er by co-sputtering, thus avoiding the costly process of ion implantation while optimizing the amount of overlap between a homogeneous Er profile and the guided light mode.

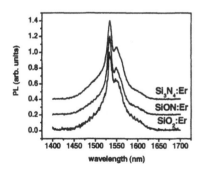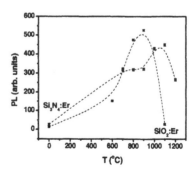

Figure 2. (a) Photoluminescence (PL), normalized to peak intensity, from sputtered SiO₂:Er, SiON:Er and Si₃N₄:Er. Peak position is at 1533 nm. (b) Plot of PL peak intensity versus annealing temperature for SiO₂:Er and Si₃N₄:Er samples. The dashed lines are meant to guide the eye.

Fig.2.(a) overlays the photoluminescence (PL) spectral profile of Er in the three sputtered hosts systems of SiO_2, SiON (refractive index n=1.6 at λ=1.55μm light), and Si_3N_4. We observe near identical PL profiles for Er within all three hosts, suggesting all three hosts have a similar degree of inhomogeneous broadening. The amorphous structure of these three hosts, even after high temperature anneal, has been confirmed by X-ray diffraction.

Efficient operation of a waveguide optical amplifier relies on two critical constraints: (i) low optical loss both in the waveguide core bulk and at the core-cladding interface, in order to ensure net gain is greater than zero; and (ii) a long spontaneous emission lifetime from the Er atom at room temperature, to allow population inversion for low pump powers. Given the upconversion gain limitation of the Er atom[6] at concentrations in excess of 10^{20} Er/cm^3, we approximate a potential upper limit on Er based gain (using literature values for Er absorption and emission cross-sections in SiO_2[6]) of ~3 dB/cm. Thus, losses due to criterion (i) must have a cumulative value significantly less than 3 dB/cm, in order to realize a waveguide amplifier with net gain.

We evaluate criterion (i) by prism coupler loss measurements and atomic force microscopy. Prism coupler results show a 2 dB/cm materials loss for reactively sputtered SiON and 0.8 dB/cm materials loss for co-sputtered SiON. The Si_3N_4 film is observed to have prohibitively high loss values of 15-20 dB/cm. The low bulk materials loss of co-sputtered SiON:Er thus makes it a feasible sputtering candidate for a waveguide amplifier.

Atomic force microscopy results show a peak-to-peak roughness of 10 nm for the reactively sputtered SiON film and 1 nm for the Si_3N_4 film. If the sputter process can not be controlled to produce smoother SiON films, we conclude a Chemical Mechanical Polish procedure will be required for SiON films (reducing peak-to-peak roughness from 10 nm to 2 nm or lower) in order to ensure the processing of buried symmetric channel SiON waveguides (with an SiO_2 cladding) that have a core-cladding interface scattering loss significantly lower than 3 dB/cm. We are currently investigating the possibility of minimizing film roughness by process control in co-sputtered SiON.

Fig.2.(b) shows the effect of annealing on PL emission from sputtered SiO_2:Er and Si_3N_4:Er (all anneals were performed for one hour). We observe that sputtered SiO_2:Er exhibits a peak luminescence at temperature T~900°C, with a steep PL roll-off for higher temperature anneals. At the doping concentrations of ~10^{20} Er/cm^3 in these samples, we suggest this roll-off may be attributed to localized Er-Er clustering at a short-range nm length scale, resulting in a reduction of the concentration of light-emitting Er. PL roll-off for high annealing temperatures has been reported in the literature for Er ion-implanted samples in both SiO_2 and Si_3N_4[7]. Er PL from annealed Si_3N_4 shows a similar behavior, with two important observations: (i) PL peak intensity occurs for samples annealed around T~1100°C, a temperature significantly higher than in SiO_2:Er. If the optimization of Er PL emission is dependent on short range structural relaxation about the Er atom, observation (i) is consistent with Si_3N_4 being a more refractory material than SiO_2 (Si_3N_4 has a higher glass transition temperature). (ii) PL intensity from annealed Si_3N_4:Er shows two peak emission plateaus. This shows two distinct optimization processes are

Host	Refractive index	$[Er]_{RBS}$ (cm^{-3})	τ_{4K} (ms)	τ_{roomT} (ms)	$\eta=\tau_{roomT}/\tau_{4K}$
CVD SiO$_2$	--	(1.04×10^{20})	27.5	24.6	0.9
SiO$_2$	1.46	0.9×10^{20}	10.8	10.1	0.93
SiON	1.6	2.4×10^{20}	10.6	4.0	0.37
Si$_3$N$_4$	2.2	1.0×10^{20}	2.4	1.2	0.50

Table 1. refractive index, 4 K and room temperature lifetime data, and Rutherford Back-Scattering(RBS) Er concentration measurements, for SiO$_2$:Er, SiON:Er and Si$_3$N$_4$:Er. 'CVD SiO$_2$' refers to a CVD grown SiO$_2$:Er sample donated by Corning, Inc. The Er concentration for the CVD sample is a nominally prepared composition as determined by Corning manufacturers.

happening in annealed Si$_3$N$_4$:Er. We interpret the two processes as a short-range and intermediate range structural relaxation within the Si$_3$N$_4$ host.

Sputtered SiO$_2$:Er is our control sample for investigating the influence of a sputtered host material on the Er lifetime, by comparing PL lifetime with a chemical vapor deposition (CVD) grown host. Our CVD grown SiO$_2$:Er reference is an SiO$_2$ majority constituent glass, optimized for EDFA application by Corning, Inc. Table 1 lists PL measurements on all three samples versus this CVD optimized sample.

We observe that sputtered SiO$_2$:Er has remarkably similar performance to CVD SiO$_2$:Er. The CVD sample contains proprietary chemical additives to minimize Er-Er clustering at a nominally claimed concentration of 1.1×10^{20} Er/cm^3. For a similar concentration of Er, sputtered SiO$_2$:Er, without any de-clustering additives, produces comparable lifetimes at λ=1533 nm. 10 ms lifetimes are the optimal range achieved in the literature for Er ion implanted samples as well. The quantum internal efficiency, defined as the ratio of room temperature to 4 K lifetimes (we are assuming a lifetime measured at 4 K is representative of the Er atom's radiative lifetime), is nearly identical for sputtered SiO$_2$:Er and the CVD sample. We conclude the sputtering process has the inherent capability to produce the optical quality requisite for an Er atom-transition based optical amplifier.

We observe that reactively sputtered SiON:Er has long lifetimes at 4 K, which quench by over a factor of two at room temperature. We conclude the host matrix for reactively sputtered SiON has a considerable concentration of temperature dependant non-radiative de-excitation sites. Similar measurements on co-sputtered SiON:Er are underway. Lastly, we observe sputtered Si$_3$N$_4$:Er lifetimes experience very little quenching between 4 K and room temperature, however the 4 K lifetime for Er is on the order of 1 ms. Polman et al.[7] reported Er lifetime values of 7 ms in ion-implanted samples annealed in the same temperature range. We thus conclude the Er concentration distribution is less disperse in sputtered Si$_3$N$_4$:Er than in sputtered SiO$_2$:Er. The <0.5 quantum internal efficiency of reactively sputtered SiON:Er and sputtered Si$_3$N$_4$:Er imply that as is, these materials are poor light emitters at room temperature, and hence very difficult material to make a laser out of (which require a high photon generation rate from spontaneous emission, within the lasing cavity). However as an optical amplifier, these materials, may still be relevant, since the Er concentration can still be population inverted in a host where its radiative emission is low.

We conclude with the report of initial gain measurements (Fig.3) on the Si_3N_4:Er doped sample. Measurements were done by the Variable Stripe Length method[8] on a film grown to a single mode (for one-dimensional confinement) thickness of 0.5 μm, using

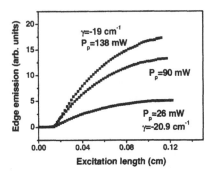

Figure 3. Variable Stripe Length method for measuring Amplified Spontaneous Emission[8] from Si_3N_4:Er. Results show a decrease in modal loss as a function of pump power.

variable pump powers at wavelength λ=488 nm. We do not observe net gain, but rather a reduction in modal loss of magnitude 1.9 cm^{-1}=7.3 dB/cm, for a pump power increase from 26 mW to 138 mW. Measurements for pump powers greater than 138 mW indicate the modal loss saturates around a value of 19 cm^{-1}. The difference between modal loss at 136 mW versus 26 mW pump powers is therefore approximately equal to 2×gain coefficient≈2ΔNσ$_{21}$ where ΔN is the concentration of population inversion (approximated as ΔN≈N) and σ$_{21}$ is the stimulated emission cross-section. If the 7.3 dB/cm=1.9 cm^{-1} reduction in modal loss is purely due to modal gain, we solve from 1.9 cm^{-1}=2Nσ$_{21}$ for σ$_{21}$ and find σ$_{21}$≅10^{-20}cm^2 for Er in Si_3N_4:Er. Literature reported values[4] for σ$_{21}$ of Er in SiO_2:Er are σ$_{21}$≅6×10^{-20}cm^2. The comparable value strongly suggests we have observed a reduction in modal loss (as a function of increasing pump power) due to the presence of modal gain. Measurements to conclusively confirm the Er emission cross-section in Si_3N_4:Er and observe net gain in co-sputtered SiON:Er are currently in progress.

CONCLUSIONS

We have investigated Er-doped SiON and Si_3N_4 materials, grown by sputter deposition, for use as waveguide core materials in SiO_2-clad waveguide optical amplifier structures. Theory modeling of the index difference of these waveguide structures (Δn=0.1↔0.75) shows significant power law improvement in device gain efficiency and areal footprint as planar waveguides. Using SiO_2:Er as a reference comparison for the sputtering process against state of the art commercial CVD processes, we conclude sputtered hosts are a high optical quality matrix for Er-doped light amplification. We compare reactively sputtered SiON:Er, co-sputtered SiON:Er, and sputtered Si_3N_4:Er; we conclude co-sputtered SiON:Er is the most promising host for net Er-based gain. Preliminary gain measurements of Si_3N_4:Er suggest the presence of optical gain under a λ=488 nm pump power of 138 mW. Comprehensive measurements are currently underway for a more detailed future publication.

ACKNOWLEDGMENTS

The authors would like to thank John Leblanc and Dr. J. Haavisto at Draper Laboratories for their facilities and assistance in helping set up a Vertical Scanning Length gain measurement system.

REFERENCES

1. ITRS (2001), International Technology Roadmap for Semiconductors: 2001, http://public.itrs.net/.
2. K. Wada, H.S. Luan, K.K. Lee, S. Akiyama, J. Michel, L.C. Kimerling, M. Popovic and H.A. Haus, "Silicon and Silica Platform for On-chip Optical Interconnection," Proc. LEOS Annual Meeting (2002).
3. L.C. Kimerling, L. Dal Negro, S. Saini, Y. Yi, D. Ahn, S. Akiyama, D. Cannon, J. Liu, J. Sandland, D. Sparacin and M. Watts, "Monolithic Silicon Microphotonics," ch.3 in *Silicon Photonics (Topics in Applied Physics, Vol.94)*, (Springer-Verlag, 2004).
4. P.C. Becker, N.A. Olsson and J.R. Simpson, Erbium-Doped Fiber Amplifiers: Fundamentals and Technology (Academic Press, 1999).
5. S. Saini, J. Michel and L.C. Kimerling, "Index Scaling for Optical Amplifiers," IEEE Journal of Lightwave Technology, **21**(10), 2368-2376 (2003).
6. W.J. Miniscalco, "Erbium-Doped Glasses for Fiber Amplifiers at 1500 nm," invited paper, IEEE Journal of Lightwave Technology, **9**(2), 234-250 (1991).
7. A. Polman, D.C. Jacobson, D.J. Eaglesham, R.C. Kistler and J.M. Poate, "Optical doping of waveguide materials by MeV Er implantation," J. Appl. Phys., **70**(7), 3778-3784 (1991).
8. L. Dal Negro, P. Bettotti, M. Cazzanelli, D. Pacifici, L. Pavesi, "Applicability conditions and experimental analysis of the variable stripe length method for gain measurements," Optics Communications, **229**, 337-348 (2004).

Optical and structural investigation on the energy transfer in a multicomponent glass co-doped with Si nanoaggregates and Er³⁺ ions

Francesco Enrichi[1], Giovanni Mattei[1], Cinzia Sada[1], Enrico Trave[1], Domenico Pacifici[2], Giorgia Franzò[2], Francesco Priolo[2], Fabio Iacona[3], Michel Prassas[4], Mauro Falconieri[5], Elisabetta Borsella[6]

[1]INFM, Dip. Fisica, Università di Padova, via Marzolo 8, 35131 Padova, Italy
[2]INFM-MATIS, Dip. Fisica e Astron., Univ. Catania, via S. Sofia 64, 95123 Catania, Italy
[3]CNR-IMM, Stradale Primosole 50, I-95121 Catania, Italy
[4]Adv. Mat. for Photonics, Corning SA, 7 bis Avenue de Vilvins B.P. No. 3, Avon, France
[5]ENEA, via Anguillarese 301, 00060 Casaccia (Roma), Italy
[6]ENEA, via E. Fermi 45, 00044 Frascati (Roma), Italy

ABSTRACT

The enhancement of the Er^{3+} ions photoluminescence (PL) emission at 1.54 μm in a Si and Er co-implanted aluminosilicate glass is investigated in details. Post-implantation annealing has been performed to recover the damage induced by the implantation process and to promote Si aggregation. It is shown that 1h treatment in N_2 atmosphere is not sufficient to induce Si precipitation for the investigated temperatures, up to 500°C. Nevertheless, the most intense Er^{3+} PL emission at 1.54 μm is achieved at 400°C. Such emission has been investigated by pumping in and out of resonance. The results suggest that good energy transfer mediators could be small Si aggregates and not only crystalline clusters. The effective excitation cross section of Er^{3+} ions has been measured in the best performing sample yielding a value of ~ 2 x 10^{-16} cm², many orders of magnitude higher than the direct absorption cross section of Er^{3+} ions: about 10^{-21} cm² in this glass. The structural and optical properties of this material are discussed and compared to those found for a standard silica substrate.

INTRODUCTION

Erbium doped materials are of great interest in optoelectronic technology due to their Er^{3+} intra-$4f$ emission at 1.54 μm [1], a standard wavelength for telecommunications since it coincides with the low-loss window of commonly used optical fibers. One of the major difficulties to overcome to realize a planar amplifier is the small cross section for Er excitation (typically ~ 10^{-21}cm²). That's why a high Er concentration and an enhancement of Er pumping efficiency are required. For the first aspect, the use of multicomponent glasses can be a good solution with respect to silica because they are characterized by a lower Er clustering behavior [2, 3]. For the second aspect, the increase of Er^{3+} pumping efficiency has been observed by incorporating Si nanocrystals in the glass [4-7]. The enhancement was attributed to an energy transfer process from the Si-nc to Er^{3+} ions. In this work we present an investigation of such an energy transfer mechanism in an aluminosilicate glass co-implanted with Si and Er, in relation to the structural properties of the material. A post-implantation annealing has been performed to recover the damage left over by the implantation process, to promote Si nucleation and to maximize the Er^{3+} emission at 1.54 μm. The comparison between properties of this specific glass and those of a standard silica host is presented, highlighting differences and discussing the implications on the realization of an optical device.

EXPERIMENTAL

A multi-component aluminosilicate glass and a common silica substrate were implanted with 80 keV Si^+ to a dose of $1x10^{17}$/cm² and 300 keV Er^{3+} to a dose of $1x10^{15}$ /cm². The implant energies of the Si and Er were chosen to optimize the overlap of their concentration profiles. A sample implanted only with Er was also prepared for comparison. Post implantation treatments were performed for 1 h in N_2 atmosphere at different temperatures, with aim of maximizing the 1.54 µm PL emission for each substrate: according to the results of previous works, at 400°C for the aluminosilicate glass [8] and at 900°C for silica [9].

PL measurements were carried out by using different excitation sources: a cw Ar laser in the range 450–520 nm, a frequency-doubled pulsed Ti:sapphire laser in the range 360–430 nm and a 1 W broadband emitting Xe lamp coupled to a monochromator in the range 250–850 nm. The excitation repetition frequency was set at 10 Hz; the signal was analyzed by a monochromator, detected by a near-infrared photomultiplier tube and sent to a lock-in amplifier using the chopper frequency as a reference. For the time resolved spectra, an acousto-optic modulator was used to ch the Ar laser beam and the PL intensity was acquired as a function of time. When pulsed Ti:sapphir laser excitation was used (pulse width 15 ns) the signal was detected by an InGaAs photodiode anc sent to an oscilloscope.

Elemental in-depth profiles were obtained by Secondary Ion Mass Spectrometry technique (SIMS) using a CAMECA IMS 4f. Structural and compositional characterization was performed with a field-emission gun (FEG) FEI TECNAI F20 SuperTwin transmission electron microscope (TEM) equipped with an EDAX energy-dispersive x-ray spectrometer (EDS) and with a Jeol 2010 FEG TEM equipped with a Gatan GIF imaging filter for energy filtered TEM analysis.

RESULTS AND DISCUSSION

The occurrence of an energy transfer between Si nanoaggregates and Er^{3+} ions in the co-doped and annealed multi-component glass is clearly shown in figure 1, which reports the intensity of the 1.54 µm PL emission as a function of the excitation wavelength. This PLE spectrum is compared t the one obtained for an only-Er implanted and annealed sample.

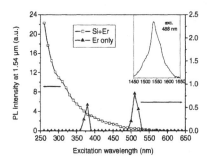

Figure 1: PLE spectra of only Er implanted and Si + Er co-implanted, 400 °C annealed samples. 1 excitation source is a 150 W Xe lamp chopped at 10 Hz. The connecting lines between the experimental points are only for guiding the eye. In the inset: the PL emission spectrum near 1.54 µm of the Si + Er + 400°C annealed sample, obtained by 488 nm Ar laser excitation.

It is worth noticing that broadband excitation is possible when Si is co-implanted in the materi with a monotonic decrease of the Er emission as a function of the excitation wavelength. On the other hand, for the only-Er doped sample, Er emission at 1.54 µm is possible only when Er^{3+} ions a directly excited (in particular at 379 nm and 524 nm). This behavior of the PLE spectrum supports the idea that Si nanograins play the role of mediators for energy transfer to Er^{3+} ions since the PLE dependence on wavelength can be ascribed to the absorption cross section of crystalline [10, 11] or amorphous Si-clusters [12]. From a technological point of view, the possibility of broadband excitation for pumping the Er^{3+} ions makes it possible to use cheap lamps in spite of the expensive 980 nm or 1540 nm lasers commonly used in commercial Er doped devices. In the inset of figure 1, the 488 nm excitation spectrum obtained by Ar laser excitation is presented. This spectrum is peaked at 1537 nm and shows a full width at half maximum of 52 nm. For Er implanted silica annealed at 900°C these values resulted respectively 1542 nm and 39 nm. The different shape of the PL emission in a multi-component glass is related to a different structural environment surrounding the Er^{3+} ions [3]. In particular the bandwidth enlargement has the technological advantage of allowing a broader amplification band when used as active material in an optical amplifier.

The reported very efficient Si-mediated pumping mechanism can be ascribed to an enhanced effective cross section for Er excitation. In figure 2 the PL intensity at 1.54 µm of the Si + Er and of the only Er implanted samples is reported as a function of the photon flux. Pulsed laser excitation at 379 nm (resonant excitation) and 390 nm (non resonant excitation) is used. The trend of the PL can be explained by taking into account that, under pulsed laser excitation, the number N^* of excited Er^{3+} ions follows (neglecting non linear effects such as upconversion):

$$N^* = N_{Er} \cdot \left(1 - \exp\left(- \sigma_{eff} \cdot \varphi \cdot \Delta t\right)\right) \qquad (1)$$

where N_{Er} is the total number of excitable Er ions, σ_{eff} is the effective excitation cross section for their excitation at the pumping wavelength, φ is the photon flux and Δt is the duration of the pulse. For the Er-only implanted sample the PL emission versus the photon flux follows a linear trend. This means that $\sigma_{eff} \cdot \varphi \cdot \Delta t \ll 1$ up to the highest fluxes. As a consequence, $\sigma_{eff} \ll 10^{-17}$ cm^2, which is expected for the direct pumping of Er ions at 379 nm. In fact, from absorption measurements we have evaluated σ_{Er} (379 nm) $\sim 10^{-20}$ cm^2. On the other hand, for the co-implanted sample the PL experimental points can be fitted with equation 1 for 390 nm excitation and adding a linear contribution for 379 nm excitation. The fitting curves are reported in figure 2, giving a cross section of about $(1.4 \pm 0.3) \cdot 10^{-17}$ cm^2 at 390 nm and $(5.2 \pm 0.8) \cdot 10^{-17}$ cm^2 at 379 nm. These values follow the wavelength dependence of the absorption cross section of Si nanograins and are in agreement with typical cross section values measured on Si-nc [10, 11]. Moreover, they are much higher than the direct absorption cross section of Er^{3+} at 379 nm. It is important to say that these numbers are a lower limit to the real cross section values since the occurrence of non linear effects such as upconversion have not been taken into account in the previous equation 1.

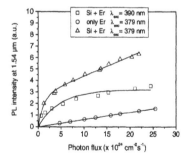

Figure 2: PL emission at 1.54 µm as a function of pump power for resonant (379 nm) and non resonant (390 nm) excitation. The continuous lines are fitting curves to the data.

Similar calculations can be done for the 476.5 nm continuous Ar laser excitation, chopped at 10 Hz, which allows the pumping of the Er ions via a Si-mediated path. When continuous pumping is used, the concentration of excited Er ions N^* can be written as:

$$N^* = N_{Er} \cdot \frac{\sigma_{eff} \cdot \phi \cdot \tau}{1 + \sigma_{eff} \cdot \phi \cdot \tau} \qquad (2)$$

In this formula τ is the spontaneous 1.54 μm lifetime resulting from radiative and non-radiative processes, while the other symbols have the same meaning as before. Again the occurrence of non linear effects and upconversion has been neglected. Equation 2 has been used to fit the experiment data (not reported), evaluating an effective cross section of $(2.9 \pm 0.8) \cdot 10^{-17}$ cm^2 at 476.5 nm by using a lifetime of 3.5 ms, which is well representative of the time decay in the range of photon fluxes considered (see figure 3). Also in this case the effective excitation cross section is many orders of magnitude higher than the direct absorption cross section of Er^{3+} ions, of the order of 10$^-$ cm^2 in this glass at 488 nm [13].

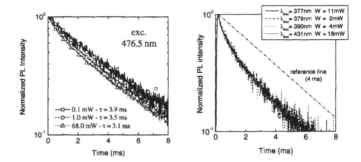

Figure 3: Time decay curves of the 1.54 μm PL emission for the co-implanted sample annealed at 400°C. The curves have been obtained by pumping with a continuous Ar laser (left panel) or a pulsed Ti:sapphire laser (right panel). Data are shown for different pumping conditions.

In figure 3, the 1.54 μm PL curves are reported, normalized to their maximum values. The curves on the left panel were obtained by using the continuous 476.5 nm Ar laser at different pump powers, chopped at 10 Hz, while the ones on the right panel were obtained by using the Ti:sapphire pulsed laser at different wavelengths and pump powers and compared with a reference single exponential decay curve (lifetime 4 ms). Looking at the shapes of the curves, we can notice that their behavior is quite different under continuous or pulsed laser conditions. In the first case (left panel) they follow a non-exponential decay which is more pronounced when the pump power is increased. If we evaluate the lifetime as the time at which the PL intensity is 1/e of its initial value, decreases from 3.9 to 3.1 ms as the excitation power on the sample increases from 0.1 mW to 68 mW. This non-exponential behavior is typical of the occurrence of cooperative upconversion mechanisms [14]. It is worth noticing that the decay curve obtained from the only-Er implanted sample (not reported) is single exponential even for the highest pump powers. The reason is that the effect of upconversion can be observed if the density of excited ions is sufficiently high. This is the case of the co-implanted sample due to the much higher effective excitation cross section of Er^{3+} ic with respect to the only-Er implanted sample. In the right panel of figure 3 the time resolved PL curves obtained by a Ti:sapphire pulsed laser excitation is used to excite the system (right panel). I this case the PL curves do not vary by changing the excitation wavelength or by increasing the pump power on the sample. This behavior agrees with the idea that Er^{3+} ions are mainly excited through energy transfer from Si aggregates, so they can be pumped independently from the used wavelength. Moreover, since $\sigma_{eff} \cdot \varphi \cdot \Delta t >> 1$ (equation 1) even for the lower pump powers, most of the available nanograins are saturated. In these conditions the use of higher pump powers cannot increase the number of excited Er^{3+} ions since the pulsed excitation of about 15 ns is much shorter than the energy transfer time, estimated in about 1 µs [15]. This means that under pulsed laser excitation only one Er ion can be excited per nanograin and conversely that the maximum number of emitting Er ions corresponds to the number of sensitizing centers (or to the total number of Er ions if they are less than the number of sensitizers).

Investigation of the structural and compositional properties of the implanted and annealed material in determining the described optical properties evidenced that a post implantation thermal treatment does not induce significant compositional modification in the glass for temperatures up to 500 °C. Moreover, HRTEM measurements on the 400 °C annealed sample (figure 4, left panel) showed no Si nanocrystals, at least of size greater than 1 nm ("threshold visibility" in TEM for Si nc embedded in silica glasses). Energy-filtered TEM (EFTEM) further corroborate this picture (figure 4, right panel). Indeed, EFTEM is able to form images of the sample by selecting only those electrons that suffered a certain energy loss when passing through the sample. In the present case, we used electrons that have lost about 16 eV (plasmon loss of Si) and compared the resulting image with that obtained by using electrons with about 23 eV of energy loss (plasmon loss of silica). The analysis shows an enhanced contrast in the implanted region due to the onset of Si-Si bonding, but no evidence of Si precipitation (either in an amorphous or crystalline phase) was found. Considering that in all of the samples an efficient energy transfer was detected by PL measurements, we can conclude that Er^{3+} ions can be excited by energy transfer from amorphous Si nanograins composed of few Si atoms.

Finally, comparing the annealing temperatures required to achieve the best PL emission in the examined aluminosilicate glass and in silica (about 400°C and 900°C, respectively), it is possible to evidence that working with a multi-component glass has the obvious potential technological advantage to require lower thermal budgets for producing the same sensitizing effect in comparison to silica.

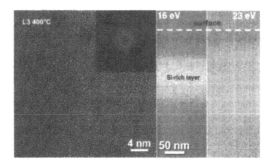

Figure 4: Cross-sectional TEM images on the Si + Er implanted aluminosilicate glass annealed at 400°C. In the left panel the high resolution image shows no detectable nanoclusters (in the inset the corresponding FFT is reported). In the right panel a comparison of two energy-filtered images of the same sample evidence the presence of a Si-rich layer.

CONCLUSIONS

In this work we have investigated the energy transfer process between Si nanoaggregates and Er^{3+} ions in a co-implanted aluminosilicate glass annealed at 400°C. This temperature is well belo~ the threshold for the formation of Si nanocrystals, as confirmed by structural analysis, suggesting that the energy transfer is efficiently mediated by amorphous Si aggregates and not necessarily by crystalline structures. It is worth noticing that in the co-implanted material the excitation of Er^{3+} ic is possible in a wide range of wavelengths and its efficiency increases with decreasing the wavelength. The basic reason for such an efficient pumping of the Er^{3+} ions is the much higher absorption cross section of Si nanograins with respect to the direct absorption cross section of Er^{3+} ions. The effective excitation cross section of Er^{3+} ions through a Si-mediated path has been evaluated to be higher than 10^{-17} cm^2 at 379 nm, 390 nm and 476 nm. This value is much higher th~ the one for direct excitation of Er^{3+} ions. Such an efficient pumping shifts the limitations of the maximum 1.54 μm intensity attainable in the material to the maximum number of available Er^{3+} ic that can interact with Si nanograins and makes it possible to observe upconversion mechanisms at much lower photon fluxes on the sample. The investigated material could be a good candidate for ~ realization of a cheap and high-performing optical amplifier. Moreover, it is perfectly compatible with today's industrial production processes and can be realized by using relatively low thermal budgets, allowing a significant improvement of the device performances at very low cost.

ACKNOWLEDGMENTS

This work was possible by the support of the EU - IST SINERGIA project. We wish to thank~ Boninelli, C. Bongiorno and C. Spinella for collaboration with the TEM analyses and A. Marino expert technical assistance.

REFERENCES

1. S. Hüfner, Optical spectra of transparent rare earth compounds (Academic, New York, 1978)
2. R. S. Quimby, W. J. Miniscalco, B. Thompson, J. Appl. Phys. **76**, 4472 (1994)
3. P. M. Peters, S. N. Hounde Walter, Journal of Non Crystalline Solids **239**, 162 (1998)
4. M. Fujii, M Yoshida, S. Hayashi, K. Yamamoto, J. Appl. Phys. **81**, 4525 (1998)
5. F. Priolo, G. Franzò, D. Pacifici, V. Vinciguerra, F. Iacona, A. Irrera, J. Appl. Phys. **89**, 264 (20C
6. C. E. Chryssou, A. J. Kenyon, T. S. Iwayama, C. W. Pitt, D. E. Hole, Appl. Phys. Lett. **75**, 2011 (1999)
7. P. G. Kik, M. L. Brongersma, A. Polman, Appl. Phys. Lett. **76**, 2325 (2000)
8. F. Enrichi, G. Mattei, C. Sada, E. Borsella, D. Pacifici, G. Franzò, F. Priolo, F. Iacona, M. Prassa~ ECOC-IOOC Proceedings vol. 3, 426 (2003)
9. G. Franzò, S. Boninelli, D. Pacifici, F. Priolo, F. Iacona, C. Bongiorno, Appl. Phys. Lett. **82**, 387~ (2003)
10. D. Kovalev, J. Diener, H. Heckler, G. Polisski, N. Kunzner, F. Koch, Phys. Rew. B **61**, 4485 (2000)
11. C. Garcia, B. Garrido, P. Pellegrino, R. Ferre, J. A. Moreno, J. R. Morante, L. Pavesi, M. Cazzanelli, Appl. Phys. Lett. **82**, 1595 (2003)
12. G. Allan, C. Delerue, M. Lannoo, Phys. Rew. Lett. **78**, 3161 (1997)
13. F. Enrichi, E. Borsella, Mat. Sci. & Eng. B **105**, 20 (2003)
14. G. N. van Den Hoven, E. Snoeks, A. Polman, C. van Dam, J. W. M. van Uffelen, M. K. Smit, J. Appl. Phys. **79**, 1258 (1996)
15. D. Pacifici, G. Franzò, F. Priolo, F. Iacona, L. Dal Negro, Phys. Rew. B **67**, 1 (2003)

Mat. Res. Soc. Symp. Proc. Vol. 817 © 2004 Materials Research Society   L6.5

The Investigation of Erbium Complexes and Erbium Doped Materials

Seunghoon Lee, Ung Kim, Juntae Kim and Sang Man Koo
Department of Chemical Engineering, College of Engineering, Hanyang University
Seoul 133-791, Korea

ABSTRACT

Erbium ion (Er^{3+}) doped materials are of great interest for their optical amplification, lasing and frequency up-conversion properties. When preparing such materials, a major problem that often arises is the formation of Er-rich oxide clusters inducing optical quenching. The materials prepared at low Er^{3+}-ion concentrations to overcome the problem have severely reduced the optical yield. Such clustering might be avoided by preparing suitable precursors. If Er is encapsulated with proper materials, clustering can be avoided and higher doping levels can be achieved. In this study, erbium phenoxide complex was obtained by metathesis reaction of erbium chloride ($ErCl_3$) with potassium phenoxide (KOPh). And heterometallic complexes were also synthesized by encapsulation of the Er with Al or Ti derivatives. The complexes were characterized by elemental analysis, infrared and nuclear magnetic resonance spectroscopic analysis. Their crystal structures were determined by X-ray single crystal diffraction analysis. In addition, the Er-doped organic-inorganic matrices using the erbium complexes were investigated about their optical properties.

INTRODUCTION

In recent years, Er^{3+} ion doped optical materials have received growing interest for their applications to the fields of integrated lasers or amplifiers for telecommunication. In these applications, materials with high Er^{3+} concentration are required to maximize the amplification. However, high Er^{3+} concentration in the devices often leads the formation of cluster type Er compounds that induce optical quenching due to energy transfer. These problems can be solved by employing suitable Er precursors either containing bulky ligands or being encapsulated with proper materials to prevent agglomeration and ensure the homogeneous dispersion. In this study, new types of Er precursors were synthesized by first introducing bulky phenoxide ligands and secondly encapsulating this complex with other metal alkoxides. The absorption and photoluminescence spectra of these complexes have been investigated and well-defined fluorescence peak with a wide bandwidth were observed.

EXPERIMENTAL DETAILS

All preparations were performed in a glovebox containing dry, oxygen-free nitrogen atmosphere. The solvents were purified by distillation. Commercial ErCl$_3$ (Strem Chemicals), AlCl$_3$ (Aldrich), Ti(OPr^i)$_4$ (Aldrich), Phenol (Aldrich) and Dibenzoylmethane (Aldrich) were used

Er(OPh)$_6$ – Erbium phenoxide was prepared by the addition of erbium chloride solution in acetonitrile to 6 equivalent of potassium phenoxide solution in acetonitrile under stirring, whereafter the reaction was allowed to proceed for 2 days at 70°C, yielding a brown solution with white precipitates (KCl). The solution was removed and evaporated to dry. The dried alkoxide was dissolved in DMF and layered by diethyl ether, yielding brown crystals. These crystals were soluble in THF, MeCN, DMF and EtOH. ^{1}H NMR data : δ 6.75 (s, 2H, CH in OPh.), δ 6.15 (s, 2H, CH in OPh), δ 5.81 (s, 1H, CH in OPh), δ 7.98 (s, 1H, CH in DMF), δ 2.75 (s, 3H, CH$_3$ in DMF), δ 2.91 (s, 3H, CH$_3$ in DMF) ppm.

Er$_2$Al$_4$(OPh)$_{16}$Cl$_2$ – Aluminum chloride was buffered by hexane and dissolved in THF. This solution was added into 4 equivalent of potassium phenoxide solution in THF. After 4 hours at room temperature, a brown solution with white precipitates (KCl) was obtained. The solution was removed and evaporated to dry. The 6 equivalent of alkoxide was dissolved in toluene and added into 2 equivalent of erbium chloride solution in toluene under stirring. After 2 days at 100°C, a brown solution with white precipitates (KCl) was obtained. The solution was removed and evaporated to dry. The dried complex was dissolved in THF and layered by hexane, resulting in brownish pink crystals. These crystals were soluble in THF, MeCN, DMSO and DMF, slightly soluble in toluene and alcohol. ^{1}H NMR data: δ 7.21 (d, 2H, CH in OPh.), δ 7.05 (d, 2H, CH in OPh), δ 6.69 (d, 1H, CH in OPh) ppm.

Er$_2$Ti$_4$(DBM)$_4$(OPh)$_{16}$Cl$_2$ – Titanium isopropoxide and dibenzoylmethane were mixed in THF under stirring (1:1). After 4 hours at room temperature, a red solution and yellow precipitates were obtained. The yellow powder was removed and added into 4 equivalent of phenol dissolving THF solution, and then 1 equivalent of potassium metal was added into the mixture solution. After 4 hours at room temperature, a yellowish black solution was obtained. The yellow powder was obtained from solvent extraction and washing with diethyl ether. The 6 equivalent of alkoxide was dissolved in MeCN and added into 2 equivalent of erbium chloride solution in MeCN under stirring. After 2days at 60°C, a reddish orange solution with white precipitates was obtained. The yellow powder was obtained from the collected solution through evaporation, washing by diethyl ether, and drying. This powder was soluble in toluene, benzene, THF, DMF, and slightly soluble in diethyl ether, MeCN and alcohol. ^{1}H NMR data: δ 7.22 (d, 8H, CH in OPh.), δ 6.92 (d, 8H, CH in OPh), δ 6.41 (d, 4H, CH in OPh), δ 7.91 (d, 4H, CH in DBM), δ 7.42 (d, 6H, CH in DBM), δ 6.69 (d, CH in DBM) ppm.

DISCUSSION

FTIR spectra of each complex are shown in figure 1. Absorption bands near 3050 cm^{-1} were due to C-H in aromatic rings. Strong C=C bands near 1480 cm^{-1} and 1595 cm^{-1} were assigned to the C=C stretching vibration of aromatic rings. The bands near 695 cm^{-1} and 760 cm^{-1} were due to aromatic =C-H out-of-plane bending. The bands near 1250 cm^{-1} arose from the vibrations of C-O bonds. Because of solvent effect in recrystallization, Er(OPh)$_6$ crystals involved DMF molecules in the crystal lattice, causing to the band of amide C=O vibrations observed near 1658 cm^{-1}. In case of Er$_2$Ti$_4$(DBM)$_4$(OPh)$_{16}$Cl$_2$, strong band near 1523 cm^{-1} arose from the vibrations of conjugated C=O in dibenzoylmethane.

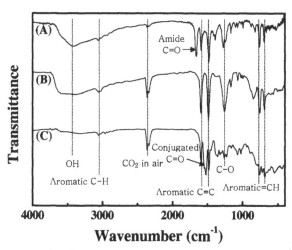

Wavenumber (cm^{-1})

Figure 1. FTIR spectra of Er(OPh)$_6$ (A), Er$_2$Al$_4$(OPh)$_{16}$Cl$_2$ (B) and Er$_2$Ti$_4$(DBM)$_4$(OPh)$_{16}$Cl$_2$ (C).

	Concentration of elements (ppm)	Molar ratio
Er	131306	Er : Al = 1 : 1.986
Al	42087.6	
Er	103420	Er : Ti = 1 : 2.346
Ti	69510.2	

Table 1. ICP-AES data of Er$_2$Al$_4$(OPh)$_{16}$Cl$_2$ and Er$_2$Ti$_4$(DBM)$_4$(OPh)$_{16}$Cl$_2$

Heterobimetallic alkoxides were confirmed by ICP-AES (see table 1).

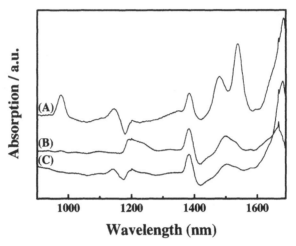

Figure 2. Absorption spectra of Er(OPh)$_6$ (A), Er$_2$Ti$_4$(DBM)$_4$(OPh)$_{16}$Cl$_2$ (B) and Er$_2$Al$_4$(OPh)$_{16}$Cl$_2$ (C) in THF.

Figure 2 shows the absorption spectra of each complex in THF. The absorption peaks at 980 nm and around 1500 nm wavelength were observed.

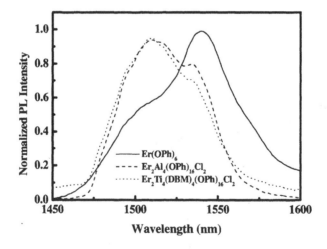

Figure 3. Photoluminescence spectra of erbium complexes in PMMA film (15 wt%, thickness: 100 μm).

Figure 3 shows the normalized photoluminescence spectra (PL) of erbium complexes doped polymers. An argon laser operating at 980 nm wavelength was used as the excitation source, and fluorescence peaks at around 1540 nm due to the $^4I_{13/2}-\,^4I_{15/2}$ transition were observed. In case of Er(OPh)$_6$, the full width at half maximum (FWHM) is measured to be ~70 nm.

CONCLUSIONS

New types of monomeric and heterobimetallic Erbium complexes were synthesized as suitable precursors for EDWA (Erbium Doped Waveguide Amplifier) materials. The use of bulky ligands or encapsulation with other metal alkoxides around Er^{3+} center will prevent the agglomeration of Er species with homogeneous phase formation in devices requiring a high level of Er-doping to maximize its high throughput. In PL study, Er complexes synthesized in this study exhibited fluorescence peak at around 1540 nm due to the $^4I_{13/2}-\,^4I_{15/2}$ transition and the full width at half maximum (FWHM) was measured to be ~70 nm. The optical properties of these complexes in inorganic and hybrid matrices are under investigation.

REFERENCES

1. P. C. Becker, N. A. Olsson and J. R. Simpson, "Erbium-Doped Fiber Amplifiers" (Academic Press, 1999).
2. M. J. F. Digonnet, "Rare Earth Doped Fiber Lasers and Amplifiers" (Marcel Dekker, INC., 1993).
3. G. Westin, M. Wijk, M. Moustiakimov and M. Kritikos, *J. Sol-Gel Sci. Tech.* **13**, 125 (1998).
4. G. Westin, 1 M. Kritikos and M. Wijk, *J. Solid State Chem.* **141**, 168 (1998).
5. M. Wijk, R. Norrestam, M. Nygren and G. Westin, *Inorg. Chem.* **35**, 1077 (1996).
6. S.W. Magennis, A.J. Ferguson, T. Bryden, T.S. Jones, A. Beeby and I.D.W. Samuel, *Synth. Metals* **138**, 463 (2003).
7. O.-H. Park, S.-Y. Seo, J.-I. Jung, J. Y. Bae and B.-S. Bae, *J. Mater. Res.* **18**, 1039 (2003).
8. H. Suzuki, Y. Hattori, T. Iizuka, K. Yuzawa and N. Matsumoto *Thin Solid Films* **438 –439**, 288 (2003).
9. G. Westin, R. Norrestam, M. Nygren and M. Wijk, *J. Solid State Chem.* **135**, 149 (1998).
10. H. Ma, A. K.-Y. Jen and L. R. Dalton, *Adv. Mater.* **14**, 1339 (2002).

Mat. Res. Soc. Symp. Proc. Vol. 817 © 2004 Materials Research Society   L6.18

Structure analysis of terbium aluminosilicate glass

Xiaoyuan Qi, Sang-Yeob Sung, Samir K. Mondal and Bethanie J. H Stadler
Department of Electrical and Computer Engineering, University of Minnesota, Minneapolis, MN 55455, USA

Glasses rich in rare-earth ions have attracted a lot of interest due to their applications in optical isolators and optical amplifiers. Integrating optical isolators with various optoelectronic devices allows sources to be integrated with lower costs, easier alignment and longer lifetimes. Glasses rich in rare-earth ions have large Verdet constants, so large Faraday rotations. Among the rare-earth ions used in paramagnetic glasses, Tb^{3+} ions have largest Faraday rotation per ion and the glasses are transparent down to 1.6um. These glasses can also avoid the temperature incompatibility and lattice match problems which are encountered when magneto-optical garnets are used for integration. The Tb^{3+} doping is also widely used in optical amplifiers. In this paper we have explored the metastable phases present in the sputtered Tb-Al-Si-O system in order to fabricate paramagnetic films with the highest possible Faraday rotations, lowest optical losses and that are easily integrated with semiconductors. A broad peak was observed in the microdiffraction pattern around $2\theta = 30$deg. This peak corresponded with the close-packed Tb-O plane spacing (111 for FCC Tb_4O_7 or 002 for HCP Tb_2O_3), but it was an "amorphous" peak with a 5 deg FWHM. Amorphous films were obtained even when the Tb concentrations were very high. Since high concentrations of Tb are known to devitrify glasses, the discovery of a high-Tb concentrated glass is exciting.

I. INTRODUCTION

Glasses rich in rare-earth ions have been widely used in optical isolator applications. In recent years, integrated optical isolators have attracted a lot of interests due to their low cost, convenience and long lifetimes. They are crucial parts of optical fiber communication systems and photonic integrated circuits. Isolators using Faraday rotation have been widely used in both bulk and thin film forms. Faraday rotation is a nonreciprocal effect. In the isolator application, the light is rotated about 45 degrees when it passes through the Faraday rotator in the forward direction and another 45 degrees when it is reflected in the backward direction. A polarizer is put in front of the Faraday rotator, and this will only allow the forward-traveling light to propagate. The backward-traveling light is blocked as it is 90 degrees to the polarizer, and in such a way, the laser or other light source is isolated from back reflections.

Yttrium iron garnets (YIG) and substituted iron garnets, e.g. bismuth-substituted YIG (BiYIG), are mainly used for Faraday rotator applications. Liquid phase epitaxy (LPE) is usually used to grow these garnets. But the high temperature and epitaxial requirements are not compatible with semiconductors or other common substrates. Many efforts have been made to overcome these problems [1]-[6]. The replacement of garnets with other materials that are amorphous, that can be made at lower temperatures, and that have large Faraday rotations would be one way to overcome these problems.

Faraday effect can be quantified as $\theta_F = VBL$ where θ_F is the rotation angle, V is the material property indicating strength of rotation, B is the magnetic field, and L is the pathlength. The magnetic field is parallel to the propagation direction of the light. Glasses rich in rare-earth ions have large Verdet constants (V's) and so large Faraday rotations [7]. Tb^{3+} has been shown to have the largest Faraday rotation per ion and the glasses are transparent down to 1.6um [8-10]. Glasses with high Tb^{3+} composition should have higher rotations. But the Tb^{3+} concentration is limited by its glass-forming abilities, or crystallinity, usually below 30mol %. As far as we know, Tb^{3+} doped glass films have never been used in integrated isolator applications and the phase diagram of the terbium aluminosilicate glass films has not been studied yet. Tb^{3+} doping has also been used in fiber amplifiers, but lower concentrations are desired for this application. [11]-[12]

II. EXPERIMENTAL

Terbium aluminosilicate glass films were deposited on both glass and quartz substrates at room temperature by RF magnetron sputtering using an Ar pressure of 2.3 mTorr and varied O_2 pressures. The background pressure was 3×10^{-6} Torr and the target-to-substrate distance was

150mm. The 2-inch diameter terbium, aluminum and silicon targets were cosputtered and their ratios were varied by adjusting the corresponding sputtering powers and bias voltages.

A Bruker AXS Microdiffractometer was used for X-ray diffraction and energy dispersive X-ray analysis (EDX) was used to analyze the compositions of the films. The magnetic properties of the films were measured using a Vibrating Sample Magnetometer (VSM) and the Faraday rotations were also measured using the setup shown below.

III. RESULTS AND DISCUSSION

A series of terbium aluminosilicate glass films with varied Tb, Al and Si ratios were made. The microdiffraction data showed four different types of diffraction patterns, shown in Fig. 1. The peaks observed in the microdiffraction patterns around $2\theta=30deg$ corresponded with the close-packed Tb-O plane spacing (111 for FCC Tb_4O_7 or 002 for HCP Tb_2O_3). Experiments also indicated that the crystallinity of the films mainly depended on the ratios of Tb, Al and Si as long as there was enough oxygen.

There are mainly two phases corresponding to two distinguished peaks at $2\theta=22deg$ and $2\theta=30deg$. Both of these peaks can be called "amorphous peaks" due to their breadth (FWHM ~5deg) which would correspond to atomic scale "grain sizes." The trinary phase diagrams can be divided into three regions as shown in Fig. 2. The middle region is a two phase field that contained both Phase 1 and Phase 2. It was also found that the susceptibility correlated with phases in the glasses. The plot of susceptibility vs. terbium concentration is shown in Fig. 3. The susceptibility increases with the Tb composition and also an increase in Phase II ratio. Fig. 4 shows the M-H curves of four samples corresponding to four different patterns in Fig. 1. They are all paramagnetic films.

49

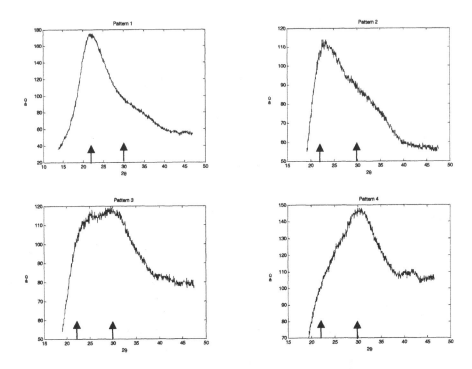

Fig. 1. Two phases corresponding to four types of microdiffraction patterns

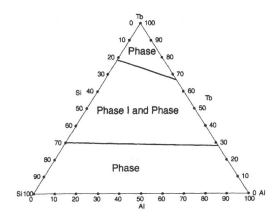

Fig. 2. Trinary Phase Diagram of terbium alumonium silicate glass films.

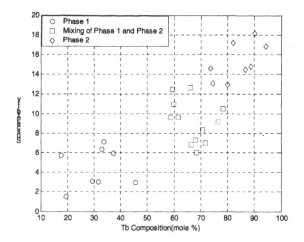

Fig. 3. Susceptibility of terbium alumonium silicate glass films vs. terbium

composition(mole %)

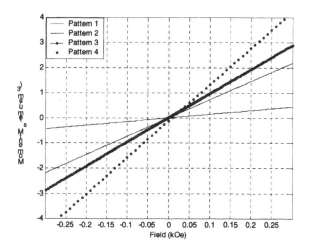

Fig. 4. M-H curves of terbium alumonium silicate glass films.

The Faraday rotations of these films were also measured with the setup shown in Fig. 5.

The films were around 5000Å thick. The rotation was measured in a waveguide configuration since these films are of interest for integrated waveguide isolators where the beam will propagate through the film in parallel to the film surface. In the high-Tb films, rotations up to $0.2/18mm=1.1x10^{-5}deg/um$ were measured at 300 Oe (78.35 at. %Tb). Due to the difficulty of waveguiding coupling and large noise, this first attempt is a rough approximation and more accurate measurements will be done in the future. In particular, the waveguide dimensions must be optimized to optimize the rotation.

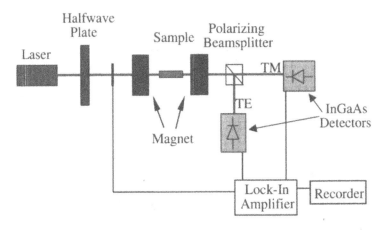

Fig. 5 Faraday Rotation measurement setup.

IV. Conclusions

Tb^{3+} doped glasses have been fabricated using integration-friendly processes. The ternary phase diagram indicated that the films with high terbium concentration may still remain amorphous. If these films could be used in photonic isolators, they would solve the lattice mismatch and temperature incompatible problems. We also correlated phase composition with terbium concentrations and with susceptibility. Faraday roation measurements indicated there was a small amount of rotations. Further study was needed.

ACKNOWLEDGEMENTS

This project was supported by the National Science Foundation under the CAREER program (ECS-0134544).

REFERENCES:

[1] M. Levy, "Epitaxial liftoff of thin oxide layers: Yttrium iron garnets onto GaAs," *Appl. Phys. Lett.,* **71** 2617-9 (1997).

[2] B. J. H. Stadler and A. Gopinath, "Magneto-optical garnet films made by reactive sputtering," *IEEE Trans,. Magn.,* **36**, 3957-61, 2000.

[3] S. A. Oliver, M. L. Chen, I. Kozulin, S. D. Yoon, X. Zuo, and C. Vittoria, "Growth and characterization of thick oriented Barium Hexaferrite films on MgO(111) substrates," *Apppl. Phys. Lett.,* **76**, 3612-14, 2000.

[4] B. Stadler, P. Yip, Y. Li, M. Cherif, K. Vaccaro and J. Lorenzo, *IEEE Trans. Mag.,* **38**(3), 1564-7, 2002.

[5] L. J. Cruz Rivera, S. Pandit, S. Pieski, R. Cobian, M. Cherif, and B. J.H. Stadler, *Proceedings of SPIE,* **4284**, 29-42, 2001.

[6] Sang-Yeob Sung, Na-Hyoung Kim, and B. J. H. Stadler, *Mat. Res. Sco. Symp. Proc.,* **768**, G4.6.1, 2003.

[7] Katsuhisa Tanaka, Koji Fujita, Nobuaki Matsuoka, Kazuyuki hirao, and Naohiro Soga, "Large Faraday effect and local structure of alkali silicate glasses containing divalent europium ions," J. Mater. Res., 13, 1989-1995, 1998.

[8] N. F. Borelli, "Faraday rotation in glass", *J. Chem. Phys.* **41**, 3289-3293, 1964.

[9] John Ballato and Elias Snitzer, "Fabrication of fibers with high rare-earth concentrations for Faraday isolator applications," *Applied Optics,* **34**, 6848-6853, 1995

[10] Katsuhisa Tanaka, Kazuyuki Hirao and Naohiro Soga, " Large Verdet constant of $30Tb_2O_370B_2O_3$ glass," *Jpn. J. Appl. Phys.,* **34**, 4825-4826, 1995.

[11] B. N. Samson, T. Schweizer, D. W. Hewak, and R. I. Laming, "Properties of dysprosium-doped gallium lanthanum sulfide fiber amplifiers operating at 1.3um", *Optics Letters,* Vol. 22, No.10, 703-705, 1997.

[12] Dong Jun Lee, Jong Heo, Se Ho Park, "Energy transfer and 1.48 emission properties in chalcohalide glasses doped with Tm^{3+} and Tb^{3+}", *Journal of Non-Crystalline Solids,* 331, 184-189, 2003

Mat. Res. Soc. Symp. Proc. Vol. 817 © 2004 Materials Research Society   L6.22

Fabrication of Tungsten-Tellurite Glass Thin Films using

Radio Frequency Magnetron Sputtering Method and Optical Property Characterization

Ki-Young Yoo, Sanghoon Shin, Youngman Kim, Jong-Ha Moon, and Jin Hyeok Kim

Center for Photonic Materials and Devices
Department of Materials Science and Engineering, Chonnam National University
300 Yongbong-Dong, Puk-Gu, Kwangju 500-757, South Korea

ABSTRACT

Tungsten-tellurite glass thin films were fabricated by radio-frequency (rf) magnetron sputtering method at various processing parameters such as substrate temperatures, Ar/O_2 processing gas flow ratio, processing pressure, and rf power from a $70TeO_2$-$30WO_3$ target fabricated by solid–state sintering method. The effects of processing parameters on the growth rate, the surface morphologies, the crystallinity, and refractive indices of thin films were investigated using atomic force microscopy, X-ray diffractometer, scanning electron microscopy, and UV spectrometer. Amorphous glass thin films with a surface roughness of 4~6 nm were obtained only at room temperature and crystalline phase were observed in all as-deposited thin films prepared at above the room temperature. The deposition rate strongly depends on the processing parameters. It increases as the rf power increases and the processing pressure decreases. Especially, it changes remarkably as varying the Ar/O_2 gas flow ratio from 40sccm/0sccm to 0sccm/40sccm. When the films were formed in pure Ar atmosphere it shows a deposition rate of ~0.2 μm /h, whereas ~1.5 μm/h when the films was formed in pure O_2 atmosphere.

INTRODUCTION

Erbium doped fiber amplifiers (EDFAs) have used silica or silicate glasses based fibers as host materials because the characteristics of the materials for optical fibers are similar to those of the silicate glasses. However, the full width at half maximum (FWHM) of these materials is about 45 nm with a non-flat gain profile at 1550 nm. This value of FWHM is insufficient to support the number of channels for WDM EDFA. Therefore, recently, tellurite glass has received great interest as a host material for broadband EDFAs[1-4] because it has the FWHM of about 80 nm at 1550 nm [2-5]. In addition to the large FWHM, tellurite glasses have a wide transmission region (0.35~6μm), good glass stability, the lowest vibration energy (about 780 cm^{-1}) among glass formers, high refractive index, low processing temperature, high nonlinear refractive index, and good chemical durability [6-9]. However, tellurite glasses have two disadvantages as a host for EDFAs. First, the phonon energy of the glass is relatively low at 770 cm^{-1}, so the $^4I_{11/2}$ → $^4I_{13/2}$ nonradiative decay is too slow to allow pumping at 980 nm. Second, the softening point of tellurite glass is at 290 °C, which makes it exposed to thermal damage at high optical intensities [10-11].

Tungsten-tellurite glasses have been proposed to overcome these problems. Tungsten-tellurite glasses differ from the conventional tellurite glasses in that they contain two glass-forming components, WO_3 and TeO_2. Tungsten oxide (WO_3) is concern with the higher phonon energy of tungsten-tellurite glass (920 cm^{-1}) as compared with tellurite glass (770 cm^{-1}), which makes it possible to employ pumping at 980 nm. Moreover, the stronger bonding character of the WO_3 network increases the glass transition and melting temperatures. As a result, the softening point of tungsten-tellurite glass is 370 °C that is higher than that of tellurite(290 °C)[10].

The purpose of this study is preparation of tungsten-tellurite glass thin films by RF magnetron sputtering method. The effects of deposition parameters on the growth and optical properties of thin films while changing substrate temperature, working pressure, input power, and target composition have been investigated.

EXPERIMENTAL DETAILS

For the preparation of thin films, a $70TeO_2$-$30WO_3$ (wt%) 2" target was fabricated by solid-state sintering method using chemicals (Tellurium (VI) oxide, 99.9 % , Tungsten (VI) oxide, 99.99 %) of High Purity Chemicals. The Si (001) and corning glass substrates were cleaned using acetone, methylalcohol, ethylalcohol, and DI water for 5min each. Thin films of Tungsten-Tellurite glass have been deposited by radio-frequency (rf) magnetron sputtering technique in the presence of argon (Ar) and oxygen (O_2) gases using the target in turbo-pumped vacuum chamber with base pressure of approximately 3×10^{-6} torr. The target-to-substrate distance was approximately 7 cm. Films were fabricated at various processing parameters such as substrate temperature, Ar/O_2 processing gas flow ratio, processing pressure, and rf power as shown in Table 1. Among them, the optimized sputtering parameters were the working pressure of 5 mTorr and the rf power of 60 W.

Table 1. Various deposition parameters of the tungsten-tellurite glass thin films.

Temperature (°C)	room temp., 50, 70, 85, 100
Working pressure (mTorr)	5, 10, 15
RF power (W)	30, 60, 90, 120
Ar/O$_2$ flow ratio (sccm)	0/40, 10/30, 20/20, 30/10, 40/0
Deposition time (min)	60

The crystallinity, microstructure, and surface roughness of glass films were investigated using X-ray diffraction (XRD) (X'pert-PRO, Phililps, Netherland), scanning electron microscopy (SEM) (S-4700, Hitachi co., Japan) and atomic force microscopy (AFM) (nanoscope IV, Digital Instrument, USA). The film thickness was measured using a cross-sectional SEM image of a deposited film. The optical transmittance was determined with a UV-VIR-IR spectrometer (U-3500, Hitachi co., Japan). The transmission spectrum of the thin film was measured in the wavelength range from 300 nm to 1100 nm. Measured transmittance spectra were used to determine the dispersive curves of the refractive indices.

RESULTS AND DISCUSSION

First, we investigated the effect of substrate temperature on the crystallinity of thin films, while keeping the other processing conditions (60W, 5mTorr, 1h) fixed. Figure 1 shows XRD patterns of thin films deposited at various substrate temperatures from room temperature (b) to different temperatures (a). Indexing of patterns shows that there are peaks from WO$_3$ crystalline phase in all patterns except the pattern from the sample deposited at room temperature that have peaks only from the substrate. This result indicates that the tungsten-tellurite films have crystalline phase when they are prepared at above the room temperature.

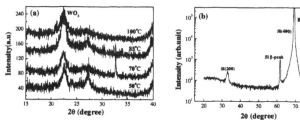

Figure 1. XRD patterns of as-deposited TeO$_2$-WO$_3$ thin film prepared at various substrate temperatures from 50 to 100 °C (a) and at room temperature (b) with rf power of 60W and working pressure of 5mTorr.

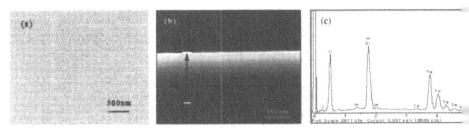

Figure 2. Plan-view (a) and (b) cross-sectional SEM images of as deposited TeO_2-WO_3 films deposited at room temperature 60W, 5mTorr, and Ar:O_2 (20:20), and (c) EDXA spectrum of the sample.

The microstructure and chemical information of as-deposited TeO_2-WO_3 thin films were also investigated using SEM. Figure 2 shows a SEM plan-view image (a), a cross-sectional image (b), and EDXA spectrum (c) of the as-deposited TeO_2-WO_3 thin film prepared at room temperature 60 W, 5 mTorr, and Ar:O_2 (20:20). There are no indication of any crystalline phases in both the plan-view and cross-sectional images. It is confirmed that the film consists of Te, W, and O elements from the EDAX spectrum. Further quantitative study may be needed to see the effect of the composition on the luminescent property of glass thin films in future.

Figure 3 shows AFM images showing the surface morphologies of tungsten-tellurite glass thin films prepared at room temperature with the various Ar/O_2 gas flow ratios while fixing rf power of 60W and working pressure of 5mTorr. All the AFM images show that films have very uniform microstructure and very smooth surface structure with surface roughness RMS of about 4~6 nm.

Figure 4 shows the effect of the processing parameters on the deposition rate of tungsten tellurite glass thin films prepared at room temperature by changing other processing parameters, such as, processing power (a), working pressure (b), and Ar/O_2 flow ratio (c). It is clearly observed that the deposition rate strongly depends on the processing parameters. It increases as

Figure 3. Atomic force microscopy images of tungsten-tellurite glass thin films prepared at room temperature with various Ar/O_2 flow ratios (a) Ar:O_2 (0:40) RMS=4.61 nm, (b) Ar:O_2 (10:30) RMS=6.53 nm, and (c) Ar:O_2 (20:20) RMS=4.77 nm.

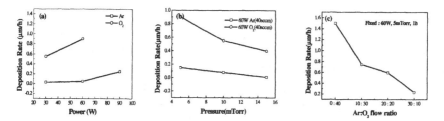

Figure 4. The effects of the processing parameters on the deposition rate of tungsten tellurite glass thin films prepared at room temperature by changing other processing parameters, such as, processing power (a), working pressure (b), and Ar/O₂ flow ratio (c).

the rf power increases and the processing pressure decreases. Especially, it changes remarkably as the Ar/O₂ gas flow ratio is changed from 40sccm/0sccm to 0sccm/40sccm. When the film was formed in pure Ar atmosphere it shows a deposition rate of ~0.2 μm /h, whereas ~1.5 μm/h when the film was formed in pure O₂ atmosphere.

The optical transmittance of the as-deposited films prepared on corning glass substrates was measured in the wavelength UV-VIR-IR spectrometer to calculate the refractive indices of tungsten-tellurite glass thin films. Figure 5 shows the transmission spectrum (a) and the calculated refractive index (b) of the tungsten-tellurite glass thin film deposited at deposited at 60 W, 5 mTorr, and Ar:O₂ (0:40)

We monitored the data from 300 nm to 1100 nm. The development of interference fringes indicates that the film thickness is uniform. The refractive index of the substrate can be determined by measuring the transmission spectrum of the substrate. The refractive indices range from 1.8 to 2.3 which are higher that than that of silicate glasses (about 1.45). It is observed that the refractive index decreases as the wavelength increases.

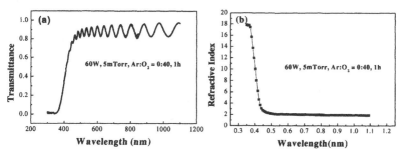

Figure 5. The transmission spectrum (a) and the calculated refractive index (b) of the tungsten-tellurite glass thin film deposited at 60 W, 5 mTorr, Ar:O₂ (0:40)

CONCLUSIONS

The effects of deposition parameters on the growth and properties of tungsten-tellurite glass thin films prepared by an rf magnetron sputtering technique on Si (001) and corning glass substrates were investigated. Amorphous films could be observed only at room temperature. Higher deposition rate could be obtained by increasing the rf power, decreasing the working pressure, and decreasing the Ar/O_2 ratio. Using the transmittance spectrum, the refractive indices of glass thin films could be measured to be about 1.8~2.3 depending on the wavelength. Further study is needed to investigate the optical properties about Er doped tungsten-tellurite glass thin films.

ACKNOWLEDGEMENTS

Authors are grateful to the financial support from Korea Ministry of Science and Technology through the Center for Photonic Materials and Devices

REFERENCES

1. A. Mori, Y. Ohishi, M. Yamada, H. Ono, Y. Nishida, K. Oikawa, S. Sudo, in : *OFC' 97*, Washington, DC, USA, 1997, PD1.
2. Y. Ohishi, A. Mori, M. Yamada, H. Onon, Y. Nishida and K. Oikawa Opt. *Lett.* **23,** 97(1998).
3. M. Yamada, A. Mori, H. Ono, K. Kobayashi, T. Kanamori and Y. Ohishi Electron. *Lett.* **34,** 370 (1998).
4. Nakai, Y. Noda, T. Tani, Y. Mimura, S. Sudo and S. Ohno OSA TOPS **25,** 82 (1998).
5. M.J.F. Digonnet (Ed.), *Rare-Earth-Doped Fiber Laser and Amplifiers*, Marcel Dekker, 2001
6. J.S. Wang, E.M. Vogel, E. Snitzer, Opt. Mater. **3,** 187 (1994).
7. R. Rolli, K. Gatterer, M. Wachtler, M. Bettinelli, A. Speghini, D. Ajo', Spectrochimica Acta Part A **57,** 2009 (2001).
8. L. Le Neindre, S. Jiang, B.C. Hwang, T. Luo, J. Watson, N. Peyghambarian, *J. Non-Cryst. Solids* **255,** 97 (1999).
9. A. Narazaki, K. Tanaka, K. hirao, N. Soga, *J. Applied Physics*, **85**(4), 2046 (1999).
10. Shaoxiong Shen, Mira Naftaly, Animesh Jha, *optics Communication* **205,** 101-105(2002).
11. S.K.J. Al-Ani, C.A. Hogarth, *Int. J. Electron.* **58,** 123(1985).

Mat. Res. Soc. Symp. Proc. Vol. 817 © 2004 Materials Research Society   L6.21

Eu-doped Yttria and Lutetia Thin Films Grown on Sapphire by PLD

S. Bär[1], H. Scheife[1], G. Huber[1], J. Gonzalo[2], A. Perea[2], A.Climent Font[3], F. Paszti[3], M. Munz[4]

[1]Institut für Laser-Physik, Universität Hamburg, Luruper Chaussee 149, 22761 Hamburg, Germany
[2]Instituto de Optica, CSIC, Serrano 121, 28006 Madrid, Spain
[3]Departamento de Física Aplicada y ICMAM, Universidad Autonoma de Madrid, 28049 Cantoblanco, Madrid, Spain
[4]Bundesanstalt für Materialforschung und –prüfung (BAM), Unter den Eichen 87, 12205 Berlin, Germany

ABSTRACT

This paper focuses on the preparation and characterization of crystalline thin films of rare-earth-doped sesquioxides (Y_2O_3 and Lu_2O_3) grown by pulsed laser deposition on single-crystal (0001) sapphire substrates. The crystal structure of the films (thicknesses between 1 nm and 500 nm) was determined by X-ray diffraction and surface X-ray diffraction analysis. These measurements show that the films were highly textured along the $\langle 111 \rangle$ direction. Using Rutherford backscattering analysis the correct stoichiometric composition of the films could be proved. The surface morphology of the thin films has been studied using atomic force microscopy. Crystalline films show a triangular surface morphology, which is attributed to the $\langle 111 \rangle$ growth direction. The emission and excitation spectra of the Eu-doped films down to a thickness of 100 nm look similar to those of the corresponding crystalline bulk material, whereas films with a thickness ≤ 20 nm show a completely different emission behaviour.

INTRODUCTION

The development of integrated optic devices demands the fabrication of high quality optically active thin films. This work focuses in particular on thin sesquioxide films, which are promising materials because they are well-known hosts for rare-earth-doped luminescent materials and solid-state lasers, e. g. $Yb:Y_2O_3$ [1]. Other potential applications are phosphor materials, high-temperature corrosion protection, and, due to the large band gap and a high dielectric constant, these materials can be used in semiconductor devices, e.g. MIS diodes, transistor gates, MOS capacitors, and DRAM [2,3,4]. Additionally, the fabrication of planar waveguide structures can be envisioned. These devices can be passive elements as well as active waveguides, where the larger emission and absorption cross sections available in crystalline matrices compared to fibres become accessible, and the confine-ment of light inside the waveguide generates a larger intensity-length product. Additionally, the guiding of the pump mode as well as the signal mode leads to an excellent overlap of the modes resulting in lower laser thresholds.

For the selection of a substrate for the growth of thin films with optical quality several require-ments must be considered: good transmission in a wide range of wavelengths, a refractive index lower than that of the growing film, low cost, and a good thermal and lattice match with the chosen film. As all these conditions are widely fulfilled, sapphire is an attractive substrate can-didate for the growth of Y_2O_3 films. The considerable small lattice mismatch, required for growing epitaxial films, is the main reason for choosing alumina substrates. Lattice matching is achieved by growth of the sesquioxides along the $\langle 111 \rangle$ direction on [0001] α-Al_2O_3. This leads to the relation $3 \times a(Al_2O_3) \approx \sqrt{2}\, a(RE_2O_3)$ which results in the case of Y_2O_3 in a mismatch of 4.8%. For the sesquioxides Lu_2O_3 and Sc_2O_3 this leads to a lattice mismatch of 2.8% and -2.5%, respectively. Another advantage of an alumina substrate is the thermal conductivity which is of the same order as that of the sesquioxides.

EXPERIMENTAL DETAILS

The targets used for ablation were sintered sesquioxide powders. The preparation was as follows: Rare-earth-doped sesquioxide powders were cold-pressed into a pellet of 2.54 cm in diameter. To increase the density, the pellets were sintered for 72 h in air at a temperature of 1700°C. The estimated density of these pellets was around 4.53 g/cm^3 which is 90% of the single-crystal density. All targets were cleaned prior to deposition by ablation with 300-500 pulses under the deposition atmosphere to ensure a homogeneous surface morphology. During the deposition process, the targets were continuously rotated to reduce the influence of crater formation on the target surface and to avoid stoichiometric changes in the target material. The α-Al_2O_3 substrates had a size of 10 mm $\times$ 10 mm $\times$ 0.5 mm and were polished on both sides with an RMS-roughness of < 0.4 nm. Prior to deposition, they were ultrasonically cleaned in a sequence of trichlorethylene, acetone, and alcohol. The substrates were placed at a fixed distance of 3.5 cm in front of the target holder. Prior to ablation, the chamber was evacuated down to a base pressure of 2.1×10^{-6} mbar. Then the chamber was refilled with oxygen. The optimum partial pressure of oxygen for the deposition process of Y_2O_3 was around 5×10^{-2} mbar. Using an ArF excimer laser (LAMBDA PHYSICS LPX 210i fluorine) operating at a wavelength of 193 nm, a pulse duration of 15 ns and a repitition rate of 10 Hz, Y_2O_3 targets were irradiated with a laser fluence of 2.1 J/cm^2. The threshold fluence of yttria upon nanosecond laser irradiation at 193 nm is $\sim$0.4 J/cm^2. The approximate growth rates of Y_2O_3 and Lu_2O_3 were 0.12 Å/pulse and 0.15 Å /pulse, respectively.

RESULTS AND DISCUSSION

The analytical tools for examining the structure of the films are X-ray diffraction and surface X-ray diffraction, Rutherford backscattering, and atomic force microscopy. To study the optical properties of the rare-earth-doped sesquioxide films, fluorescence and excitation measurements have been performed.

X-ray diffraction

To investigate the grade of crystallinity for different growth conditions, XRD measurements have been performed on films grown at different substrate temperatures and oxygen pressures, because these parameters have been found to be the most important. For a 500 nm thick film grown from a sintered Y_2O_3 target at a O_2 pressure of 5×10^{-2} mbar and a substrate temperature of 20°C, no Y_2O_3 diffraction peaks are visible – the film is completely amorphous. At a substrate temperature of 300°C, the yttria {222} reflection peak appears at 29.1°. At 60.26°, a higher order of this reflection, the {444} peak, arise. Additionally, the {400}, the {440}, and the {622} reflection peaks are visible. These peaks have less intensity than the {222} and {444} peaks, indicating a preferred growth of yttria in the $\langle 111 \rangle$ direction. This effect becomes stronger at a substrate temperature of 700°C (see figure 1, left). In this case, the {222} peak at 29.19° (FWHM = 0.163°) dominates the spectrum and is even stronger than the α-Al₂O₃ peak (FWHM = 0.1°). The {400} peak is also visible, but nearly four orders of magnitude smaller. This film has to be termed polycrystalline because there is no evidence of laterally connected crystallites forming the film. From these measurements it can be seen that the 500 nm thick films grown at substrate temperatures > 300°C are uniaxially textured along the ⟨111⟩ direction, because the X-ray diffraction patterns reveal mainly the {222} oriented diffraction line.

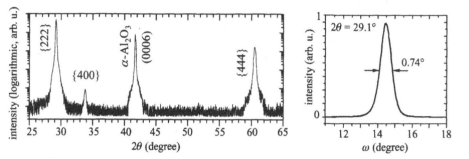

Figure 1. X-ray diffraction pattern of an Y₂O₃ film (left) grown at $T_{sub} = 700$°C and $p(O2) = 0.05$ mbar, rocking curve of the {222} reflection peak.

The same effect (preferred growth in ⟨111⟩ direction) is observed when the oxygen partial pressure is varied at constant temperature. If the pressure increases from 4×10^{-3} mbar the FWHM of the {222} reflection peak decreases till it reaches a minimum of 0.163° at 5×10^{-2} mbar. Further increase of the pressure results in a strong increase of the FWHM. Additionally, it was observed that the {400} peak becomes stronger at higher oxygen content. This crystallinity of the Y₂O₃ films was observed down to a film thicknesses of 20 nm. The width of the rocking curve (see figure 1, right) provides information about the out-of-plane-orientation and was determined to 0.743° in case of the {222} direction (FWHM of (0006) peak rocking curve is 0.3°). In the 20 nm thick Y₂O₃ film, the width of the {222} peak at 29.19° is 0.573° resulting in an estimated crystallite size of 18 nm.

Rutherford Backscattering

To confirm the composition of the films Rutherford backscattering (RBS) experiments have been performed. In this case, a 20 MeV ⁴He⁺ beam with a spot size of 1 mm² was used to bombard the samples. The backscattered particles have been detected with a PIPS silicon detector. By this measurement, it could be shown that the films have the correct stoichiometric composition and contain no impurities. At optimum growth conditions a channelling effect was observed, which can be explained as epitaxial growth of the Y₂O₃ film along the ⟨111⟩ direction on the [0001] sapphire. The epitaxial growth means that lattice matching between the substrate and the Y₂O₃ film can be achieved. This is an important result for possible waveguide applications because dislocations and defects at the interface are minimized.

Atomic Force Microscopy

Figure 2 (left) shows the surface morphology of a 500 nm thick crystalline yttria film The triangular (2D) or pyramidal (3D) shaped crystallite structure observed can be assigned to the ⟨111⟩ growth direction. The crystallites have edge lengths of the order of 120 nm–160 nm. A surface morphology with pyramidally shaped grains was also observed by [5] and [6]. A similar structure was occurs at the cleavage of a ⟨111⟩ grown Y_2O_3 bulk crystal. This leads to the conjecture that the ⟨111⟩ growth direction is connected with a pyramidal structure of the crystallites. In contrast to the cleavage structure in the bulk crystal, the film surface shows distorted triangles which are not parallel to the surface, but show a small angle with the substrate. This effect can be explained by the lattice mismatch between the grown yttria film and the corundum substrate. The RMS-roughness of the crystalline film is 2.4 nm. In general, the roughness of thick films, amorphous as well as crystalline, is comparably small, which clearly shows that these films can have optical quality.

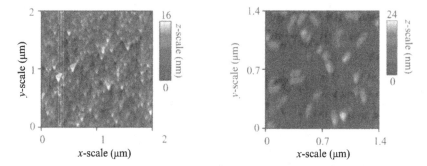

Figure 2. Surface of a 500 nm crystalline Y_2O_3 film (left). Surface of 5 nm Y_2O_3 film (right).

Additionally, in figure 2 (right), the surface of a 5 nm thick yttria film is presented. It can be seen that at this early stage of film growth, there is no complete film covering the substrate surface. Instead, three dimensional island growth of small crystallites occurs. The size of the crystallites can also be seen in figure 3. The in-plane dimensions are between 50 nm and 200 nm, whereas the height ranges from 15 nm to 25 nm. 3D-island growth of yttria on α-alumina was already observed by [6]. An interesting, however, until now not observed feature can be seen in the shape of the grains. Most of the edges of the crystallites have a 60°/120° angle, which is, together with the AFM result of a 500nm thick crystalline film, an additional proof of the ⟨111⟩ growth direction.

Optical Spectroscopy

Fluorescence and excitation measurements of the Eu(4%):Y_2O_3 films in the ultraviolet and visible region have been performed at room temperature with a modular fluorescence spectrometer (Yobin Yvon FL 321 FLUOROLOG-3). All films were grown at a substrate temperature of 700°C and an oxygen partial pressure of 5×10^{-2} mbar. The excitation spectra ($\lambda_{em} = 611$ nm) shown in the left part of figure 3 are characterized by the onset of the strong charge transfer band starting around 300nm and peaking around 240 nm. The excitation spectrum of the 100 nm film has a structure similar to the equivalent bulk excitation spectrum (see figure 3 (a)) with the exception that the spectrum presented in figure 3 (b) shows a strong broadening of the lines compared to the bulk spectrum. Down to a film thickness of 5 nm, the spectra have the same peak structure (figure 3 (c) the peaks appear very weak due to a short integration time). Additionally, a broad background from 370 nm to 450 nm appears in the 20 nm and 5 nm film. At a film thickness of 1 nm, only this broad structure is visible and no sharp lines can be found.

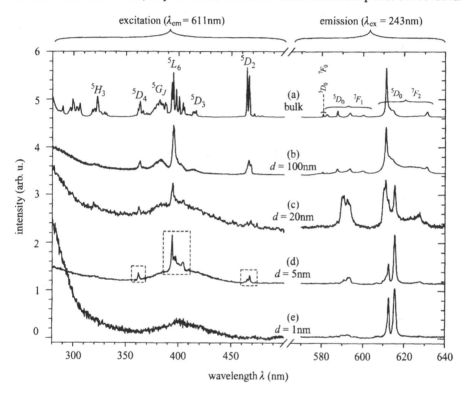

Figure 3. Excitation (left) and emission spectra (right) of Eu(4%):Y_2O_3 films with different thicknesses. The regions marked with a rectangle were measured with a longer integration time.

A more distinct, thickness dependent luminescence behavior can be observed in the emission spectra (λ_{ex} = 243 nm). The emission spectrum of the 100 nm film is similar to that of a corresponding crystalline bulk sample (figure 3 (b), right). The intense, narrow peak at 611 nm originates from the $^5D_0 \rightarrow {}^7F_2$ transition of Eu^{3+} in the C_2 site whereas the less intense features at 586 nm, 592 nm, and 599 nm belong to the $^5D_0 \rightarrow {}^7F_1$ transition of Eu^{3+} ions occupying the C_2 as well as the $C3i$ site. However, when the film thickness is reduced below 100 nm, the spectrum of Eu^{3+}:Y_2O_3 shows drastic changes as it can be seen in the right part of figure 3 (c) - (e). The peak intensity at 611 nm decreases with thinner films and completely vanishes in the case of the 1 nm film. The peaks at 612.6 nm and 615.5 nm, which appear in the shoulder of the 611nm peak in the 500 nm and 100 nm thick films, can now be resolved, and their intensity increases with decreasing film thickness. These peaks can be assigned to $^5D_0 \rightarrow {}^7F_2$ transitions of the C_2-sites which are normally very weak. The features around 580 nm, the transition $^5D_0 \rightarrow {}^7F_0$ of the C site at 579.4 nm, and the transition $^5D_0 \rightarrow {}^7F_1$ of the C_{3i} site at 581.2 nm vanish completely in films with a thickness smaller than 100 nm. The peak at 630 nm ($^5D_0 \rightarrow {}^7F_2$), clearly visible in the spectra of the 100 nm and 20 nm films, decreases drastically in intensity in the 5 nm film and vanishes completely in the 1 nm film. In addition, a blue shift of this peak is observed.

Similar spectroscopic results have been obtained in thin films of Eu^{3+}:Lu_2O_3. Independen of CT or direct excitation, the main emission line at 611 nm vanishes and the two peaks a 613 nm and 616 nm gain intensity. A similar change has been observed in the broad emission characteristics of nanocrystalline Eu^{3+}:Y_2O_3 [7]. However, in that case the emission was much broader than in the spectra presented here. The different luminescence behaviour of very thin films and nanoparticles can be explained by surface effects because the presented thin films are not closed films by means of a completely covered substrate surface, but single 3D islands. Due to a large surface-to-volume ratio of these islands, the number of defects, such as unsaturated bonds or varying bond lengths, which have a significant influence on the spectroscopic behavior [8], is strongly increased. The sharp lines obtained by the optical spectroscopy show clearly the crystalline behaviour of the films. Additionally, subplantation of Y_2O_3 into the Al_2O_3 matrix caused by high-energy particles in the plasma plume, can lead to the formation of $Y_3Al_5O_{12}$ or $YAlO_3$, and thus result in a change of spectroscopic beahvior.

CONCLUSION

In this work, we present good-quality crystalline thin films, by means of XRD, RBS and optical spectroscopy, of europium-doped sesquioxides (Y_2O_3, and Lu_2O_3) grown by pulsed laser deposition on single-crystal {0001} sapphire substrates. At the optimum growth conditions the RE_2O_3 films are highly textured along the $\langle 111 \rangle$ direction. Emission spectroscopy reveals great differences between thick films (100 nm and 500 nm), which have a bulk-like emission characteristic, and thin films (20 nm, 5 nm, and 1 nm). This behaviour can be explained by a change in the local structure around the Eu^{3+} ions near the interface and the surface.

REFERENCES

1. L. Fornasiero, E. Mix, V. Peters, K. Petermann and G. Huber, Cryst. Res. Tech. **34** (2), 255 (1999).
2. S. L. Jones, D. Kumar, R. K. Singh, and P. H. Holloway, Appl. Phys. Lett. **71** (3), 404 (1997)
3. A. C. Rastogi and R. N. Sharma, J. of Appl. Phys. **71** (10), 5041 (1992).
4. S. Zhang and R. Xiao, J. of Appl. Phys. **83** (7), 3842 (1998).
5. K. G. Cho, D. Kumar, D. G. Lee, S. L. Jones, P. H. Holloway and R. K. Singh, Appl. Phys. Lett. **71** (23), 3335 (1997).
6. M. B. Korzenski, P. Lecoeur, B. Mercey, D. Chippaux, B. Raveau and R. Desfeux, Chem. Mat. **12**, 3139 (2000).
7. Z.Qi, C. Shi, W. Zhang, W. Zhang, and T. Hu, Appl. Phys. Lett. **81**, 15 (2002)
8. P. Burmester, Optisch active, kristalline, Selten Erd-dotierte Y_2O_3-PLD-Schichten auf α–Al_2O_3, PhD thesis, University of Hamburg, Shaker Verlag, Aachen (2003)

Photonic Bandgap Crystals

Mat. Res. Soc. Symp. Proc. Vol. 817 © 2004 Materials Research Society   L2.3

Compact All Pass Transmission Filter using Photonic Crystal Slabs

Wonjoo Suh and Shanhui Fan

Department of Electrical Engineering,

Stanford University, Stanford, CA 94305

Abstract

We show that both the coupled photonic crystal slab and the single photonic crystal slab structure can function as an optical all-pass transmission filter for normally incident light. The filter function is synthesized by designing the spectral properties of guided resonance in the slab. We expect this compact device to be useful for optical communication systems.

Introduction

Compact optical filter structures are of great interest for optical communication applications. In particular, optical all-pass *transmission* filters, which generate significant delay at resonance, while maintaining 100% transmission both on and off resonance have been useful for applications such as optical delay or dispersion compensation[1][2] and the demand for making compact optical filters with these characteristics is currently increasing. In all pass reflection filters such as Gires-Tournois interferometers, proper signal processing is needed in signal extraction. Also, cascading multiple devices to obtain a high capacity delay line remains a challenge in this reflection mode. Here, we introduce a new type of optical all-pass transmission filter based upon guided resonance in photonic crystal slabs, which consist of a periodic lattice of air holes introduced into the dielectric slab.

Operating Mechanism

Guided resonances in photonic crystal slabs [3]-[10] provide a very compact way to generate useful spectral functions for externally incident light.

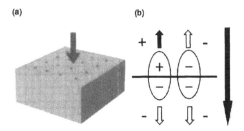

Figure 1. (a) Schematic of a photonic crystal filter consisting of a single photonic crystal slab. The arrow represents the direction of the incident light. (b) Schematic of a theoretical model for a resonator system that supports two resonant states with opposite symmetry with respect to the mirror plane perpendicular to the incident light.

An example of a photonic crystal slab consists of a periodic array of air holes introduced into a high index dielectric slab, as shown in Figure 1(a). Wang and Magnusson showed that a slab can function as a notch filter with a Lorentzian reflection line shape, when the slab thickness is appropriately chosen and a single resonance is placed within the vicinity of the signal frequency[11]. However, the spectral response has a strong variation in the intensity and therefore the filter is in the reflection mode. Therefore, in order to obtain 100% transmission while maintaining the group delay on resonance, we need two resonances that are degenerate. With two degenerate resonances with opposite symmetry, we are able to cancel out the effect of the strong variation in the transmission intensity. This is schematically shown in Figure 1(b), and this is referred to as accidental degeneracy.

Coupled Slab All Pass Filter

First we demonstrate a coupled photonic crystal slab configuration as a physical realization of creating an even and odd mode to generate a large resonant group delay while maintaining complete transmission both on and off resonant frequencies, using finite-difference time-domain (FDTD) simulations[12]. By cascading two photonic crystal slabs, we are able to obtain two resonances since each slab provides a resonance. In order to create accidental degeneracy in the two resonances, we need to balance the interference between the evanescent coupling and the

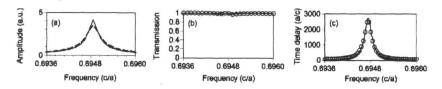

Figure 2. Spectral responses for the coupled slab structure. (a) Resonance amplitudes of even (dashed line) and odd (solid line) mode. (b) Transmission spectrum. (c) Group delay. In both (b) and (c), the solid line represents the theory and the open circles correspond to FDTD simulations

propagating coupling between the slabs. This can be done by choosing the appropriate displacement between the slabs to be $0.4a$, where a is the lattice constant. Within each slab of thickness $1.05a$, there are square lattice of air holes with radius $0.1a$, and the dielectric constant was chosen to be $11.4[13]$. For this design, we can see that there is accidental degeneracy in the even and odd resonances (Figure 2(a)). Therefore, the transmission spectrum shows near 100% transmission over the entire bandwidth both on and off resonance (Figure 2(b)), and yet there is a significant group delay in the vicinity of the resonance frequency (Figure 2(c)).

Single Slab All Pass Filter

Now, we show that a *single photonic crystal slab* can also function as an all-pass transmission filter, thus providing an extremely compact way of generating useful filter functions, and further demonstrating the versatility of photonic crystal structures.

To generate either an all-pass transmission filter function, we already know that one would need to have two resonant modes in the vicinity of the signal frequencies, which possess opposite symmetry with respect to the mirror plane perpendicular to the propagating direction. In order to design a single photonic crystal slab as shown in Figure 1(a) to have two resonant modes, we step back and probe the operating mechanism of these devices, which is the guided resonance. A guided resonance originates from the guided modes in a uniform dielectric slab, and is therefore strongly confined within the slab. And yet the periodic index contrast provides the phase matching mechanisms that allow these modes to couple into free space radiations in the vertical direction. Since a dielectric slab structure supports TE or TM guided modes that are even or odd with respect to the mirror plane at the center of the slab, a guided resonance could also be

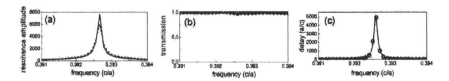

Figure 3. Spectral response functions for the one slab structure (a) The spectra of resonance amplitudes for the even mode (dashed line) and the odd mode (solid line). (b) Transmission spectrum for normally incident light. (c) Group delay spectrum. In both (b) and (c), the solid line represents the theory and the open circles correspond to FDTD simulations.

designed to have either even or odd symmetry. By appropriately choosing the structural parameters, it is then possible to place both an even resonance and an odd resonance in the vicinity of the signal frequency.

Again, in a FDTD simulation, we excite the resonant modes by a pulse of a normally incident plane wave. The line shapes of even and odd modes can then be obtained by Fourier-transforming the temporal decay of the resonance amplitudes. When the structure is chosen to have a thickness of $2.05a$ where a is the lattice constant, a radius of air holes of $0.12a$, and a dielectric constant of 10.07, which corresponds to that of AlGaAs in optical frequencies[14], both the even and odd mode have the same frequency and widths as shown in Figure 3(a). The transmission spectrum therefore shows near 100% transmission over the entire bandwidth both on and off resonance as can be seen in Figure 3(b), while a large resonant delay is generated in the vicinity of the resonant frequency (Figure 3(c)). The peak delay of $5000(a/c)$ corresponds to $10.14ps$, when the operating wavelength is at $1550nm$. For such a delay, the structure is only $1.2\mu m$ thick.

Conclusion

We note that the all pass filter proposed here can be readily cascaded to create optical delay lines since the filter operates in a transmission mode. In such an optical delay line, it has been shown that the maximum capacity is inversely related to the dimension of each stage[15]. Consequently, our filter structure, which is extremely compact, is useful for increasing the capacity of such delay lines. Also, unlike many single-mode integrated optical devices, both filter structures proposed here couples easily with optical fibers, since the mode of a fiber is typically far larger than the periodicity of the crystal. With a square lattice, the structures are inherently polarization independent, which is required for most communication applications. Polarization- selective dispersion characteristics, on the other hand, can also be readily designed by simply choosing a crystal lattice with less symmetry. Finally, these structures are far more compact than conventional multi-layer thin film devices commonly used. We therefore expect these novel and compact devices to be useful in optical communication systems.

Acknowledgement

This work was partially supported by the US Army Research Laboratories under Contract No. DAAD17-02-C-0101, and by the National Science Foundation (NSF) grant ECS-0200445. The computational time was provided by the NSF NRAC program.

References

[1]   G. Lenz and C. K. Madsen, *J. Lightwave Technol.* **17**, 1248 (1999).

[2]   C.K. Madsen, J.A. Walker, J.E. Ford, K.W.Goossen, T.N. Nielsen and G. Lenz, IEEE Photon. Techno. Lett. **12**, 651 (2000).

[3]   M. Kanskar, P. Paddon, V. Pacradouni, R. Morin, A. Busch, J. F. Young, S. R. Johnson, J. Mackenzie, and T. Tiedje, *Appl. Phys. Lett.* **70**, 1438 (1997)

[4]   V. N. Astratov, I. S. Culshaw, R. M. Stevenson, D. M. Whittaker, M. S. Skolnick, T. F. Krauss, and R. M. De La Rue , *J. Lightwave Technol.* **17**, 2050 (1999).

[5]   S. Fan and J. D. Joannopoulos, *Phys. Rev. B*, **65**, 235112 (2002).

[6]   M. Boroditskky, R. Vrijen, T. F. Krauss, R. Coccioli, R. Bhat, and E. Yablonovitch, *J. Lightwave Technol.* **17**, 2096 (1999).

[7]   A. Erchak, D. J. Ripin, S. Fan, P. Rakich, J. D. Joannopoulos, E. P . Ippen, G. S. Petrich and L. A. Kolodziejski, *Appl. Phys. Lett.* **78**, 563 (2001).

[8]   H. Y. Ryu, Y. H. Lee, R. L. Sellin, and D. Bimberg, *Appl. Phys. Lett.* **79**, 3573 (2001).

[9]   M. Meier, A. Mekis, A. Dodabalapur, A. A. Timko, R. E. Slusher and J. D. Joannopoulo Appl. Phys. Lett. **74**, 7 (1999).

[10]   S. Noda, M. Yokoyama, M. Imada, A. Chutinan, and M. Mochizuki, *Science*, **293**, 1123 (2000).

[11]   S.S. Wang and R. Magnusson , *Opt. Lett.* **19**, 919 (1994).

[12]   K. S. Kunz and R. J. Luebbers, *The Finite-Difference Time-Domain Methods for Electromagnetics* (CRC Press, Boca Raton, FL, 1993); A. Taflove and S. Hagness, *Computational Electrodynamics: The Finite-Difference Time-Domain Methods* (Artech House, Boston, 2000).

[13]   W. Suh and S. Fan, *Opt. Lett.* **28**, 1763, (2003)

[14]   Edward D. Palik, *Handbook of optical constants of Solids* (Academic Press, San Diego, Calif., 1985).

[15]   Z. Wang, S. Fan, *Phys. Rev. E*, **68**, 066616, (2003)

Mat. Res. Soc. Symp. Proc. Vol. 817 © 2004 Materials Research Society   L2.5

Omnidirectional reflectance and optical gap properties of Si/SiO$_2$ Thue-Morse quasicrystals

L. Dal Negro[1], M. Stolfi[1,2], Y. Yi[1], J. Michel[1], X. Duan[1], L.C. Kimerling[1]
J. LeBlanc[2], J. Haavisto[2]

[1]Massachusetts Institute of Technology, 77 Massachusetts Avenue, Cambridge, MA 02139
[2]Charles Stark Draper Laboratory, 555 Technology Square, Cambridge, MA 02139

ABSTRACT

Aperiodic one dimensional Si/SiO$_2$ Thue-Morse (T-M) multilayer structures have been fabricated, for the first time, in order to investigate both the band-gap behavior, with respect to the system size (band-gap scaling), and the omnidirectional reflectance of the fundamental optical band-gap. Variable angle reflectance data have experimentally demonstrated a large reflectance band-gap in the optical spectrum of a T-M quasicrystal, in agreement with transfer matrix simulations. We have explained the physical origin of the T-M omnidirectional band-gap as a result of periodic spatial correlations in the self-similar T-M structure, as revealed by Fourier Transform and Wavelet analysis. The unprecedented degree of structural flexibility showed by T-M systems can provide an attractive alternative to photonic crystals for the fabrication of photonic devices.

INTRODUCTION

Photonic quasicrystals (PQ's) are deterministically generated dielectric structures with non-periodic refractive index modulation. PQ's represent an intermediate organization stage between periodic dielectric materials, namely photonic crystal structures[1-3], and random media[4-7]. One dimensional PQ's can be generated by stacking together layers of different dielectric materials, A and B, according to simple rules, encoding a fascinating complexity. PQ's show peculiar physical properties like the formation of multiple frequency band-gap regions, called pseudo band-gaps[8,9], the presence of fractal transmission resonances[10,11] and the occurrence of critically localized states[12,13] (field states that decay weaker than exponentially, typically by a power law, and have a rich self similar structure). The presence of large band-gap regions in photonic structures without translational symmetry shows close analogies with the electron behavior in amorphous semiconductors or glass materials. Since the first experimental realization by Gellermann et al. [14] of an optical Fibonacci quasicrystal, the Fibonacci system has been predominantly investigated leading to the experimental demonstration of transmission scaling[14,15], symmetry induced resonances[16], complex light dispersion[17] and strong band-edge group velocity reduction[18]. Fibonacci quasicrystals are an example of quasi-periodic structures with delta-like Fourier power spectrum (FPS) characterized by non-periodic self-similar Bragg peaks which are responsible for the location and width of the energy pseudo band-gaps[19]. However, there are other classes of quasicrystals that exhibit a "more complex" structure than the Fibonacci ones. In particular, deterministic aperiodic structures are characterized by singular continuous Fourier spectrum[20]. The principal example of an aperiodic structure is given by the Thue-Morse (T-M) sequence[20-22], generated by the simple inflation rule σ_{T-M}: A→AB, B→BA[23]. The lower order T-M sequences are given by the strings: S_0=A, S_1=AB, S_2=ABBA, S_3=ABBABAAB, etc. An extensive theoretical literature describes both the localization and the intriguing scaling of T-M resonant transmission states[23] in terms of a generalized trace map approach[23,24], originally introduced by Kohmoto et al. [25] to explain the Fibonacci case. In addition, electron scattering from a-periodic potentials has been investigated by Fourier transform methods showing a close relationships between the geometry of the potential and the physical properties of the corresponding energy spectra[19]. However, the band-gap properties of T-M structures have not been investigated experimentally.

EXPERIMENTAL RESULTS

Here we report on the first experimental realization and study of the optical band-gap properties of Si/SiO_2 T-M quasicrystals up to 32 layers thick. For simplicity, the thickness $d_{A,B}$ of the two materials, SiO_2 (layer A) and Si (layer B), has been chosen to satisfy the Bragg condition, $d_A n_A = d_B n_B = \lambda_0/4$, where n_A and n_B are the respective refractive indices and λ_0=1550nm. The samples were fabricated on transparent fused silica substrates through RF-

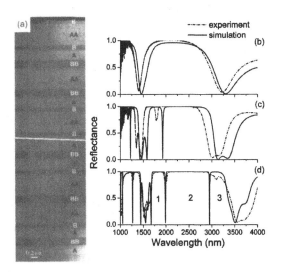

Figure 1. (panel a) Transmission Electron Microscope (TEM) cross section of the 32 layer (S_5) T-M structure. The light and dark layers correspond to SiO_2 and Si respectively. The letters on the different layers indicate all the components of the S_5 sequence. (panel b) Experimental (dash-dot line) and calculated (solid line) reflectance for the 8 layer (S_3) T-M structure. (panel c) Experimental (dash-dot line) and calculated (solid line) reflectance for the 16 layer (S_4) T-M structure. (panel d) Experimental (dash-dot line) and calculated (solid line) reflectance for the 32 layer (S_5) T-M structure. For all the simulations we have considered n_A=1.46 (SiO_2), n_B=3.53 (Si) and an incidence angle θ=30°, which approximately compensates for the large microscope objective numerical aperture (N.A.=0.6). The layer thickness is defined by the Bragg condition at λ_0=1.55μm.

magnetron sputtering in a Kurt J. Lesker CMS 18 UHV sputtering system using Si and SiO_2 targets. The Si and SiO_2 were sputtered in an Ar plasma at a power of 300 W and a deposition pressure of 3×10^{-3} Torr. The reflectance measurements were performed using a Nicolet Magna 860 Fourier Transform Infrared (FTIR) spectrometer coupled to a Nic-Plan IR Microscope for better spatial resolution.

For variable angle measurements we used the main chamber of the same FTIR equipped with a VeeMax variable angle specular reflectance accessory and a ZnSe polarizer.
In Fig. 1 (a) we show the TEM picture of the 32 layer T-M structure (S_5). In Fig. 1 (b-d), we plot the measured and calculated reflectance spectra of three T-M samples with different numbers of layers. The reflectance spectrum of 8 layer T-M sample (S_3), Fig.1 (b), has a large photonic band-gap around 2300nm. In the spectrum of the 16 layer T-M sample (S_4), Fig.1 (c), the original band-gap of Fig. 1 (b), is split into two distinct adjacent band-gaps by a narrow transmission band. The S_5 case, Fig. 1 (d), has three adjacent band-gaps (numbered 1,2,3) separated by two distinct narrow transmission regions.

DISCUSSION

The patterns revealed in the reflectance spectra can be related to the intrinsic properties of the T-M structures through Wavelet Decomposition and Fourier Transform analysis. Wavelet analysis maps one dimensional time signals into a two dimensional time-frequency domain. Since this approach is able to catch the fine structure of complex signals by performing local correlation analysis on a multiscale level[26-28], it is an ideal tool to highlight self-similarity and multifractality[28]. The basic idea of Wavelet decomposition is to represent complex signals as the sum over all time (or space) of the signal itself multiplied by scaled, shifted versions of the basis wavelets (complex conjugate $\overline{\varphi}$), simply defined as localized oscillation with zero average[26-28]. This process produces the wavelet coefficients $C(a,b)$ that are local functions of scale a and position b, and can be expressed as[28]:

$$C(a,b) = \langle \varphi_{a,b} | f \rangle = \int_{-\infty}^{+\infty} \overline{\varphi(a,b,t)} f(t) dt \qquad (1)$$

The wavelet coefficients describe how, on different length scales, the considered wavelet function φ can approximate the detailed structure of the analyzed signal $f(t)$. The information contained in the wavelet coefficients can be structured in the form of a two dimensional state map of the scale parameter a and position parameter b. In order to highlight the fractal pattern of an ideal, infinite T-M optical transmission we represented the T-M transmission spectrum of a very large structure (containing 256 layers) as a "time varying" one-dimensional signal and calculated the Wavelet decomposition map shown in Figure 2.

This analysis shows a clear self similar structure with a remarkable triplication pattern (Fig.2). This symmetry in the transmission spectrum is the origin of the experimentally observed band-gap triplication pattern already observed in the 32 layer T-M structures.
To achieve a further understanding of the T-M optical properties, we compare the transmission spectra with the Fourier power spectra obtained through discrete Fourier analysis. In Fig 3(a) we show the calculated T-M transmission spectra versus the normalized frequency $\delta = \omega / \omega_0$ from 0 to 2 ($\omega_0 = 2\pi / \lambda_0$) for the S_3, S_4 and S_5 generations respectively.

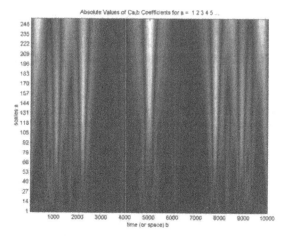

Figure 2. Wavelet decomposition map of a continuous signal given by the optical transmission of a 256 layer Thue-Morr aperiodic structure. Symmetric Mexican Hat wavelets are used as a basis set. This function is proportional to the secon derivative of the Gaussian probability density function. The horizontal axis shows the components of the frequency vector a represents an effective spatial dimension swept by the Wavelets. The vertical axis shows the different scales (from 1:1 to 1:256 that are considered in the calculation. The smaller the vertical scale, the finer the details that can be studied in the signal.

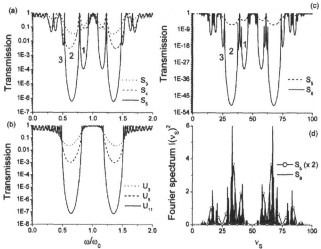

Figure 3. (panel a) Normalized frequency plot of the calculated transmission spectra of S_3 (dot-line), S_4 (dashed-line) and S_5 (solid-line) T-M structures. The numbers in the three band-gap regions of the S_5 structure correspond to the band-gap nomenclature shown in Fig. 1 (d). (panel b) Normalized frequency plot of the calculated transmission spectra of U_3 (dot-line), U_5 (dashed-line) and U_{11} (solid-line), the periodic approximant units for the fundamental band-gaps of the T-M structures. (panel c) Rescaled (horizontal scale in the phase plots of Fig. 2 (a) and (b) has been rescaled by the constant factor of 50) transmission spectrum of a 256 (solid line) layer (S_8) and 32 (dashed line) layer (S_5) T-M structure respectively. (panel d) Fourier Power Spectra (FPS) of the associated T-M strings with 32 (S_5) and 256 (S_8) characters respectively. The FPS of the S_5 string has been expanded by a factor of 2.

78

The three band-gaps in S_5 are labeled as the experimental data in Fig 1 (d). At the normalized frequency around 0.3, the smaller adjacent band-gaps of S_5 split into two separate band-gaps following the same pattern occurring at the two fundamental band-gaps in the T-M generation S_4 (see dashed line, central band-gaps).

Discrete Fourier analysis [19-25,29] was performed on T-M strings with 32 characters (S_5) and 256 characters (S_8) by assigning the numerical value 1 and 0 to the A and B symbols respectively. The FPS of S_8 and S_5 are plotted in Fig. 3 (d), as a function of the percentage of sampling frequency v_s. Comparing the two spectra in Fig. 3 (d), the FPS of the shorter S_5 string collapses, for the S_8 case, into a hierarchy of narrow frequency peaks connected by "small bands" of secondary frequency contributions. The fine structure of the S_8 case around the broader S_5 bands realizes the same triplication pattern previously demonstrated through Wavelet

Decomposition Analysis.

Since all the information contained in the optical transmission of Fig. 3 (a) falls between 0 and 2 in a normalized frequency scale and reflects all the frequencies in the associated FPS, we can establish a natural correspondence between Fig. 3 (d) and Fig. 3 (a) by rescaling the transmission data, as shown in Fig. 3 (c). There is high correlation between the Fourier peaks of the S_8 string in Fig. 3(d) and the position and width of the corresponding optical band-gaps in Fig. 3 (c). In particular, the physical origin of the fundamental T-M band-gaps can be attributed to local correlations in the form of periodic strings with the corresponding frequency (~ 30%) in the associated FPS (see Fig. 3 (d)). As an example, we consider the periodic approximant for the T-M structure, strings of the kind $U_n=(ABB)^n$, ($U_n=ABBABBABB...n$ times). The calculated transmission for U_3, U_5 and U_{11}, shown in Fig. 3 (b), qualitatively reproduces the two T-M fundamental band-gaps. It is interesting at this point to ask if the T-M central band-gaps also share the distinctive band-gap physical properties of their periodic approximants, in particular omnidirectional reflectance. We answer the question affirmatively, investigating for the first time the wide angle reflectance behavior of a T-M structure.

In Fig. 4 (a) we show the calculated reflectance spectra for both Transverse-electric (TE) and Transverse-magnetic (TM) modes versus wavelength for different incidence angles up to 89°. The grey shaded area shows unambiguously the occurrence of a large omnidirectional band-gap region corresponding to the central band-gap and determined only by the TM polarization. The experimental variable angle reflectance data is plotted in Fig. 4 (b) along with the calculated variable angle reflectance of the U_{11} approximant structure up to 70° incidence for the TM polarization. A large region of wide angle reflectivity is experimentally demonstrated at the central band-gap of the T-M structure, which corresponds with the angular behavior of the periodic U_{11} structure. The comparison with the simulation data in Fig. 4 (a) demonstrates the omnidirectional reflectivity of the experimental T-M structure.

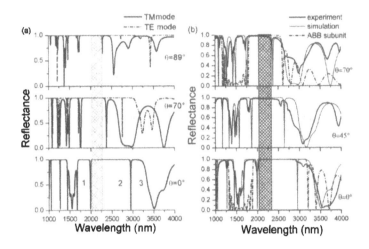

Figure 4. (panel a) Calculated reflectance of the 32 layer (S_5) T-M structure for both TM (solid line) and TE (dashed-dot line) modes at different incidence angles specified in the Figure.(panel b) Measured variable angle reflectance data (thick solid line) and transfer matrix simulation (thin solid line) for the TM polarization and incidence angles specified in the Figure. At the incidence angles θ=0° and θ=70° the calculated reflectance of the periodic U_{11} T-M approximant is also shown for comparison.

CONCLUSIONS

In conclusion, we have reported the first experimental study of the band-gap properties in Si/SiO$_2$ T-M quasicrystals, showing the remarkable scaling properties of the transmission spectra both with Wavelet Decomposition and Fourier Transform Analysis. In addition, we demonstrated large omnidirectional reflectance in a 32 layer T-M structure for the first time. We related the physical properties of the T-M optical spectrum with the complex geometry of the T-generating sequence, suggesting that local periodic correlations can be at the origin of the T-M large optical band-gaps. The unprecedented degree of structural flexibility of T-M systems can provide an attractive alternative route to regular photonic crystals for the fabrication of multi-frequencies devices, sensors and optical filters.

ACKNOWLEDGMENTS

This work has been supported by the Draper Laboratory Incorporated subcontract, No. DL-H-546257. We acknowledge L. Mayes from National Semiconductors for technical support during TEM sample preparation and T. McClure from MIT – CMSE for valuable technical support.

REFERENCES

1. E. Yablonovitch, Phys.Rev.Lett. **58**, 2059 (1987)
2. S. John, Phys.Rev.Lett. **58**, 2486 (1987)
3. J.D. Joannopoulos, P.R. Villeneuve, S. Fan, Nature (London) **386**, 143 (1997)
4. P.W. Anderson, Philos. Mag. **52**, 505, (1985)
5. S. John, Phys.Rev.Lett. **53**, 2169 (1984)
6. M.P. van Albada, A. Lagendijk, Phys.Rev.Lett. **55**, 2692 (1985)
7. D.S. Wiersma, P. Bartolini, A. Lagendijk, R. Righini, Nature (London) **390**, 671 (1997)
8. F. Nori, J.P. Rodriguez, Phys.Rev.B, **34**, 2207 (1986)
9. R.B. Capaz, B. Koiller, S.L.A. de Queiroz, Phys.Rev.B, **43**, 6402 (1990
10. T. Fujiwara, M. Kohmoto, T. Kokihiro, Phys.Rev.B, **40**, 7413 (1989)
11. C.M. Soukoulis, E.N. Economou, Phys.Rev.Lett.,**48** 1043 (1982)
12. M. Kohmoto, B. Southerland, C. Tang, Phys.Rev.B, **35**, 1020 (1987)
13. E. Maciá, Phys.Rev.B, **60**, 10032 (1999)
14. W. Gellermann, M. Kohmoto, B. Southerland, P.C. Taylor, Phys.Rev.Lett., **72**, 633 (1993)
15. M. Dulea, M. Johansson, R. Riklund, Phys.Rev.B., **47**, 8547 (1993)
16. R.W. Peng, X.Q. Huang, F. Qiu, A. Hu, S.S. Jiang, M. Mazzer, Appl.Phys.Lett., **80**, 3063 (2002).
17. T. Hattori, N. Tsurumachi, S. Kawato, H. Nakatsuka, Phys.Rev.B., **50**, 4220 (1994)
18. L. Dal Negro, C.J. Oton, Z. Gaburro, L. Pavesi, P. Johnson, A. Lagendijk, M. Righini, L. Colocci, D. Wiersma, *Phys.Rev.Lett.*, **90**, 5, art.num.055501-1 (2003)
19. J.M. Luck, Phys.Rev.B, **39**, 5834, 1989; R.W. Peng, M. Wang, A. Hu, S.S. Jang, G.J. Jin, D. Feng, Phys.Rev.B **52**, 13310 (1995)
20. Z.M. Cheng, R.Savit, R.Merlin, Phys.Rev.B, **37**, 4375 (1988);
21. M. Queffélec, *Substitution Dynamical Systems-Spectral Analysis, Lecture Notes in Mathematics*, vol.1294, (Springer, Berlin, 1987).
22. F. Axel, J.P. Allouche, M. Kleman, M. Mendès-France, J. Peyrière, J.Phys., (Paris), Colloq.47, C3-181 (1986)
23. N.Liu, Phys.Rev.B, **55**, 3543, (1997)
24. X. Wang, U. Grimm, M. Schreiber, Phys.Rev.B, **62**, 14020 (2000); S. Cheng, G. Jin, Phys.Rev.B, **65**, 134206-1 (2002)
25. M. Kohmoto, B. Southerland, K. Iguchi, Phys.Rev.Lett. **58**, 2436 (1987)
26. D.F.Walnut, *An Introduction to Wavelet Analysis*, Birkhauser Boston; (September 27, 2001)
27. B.Burke Hubbard, *The World According to Wavelets: The Story of a Mathematical Technique in the Making*, AK Peters Ltd; 2nd edition (May 1998)
28. M.Holschneider, *Wavelets, An Analysis Tool*, Clarendon Press, Oxford, (1995) C.Godrèche, J.M. Luck, Phys.Rev.B, **4**

at. Res. Soc. Symp. Proc. Vol. 817 © 2004 Materials Research Society L2.6

Preparation of opal-based PBG crystals to develop multiple stop bands

Yen-Tai Chen and Leo Chau-Kuang Liau
Department of Chemical Engineering, Yuan Ze University,
135 Yuan-Tung Rd., Chung-Li, Taiwan 320
email: lckliau@saturn.yzu.edu.tw

ABSTRACT

Opal-based photonic band gap (PBG) crystals with multiple stop bands were prepared utilizing a sol-gel method. The fabricating procedure includes colloidal crystal syntheses, dispersion, sedimentation, coating, and thermal treatments. Each of the steps can affect the PBG properties, such as stop band locations and ranges. Different stop bands of the photonic crystals can be produced by controlling the particle sizes prepared by the colloidal crystal syntheses. A PBG crystal film with a certain stop band was formed using a particular size of the colloidal crystals coated on glass substrates. In this study, two layers of different particle sizes of PBG crystal were fabricated by different deposition conditions to demonstrate the feasibility of producing multiple stop bands. These conditions can affect the stack layers and structural regularity for forming the PBG layers. In addition, the stop band intensity of the PBG layer can be further improved by the step of thermal treatments. Results imply that multiple stop bands can be feasibly designed and produced as multiple PBG layers coating with a certain SiO_2 particle size for each layer.

INTRODUCTION

Nowadays, photonic band gap (PBG) technology has been paid attention to many researchers, especially the field of opto-electronics. This technology has been applied on promoting the modern optical communication and improving efficiencies of light emitted diode (LED) and laser devices. Furthermore, the PBG elements have been designed as photonic integrated circuits developed to take part in the conventional microelectronics for the purpose of miniaturizing devices [1].

The theory of PBG is that certain frequency ranges of light wave propagation can be forbidden due to the periodical crystal structure of PBG materials. The ability of PBG to control the propagation of certain light frequency described as a photonic stop band is similar to manipulating electrons in semiconductor as an on-off signal. The structure of PBG materials were first proposed by Yablonovitch [2] to improve an efficiency of laser device and John [3] to create light localization both in 1987. The typical photonic crystals are fabricated with face-centered cubic (fcc) structure of silica as called opal PBG. However, more different photonic crystal materials, such as

PS [4], CdSe [5], ZnO [6], TiO2 [7], and structures were successfully established, i.e. rod-matrix structure [8], inverse opal structure [9] in recent years.

There are many methods to produce photonic crystals with different structures, such as such as bar-matrix structure using a lithography method [8]. One of the practical methods adopts sol-gel technique to synthesize colloidal crystals as materials to fabricate PBG crystals. The fabricating procedure of this method starts from a step of colloidal particle synthesis, following with powder dispersion, coating, self-assembling by sedimentation [9], and then processing thermal treatments. Each of these steps can greatly affect the PBG properties, such as stop band intensity, location, etc. For instance, self-assembling by sedimentation was taken as a critical step to influence the growth of PBG layers. Therefore, understanding the mechanism of particle sedimentation is essential if multiple stop bands of PBG crystals are to be made from the colloidal particles.

In this work, the feasibility of fabricating opal-based PBG with multiple stop bands was to be evaluated according to our developed processes. First of all, one size of colloidal particle size was first synthesized utilizing the sol-gel method by controlling the operating conditions. The sample films of PBG crystal layers were then produced based on the fabricating processes. The surface structures and photonic properties of these samples can be measured by analytical instruments. The development of PBG crystals with multiple stop bands was discussed and evaluated according to the results.

EXPERIMENT DETAILS

The experimental procedure includes colloidal particle synthesis, powder dispersion, deposition, self-assembling by sedimentation, and then processing thermal treatments. In the first step, a desired SiO_2 particle size used to fabricate photonic crystal is synthesized by a sol-gel method. TEOS, purchased from ACROS Corp., was dissolved in ethanol and stirred for 5 min; then ammonia (30 wt%) was added into the solution drop by drop. The mixture solution was stirred at 25℃ to synthesize SiO_2 colloidal solutions for several hours (aging time) depending on the preparation of desired particle sizes. This sol-gel reaction was terminated by adding certain amount of deionized (DI) water.

The SiO_2 powder was obtained by centrifuging the colloidal particle solution at 4000 rpm for 20 minutes. The prepared powder was washed three times with DI water using ultrasonic cleaner (ULTRAsonik model 104H) about 20minutes. After the sample was dried at 40℃ for an hour, the powder was then sintered for 12 hours at 500℃. The prepared powder was dispersed in DI water by ultrasonic cleaner for 1 hour as 100 ml water per gram of the powder. The solution was coated on glass substrates washed with DI water by ultrasonic cleaner. After being stayed for 3 days at room temperature, the PBG samples were sintered for 24 hours at 105°C.

The prepared PBG sample properties were analyzed using several instruments. The stop bands of the PBG samples were collected using a UV-vis-NIR spectrometer (Perkin Elmer Lambda 900). In addition, the structure and surface of the PBG samples were analyzed by the microscopy images performed using a SEM by HITACHI S-3000H.

DISCUSSION

Two different colloidal particle sizes, 550 nm and 190 nm, were prepared using sol-gel methods to fabricate two layers of PBG crystals, respectively. The wavelengths of the stop bands corresponding to each sample were estimated around 1200nm and 400nm, respectively, from a theoretical equation [11]. A single layer of a PBG film using each sample size was deposited on a glass substrate by the experimental procedure. Figure 1 shows the stop bands for the two sample films. Figure 1(a) elucidates the significant stop band at wavelength 1200 nm for the sample of larger particle size. The intensity of the stop band increases after the PBG film was thermally treated. Figure 1(b) shows the transmittance for the sample films of the 190 nm particle size. However, the stop band around 400 nm can not be clearly found in this figure.

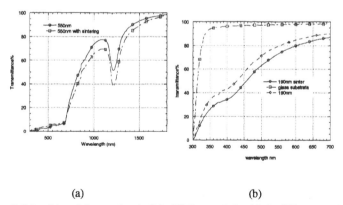

(a)   (b)

Figure 1. Intensities of the stop band of the PBG crystals for (a) the 550nm sample and (b) the 190nm sample.

A simple design of experiment was carried out using these particle sizes to test the preparation of the multiple stop bands as shown in Table 1. Two layers composed of two different particle sizes were stacked on the substrates layer by layer. The objective is to evaluate the feasibility of producing the multiple stop bands by different particle sizes within a PBG film. The deposited samples were sintered at 105 ℃ for 24 hours to evaluate the effect of the thermal treatment on PBG formation.

Table 1. Design of experiment for different samples

sample	A	B	C	D	E
Particle size of the top layer (nm)	550	550	190	190	-----
sintered	NO	YES	NO	YES	-----
Particle size of the bottom layer (nm)	190	190	550	550	mixing particles

Figure 2 demonstrates the results of the measurements in different wavelength for these samples. In figure 2(a), the intensities of samples A and B are more obvious than samples C and D at 1200 nm. This indicates that the intensity is affected by the combination of the layer preparation by different particle sizes. In addition, samples after the thermal treatment show stronger intensity of the stop bands. This implies that the sintering process can affect the stop band intensity. However, the stop band around 400nm for these samples is detected very weakly due to the glass substrate absorption as shown in figure 2(b).

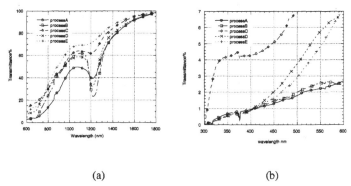

(a)   (b)

Figure 2. Intensities of the stop band of the PBG for different sample (a) wavelength range from 600~1800nm (b) wavelength range from 600~300nm

The particle size distributions of the two layer samples can be depicted using SEM pictures. The structure of sample B is shown in figure 3 by SEM as the larger particles are on the top layer and the smaller ones are on the bottom layer. On the contrary, if the samples C and D were made as the larger particles on the bottom and the smaller particles on the top, the distribution is quite different from that of sample B as shown in figure 4. From these figures, it appears that the smaller particles can diffuse down to the bottom layer as they distribute among the larger particles. This structure can further change the PBG properties and affect the stop band intensity and ranges as shown in figure 2.

(a)   (b)

Figure 3. The SEM pictures of sample B (a) top view (b) side view

Figure 4. The SEM picture of sample C

For sample E, one layer of the film was produced by mixing the two different particle sizes together to evaluate the particle interactions during the sedimentation. It is found that the intensity of the stop bands at 1200 nm becomes weaker and shifted as shown in figure 2. This is because the larger particles (550 nm) are surrounded by the smaller particles (190 nm) as illustrated in figure 5. Therefore, the stop band of sample E can be interfered by the distribution of these particles.

Figure 5. The SEM picture of sample E

CONCLUSIONS

The procedure of preparing PBG samples with multiple stop bands were proposed and developed in this work. Two layers with different particle sizes were produced to demonstrate the feasibility of fabricating these samples. The difficulties of developing multiple stop bands within a film are increased due to the complex fabricating procedures. In addition, the operating conditions of the process can affect the PBG sample properties from the analysis, such as intensity of stop bands and the structure uniformity. Results indicate that this proposed method can be applied to prepare and produce PBG samples with multiple stop bands.

REFERENCES

1. E. Yablonovitch, Scientific American, p47 (2001).
2. Yablonovitch, E. Phys. Rev. Lett. 58, 2059~2062 (1987).
3. John, S. Phys. Rev. Lett. 58, 2486~2489 (1987).
4. Rogach, A. L.; Kotov, N. A.; Koktysh, D. S.; Ostrander, J. W.; Ragoisha, G. A.; Chem. Mater., 12(9), 2721-2726 (2000).
5. Sacks, M. D., Tseng, T. Y., J. Am. Ceram. Soc, 67, 526-532 (1984).
6. C. C. Kao, 2002 Hsinchu Materials Nanotechnology Forum, Hsinchu, Taiwan.
7. Wijnjohn J. E. G. J. and Vos W. L., 281 802-584 (1998).
8. Lin, S.Y., Fleming, J.G., Hetherington, D. L., Smith, B.K., Biswas, R,. Ho, K.M., Sigalas, M.M., Zubrzycki, W., Kurtz, S. R., Bur, J., Nature, 394, 251-253 (1998).
9. Rui M. Almeida , Sabine Portal, Current Opinion in Solid State and Materials Science 7, 151–157(2003).
10. N. Stefanou, V.Yannopapas, A. Modinos, Computer Physics Communication, 13 , 49-47 (1998).
11. H. Miguez, H.,Blanco,A.,Meseguer, F., Lopez, C.,Phy. Rev. B, 59, 1563, 1999.

Mat. Res. Soc. Symp. Proc. Vol. 817 © 2004 Materials Research Society   L2.8

Optical Spectroscopy of Silicon-On-Insulator Waveguide Photonic Crystals

D. Bajoni, M. Galli, M. Belotti, F. Paleari, M. Patrini, G. Guizzetti, D. Gerace, M. Agio, L.C. Andreani
INFM and Dipartimento di Fisica "A. Volta", via Bassi 6, I-27100 Pavia, Italy

Y. Chen
Laboratoire de Photonique et de Nanostructures, CNRS, Route de Nozay, 91460 Marcoussis, France
Département de Chimie, Ecole Normale Supérieure, 24 Rue Lhomond, 75231 Paris Cedex 05, France

ABSTRACT

We report on a complete optical investigation on two-dimensional silicon-on-insulator (SOI) waveguide photonic crystals obtained by electron beam lithography and reactive ion etching. The dispersion of photonic modes is fully investigated both above and below the light-line by means of angle- and polarization-resolved micro-reflectance and attenuated total reflectance measurements.
The investigated samples consisted in a) large area (300 x 300 μm^2) two-dimensional (2D) triangular lattices of air holes containing repeated line–defects; b) small area triangular lattices of holes with different number of periods and /or line defects integrated in a ridge type waveguide.
In the case of large area samples, variable-angle reflectance and ATR is measured from the sample surface in a wide spectral range from 0.2 to 2 eV both in TE and TM polarizations. The sharp resonances observed in the polarized reflectance and ATR spectra allow mapping of the photonic dispersion of both radiative and guided modes. Experimentally determined and compared to those calculated by means of an expansion on the basis of the waveguide modes.
In the case of ridge type waveguide-integrated photonic crystals, transmission is measured in the 0.9-1.7 eV spectral range by an edge-coupling technique. Transmission spectra exhibit significant attenuation corresponding to the photonic gaps along the Γ–M and Γ–K directions respectively, even when a small number of hole periods is integrated in the ridge waveguide. Good agreement is obtained by comparing the measured transmission spectra with the calculated photonic bands.

INTRODUCTION

Since the pioneering works from Joannopoulos [1] and Yablonovitch [2], great interest has arosen in photonic crystals (PCs) for their peculiar optical properties. Periodically patterned planar waveguides are emerging as one of the best performing structures for the control of light propagation in three dimensions (3D) [3]-[5]. This is obtained by means of a two-dimansional (2D) photonic lattice embedded in a slab waveguide, which provides additional confinement in the vertical direction by means of total internal reflection. Nanoscale waveguides may be obtained by introducing linear defects in a periodic structure. A common design is the so called W1 waveguide, which consists in a missing row of holes in a triangular lattice along the Γ-K symmetry direction. The $\omega(k)$ dispersion of the defect- and bulk- photonic modes lies partly above and partly below the dispersion line of light in air. As in conventional dielectric slabs, the former modes can couple with external radiation being quasi-guided modes with intrinsic finite linewidths, while the latter are truly guided modes with very low propagation losses which are more suitable for photonic applications. The existence of truly-guided modes in PC-slabs requires a high dielectric contrast between core and cladding materials. In this respect, silicon-on-insulator (SOI) patterned waveguides has become one of the preferred systems for the study and design of micro-optical devices, due to the very well developed silicon technology. The experimental determination of photonic bands is extremely important for fundamental research as well as for applications, as it contains full information on the propagation properties of these systems. In the following, we show that a deep understanding of photonic mode dispersion over the whole reciprocal space can be obtained by optical spectroscopy means, especially by attenuated total reflectance (ATR) which was never applied before to these systems.

SAMPLE FABRICATION AND EXPERIMANTAL TECHNIQUES

Several different photonic crystal structures were fabricated by means of Electron Beam Lithography (EBL) followed by Reactive Ion Etching (RIE) on SOI wafers (SOITEC). The waveguides consisted in a 260 nm thick Silicon core on a 1 μm thick SiO_2 cladding grown on a Silicon substrate. EBL was performed on PMMA resist using a JEOL JBX5D2U vector scan generator at 50 keV energy. After developing the PMMA, pattern transfer to the waveguide core was realized through three-layer process: a 500 nm-thick S18 bottom layer, a 50 nm-thick

---

The work reported in this paper was supported by MIUR through COFIN and FIRB programs.

Figure 1. SEM micrographs of (a) W1_BULK and (b) patterned ridge-waveguide with 6 periods lattice along Γ–M direction.

Ge middle layer and a 150 nm-thick PMMA top layer. The top and middle layer were etched by standard RIE techniques, while the silicon top layer is etched by RIE using a SF_6 and CHF_3 gas mixture. The RIE parameters were optimised to obtain steep sidewalls in silicon [6].

Bulk samples (Fig. 1a) and samples containing W1 line-defects were realized in the triangular array of air holes with lattice constant $a = 500$nm and hole radius $r = 0.34a$. The patterned area was 300x300 μm^2, and W1 defects were repeated with different supercell periodicities $d = m\sqrt{3}a$ ($m = 4, 5, 6$) along the Γ–M direction. Triangular lattices of air holes, with lattice constant $a = 400$nm and hole radius $r=0.3a$, were also etched in 6 μm wide ridge-waveguides (Fig. 1b). This samples were realized for both lattice directions and different number of periods ($n=2,4,6$).

The optical response of the samples has been probed by means of angle-resolved specular reflectance. This technique has been applied before to measure the band dispersion of 1D and 2D PCs [7]-[9]. When the energy and the wave vector of the incoming beam match those of a photonic mode of the system, a resonant feature appears in the reflectance spectrum which marks the excitation of the photonic mode. By varying the angle of incidence and energy of the light, the band structure of the sample can be experimentally plotted in a large part of the Brillouin zone. The major drawback of this technique is that only radiative modes which lie above the light-line can be excited, while it is not suitable to probe the dispersion of the truly-guided modes which extend in the cladding regions and are evanescent in air. This limit can be overcome by means of angle-resolved Attenuated Total Reflectance (ATR) [10]. By putting a high index prism at a small distance from the surface of the sample, efficient coupling between the evanescent fields at the air-prism-sample interfaces can be achieved, therefore allowing the excitation of truly-guided modes of the system.

Angle-resolved specular reflectance and ATR from the sample surface are measured in the spectral range 0.3-1.8 eV, at a spectral resolution of 0.5 meV, by means of a micro-reflectometer coupled to a Fourier-transform (FT) spectrometer (Bruker IFS66s). The angle of incidence θ is varied in the range 4° - 75° with an angular resolution of 0.5° set by the very small but finite numerical aperture of the beam that is focused on the sample. Transverse-electric (TE) and transverse-magnetic (TM) polarizations with respect to the plane of incidence are selected by means of KRS5 wire-grid and calcite Glann-Taylor polarizers. ATR measurements were carried on with a ZnSe prism (refractive index 2.4, angle of total internal reflectance >23°) that was kept close to the sample surface at a distance of ~ 250 nm by means of piezoelectric actuators.

The transmission properties of patterned ridge-waveguides were investigated in the 0.9-1.7 eV spectral range. The white-light from a Hg arc-lamp is focused to a 10μm spot on the cleaved edge of the sample by means of a 25x reflecting microscope objective. Transmitted light is then collected from the other cleaved edge of the waveguide by means of an identical objective and sent to the FT spectrometer equipped with a InGaAs detector for spectral analysis. The sample and the collecting objective were mounted on a high precision nanopositioner. Transmittance spectra of patterned ridge-waveguides were calibrated using an identical unpatterned waveguide as a reference. Due to the very strong dependence of the coupling and collecting efficiency on quality of the cleaved facets, each patterned waveguide was etched close (10 μm apart) to a reference one.

RESULTS AND DISCUSSION

Angle-resolved reflectance spectra were measured for W1_BULK along both symmetry orientations and for both polarizations. Results obtained for the Γ–K orinetaiotn and TE polarization are shown in Fig. 2a. The curves

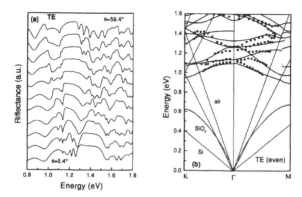

Figure2. (a) Reflectance spectra for the W1_BULK sample measured for TE polarization along the Γ–K orientation. (b) Measured dispersion of the photonic bands as derived from reflectance spectra (circles), compared to calculated dispersion (lines).

display a prominent interference pattern due the multiple interference occurring at the core-cladding and cladding-substrate interfaces. Superimposed to the interference fringes, several sharp features are clearly visible, whose energy position varies with angle, polarization and lattice direction. We ascribe these structure to resonant coupling of the incident radiation to quasi-guided modes of the patterned waveguide. The line-shape of these features may change with their position with respect to the to the interference pattern: they may appear as maxima, minima or even with dispersive-like shape. This leads to some uncertainty in the absolute value of the experimentally determined energy bands, which is however estimated to be in the range of a few meV.

By plotting the energy position of the resonant structures versus the in-plane component of the wavevector of light the photonic band dispersion is determined, as shown in Fig. 2b. The experimental band structure is

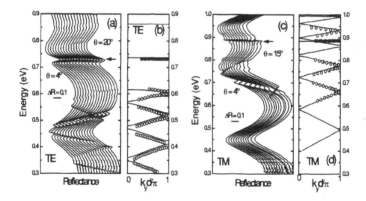

Figure 3. (a,c) Reflectance measurements for the sample W1_4 along the Γ–M orientation for TE and TM polarizations. The curves are slightly shifted for clarity. The defect modes are indicated by arrows. (c,d) Corresponding photonic bands folded into a reduced Brillouin zone due to supercell periodicity: experiment (closed circles for defect modes, open circles for bulk modes) and theory (lines).

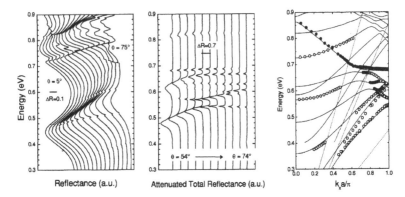

Figure 4. Experimental reflectance (a) and ATR (b) measured along the Γ–K orientation for TE polarization (curves are shifted for clarity). (c) Measured dispersion of the photonic bands as derived from reflectance and ATR spectra for TE polarization (closed circles for defect modes, open circles for bulk modes), compared to calculated dispersion (lines). Dotted lines: light lines of air, SiO₂ and Si.

compared to a full 3D calculation, as obtained by expanding the magnetic field on the basis of the guided modes of an unpatterned waveguide and assuming an average refractive index in each layer [11]-[12]. Rather good agreement between theory and experiment is obtained for dispersion and energy of photonic bands both TE and TM polarizations (not shown in Fig 2b). The dispersion of the lowest bands up to ~ 1 eV is particularly well reproduced, whereas small discrepancies between experiment and calculations can be seen for higher energy bands. This may be attributed to the dispersion of the refractive index of the layers, which has not been included in calculations. Figures 3a and 3c experimental reflectance spectra of sample W1_5 measured along the Γ–M direction, i.e. orthogonal to the line defects, and for both polarization. Also in this case, resonant features corresponding to the excitation of photonic modes which exhibit a marked dispersion with the incidence angle are clearly observed. Besides these, two dispersionless resonances occur at 0.73 eV and 0.89 eV for TE and TM polarizations, respectively. These correspond to the defect modes localized at the missing rows of holes, which act as cavities. A comparison with calculated dispersion is presented in Fig. 3b and 3d for TE and TM polarizations, respectively. Excellent agreement is found between experimental and calculated bands. Notice that along for this sample orientation the wave vectors of the modes are perpendicular to the line-defects, and the photonic bands are folded into a reduced Brillouin zone due to the new, longer periodicity of the system. As a result, most of the bands fall above the light line and are observed in angle-resolved reflectance spectra.

The optical response of sample W1_5 along the Γ–K direction were thoroughly studied by means of both angle resolved reflectance and ATR. The wave vectors of the photonic modes are in this case parallel to the line defects, which provides additional waveguiding of radiation along the defect: no localization of light is expected and the defect mode is therefore dispersive due to its finite group velocity, as observed from reflectance spectra shown in Fig. 4a. Figure 4b show ATR spectra measured for the same sample orientation. In this case, excitation of guided modes appears as "absorption-like" dips on a flat, almost unitary background, which correspond to the total internal reflection regime. Notice that while resonances observed in reflectance spectra have amplitudes of the order of 0.05, those observed in ATR geometry can be deep as 0.8. This means that as much as the 80% of the incoming energy can be transferred to the PC slab by such a prism-coupling. Such a highly efficient energy transfer is attributed to the overlap of the evanescent tails of the modes freely propagating in the ZnSe prism with those of the PC guided modes. This overlap mostly occurs in air spacer between the prism and the surface of the sample and the coupling is thus highly dependent on their distance. When the prism is placed very close to sample surface, the modes are strongly affected by its presence and broad features appear in the spectra. The band dispersion is also affected and the energies of the features are slightly shifted. By increasing the width of the air spacer, the resonances become sharper and their energy positions stick with the values expected for the SOI PC slab. For the measurements of Fig. 4b the separation distance is ~250nm.

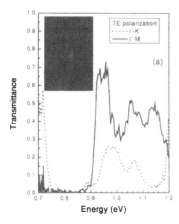
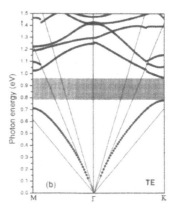

Energy (eV)

Figure 5. (a) Transmittance spectra of the patterned waveguides with six periods along the Γ–K and Γ–M directions. The inset show the light coming out from the ridge recorded with an infrared vidicon camera. (b) Calculated photonic bands for the corresponding triangular lattice for TE polarization. Shaded area indicate the overlap between photonic band-gaps along the Γ–K and Γ–M directions

The combination of results obtained from normal reflectance and ATR spectra yields an very precise experimental reconstruction of the band dispersion of the sample, along the whole extent of the Brillouin Zone. As a matter of fact, in the considered spectral range, the value of wave vectors obtained by means of the high-index prism are greater than π/a and the experimental points must therefore be reported in the first Brillouin zone by subtracting a reciprocal lattice vector. Excellent agreement is found between the experimental dispersion and the full 3D calculations for both TE and TM polarizations, as shown in Fig. 4c (TM bands are not reported).

The dispersive behaviour of the defect modes is traced in the reciprocal space from the low-k region across the light-line up to the edge of the reduced Brillouin zone. We observe that the TE-like defect mode is characterized by a very unusual behaviour of the dispersion, which is pronounced in the radiative region and become very smooth in the guided region, where the mode propagates with low group velocities.

Transmission spectra of ridge-waveguide photonic crystals having 6 periods lattice are shown in Fig. 5a. Light propagation along the Γ–M and Γ–K directions was measured for TE polarization only. Considerable attenuation of light transmission due to the presence of the photonic band gap is clearly observed in the spectra, as well as the expected Fabry-Perot interference fringes caused by the finite size of the system. We notice that transmittance values as low as 0.005 are achieved within the stop-band even for this small number of periods. This is attributed to the high dielectric contrast of the SOI system. On the other hand, transmittance values for energies falling outside the band-gap reach values of ~0.5, thus implying a total attenuation ratio between rejected and transmitted light of the order of -20 dB. Calculated bands for TE polarization are shown in Fig. 5b. Good agreement is found for both energy position and bandwidth of the photonic band-gap.

CONCLUSIONS

2D Silicon-on-Insulator photonic crystal waveguides with periodically repeated W1 line-defects and patterned ridge-waveguides have been successfully fabricated by means of Electron Beam Lithography and Reactive Ion Etching. The photonic band dispersion of the samples has been directly probed by means of angle- and polarization- resolved reflectance and ATR measurements. The propagation characteristics of linear defects have thus been investigated in details both above and below the light-line. The resulting dispersion is in very good agreement with full 3D calculation based on the expansion of the magnetic field on the guided modes. Transmittance measurements performed on the patterned ridge-waveguides with a small number of periods confirmed the formation of a complete stop-band for TE polarized light.

REFERENCES

[1] J.D. Joannopoulos, R.D. Meade, J.N. Winn, *Photonic Crystals Molding the Flow of Light*, Princeton: Princeton University Press, 1995.

[2] E. Yablonovitch, T.J. Gmitter, K.M. Leung, *Phys. Rev. Lett., vol. 67.p. 4753, 1990.*

[3] K. Sakoda, *Optical Properties of Photonic Crystals*, Springer Verlag , 2001.

[4] See papers in IEEE J. Quantum Electron. *vo.l 38*, Feature section on photonic crystal structures and applications, T.F. Krauss and T. Baba, Ed., 2002.

[5] H. Johnson, J.D. Joannopoulos, *Photonic Crystals: the Road from Theory to Practice*, Kluwer Academic Publishers, 2002.

[6] D. Peyrade, Y. Chen, A. Talneau, M. Patrini, M. Galli, L.C. Andreani, E. Silberstein, P. Lalanne, *Microelectron. Engin., vol 61-62 p. 529*, 2002.

[7] V. N. Astratov, D. M. Whittaker, I. S. Culshaw, R. M. Stevenson, M. S. Skolnick, T. F. Krauss and R. M. De La Rue, *Phys. Rev. B, vol. 60 p. R16255,*1999.

[8] AV. N. Astratov, I. S. Culshaw, R. M. Stevenson, D. M. Whittaker, M. S. Skolnick, T. F. Krauss, and R. M. De La Rue, *J. Lightwave Technol., vol. 17 p. 2050*, 1999.

[9] V. Pacradouni, W. J. Mandeville, A. R. Cowan, P. Paddon, J. F. Young, and S. R. Johnson, *Phys. Rev. B vol. 62 p. 4204*, 2000.

[10] M. Galli, M. Belotti, F. Paleari, D. Bajoni, M. Patrini, G. Guizzetti, D. Gerace, M. Agio, L.C. Andreani, Y. Chen, *submitted to Phys. Rev. B.*

[11] L.C. Andreani and M. Agio, in [4], *p. 891.*

[12] L.C. Andreani and M. Agio, Appl. Phys. Lett., *vol. 82 p. 2011*, 2003.

Mat. Res. Soc. Symp. Proc. Vol. 817 © 2004 Materials Research Society   L5.8

Optical and Nanomechanical Characterization of an Omnidirectional Reflector Encompassing 850 nm Wavelength

Manish Deopura, Yoel Fink and Christopher A. Schuh
Department of Materials Science and Engineering, Massachusetts Institute of Technology
77 Massachusetts Avenue, Cambridge, Massachusetts, USA 02139

ABSTRACT

We demonstrate that multilayers composed of nineteen alternating layers of tin sulfide and silica can function as omnidirectional reflectors. These materials exhibit omnidirectional reflectivity for a range of frequencies in the near infra-red (NIR) encompassing the 850 nm wavelength. A refractive index contrast of 2.7/1.46 is achieved, one of the highest values demonstrated until now in NIR photonic bad gaps. In addition, new nanoindentation procedures have been developed to measure mechanical properties of these fine laminate materials, and demonstrate that tin sulfide-silica multilayers are mechanically stable for practical applications.

INTRODUCTION

Optical materials design requires knowledge from several fields including the physics of devices, optical properties of materials, processing methods, and characterization for robustness in applications. These various facets provide many design constraints, and opportunities at the same time. Most optical materials design in the past has been empirical, with researchers in academia and industry focusing on physics of devices and development of materials with low optical loss, since these are key requirements in any optical device. In contrast, systematic procedures for processing these optical materials and corresponding characterization techniques (particularly mechanical characterization) have been developed only to a limited extent.

More recently, a new trend in optical materials design has emerged. With the advent of photonic crystals in the last decade [1] and the generalization of some optical device principles, researchers from other fields (i.e., polymer synthesis or materials mechanics) have begun to be actively involved in shaping material design schemes. In photonic crystals, the omnidirectional reflector [2,3], a 1-D photonic band gap structure, has recently been reported for 700 nm wavelengths. When properly shaped in an appropriate geometry, this structure can provide optical performance similar to 2-D and 3-D photonic band gap structures without the associated processing costs and complexity.

The purpose of the present paper is to extend the concept of the 700 nm omnidirectional reflector reported in Ref. [3], to develop a 1-D photonic crystal for the 850 nm wavelength. This wavelength is significant for optical data transmission technologies, however until now it has been limited to generally short distance operation. To extend the 850 nm technology to long distance communications, practical limitations exist, most significantly the high attenuation and dispersion of the transmitted signal. Traditional fibers have already reached their attenuation and dispersion limit, and hence new materials systems for use as photonic crystal fibers need to be pursued. Recently photonic crystal fibers have been fabricated for use at other frequencies [4]. The present effort combines therefore two aspects: (a) illustration of a general systematic method for optical materials design, and (b) specific application of this approach to the design of an omnidirectional reflector for the NIR encompassing the 850 nm wavelength.

MATERIALS SELECTION

For photonic band gap structures like the 1-D omnidirectional reflector, obtaining high refractive index contrast is essential. At the visible and NIR frequencies, however, the electronic and ionic polarizability of most materials dictate that the refractive index is usually low. Recently, tin sulfide, which has a relatively high refractive index, has emerged as a possible candidate for practical applications as a photonic material for both 1-D [3,5] and 3-D [6] structures. Tin sulfide is optically transparent above its band edge of 590 nm [7] and has measured refractive index of n = 2.7 [8]. In combination with tin sulfide as the high refractive index material we have chosen silica as a low refractive index material (with n = 1.46). Omnidirectional reflectors for visible frequencies have already been successfully fabricated using this materials combination [3].

MATERIALS PROCESSING

The target structure for reflection in the 750-900 nm range was identified as a 19 layer structure with layer thicknesses of h_1= 75 nm for tin sulfide and h_2 = 135 nm for silica [8,9]. To fabricate this multilayer structure, the two materials were deposited sequentially using separate vacuum deposition chambers, on a square glass substrate 22 mm on each side. Tin sulfide layers were deposited using a thermal evaporator (CVC) operating at 10^{-6} torr and 10 A; the layer thickness and deposition rate were measured in-situ with a crystal thickness monitor. Silica layers were vacuum deposited using an electron beam evaporator (NRC 119) at 5×10^{-7} torr operating at 10 kV and 50 A; again the layer thickness and deposition rate were monitored in-situ with a sensor. The substrate was maintained at room temperature during all deposition procedures.

ANALYTICAL CHARACTERIZATION

Scanning electron microscopy (SEM) was used to characterize the thickness of the individual layers of the multilayer structure. In the SEM micrograph in Fig. 1, the bright regions correspond to tin sulfide and the dark regions correspond to silica; clearly the layers are well defined and relatively uniform. Root mean square (RMS) surface roughness was measured after each stage of the muti-step deposition process, to verify that roughness values were well within the low-loss optical scattering limit. In general, the surface roughness after any one step was less than ~1%. Rutherford backscattering was also used to verify that the stoichiometry of the tin sulfide thin films was reasonably close to the target value of SnS_2, being in the present case near $SnS_{1.85}$. Further detailed information on analytical characterization is presented in Ref. [8].

OPTICAL CHARACTERIZATION

Using the structure profile of the multilayer and the appropriate refractive index values for tin sulfide and silica (as described previously), we have constructed the expected band diagram for the multilayer, as shown in Figure 2. Here the black shaded areas highlight regions of propagating states, whereas the white areas represent regions containing evanescent states. The grey region represents the omnidirectional photonic band gap. The theoretically calculated values of ω_h and ω_l which define the omnidirectional band edges correspond to 730 nm and 870 nm respectively. The characteristic dimensionless parameter $\eta = 2(\omega_h - \omega_l)/(\omega_h + \omega_l)$, which quantifies the extent of the omnidirectional range, has a value of $\eta \sim 17\%$.

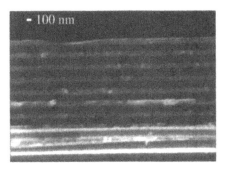

Figure 1 Cross sectional scanning electron micrograph for the 19 layer sample. Bright regions correspond to tin sulfide and dark regions correspond to silica.

To experimentally measure the optical response of the multilayer specimens a UV-VIS-NIR spectrophotometer was used. Normal incidence reflectivity measurements were done using a spectral reflectance accessory (SRA) and the oblique angle measurements up to 70° were done using a variable angle reflectivity stage (VASRA). A freshly evaporated aluminum mirror was used as background for the reflectance measurements.

The measured reflectance spectra for normal (0°), 45° and 70° for both the transverse electric (TE) and transverse magnetic (TM) modes of light polarization are presented in Figure 3. Around the target wavelength, nearly 100% reflectivity is observed. The shaded region in this figure shows the overlap of regimes of high reflectivity, which denotes the range of the omnidirectional band. Although it is not shown explicitly in Fig. 3, the measured response in Fig. 3 matches quite closely with the expected values obtained from theoretical calculations.

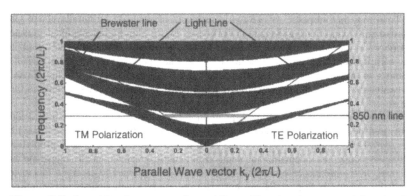

Figure 2 Projected band structure of a multilayer film showing an omnidirectional reflectance range in the first harmonic. Propagating states, black; evanescent states, white; and omnidirectional reflectance range, grey. The film parameters are $n1 = 2.7$ and $n2 = 1.46$ with a thickness ratio $h2/h1 = 145/85$ for the tin sulfide-silica system.

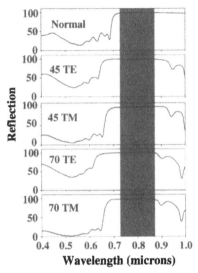

Wavelength (microns)

Figure 3 Measured reflectance spectra as a function of wavelength for TM and TE modes at normal, 45°, and 70° angles. The semi-transparent region shows the experimentally observed omnidirectional band-gap.

NANOMECHANICAL CHARACTERIZATION

In order to assess mechanical properties of these omnidirectional multilayer structures nanoindentation has been used. In general, nanoindentation techniques are not as well developed for multilayered films, as compared to the relatively common techniques used for individual thin films.

To ensure that properties of the two constituent materials of the multilayer are sampled in their true bulk proportions, upper and lower bounds for the allowed indentation depths need to be prescribed. In order to determine the upper bound indentation depth we make use of the well known 10% rule, which dictates that maximum indenter penetration should be 10% of the film thickness to avoid substrate artifacts. Since the total thickness of the multilayer film is approximately 2.5 μm, the upper bound indentation depth translates to ~ 250 nm in the present case.

It is a unique feature of the present multilayer structures that a lower-bound indentation depth need be prescribed; for very shallow indentations only the properties of the uppermost layer will be sampled, and in fact the true phase-averaged mechanical properties of the multilayer require an indentation of depth significantly greater than the repeat period of the structure. In the present case a simple model has been developed to ascertain this lower bound indentation depth. In this model, we have adopted the relatively common assumption that the indenter samples a volume of material which is roughly spherical, with a radius an order of magnitude greater than the indentation depth. Based on these geometrical approximations we have calculated the effective volume fraction of tin sulfide that the indenter samples, as a function of the indentation depth (details are available in Ref. [8]). Figure 4 shows the results for the 19 layer samples; as expected, for very shallow indentation depths, none of the silica is sampled since the tin sulfide is the top exposed surface. For larger values of indentation depth (>150 nm), the effective volume fraction converges to the true volume fraction of tin sulfide in the multilayer (~35%).

Based on the above analyses we estimate that to measure the true properties of the multilayer specimens, nanoindentation experiments should be conducted to depths in 150 – 250 nm range. Indentation tests were performed using a Hysitron Triboindenter with a calibrated diamond Berkovich tip, to depths in the specified range above. The standard Oliver-Pharr analysis was used to extract Young's modulus and hardness values. Ten independent experiments were conducted at any given indentation depth. In addition, relatively thick films of both silica and tin sulfide were used to develop baseline expectations for the mechanical properties of the multilayer structures.

The modulus and hardness values obtained for both the multilayer and the individual tin sulfide and silica films are presented in Table I. For the silica films, both the modulus and hardness values are lower than the bulk values reported in literature (70GPa and 18GPa, respectively). This is almost certainly attributable to the relatively low density of thin film silica prepared by e-beam evaporation. As we have shown elsewhere [8], the low density of the present silica dramatically reduces the modulus and hardness compared to cast silica glass [10].

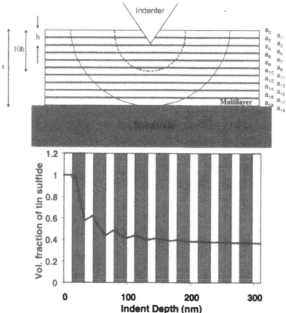

Figure 4(a) (top) Illustration of spherical cavity model. For calculation, x coordinate values of the layers are denoted by $a_0, a_1, \ldots a_{19}$. Depth of the indent is denoted by h. 10h denotes the depth to which the plastic deformation is sensed. The total thickness of the multilayer sample is denoted by t. (Note: diagram is not to scale)
Figure 4(b) (bottom) The volume fraction of tin sulfide enclosed in the plastic deformation zone for the mutilayers.

Table I. Modulus and Hardness values for silica, tin sulfide and multilayer specimen

Sample	Modulus (GPa)	Hardness (GPa)
Silica	13.3 (±1.2)	2.87 (±0.19)
Tin Sulfide	39.6 (±2.3)	2.14 (±0.12)
Multilayer	35.8 (±3.2)	3.75 (±0.18)

Examining the modulus value of the multilayer system, we find it lies between the moduli c silica and tin sulfide, roughly in line with the rule of mixtures. On the other hand, it is seen tha the hardness value of the multilayer falls slightly outside the range defined by tin sulfide an silica, the constituent materials. In principle, we would expect that the mechanical properties c these dielectric multilayer systems are primarily dictated by the intrinsic properties of thei constituents, and not by any special aspects of their engineered structure. We believe that th slightly increased hardness of the multilayer structure is likely attributable to fluctuations in th deposition rates of the two constituent materials, which consequently affects the density of th silica phase. As we discussed briefly above, the hardness of deposited silica is quite sensitive t density fluctuations, and small process variations can quite readily account for the unexpectedl high hardness of the multilayer.

It is encouraging to see that in general, the strength and hardness of multilayer mirrors ca be anticipated based on the properties of the constituents used to form them. Perhaps mor importantly, the present nanoindentation results suggest that the tribological characteristics o these multilayers can be improved *independently of the optical response* by adjusting th deposition parameters to promote higher density in the silica phase.

CONCLUSIONS

We have demonstrated that tin sulfide-silica multilayers exhibit omnidirectionality for nea infra red wavelengths encompassing the 850 nm communications wavelength; a omnidirectional band gap of over 140 nm has been achieved. Using standard vacuum depositio techniques in combination with a variety of analytical and nanomechanical characterizatio techniques, we have devised a systematic procedure for optical materials design.

REFERENCES

1 E. Yablonovitch, Phys. Rev. Lett. 58, 2059 (1987).
2 Y. Fink, J. N. Winn, S. Fan, C. Chen, J. Michel, J. D. Joannopoulos, and E. L. Thomas, Science 282, 1679 (1998).
3 M. Deopura, C.K. Ullal, B. Temelkuran and Y. Fink, Optics Letters, 26, 17, 1370-1372 (2001).
4 S. D. Hart, G. R. Maskaly, B. Temelkuran, P. H. Prideaux, J. D. Joannopoulos, Y. Fink., Science 296, 511-513 (2002).
5 M. Deopura, Y. Fink and C. A. Schuh, MS&T 2003 Proceedings, Chicago Meeting.
6 M. Müller, R. Zentel, T. Maka, S. G. Romanov, and C. M. Sotomayor Torres Advanced Materials 12 20 (2000).
7 P. A. Lee, G. Said, R. Davis, and T. H. Lim, J. Phys. Chem. Solids 30, 2719 (1989).
8 M Deopura, SM thesis, Massachusetts Institute of Technology (2003).
9 F. Abeles, Ann. Phys. 5, 706 (1950).
10 K. N. Rao, L. Shivlingappa and S. Mohan, Mater. Sci. & Engg., B8, 38 (2003).

at. Res. Soc. Symp. Proc. Vol. 817 © 2004 Materials Research Society

Fabrication of Photonic Crystals in Microchannels

Chun-Wen Kuo[1,2], Hui-Mei Hsieh[1,2], Jung-Chuan Ting[1]
Yi-Hong Cho , Kung Hwa Wei and Peilin Chen *[1]

[1]Center for Applied Sciences, Academia Sinica, 128, Section 2, Academia Road, Nankang, Taipei 115, Taiwan
[2]Department of Material Science and Engineering, National Chiao Tung University, Hsin Chu 300, Taiwan.
*Corresponding author: E-mail: peilin@gate.sinica.edu.tw, Tel:+886-2-2789-8000, Fax:+886-2-2782-6680

Abstract

We have developed a fabrication procedure for growing photonic crystals in the lithographic defined microchannels, which enables easy integration with other planar optical components. This technique is based on the directed evaporation induced self-assembly of nanoparticles in the microchannels.   Substrates with pre-patterned microchannels (30-100 μm wide) were dipped into solution of nanoparticles for several days.   By controlling the evaporation rate, the meniscus contacting the microchannels will undergo evaporation-induced self-assembly. The capillary forces cause nanospheres to crystallize within the microchannels forming colloidal photonic crystals in the microchannels. Two types of colloidal particles, polystyrene and silica, have been employed to fabricate colloidal photonic crystals in the microchannels. Both types of colloidal particles were found to form large-area well-ordered colloidal single crystals in the microchannels.   The optical reflection spectra from the (111) surfaces of the colloidal crystals formed by various sizes of nanoparticles have been measured.   And the measured reflection peaks agree with the photonic bandgap calculated by the plane wave expansion method.

Introduction

The fabrication of three-dimensional periodic dielectric nanostructures, which can work as photonic crystals, has drawn lots of research attention lately, because of their potential capability in controlling the propagation of light [1-3].   There are many approaches to fabricate photonic crystals. However, those utilized self-assembly of monodisperse nanospheres remain to be the simplest and the most inexpensive route to construct three-dimensional periodic nanostructures [4-5].   Standard practice of using self-assembly process to fabricate photonic crystals is to place the substrates into a dilute suspension of nanosphere particles in ethanol [6]. As the solvent dries, the particles self-assemble at the receding ethanol meniscus and form close packed nanostructures. In many applications, it has been suggested by the theoretical calculation [5] that the inverse opal structure is a better photonic material. The inverse opals can be constructed by filling the void of the opal with high dielectric constant materials and dissolving away the nanoparticles.

In order to integrate photonic crystals into communication devices, it is has been proposed [7-9] to fabricate photonic crystals in the microchannels, which can work as or interconnect with functional optical components such as switches, mirrors, filters and waveguides. We have developed a fabrication procedure to synthesize photonic crystals in the lithographic defined microchannels, which enables easy integration with other optical components. This technique is based on the directed evaporation induced self-assembly of nanoparticles. Substrates with pre-patterned microchannels (30-100 mm wide) were dipped into the solution of nanoparticles for several days. By controlling the evaporation rate, the meniscus contacting the microchannels will undergo evaporation-induced self-assembly, which causes nanospheres to crystallize within the microchannels forming photonic crystals. Another advantage of growing photonic crystals in the microfluidic system is that the microfluidic channels have been demonstrated to be one of the most promising miniature integrated platform for many applications such as biosensing, display and lab-on-a-chip [10-11]. Microfluidic channels systems using soft materials have drawn lots of research attention latterly because of their low cost and ease of fabrication. It has been shown [12] that it is possible to manipulate microfluidic with very complicated LSI layout, which allows addressing large arrays of microchannel compartments. In this experiment we have utilized soft materials, polydimethylsiloxane (PDMS), to construct microchannels.

Results

To fabricate PDMS microchannels, silicon substrates (1×2 cm) coated with 8 μm thick negative photoresist SU8-2015 (MicroChem) were first used to create microchannel patterns, which were 20-50 μm wide lines separated by 50-100 μm. PDMS microchannels were formed by casting the two component PDMS mixtures (Sylgar 184, 8:1) into the patterned silicon substrates and cured at 90 ^{0}C for two hours. Free-standing PDMS microchannels have been obtained after the peel-off process.

To integrate periodic particle arrays with other planar optical components, it is important to form photonic crystals inside the microchannels. To grow photonic crystals, monodisperse polystyrene and silica beads of various diameters purchased from Bangs Laboratories, Inc. (Fishers, IN) were diluted in a methanol solution (1:50 to 1:100). To utilize the evaporation induced self-assembly in the microchannels, the

PDMS microchannel was placed in a 10 ml vial, which contained 200 μl of polystyrene solution. In general, it took 3-5 days to fill the whole microchannels with photonic crystals. Fig. 1a shows the SEM image of the photonic crystals formed by 350 nm polystyrene beads in the microchannel. The photonic crystal formed by 330 nm silica nanoparticles is shown in figure 1b. Both silica and polystyrene nanoparticles can form self-assembled three-dimensional periodic structures by this approach. The colloidal crystals fabricated by this scheme show very good quality and yield very large domain as seen from figure 1 c and d. Because of the long range ordering of the photonic crystals, they exhibit very unique optical property. To investigate the optical properties of individual microchannels, the photonic crystals in microchannels was monitored by a microscope (Olympus CX41), which was equipped with a CCD camera (Sony Exwave HAD) and a fiber spectrometer (Ocean Optics, USB 2000).

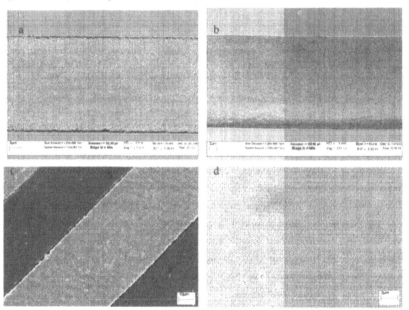

Figure 1. SEM images of photonic crystals formed in the microchannels using a)350 nm polystyrene beads, b) 330 silica nanoparticles, c) and d) 550 nm polystyrene nanoparticles

The evaporation induced self-assembly process depends on the capillary force, which grows photonic crystals very slowly in the unsealed microchannels. An alternative approach is to use spin-coating process, which have been demonstrated capable of producing photonic crystals in the microchannel. To conduct spin-coating experiment, surfactant (Triton X-100, Alrich) was added in the polystyrene solution to wet the microchannels. Depending on the diameter of the polystyrene beads, the speed of the spin-coater has been varied between 2000-3000 RPM. The spin-coating process provides a rapid technique to fabricate photonic crystals in the microchannels. However, the photonic crystals produced by this approach have more cracks and it was not easy to obtain good photonic crystals over large area.

Figure 2. SEM images of CdSe inverse opal along a) (111) and b) (100) direction.

Discussions

Because only inverse opal structures exhibit complete bandgap, we have also developed a technique to grow inverse opal in the microchannels. In this case, the microchannels have been fabricated on the ITO glasses, and the photoresist was used to define the microchannels. To form inverse opal, we have utilized electrochemical deposition to fill the void of the opal formed by polystyrene nanoparticles and the polystyrene beads were removed by dipping the microchannels into the THF solution. Figure 2 shows the SEM images of CdSe inverse opal, which was obtained by electrochemical deposition using $CdSO_4$ and SeO_2 solution.

One of the most important features of the photonic crystals is their optical property, which depends on the dimension of their periodic structure. We have measured the reflection spectra at normal direction of the photonic crystals formed in the microchannels using different sizes of polystyrene beads. Figure 3a shows the reflected spectra of photonic crystals in the microchannels using different diameter polystyrene beads. It is clear that that as the diameter of the polystyrene increases the reflection maximum exhibit red-shift. The whole visible region is covered by the reflection peaks of photonic crystals formed by 200-300 nm polystyrene beads. In another word, the optical property of photonic crystals can be tuned in the visible region by using different diameter of polystyrene in the range of 200-300 nm. The reflection maximum can be modeled by the photonic bandgap (PBG) theory, which treats the propagation of photons in the periodic dielectric materials the same way as the electrons in the periodic atomic lattices. A software package based on the plane wave expansion method can be downloaded from a website [13]. The calculated results are depicted in figure 3b. The experimental measurement agrees well with the prediction of PBG theory.

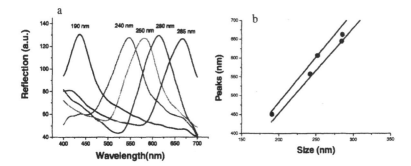

Figure 3. a) The reflection spectra of photonic crystals formed by various diameters of polystyrene beads. b) The maximum of the reflection peaks and the calculated upper and lower band of the photonic bandgaps.

Conclusion

In summary we have fabricated photonic crystals in the microchannels using two different types of nanoparticles, polystyrene and silica. Both nanoparticles are capable of forming well-ordered three-dimensional periodic nanostructures inside the microchannels. The reflection spectra from these photonic crystals match well with the photonic band gap calculation using the plane wave expansion method.

Acknowledgement

This research was supported in part by National Science Council, Taiwan, under Contract 92-2113-M-001-036.

References

1. E. Yablonovitch, *Phys. Rev. Lett.* , **58**, 2059 (1987)
2. R. Maede, A.M. Rappe, K.D. Rommer, J.D. Joannopoulos, *Phys. Rev. B*, **48**, 8434 (1993).
3. S. G. Johnson and J. D. Joannopoulos, *Photonic Crystals: The Road from Theory to Practice* (Kluwer, Boston, 2002).

4. Y. Xia, B. Gates, Y. Yin, Y. Lu, *Adv. Mater.,* **12**, 693 (2000).

5. C. Lopez, *Adv. Mater.*, **15**, 1679 (2003).

6. P. Jiang, J.F., Bertone, K.S. Hwang, V.L. Colvin, *Chem. Mater.*, **11**, 2132. (1999)

7. S.M. Yang, G.A. Ozin, *Chem. Commun.*, 2507 (2000).

8. G.A. Ozin, S.M. Yang, *Adv. Funct. Mater.*, **11**, 95 (2001).

9. S.M. Yang, H. Miguez, G. Ozin, *Adv. Func. Mater.*, **12**, 425 (2002).

10. S.R. Quake, A. Scherer, A. *Science*, **290**, 1536 (2000).

11. M.A. Unger, H.P. Chou, T. Thorsen, A. Scherer, S.R. Quake, *Science*, **288**, 113 (2000).

12. T. Thorsen, S.J. Maerkl, S.R. Quake, *Science*, **298**, 580 (2002).

13. Software was downloaded from http://jdj.mit.edu/mpb/download.html

Nanocrystals

Mat. Res. Soc. Symp. Proc. Vol. 817 © 2004 Materials Research Society   L4.1

Low loss silica waveguides containing Si nanocrystals

C. Garcia[1], B. Garrido[1], P. Pellegrino[1], J.R. Morante[1], M. Melchiorri[2], N. Daldosso[2], L. Pavesi[2], E. Scheid[3], G. Sarrabayrouse[3]
[1]Electronics, University of Barcelona, Barcelona, Spain;
[2]Physics, University of Trento, Trento, Italy;
[3]LAAS - CNRS, Toulouse, France;

ABSTRACT

We study the optical and structural properties of rib-loaded waveguides working in the 600-900 nm spectral range. A Si nanocrystal-rich SiO_x with Si excess nominally ranging from 10 to 20% forms the active region of the waveguide. Starting materials were fused silica wafers and 2 μm-thick SiO_2 thermally grown onto Si substrate. Si nanocrystals were precipitated by annealing at 1100°C after quadruple Si ion implantation to high doses in a flat profile. The complete phase separation and formation of Si nanocrystals were monitored by means of optical tools, such as Raman, optical absorption and photoluminescence. Grain size distribution was obtained by electron microscopy. The actual Si excess content was obtained by X-ray photoelectron spectroscopy. The rib-loaded structure of the waveguide was fabricated by photolitographic and reactive ion etching processes, with patterned rib widths ranging from 1 to 8 μm. M-lines spectroscopy measurements provided a direct measurement of the refractive index and thickness of the active layers versus Si excess. When coupling a probe signal at 780 nm or 633 nm into the waveguide, an attenuation of at least 11 dB/cm was observed. These propagation losses have been attributed to Mie scattering, waveguide irregularities and direct absorption by the silicon nanocrystals.

INTRODUCTION

Si nanocrystals in amorphous matrices have been subject of intensive studies as a suitable optical active medium for CMOS-compatible photonic applications. The optical and structural properties of nanostructured silicon have been exhaustively investigated, and are relatively well understood [1, 2]. More recently, the important observation of optical gain from nanocrystalline Si has been reported [3, 4]. In this contribution we focus the attention in analysing the loss mechanisms in waveguide structures built with this material.

EXPERIMENTAL DETAILS

In order to introduce Si excess inside the oxide matrix, multiple Si^+ ion implantations have been performed in both 2 μm thick SiO_2 films thermally grown on Si and pure fused quartz wafers. Ion energies and relative doses of the multiple implantations were chosen to give rise to a "box-like" Si super-saturation down to a depth of about 0.4 μm. In such a way, the processing of the layers will allow to fabricate channel waveguides with a well defined homogeneous in-depth active media providing refractive index contrast with the claddings. The total ion doses were tailored to introduce a Si excess nominally ranging from 10% to 20% referred to the silica atomic concentration before implantation. Then the samples were annealed at 1100 °C in N_2 atmosphere for different durations, ranging from 1 min up to 16 h.

Conventional Transmission Electron Microscopy (TEM) in dark field was used to image the different multilayer structure, while Energy-Filtered TEM (EFTEM) was employed to monitor the size distribution of nanocrystals and its evolution with the implanted dose. Systematic photoluminescence (PL) measurements were performed in all samples by using an Ar laser for ultraviolet excitation at 365 nm. The luminescent emission was analysed in backscattering mode by a 0.6 m monochromator and detected with a GaAs Hamamatsu photomultiplier.

The rib-loaded waveguides (WGs) have been first modelled, using the experimentally determined refractive index of the active layer. Simulations were performed in order to ensure that the chosen geometry allows the propagation inside the active layer with a good confinement factor. A set of rib widths was considered, with rib widths ranging from 1 to 8 μm, with over 65% of the energy confined inside the active layer. Simulations also proved that the WGs supported only the fundamental mode for rib widths under 4 μm, and started being multimode after that.

A 1μm-thick PECVD SiO₂ layer was deposited over the 4" wafer with the 15% nominal (9.5% measured) Si excess. Deposition was performed at 300 °C. Ridge structures were then formed in this layer by using standard photolithography and reactive ion etching to a depth of 0.7μm. An unetched cladding layer of 0.3μm was preserved, partly to act as a protection and partly to ensure single mode behaviour and mode symmetry. The mask used in the photolithographic process was patterned with 1 cm² squares, each containing eight groups of waveguides 1 to 8 μm-wide, separated by 240 μm in order to ensure proper mode isolation.

The waveguides have been characterized by coupling-in visible light from a laser (780nm – about 7 mW; 633nm – few mW) through a tapered fiber and coupling-out the light with a microscope objective (40x) matched to a zoom mounted on a high performance CCD camera. Top-view observation has been performed by means of an optical microscope and a CCD camera mounted on top of the measuring system. A prism beam splitter allows splitting the transmitted light in two parts: one is directed to the CCD camera, the other is directed to a calibrated photodiode, which yields power measurement of the transmitted light.

DISCUSSION

To characterize the material used to fabricate the waveguides, several experimental techniques have been employed. A typical dark field TEM picture of the Si-nanocrystal-rich layer, overlapped with the Si ion implanted profile as obtained by SIMS is shown in Fig. 1. From these measurements, the thickness of the implanted layer has been estimated to be around 300 nm, independent on the implanted ion dose. XPS analysis provides a quantitative evaluation of the silicon and oxygen content inside the implanted region, and then a direct measure of the introduced Si excess. For the three implantation schemes, a Si excess of 13, 9.5 and 7% have been found for the nominal 20, 15 and 10%, respectively. The variation with respect to the expected values, calculated with the help of the SRIM simulator, can be mainly assigned to the expansion of the layer to accommodate a number of ions, which is comparable to the number of target atoms. A further refinement of the calculation, including swelling of the implanted layers, brings to an almost perfect agreement between the measured and the calculated values.

The thermal evolution of the implanted silicon ions have been followed after each annealing step by means of Raman and visible absorption measurements. In all the as-implanted samples the Raman spectrum is dominated by a broad structure peaked around 480 cm⁻¹, signature of the transverse optical mode of amorphous Si aggregates. From the similarity of areas under the Raman bands before and after annealing we can conclude that most of the Si excess is already

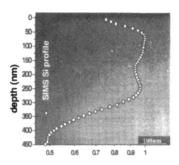

Figure 1. SIMS profile of the implanted silicon as a function of sample depth. The nanocrystal-rich layer, as observed in dark-field TEM, clearly overlaps with the flat implantation.

clustered in amorphous aggregates even in the as-implanted state. After annealing the wafers at 1100 °C for 20 minutes, the Raman spectrum is converted into a narrow asymmetric peak centred at about 521 cm^{-1}, indicating a rapid and complete crystallization of the amorphous clusters. Further annealing at the same temperature does modify neither the shape nor the intensity of the Si crystalline signal. Moreover, the integrated signal of the crystalline Si peak perfectly scales with the implanted dose. Similar results have been obtained from optical transmission in the whole visible range: the absorption spectrum is essentially the same for all the samples, once rescaled to the implanted dose, and it does not change after 20 min. annealing. We can then infer that the crystalline fraction of precipitated Si is proportional to the implanted dose and for an annealing step of only 20' at 1100 °C a complete crystallization occurs.

By varying the implanted dose a change in the average size and density of precipitated nanocrystals is expected. EFTEM analysis provides valuable information to this regard, allowing building up a statistical histogram of the nanocrystal size as a function of the introduced Si excess. A systematic increase of the average size of the Si nanograins has been observed when increasing the implantation dose, together with a rather small dispersion around the mean value (0.3-0.5 nm typical). The typical diameters range from 3.6 nm obtained for the sample with the lowest Si excess (7%), to 4.1 nm for the intermediate Si excess (9.5%) and up to 4.6 nm for the largest Si excess (13%). By combining the values of both the measured Si excess and average particles size, an estimate of the density of nanocrystals can be performed., We obtain a density of about 4x10^{18} Si nc/cm^3 for all the samples studied here (all the Si excess). The observation by EFTEM imaging that the superficial density of the nanoparticles is quite similar in all kind of samples provides an additional support to this evaluation. Then we can conclude that in the present conditions we have obtained a set of different samples in which the increasing amount of implanted Si has been converted in a population of nanocrystals with the same density but slightly larger sizes.

The characterization of the PL emission of the samples has been also performed. Typical PL spectra are presented in Fig. 2, with an inset showing the evolution of the total integrated PL intensity with the ion dose. The red-shift of the maximum is perfectly consistent with the increase in size that has been measured from EFTEM and is in agreement with the emission coming from band-edge recombination of excitons quantum confined in Si-nc.

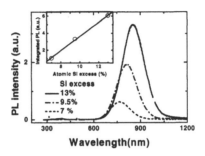

Figure 2. PL spectra of the different implanted and annealed samples. In the inset, the integrated intensity has been plotted as a function of the introduced Si excess.

The values obtained for the maximum of the emission are consistent with values obtained by us in the past from single Si+ ion implants in thermal oxides [5]. Similarly to the results reported in such papers, the PL emission energy peak ranges for the samples studied here between 1.62 eV (lowest dose, 7%) and 1.45 eV (highest dose, 13%). It is also remarkable that the integrated intensity perfectly scales with the Si excess and that higher dose emission spectra overlap those for lower doses in the low wavelength side (high energy). This is apparently due to the fact that the excitation cross-section in the blue-UV-range perfectly scales also with Si excess. So, this scaling is due to the fact that the density of states available for excitation at high energies is proportional to the number of Si excess atoms and not to the density of Si nanocrystals.

In order to evaluate the thickness and refractive index of the active layer, their dependence with the Si excess and original matrix, we performed m-lines measurements. The thickness of th active layer found from m-lines does not depend of the Si excess and is around 340 nm, in good agreement with the values shown before. The refractive index values are presented in Fig. 3 both for pure silica and thermal oxide samples. Thus, by taking the values of refractive index of the composites at 632 nm and the extreme value for $x=2$ (SiO$_2$, $n=1.46$) we can interpolate a dependence of the refractive index with Si excess. This is shown in Fig. 3, which provides very useful data to prepare composite layers with a targeted refractive index.

Efficient propagation of light was observed by coupling a signal both at 633 and 780 nm into the waveguide. Propagation losses measurements have been performed by using both the top-view (light scattering) and the insertion losses technique. The former one is based on the assumption (right in presence of homogeneous surface without irregularities) that the scattered light intensity observed from the top of the waveguide is directly proportional to the light intensity propagating in the waveguide. Hence the decrease of the scattered light as a function of the position along the waveguide is representative of the attenuation of the propagating optical mode and can be easily detected by looking at the top surface of the waveguide with a CCD camera. The image is then acquired by a frame grabber and analysed to obtain the intensity profile as a function of the waveguide length. This approach is not the best for precise measurements of losses coefficients if compared to the usual cut-back method or to insertion losses technique. However, it is a simple, rapid and not sample-consuming technique to have an upper estimation of propagation losses. The latter one, insertion losses technique, is based on the measurement of the power transmitted from the waveguide output facet as a function of the input signal power and needs an

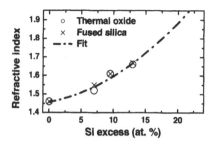

Figure 3. Refractive index of the active layer as a function of the introduced Si excess. The values obtained by M-lines measurements have been interpolated by a polynomial fit.

independent estimation/measurements of the coupling in and collection losses. The losses through the optical collection system have been determined by measuring the transmitted power of the tapered fiber in front of the microscope objective: the value was about 5 dB. Also the tapered fiber losses have been measured in order to know the real power injected into the waveguide. The coupling losses are mainly due to the different numerical aperture between tapered fiber and waveguide sample, to the tapered fiber's focused spot size against waveguide geometrical dimensions, to reflections and edge defects or irregularities. An experimental evaluation of about 15-16 dB has been obtained by a comparison with similar waveguides for which it was possible to determine the propagation losses by cut-back technique, and hence to deduce the coupling losses. An estimation of coupling losses has been made also by calculating the contribution of the various effects, resulting in losses of about 17 dB for the largest waveguide (8 μm), increasing up to 23 dB for the smallest (2 μm).

Fig. 4a shows, as example, the scattered light intensity along the sample length of a 8 μm wide waveguide with 9.5% Si excess. An exponential fit of the data yields a propagation loss coefficient of about 11 dB/cm. Propagation losses coefficients as a function of the rib width have been obtained as mean values of four different waveguides. The values obtained by the top-view measurements have been validated by the insertion losses measurements: the results are in agreement within 1 dB/cm.No significant differences in the propagation loss coefficient have been found by changing the wavelength of the light from 633nm to 780nm. Remarkable differences between the various rib widths can be observed, as shown in Fig. 4b: larger losses, around 20 dB/cm, are present in the small (1 and 2 μm) waveguides, whereas they stabilize around 11 dB/cm in the large ones (4, 6 and 8 μm). This can be mainly attributed to roughness scattering of the side walls of the ridge, together with the loss of confinement, both terms being more predominant when reducing the channel width.

The asymptotic value of about 11 dB/cm, which represents the intrinsic loss due to propagation in the active layer, has been analysed further. Scattering losses from the ensemble of Si previously determined, and result in about 2 dB/cm at 633 nm and 1 dB/cm at 780 nm. If we assume that the remaining 9 dB/cm losses are absorption losses we can get an upper limit estimate to the absorption cross-section for Si-nc, $\sigma_{abs} \sim 5 \times 10^{-19}$ cm^2. This low value of absorption cross-section

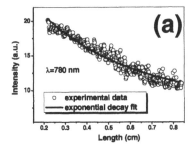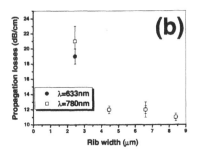

Figure 4. (a) Scattered light intensity (top-view measurements) along the sample length of a μm rib width waveguide at 780 nm. (b) Variation of the propagation losses with rib width at 633nm and 780nm.

is compatible with other reported data and supports the 4-level model for the stimulated emission in Si-nc. If for the stimulated emission cross-section a value of about 1×10^{-17} cm^2 is assumed [6] we can get an estimate of the gain coefficient in these waveguides, attainable when the system is pumped with suitable high energy radiation. The ultimate performances of similar, optimized waveguide amplifier under optical pumping should be extremely promising, being just 11 dB/cm losses and a gain coefficient of about 160 dB/cm.

CONCLUSIONS

We have been able to get a deep insight into the formation, loss contributions and mechanisms in waveguides made with Si nanocrystals-rich oxides. An exhaustive characterization of the active layer has been performed in order to control the formation and emission properties of the nanoparticles, and get valuable physical parameters, such as the nanocrystal density and size and refractive index of the multilayer structure. Propagation losses have been measured on the processed waveguides, and related to specific contributions. An asymptotic value, around 2 dB/cm of Mie scattering and 9 dB/cm of direct absorption was evaluated. The feasibility of a highly performance optical amplifier which makes use of similar optimized structures has been suggested.

REFERENCES

1. B. Garrido, M. López, C. García, A. Pérez-Rodríguez, J. R. Morante, C. Bonafos, M. Carrada A. Claverie, J. Appl. Phys. **91**, 798 (2002).
2. S. Ossicini, L. Pavesi, F. Priolo *Light Emitting Silicon for Microphotonics*, Springer Tracts in Modern Physics , Vol. 194 (Springer-Verlag, Berlin 2003).
3. L. Pavesi, L. Dal Negro, C. Mazzoleni, G. Franzò, F. Priolo, Nature **408**, 440 (2000).
4. J. Ruan, P.M. Fauchet, L. Dal Negro, M. Cazzanelli, L. Pavesi, Appl. Phys. Lett. **83**, 5479 (2003).
5. B. Garrido, M. López, O. González, A. Pérez-Rodríguez, J. R. Morante, C. Bonafos, Appl. Phys. Lett. **77**, 3143 (2000).
6. L. Dal Negro, M. Cazzanelli, N. Daldosso, Z. Gaburro, L. Pavesi, D. Pacifici, G. Franzò, F. Priolo, and F. Iacona, Physica E **16**, 297 (2003).

Mat. Res. Soc. Symp. Proc. Vol. 817 © 2004 Materials Research Society   L4.2

In-situ Control of Nitrogen Content and the Effect on PL Properties of SiNx Films Grown by Ion Beam Sputter Deposition

Kyung Joong Kim[1,a], Dae Won Moon[1], Moon-Seung Yang[2] and Jung H. Shin[2]
[1]Nano Surface Group, Korea Research Institute of Standards and Science (KRISS), P.O.Box 102, Yusong, Taejon 305-600, Korea
[2]Department of Physics, Korea Advanced Institute of Science and Technology (KAIST) 373-1 Kusung-dong, Yusong-gu, Taejon, Korea

ABSTRACT

Strong visible luminescence was observed in silicon nitride (SiN_x) thin films grown by ion beam sputter deposition (IBSD) using nitrogen ion as a sputtering source. Nitrogen content (x) of the films was controlled by variation of the sputtering N_2 ion flux and analysed by *in-situ* x-ray photoelectron spectroscopy (XPS). Relative sensitivity factors of Si and N peaks could be calculated by Rutherford backscattering spectroscopy. The photoluminescence (PL) spectra of the post-annealed samples showed visible luminescence at blue-green region. PL energy showed a blue-shift due to quantum confinement with decreased excess Si and intensity showed a maximum value near x = 1.1. These PL properties are well correlated with the formation of Si nanocrystals (nc-Si). We found that there is a great increase of PL energy of SiN_x thin films compared with SiO_x thin films, which indicate that the surface state of Si nanocrystals plays an important role to increase PL energy and intensity.

[a]kjkim@kriss.re.kr

INTRODUCTION

Si nanocrystal imbedded in SiO_2 matrix has been studied widely because of its luminescence in the visible range due to the quantum confinement effect.[1-3] Many studies have been focused on the effect of oxygen content and other matrix materials on the energy and intensity of the photoluminescence. Silicon rich silicon oxide (SRSO; SiO_x) and SiO_2/Si multilayers are representative two structures for the formation of Si nanocrystal (nc-Si) by annealing at high temperatures. The exact stoichiometry of SiO_x and the relative thickness of SiO_2 and Si layers in SiO_2/Si multilayers are main parameters to determine the PL properties. These materials can be fabricated by various growth methods such as implantation of Si ions in SiO_2 matrix [4], plasma enhanced chemical vapor deposition (PECVD)[3,5], rf sputtering [6], evaporation of SiO [7,8] and etc.

Recently, silicon rich silicon nitride (SRSN; SiN_x) proposed as a promising candidate as a host material for visible luminescence by formation of Si nanocrystal. SiN_x is available to passivate nc-Si without introducing the oxygen-related deep states. Visible luminescence from amorphous Si nanoclusters in SiN_x had been already reported [9-11]. However, the effect of stoichiometry and different surface passivation on visible luminescence from nc-Si in SiN_x was not studied seriously yet.

Ion beam sputter deposition (IBSD) has many advantages for low-temperature growth of Si, SiO_x and SiO_2 thin films [12,13]. Stoichiometry and thickness of the films can be controlled exactly by controlling with oxygen pressure and growth time. The hydrogen content of IBSD films is almost negligible compared to LPCVD films. Energetic ion bombardment has been shown to have an electrically damaging effect [14] on the growing films in conventional sputtering deposition, while in IBSD, the ion beam damaging effect is not significant. This also makes defect density of IBSD films relatively less.

In this paper, we report on the growth method and blue-green visible luminescence of SiN_x thin films. Nitrogen ions were used as sputtering source and ambient species. The nitrogen content of SiN_x thin films was controlled quantitatively by in-situ XPS measurement in order to study the effect of stoichiometry on PL properties. PL spectra of SiN_x thin films after post-annealing showed strong luminescence in the range of 1.77 - 2.75 eV due to the formation of Si nanocrystals. The small increase of PL energy as the excess Si decreased from 12.4 % to 5.9 % could be explained by quantum confinement effect due to the decrease of size of nc-Si. However, the PL energy of SiN_x thin films showed a great difference of with that of SiO_x with similar excess Si. This could be also understood by surface effect of nc-Si, which showed that surface environment plays a major role in determining the PL energy.

EXPERIMENTAL DETAILS

SiN_x thin films were grown by ion beam sputter deposition of a Si wafer at room temperature. However, in the case of IBSD where Ar ion is used as a sputtering source and N_2 gas is used as an ambient gas, N_2 is not so reactive to form silicon nitride. Only a weak N 1s peak was observed at high pressure of N_2 gas flow. This problem could be solved using nitrogen ions as a sputtering source. We used mixed ions of nitrogen and argon as sputtering source. In this case, nitrogen ions can act as a sputtering source and ambient ion species to react with adsorbed Si atoms.

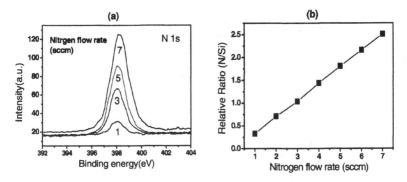

Fig. 1. XPS spectra of N 1s line showing the change of nitrogen content (a), Relative area ratio (N 1s /Si 2p) as a function of nitrogen flow rate (b).

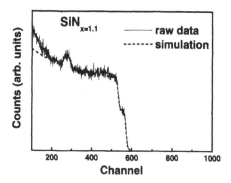

Fig. 2. A RBS spectrum of a $SiN_{x=1.1}$ thin film grown as a reference specimen.

The nitrogen content (x) was controlled by variation of relative flow rates of N_2 and Ar gases introducing to the DC ion gun. Figure 1 shows N 1s XPS data and the variation of stoichiometry of SiN_x films grown as a function of relative flow rate of Ar and N_2 gas. Nitrogen content linearly increases with the increase of N_2 flow rate from 1 to 7 sccm. Although the stoichiometric Si_3N_4 state was not able to grow, SiN_x films with enough excess Si content to form Si nanocrystals could be grown. The relative sensitivity factors of Si and N were measured by comparing the peak areas of Si 2p and N 1s from in-situ XPS analysis with RBS data of a preliminary sample as shown in Figure 2. The excess Si content of the films was defined as the excess portion of Si content under the hypothesis that the thin films are the mixture of Si and stoichiometric Si_3N_4 state. The Excess Si was controlled in range from 5.9 to 12.4 %. The layer thickness was controlled by growth rate calibrated by measurement of cross-sectional TEM image of a reference SiN_x film grown within a given time. The growth rate of this experiment was 0.3 Å/s.

The samples were annealed at optimum conditions of 950°C and 10 min in ultra-pure nitrogen ambient using a horizontal furnace to form Si-nanocrystals. The samples were hydrogenated to passivate the Si dangling bonds for 1 hour at 650°C under H_2 forming gas flow. Photoluminescence spectra were measured at room temperature using the 325 nm line of He-Cd laser as the excitation source and a charge coupled device camera as PL detector.

RESULTS and DISCUSSIONS

The PL emission is related with the phase of nanostructures. Generally, as the annealing temperature increases, nanoclusters with amorphous phase are formed initially and then fransformed to nanocrystals by more elevated annealing temperature. Figure 3 shows the room-temperature PL spectra of SiN_x thin films after annealing and hydrogenation. Broad

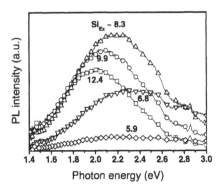

Fig. 3. PL spectra of SiN$_x$ films as a function of excess Si content.

PL spectra were observed in the range of 1.7-2.7 eV. The PL peak energy blue-shifts from 2 to 2.34 eV as excess Si content decrease from 12.4 to 5.9 at. %. The blue-shift in this small range is believed to result from quantum confinement effect with the decrease of nc-Si size. This result is in good agreement with previous reports on the effect of stoichiometry on nc-Si luminescence from nc-Si imbedded in SiO$_2$ matrix [15].

The formation of large (3~5 nm) and small(1~2 nm) sized nc-Si in the annealed SiO$_{x=1.64}$ and SiN$_{x=1.1}$ thin films were confirmed by high-resolution transmission electron microscopy (HR-TEM) measurement as shown in figure 4. The formation of nc-Si indicates that the photoluminescence is originated from nc-Si. And the increase of PL energy and intensity of SiN$_x$ thin films at higher annealing temperature shows that nano-crystalline Si is much more effective for the emission of PL than the amorphous nano-clusters formed at low annealing temperature. Figure 5 (a) shows the normalized PL intensity of SiN$_x$ thin films as a function of excess Si content. PL intensity of SiN$_x$ thin films shows a maximum value near the excess Si of 8.2% (x=1.1). PL intensity is related with the density of Si nanocrystals. In the case of high excess Si, the number of nanocrystals will decrease and the size of them increase due to the coalescence of the small nanocrystals.

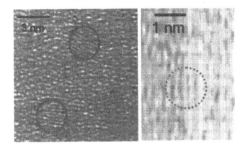

Fig. 4. HR-TEM images of a SiO$_{x=1.64}$ (left) and SiN$_{x=1.1}$ (right) thin films.

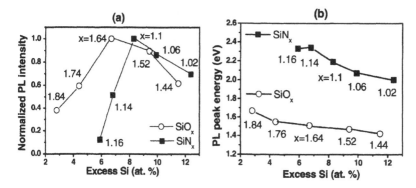

Fig. 5. Comparison of normalized PL intensity (a) and PL energy (b) of SiN_x and SiO_x as a function of excess Si content.

However, if the excess Si is highly decreased, the PL intensity decreases again because the nucleation number decreases. This tendency is very similar with that of SiO_x thin films although the excess Si is slightly different.

The PL energy of SiN_x and SiO_x thin films is compared as a function of excess Si content in figure 5(b). There is a limitation in the enlarging of the band-gap by decreasing the size of nc-Si in SiO_2 matrix because carriers of oxygen passivated nc-Si in SiO_2 matrix localized at the deep state. Therefore, the PL of SiO_x is limited in red and near infra-red region. However, the PL of SiN_x is observed in the blue and green region by controlling the excess Si and using the nitride-passivated nc-Si in Si_3N_4 matrix. The blue-shifts in the narrow wavelength region in the PL of SiO_x and SiN_x films by decreasing excess Si can be explained by quantum confinement effect due to the size decrease of nanocrystals. However, there is a big difference of PL energy about 0.8 eV between the SiO_x and SiN_x films with similar excess Si contents. This big difference is difficult to be understood as a result of quantum confinement effect due to the size decrease of nc-Si. This seems to be due to the surface state of the nanocrystals. The bandgap of Si nanocrystals affected by the surface environment of the nc-Si due to matrix effect. This result suggests that surface environment of nc-Si as well as nc-Si size is a dominant factor for efficient luminescence.

CONCLUSIONS

SiN_x thin films were successfully fabricated by ion beam sputter deposition and the stoichiometry of the films were measured and controlled with in-situ XPS analysis. The variation of visible luminescence was possible by variation of nanocrystal size and change of host material. We have shown the effect of film stoichiometry and surface passivation on luminescence by comparing the SiO_x and SiN_x thin films with similar excess Si. The PL spectra of SiN_x show strong luminescence in the range of 1.77 -2.75 eV, and the PL energy showed a blue-shift due to quantum confinement effect with the decreased excess Si content.

However, the big difference of about 0.8 eV in PL the energy of SiN_x and SiO_x films is well correlated with the interface control. Consequently, the formation of nc-Si and surface passivation of nc-Si are the major parameters to determine the PL energy and intensity.

ACKNOWLEDGMENTS

This work was supported in part by MOST, Korea, through a National Research Laboratory Program.

REFERENCES

1. Z. H. Lu, D. J. Lockwood and J. M. Baribeau, *Nature* **378**, 258 (1995)
2. V. Vinciguerra, G. Franzo, F. Priolo, F. Iacona and C. Spinella, *J. Appl. Phys.* **87**, 8165 (2000)
3. M. Zacharias, J. Heitmann, R. Scholz, U. kahler, M. Schmidt, and J. Blasing, *Appl. Phys. Lett.* **80**, 661 (2002).
4. B. Garrido, M. Lopez, O. Gonzalez, A. Perez-Rodriguez, J. R. Morante and C. Bonafos, *Appl. Phys. Lett.* **77**, 3143 (2000).
5. F. Priolo, G. Franzo, D. Pacifici, V. Vinciguerra, F. Iacona and A. Irrera, *J. Appl. Phys.* **89**, 264 (2001)
6. K. Yoshida, I. Umezu, N. Sakamoto, M. Inada and A. Sugimura, *J. Appl. Phys.* **92**, 5936 (2002)
7. M. Molinari, H. Rinnert, and M. Vergnat, *Appl. Phys. Lett.* **82**, 3877 (2003).
8. U. Kahler and H. Hofmeister, *Appl. Phys. A* **74**, 13 (2002).
9. N. M. Park, C. J. Choi, T. Y. Seong, and S. J. Park, *Phys. Rev. Lett.* **86**, 1355 (2001)
10. H. Kato, N. Kashio, Y. Ohki, K.S. Seol, and T. Noma, *J. Appl. Phys.* **93**, 239 (2003)
11. Y. Q. Wang, Y. G. Wang, L. Cao, and Z. X. Cao, *Appl. Phys. Lett.* **83**, 3474 (2003)
12. C. Saha, S. Das, S. K. Ray and S. K. Lahiri, *J. Appl. Phys.* **83**, 4472 (1998).
13. M. F. Lambrinos, R. Valizadeh and J. S. Colligon, *J. Vac. Sci. Technol.* **B16**, 589 (1998).
14. W. C. Dautremont-Smith and L. C. Feldman, *J. Vac. Sci. Technol.* **A3**, 873 (1985).
15. F. Iacona, G. Franzo, and C. Spinella, *J. Appl. Phys.* **87**, 1295 (2002)

Quantum confinement effect of silicon nanocrystals *in situ* grown in silicon nitride films

Tae-Youb Kim, Nae-Man Park, Kyung-Hyun Kim, Young-Woo Ok[1], Tae-Yeon Seong[1], Cheol-Jong Choi[2] and Gun Yong Sung
Future Technology Research Division, Electronics and Telecommunication Research Institute, Daejon, 305-350, Korea
[1]Department of Materials Science and Engineering, Kwangju Institute of Science and Technology, Kwangju 500-712, Korea
[2]Samsung Advanced Institute of Technology, Yongin Gyeonggi-do 449-712, Korea

ABSTRACT

Silicon nanocrystals were *in situ* grown in a silicon nitride film by plasma enhanced chemical vapor deposition. The size and structure of silicon nanocrystals were confirmed by high-resolution transmission electron microscopy. Depending on the size, the photoluminescence of silicon nanocrystals can be tuned from the near infrared (1.38 eV) to the ultraviolet (3.02 eV). The fitted photoluminescence peak energy as $E(eV) = 1.16 + 11.8/d^2$ is an evidence for the quantum confinement effect in silicon nanocrystals. The results demonstrate that the band gap of silicon nanocrystals embedded in silicon nitride matrix was more effectively controlled for a wide range of luminescent wavelengths.

INTRODUCTION

Because of its indirect band gap of 1.1 eV, silicon is characterized as having a very poor optical radiative efficiency and only produces light outside the visible range. Silicon nanostructures, however, which show a quantum confinement effect have an enhanced rate of electron-hole radiative recombination [1]. In recent years, a great deal of research on silicon nanocrystals embedded in a silicon oxide matrix have been conducted because of their potential for applications in silicon-based optoelectronic devices [2-3]. However, a number of groups have reported that when the crystallite size of silicon nanostructures in a silicon oxide matrix is controlled, the experimental photoluminescence energies in air are not in good agreement with values that are theoretically calculated from quantum confinement effects [4-5]. Wolkin et al. proposed that oxygen is related to the trapping of an electron (or even an exciton) by silicon-oxygen double bonds and produces localized levels in the bandgap of nanocrystals [6]. Therefore a quantum confinement effect is not good agreement with the theoretical calculation in silicon nanostructures, after exposure to air [7-8]. Even when a silicon oxide is used as a typical matrix material that hosts silicon nanostructures, a silicon oxide matrix may not provide an appropriate emission state for a quantum confinement effect in small silicon crystallites. Because of this, the focus of the present study was on an appropriate matrix material for silicon nanocrystals. There appear to be few localized states that correlate with the optical process of carriers at nanocrystal surface in a silicon nitride matrix, as shown in a previous report related t amorphous silicon quantum dot structures [9-10]. In the present work, we report on silicon nanocrystals that were *in situ* grown in a silicon nitride film by plasma enhanced chemical vapor deposition. Typically, silicon nanocrystals are obtained by the post annealing of a silicon-ric silicon oxide at 1100°C [11]. The method described here is desirable in terms of integratin silicon based optoelectronic components. This new method permits a good match with the quantum confinement effect in zero-dimensional crystalline silicon by controlling the crystal siz because this provides a good emission state in small silicon nanocrystals, when a silicon nitrid matrix is used.

EXPERIMENTAL DETAILS

The silicon nanocrystals were prepared by plasma enhanced chemical vapor deposition, in which argon-diluted 10% silane and nitrogen gas at a purity in excess of 99.9999% were used a the reactant gas sources. (100) crystalline silicon wafers were employed as sample substrates. The total pressure and growth temperature were fixed at 0.5 Torr and 250 °C, respectively. The flow rate of silane and nitrogen were used to modulate the rate of growth of the silicon nitride film and, eventually, to control the size of the silicon nanocrystals. The flow rate of silane and nitrogen was in the range from 4 to 12 sccm and from 500 to 1800 sccm, respectively. No post annealing process was required after growing the silicon nitride film. The size and microscopic structure of the silicon nanocrystals were confirmed by high-resolution transmission electron microscopy using a JEOL Electron Microscopy 2010 instrument operated at 200 kV. To demonstrate the quantum confinement effect, the photoluminescence of silicon nanocrystals with various dot sizes were measured. A charge coupled device detector was used for the photoluminescence measurements at room temperature, with a He-Cd 325 nm laser as the excitation source.

RESULTS AND DISCUSSION

Figure 1 shows high-resolution transmission electron microscopy (HRTEM) images of silicon nanocrystals embedded in a silicon nitride film. The silicon nanocrystals appear as dark spots and the silicon substrate appears as a dark region. The average size (diameter) and dot density of silicon nanocrystals was 4.6 nm and 6.0×10^{11} /cm^2, respectively. The standard deviation for the size distribution was about 0.28 nm. This sample showed a peak for red-colored light (700 nm) in the photoluminescence spectrum. The inset clearly shows that the dark spots are crystallites of silicon nanocrystals separated by an amorphous silicon nitride matrix.

Figure 1. Cross-sectional high-resolution transmission electron microscopic (HRTEM) images of silicon nanocrystals embedded in a silicon nitride film. The average size (diameter) and the dot density of silicon nanocrystals was 4.6 nm and 6.0×10^{11}/cm^2, respectively. The inset clearly shows that the dark spots are nanocrystals separated by an amorphous silicon nitride matrix.

As the size of a quantum structure decreases, the band gap of the material increases from the quantum confinement effect. A blue shift in optical luminescence is the result of this effect. To demonstrate the quantum confinement effect, we measured the photoluminescence of silicon nanocrystals with various dot sizes. The change in the photoluminescence peak energies with the nanocrystal sizes, as determined by HRTEM is shown in Fig. 2. When the crystal size was decreased from 6.1 to 2.6 nm, the photoluminescence peak energy shifted toward higher energy from 1.46 eV (850 nm) to 3.02 eV (410 nm). This figure clearly shows that the photoluminescence peak is blue shifted with decreasing nanocrystal size. Assuming an infinite

potential barrier, the energy gap, E, for three-dimensionally confined silicon nanocrystals can b
expressed as $E(eV) = E_{bulk} + C/d^2$ according to effective mass theory, where E_{bulk} is the bul
crystal silicon band gap, d the dot size (diameter), and C the confinement parameter. The best f
for the data shown in Fig. 2 is to the equation $E(eV) = 1.16 + 11.8/d^2$. Fitted bulk band gap c
1.16 is in good agreement with reported literature values for bulk crystal silicon (1.1 - 1.2 eV
and exists a great difference between bulk band gap of amorphous silicon (1.5 – 1.6 eV) [9]. Thi
result further verifies that silicon nanoclusters have a crystal structure, as shown in the HRTEN
of Fig. 1. The fitted confinement parameter of 11.8 is also very large compared with that of a
amorphous silicon quantum dot of 2.4 [9, 12-13]. In previous studies, it has been shown that th
confinement parameter by a theoretical calculation can be changed depending on the calculatio
method and is about 7 to 14 [12-17]. Our fitted parameters are consistent with the Effective Mas
Approximation. Therefore, the HRTEM measurements and the accompanyin
photoluminescence results verify the quantum confinement effect in silicon nanocrystals.

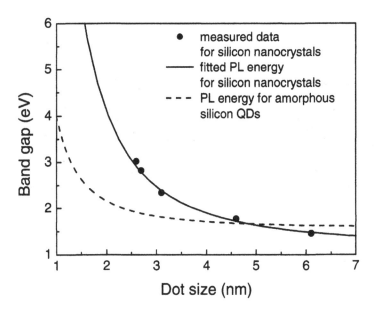

Figure 2. Photoluminescence peak energy of silicon nanocrystals as a function of crystal size
The solid line was obtained from the effective mass theory for three-dimensionally confine
Silicon nanocrystals. The dashed line was obtained from the effective mass theory for amorphou
silicon quantum dot structure in Ref. 9.

Figure 3 shows a room-temperature photoluminescence spectrum obtained from various sized silicon nanocrystals, where the tuning of the photoluminescence emission from 410 to 900 nm is possible by controlling the size of the silicon nanocrystal and, as a result, the emission color can be changed by controlling the size of the nanocrystal. For example, nanocrystal sizes corresponding to red, green, and blue emission are 4.6, 3.1, 2.7 nm, respectively.

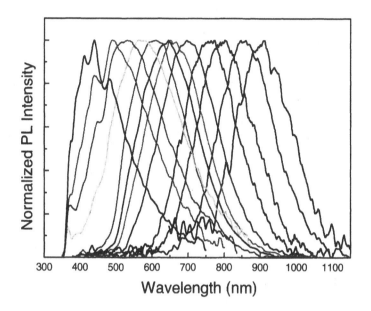

Figure 3. Room-temperature photoluminescence spectra of silicon nanocrystals. The peak position can be controlled by appropriate adjustment of the nanocrystal size.

CONCLUSIONS

well-organized silicon nanocrystals were grown in a silicon nitride film by plasma enhanced chemical vapor deposition without a post-annealing process. HRTEM measurements and photoluminescence results provided convincing evidence of a quantum confinement effect in the silicon nanocrystals, and a silicon nitride matrix was found to provide a good emission state i small silicon nanocrystals. The band gap of the silicon nanocrystals could be controlled fror 1.38 to 3.02 eV by decreasing the nanocrystal size, demonstrating the viable potential for the fabrication of effective Silicon-based light-emitting diodes in a wide range of luminescer wavelengths.

ACKNOWLEDGEMENT

This work was supported by the Ministry of Information and Communication in Korea.

REFERENCES

1. *Light Emission in Silicon : From Physics to Devices*, edited by D. J. Lockwood (Academi Press, San Diego, 1998), Chap. 1.
2. N. Lalic and J. Linnros, *J. Lumin.* **80**, 263 (1999).
3. S.-H. Choi and R. G Elliman, *Appl. Phys. Lett.* **75**, 968 (1999).
4. S. Schuppler et al, *Phys. Rev. Lett.* **72**, 2648 (1994).
5. T. van Buuren, L. N. Dinh, L. L. Chase, W. J. Siekhaus, and L. J. Terminello, *Phys. Rev. Let* **80**, 3803 (1998).
6. M. V. Wolkin, J. Jorne, and P. M. Fauchet, G Allan, and C. Delerue, *Phys. Rev. Lett.* **82**, 19 (1999).
7. J. S. Biteen, A. L. Tchebotareva, A. Polman, N. S. Lewis, and H. A. Atwater, *Mat. Res. Soc Symp. Proc.* **770** (2001).
8. A. Puzder, A. J. Williamson, J. C. Grossman, and G Galli, *Phys. Rev. Lett.* **88**, 97401 (2002).
9. N.-M. Park, C.-J. Choi, T. Y. Seong, and S.-J. Park, *Phys. Rev. Lett.* **86**, 1355 (2001).
10. Y. Q. Wang, Y. G Wang, L. Cao, and Z. X. Cao, *Appl. Phys. Lett.* **83**, 3474 (2003).
11. F. Iacona, G Franzò, and C. Spinella, *J. Appl. Phys.* **87**, 1295 (2000).
12. K. Nishio, J. Kōga, T. Yamaguchi, and F. Yonezawa, *Phys. Rev. B* **67**, 195304 (2003).
13. S. Furukawa and T. Miyasato, *Phys. Rev. B* **38**, 5726 (1988).
14. L.-W. Wang and A. Zunger, *J. Phys. Chem.* **98**, 2158 (1994).
15. S. Öğüt, J. R. Chelikowsky, and S. G Louie, *Phys. Rev. Lett.* **79**, 1770 (1997).
16. J. P. Proot, C. Delerue, and G Allen, *Appl. Phys. Lett.* **61**, 1948 (1992).
17. *Light Emission in Silicon : From Physics to Devices*, edited by D. J. Lockwood (Academic Press, San Diego, 1998), Chap. 7.

Mat. Res. Soc. Symp. Proc. Vol. 817 © 2004 Materials Research Society   L4.4

Optical Study of SiO$_2$/nanocrystalline-Si Multilayers Using Ellipsometry

Kang-Joo Lee[1], Tae-Dong Kang[1], Hosun Lee [1,a], Seung Hui Hong[1], Suk-Ho Choi[1], Kyung Joong Kim[2], and Dae Won Moon[2]

[1]Department of Physics and Institute of Natural Sciences, Kyung Hee University, Suwon 449-701, Korea
[2]Nano Surface Group, Korea Research Institute of Standards and Science, P.O.Box 102, Yusong, Taejon 305-600, Korea

ABSTRACT

Using variable-angle spectroscopic ellipsometry, we measure the pseudo-dielectric functions of as-deposited and annealed SiO$_2$/SiO$_x$ multilayers (MLs). The SiO$_2$(2nm)/SiO$_x$(2nm) MLs have been prepared under various deposition temperature by ion beam sputtering. The annealing at temperatures $\geq$ 1100°C leads to the formation of Si nanocrystals (nc-Si) in the SiO$_x$ layer of MLs. Transmission electron microscopy images clearly demonstrate the existence of nc-Si. We assume a Tauc-Lorentzian lineshape for the dielectric function of nc-Si, and use an effective medium approximation for SiO$_2$/nc-Si MLs as a mixture of nc-Si and SiO$_2$. We successfully estimate the dielectric function of nc-Si and its volume fraction. We find that the volume fraction of nc-Si decreases after annealing, with increasing x in as-deposited SiO$_x$ layer. This result is compared to expected nc-Si volume fraction, which was estimated from stoichiometry of SiO$_x$.

INTRODUCTION

In order to realize light-emitting-diode using silicon-based materials, intensive investigations on nanocrystalline silicon (nc-Si) have been carried out. Due to the confinement of electron and hole carriers in nano-scale volumes, we expect enhanced luminescence efficiency due to increasing recombination rate of carriers as well as the visible luminescence arising from quantum confinement effect [1].

Notably, superlattices composed of alternating nc-Si and SiO$_2$ layers have been given much attention due to a large volume fraction of nc-Si and the controllability of the size and density of nc-Si crystallites [2-3]. One promising method of nc-Si/SiO$_2$ MLs relies on the growth of SiO$_x$/SiO$_2$ multilayers (MLs) and subsequent heat treatment [4]. We used ion beam sputtering deposition (IBSD) method to grow SiO$_x$/SiO$_2$ MLs. IBSD has the advantage of low operational pressure of the ion sources and the precise control of the ion beam parameters, compared to plasma-based rf sputtering techniques. Defect densities are relatively low because a neutralized ion beam is used for sputtering and a substrate is not immersed in the plasma [5,6].

In order to optimize nc-Si/SiO$_2$ MLs for optoelectronic devices, we need various structural and optical characterization methods [2,3,5,7]. Using spectroscopic ellipsmetry, we can characterize the structural and optical properties at the same time. In detail, we can estimate the thickness of the nc-Si and SiO$_2$ layers and the volume fraction of nc-Si, as well as the dielectric function of nc-Si. Recently, several researchers successfully adopted Tauc-Lorentz (TL) model for the dielectric function of nc-Si of the diameters of 1 to 3 nm [8,9].

a) Corresponding author: hlee@khu.ac.kr

In this work, we measured the pseudo-dielectric function of as-deposited single thick layer SiO_x and its annealed nc-Si embedded in SiO_2 matrix, as well as as-deposited $(SiO_x/SiO_2)_{50}$ and annealed $(nc-Si/SiO_2)_{50}$ MLs, by using variable angle spectroscopic ellipsometry at room temperature. Using the T-L model for the dielectric function of nc-Si and using Bruggeman effective medium approximation (BEMA), we determined the dielectric function of nc-Si and the volume fraction of nc-Si phase.

EXPERIMENTAL DETAILS

Two sets of nc-Si/SiO_2 samples were grown. First, alternating layers composed of 2nm-thick SiO_x and 2nm-thick SiO_2 thin films with 50 periods were grown on p-type (100) Si wafers at various temperatures of room temperature, 100 °C, 200 °C, and 300 °C by reactive IBSD using a Kaufman type DC ion gun and Ar^+ beam with ion energy of 750 eV and a Si target. Second, a single layer of 80 nm-thick SiO_x was grown in the identical conditions, for comparison. The deposition chamber was evacuated to a pressure of 2.0×10^{-8} torr before introducing argon gas into the system. Details of the system are described elsewhere [5]. We controlled the relative oxygen content by varying both the sputtering rate and oxygen gas pressure. After deposition, the samples were annealed at 1100 °C for 20 minutes in a ultra-pure nitrogen ambient using a horizontal furnace to form Si nanocrystals in the SiO_x layers. In order to to passivate Si dangling bonds, the samples were hydrogenated for 1 hour at 650 °C under hydrogen gas flow. The stoichiometry of the SiO_x layers was analyzed by in-situ x-ray photoelectron spectroscopy (XPS using Al $k\alpha$ line of 1486.6 eV, thereby sidestepping the problem of surface contamination and oxidation [10]. Transmission electron microscopy images clearly demonstrate the existence of nc-Si of about 3 to 5 nm diameter for the annealed $SiO_2(2nm)/SiO_x(2nm)$ MLs and 80nm-thick single $SiO_{1.6}$ layer.

Spectroscopic ellipsometry measurements were performed on samples with incidence angles of 70 ° and 75 ° using a variable angle ellipsometer [Woollam VASE model] with and without an auto-retarder in the spectral range of 1.0 – 6.0 eV. The capability of multiple angles of incidence increases the accuracy in determining the layer dielectric function out of pseudo-dielectric functions. The pseudo-dielectric functions were fitted using non-linear Levenberg-Marquardt algorithm using WVASE32 software.

DISCUSSION

Figure 2 show the plots of the imaginary part of pseudo-dielectric function (discrete symbol) and its curve fits (solid line) for as-deposited and annealed $(SiO_2)_{50}/(SiO_x)_{50}$ ML for x= which were grown at room temperature, at 70 degree of angle of incidence. Note that the H-passivation did not change the pseudo-dielectric function of annealed MLs. We also measured and fitted the pseudo-dielectric function for the MLs for x= 1.2, 1.4, 1.6, and 1.8, which were grown at room temperature. We measured the pseudo-dielectric functions for other deposition temperatures such as 100 °C, 200 °C, and 300 °C. For simplicity, the data and fitting are not shown here. We adopted the TL lineshape model for the dielectric function of nc-Si [11]. We successfully fitted the pseudo-dielectric function by using BEMA [12]. We assumed a mixture of

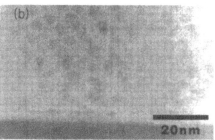

Figure 1. TEM images of the $(SiO_2)_{50}/(SiO_x)_{50}$ MLs ((a) x=1.0, and (b) x=1.2) annealed at high temperature. For visibility, a few Si nano-crystallites were marked. The large pear-shaped nc-Si in the middle is a fusion of two or three circular nc-Si's.

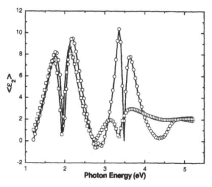

Figure 2. Plot of the imaginary part of pseudo-dielectric function (discrete symbols) and their curve fits (solid lines) for as-deposited (circles) and annealed (rectangles) $(SiO_2)_{50}/(SiO_x)_{50}$ MLs, which were grown at room temperature. The oxygen composition x was 1.0.

nc-Si and SiO_2 for the MLs. The equation for EMA is given by

$$f_{ncSi} \frac{\varepsilon_{ncSi} - \langle \varepsilon \rangle}{\varepsilon_{ncSi} + 2\langle \varepsilon \rangle} + f_{SiO2} \frac{\varepsilon_{SiO2} - \langle \varepsilon \rangle}{\varepsilon_{SiO2} + 2\langle \varepsilon \rangle} = 0 \qquad (1)$$

where ε_{ncSi} and ε_{SiO2} are the dielectric functions of, respectively, nc-Si and SiO_2, $\langle \varepsilon \rangle$ is the pseudo-dielectric function, and f_{ncSi} and f_{SiO2} are the volume fraction of nc-Si and SiO_2, respectively.

In TL model, the imaginary dielectric function ε_2 is determined by multiplying the Tauc joint density of states by the ε_2 obtained from the Lorentz oscillator model. Thus ε_2 is given by,

$$\varepsilon_2(E) = \frac{1}{E}\left[\frac{AE_0C(E-E_g)^2}{(E^2-E_0^2)^2 + C^2E^2}\right] \quad \text{for } E > E_g \quad (2)$$

$$= 0 \qquad \text{for } E < E_g$$

where A, E, E_g, and C are the four parameters to describe the spectral dependence of $\varepsilon_2(E)$ [11]

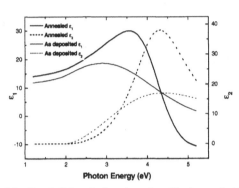

Figure 3. Plot of the fitted dielectric function of nc-Si of annealed MLs and SiO$_x$ of as-deposited MLs for x= 1 from Fig.2.

The real part of the dielectric function $\varepsilon_1(E)$ is then obtained using Kramers-Kronig transformation, with an additional parameter of $_1\varepsilon$.

Figure 3 shows the fitted dielectric function of nc-Si of annealed MLs and SiO$_x$ of as-deposited MLs for x= 1 from Fig.2. Compared to the dielectric function of bulk Si, the values of dielectric function decreased. For example, the maximum of ε_1 of bulk Si is about 45. The peak from nc-Si is much more pronounced that that of SiO$_x$, as expected because of crystallization. Note that the dielectric functions of both SiO$_x$ and nc-Si were fitted using TL model. The fitted dielectric function of nc-Si is similar to that of Ref. 8. The average diameter of their nc-Si was about 1.5nm. In general, the dielectric functions of semiconductors in the form of quantum well and bulk show main peaks associated with E$_1$ (3.396 eV) and E$_2$ (4.270 eV) critical points. However, in the case of nc-Si, the dielectric function shows only single peak in the experimental range, possibily because of large broadening of E$_1$ and E$_2$ band gaps. This phenomenon may also be attributed to the inadequacy of band structure in nano-crystallites due to their extremely small volume and increased surface effect.

The ML fitting of nc-Si/SiO$_2$ MLs such as a model of alternating SiO$_2$ and nc-Si layer of 50 periods was also successful. However, the fitting assuming a mixture of nc-Si and SiO$_2$ in a single layer gave a better goodness of fit values than assuming MLs. This may be attributed to nonuniform size and irregular ordering of nc-Si's sccording to TEM data. We also note that the BEMA fitting gave a more successful fitting for MLs rather than the same fitting for single-layer annealed nc-Si/SiO$_2$. This suggests that the size uniformity of nc-Si in annealed MLs are

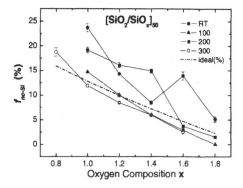

Figure 4. Plot of the fitted volume fraction of nc-Si in the MLs.

better than that of nc-Si in single layer. This is reasonable because SiO_x and SiO_2 layers are periodically distributed and SiO_x will become decomposed into SiO_2 and nc-Si phase after high-temperature annealing. Therefore, in order to get a narrow size distribution of nc-Si, ML growing of SiO_x/SiO_2 is better than the single layer growing of SiO_x as was shown in Fig.1.

Figure 4 shows the fitted volume fraction of nc-Si in the MLs. We also plotted an expected curve from the consideration of the stoichiometry of SiO_x (solid lines) in Fig. 4. In Fig. 4, we observe that the volume fraction of nc-Si decreased with increasing x in as-deposited SiO_x layer. The formation of Si nanocrystals in SiO_x thin films after annealing was studied by in-situ XPS analysis. The phase separation of excess Si in the as-deposited SiO_x layer is the driving force for the formation of Si nanocrystals.

The decrease of nc-Si with increasing x, in other words, approaching x=2.0, can be explained with the consideration of stoichiometry of SiO_x layer. The amount of the precipitated nc-Si can be estimated by assuming that all the SiO_x will decompose into SiO_2 and nc-Si after annealing. We estimated the mass of the SiO_2 layer after annealing using the equations,

$$f_{ncSi} = \frac{V_{ncSi}}{V_{SiO2} + V_{ncSi}} = \frac{t}{1+t}$$

$$t = \frac{V_{ncSi}}{V_{SiO2}} = \frac{2-x}{2+x} \left(\frac{m_{Si}}{m_{SiO2}} \right) \left(\frac{\rho_{SiO2}}{\rho_{Si}} \right),$$

where m_{Si} and m_{SiO2} are the molecular weight of Si and SiO_2 molecules, and ρ_{Si} (2.329 g/cm^3) and ρ_{SiO2} (2.2 g/cm^3) are the mass density of nc-Si and SiO_2 phase, respectively [13,14]. Note that we assumed that the SiO_2 and SiO_x layers in the as-grown MLs have the same number of Si atoms. The calculated V_{ncSi} fraction is shown in Fig. 4. The discrepancy between the calculation and experimental data can be attributed the simplified model. We assumed that SiO_2 and SiO_x layers in the as-grown MLs have the same number of Si atoms. However, the number of Si atoms in the SiO_x layer should be larger than that of the SiO_2 layer assuming the same thickness of both layers. This will increase the calculated V_{ncSi}, especially in the low x region in Fig. 4.

CONCLUSION

We measured the pseudo-dielectric functions of as-deposited and annealed SiO_2/SiO multilayers using variable-angle spectroscopic ellipsometry between 1.5 eV and 6 eV. Th $SiO_2(2nm)/SiO_x(2nm)$ MLs have been prepared under various deposition temperatures by ic beam sputtering. The annealing at temperatures larger than 1100°C leads to the formation of nanocrystals in the SiO_x layer of MLs. Using Tauc-Lorentzian lineshape for the dielectri function of nc-Si, and assuming SiO_2/nc-Si MLs as a mixture of nc-Si and SiO_2, we successfull estimated the dielectric function of nc-Si and its volume fraction. The volume fraction of nc-decreased after annealing, with increasing x in as-deposited SiO_x layer. The formation of nc-was verified by TEM. The fitted nc-Si volume fraction was compared to the estimated valt from stoichiometry of SiO_x.

ACKNOWLEDGEMENTS

This work was supported in part by Korea Research Foundation Grant No. KRF-2003-005-C00001.

REFERENCES

[1] "Light emission in silicon: From physics to devices", D. J. Lockwood, Semiconductors and semimetals Vol. 49, (Academic press, San Diego, 1998).
[2] D. J. Lockwood, Z. H. Lu, and J. –M. Baribeau, Phys. Rev. Lett. **76**, 539 (1996).
[3] P. Photopoulos, A. G. Nassiopoulou, and D. N. Kouvatsos, Appl. Phys. Lett. **76**, 3588 (2000).
[4] Y. Q. Wang, G. L. Kong, W. D. Chen, H. W. Diao, C. Y. Chen, S. B. Zhang, and X. B. Liao, Appl. Phys. Lett. **81**, 4174 (2002)]
[5] J.-S. Bae, S.-H. Choi, K. J. Kim, and D. W. Moon, J. Korean. Phys. Soc. **43**, 557 (2003),
[6] C. Saha, S. Das, S. K. Ray, and S. K. Lahiri, J. Appl. Phys. **83**, 4472 (1998).
[7] L. Khriachtchev, S. Nivikov, and O. Kilpelä, J. Appl. Phys. **87**, 7805 (2000).
[8] M. Losurdo, M. M. Giangregorio, P. Capezzuto, G. Bruno, M. F. Cerqueira, E. Alves, and M. Stepikhova, Appl. Phys. Lett. **82**, 2992 (2003).
[9] D. Amans, S. Callard, A. Gagnaire, J. Joseph, G. Ledoux, and F. Huisken, J. Appl. Phys. **93**, 4173 (2003).
[10] K. J. Kim et al., Unpublished.
[11] G. E. Jellison, Jr, and F. A. Modine, Appl. Phys. Lett. **69**, 371 (1996); **69**, 2137 (1996).
[12] D. E. Aspnes, Thin Solid Films, **89**, 249 (1982).
[13] "Silicon Processing for the VLSI Era, Volume 1 –Process Technology", S. Wolf and R. N. Tauber, (Lattice Press, 1986).
[14] "Semiconductors; Group IV Elements and III-V Compounds", Ed. By O. Madelung, in "Data in Science and technology", Ed. R. Poerschke, (Springer-Verlag, Berlin, 1991).

Mat. Res. Soc. Symp. Proc. Vol. 817 © 2004 Materials Research Society   L6.3

Active Photonic Crystal Devices in Self-Assembled Electro-Optic Polymeric Materials

J. Li[1], P. J. Neyman[2], M. Vercellino[3], J. R. Heflin[2], R. Duncan[3], and S. Evoy[1]

[1]Department of Electrical and Systems Engineering, The University of Pennsylvania, Philadelphia, PA 19104, [2]Department of Physics, Virginia Polytechnic Institute and State University, Blacksburg, VA 24061, and [3]Luna Innovations, Blacksburg, VA 24060

ABSTRACT

Photonic crystals (PC) offer novel approaches for integrated photonics by allowing the manipulation of light based on the photonic bandgap effect rather than internal-reflection mechanisms employed in traditional devices. Electro-optic polymers represent interesting possibilities for the development of devices leveraging control over the phase of a confined propagating wave. We here report on the development of such active photonic crystal technology in ionically self-assembled monolayers. The simulation of active photonic devices such as Mach-Zehnder interferometers and wavelength multiplexers is first presented. We then report on the synthesis and optical characterization of electro-optic films grown through the ISAM technique. We conclude by presenting the preliminary development of a nanofabrication platform that would enable the realization of active photonic devices in such materials.

INTRODUCTION

There is an acute need for the development of novel photonic devices that would perform on-chip functions such as signal conditioning and processing, and support the low-cost integration of optical networks. Photonic crystals (PC) offer novel approaches for such integrated technology by allowing the manipulation of light based on the photonic bandgap effect rather than internal-reflection mechanisms employed in traditional devices [1,2]. Several types of photonic crystal devices have already been reported, including waveguides and fibers offering enhanced control over light propagation [3], light emitting diodes showing good suppression of lateral emission [4], as well as laser devices realized through distributed feedback structures [5]. While these devices leveraged the photonic bandgap effect to confine wave propagation, this platform has yet to be implemented in devices that would rather leverage the manipulation of frequency or phase of the propagating wave. Such approach would enable the deployment of photonic crystals in important devices such as Mach-Zehnder Interferometers (MZIs), as well as wavelength multiplexers (MUX's) and demultiplexers (DeMUX's).

Electro-optic polymers represent interesting possibilities for such development. These flexible and robust materials offer the tunability of their refractive index under the application of electric fields, thus allowing the control over the phase of a confined propagating wave. In addition, electro-optic control of refractive index also offers the tunability over the photonic bandgap effect, and would result in a frequency-selective mechanism that could be leveraged for high-speed switching devices. Finally, an integrated photonics platform based on polymeric materials would also have other desirable characteristics such as low-cost processing and manufacturability.

Non-linear materials must possess nonzero even-order nonlinear optical susceptibility and therefore must lack a center of inversion. Several methods have been reported for the synthesis of such materials including electric-field poled polymers[6], Langmuir-Blodgett films [7] and covalent self-assembled monolayer structures [8]. Ionically self-assembled monolayers (ISAMs) have also been proposed for the synthesis of such films [9-12]. This technique involves the layer-by-layer growth of a supramolecular structure trough the alternate immersion of a substrate in aqueous solutions respectively containing negatively-charged and positively-charged polymeric dyes. The ISAM technique has been used to produce a noncentrosymmetric arrangement of nonlinear optical (NLO) chromophores that are more stable thermally and temporally than poled polymer systems [10,11] with electro-optic coefficients as high as $r_{33} > 20$ pm/V, approaching those of lithium niobate [13].

We here report on the development of an active photonic crystal technology implemented in ionically self-assembled monolayers. The simulations of active PC devices such as Mach-Zehnder interferometers and wavelength multiplexers are first presented. We then report on the synthesis and optical characterization of electro-optic films grown through the ISAM technique. We conclude by presenting the preliminary development of a nanofabrication platform that would enable the realization of active photonic devices in such materials.

ACTIVE PHOTONIC CRYSTAL DEVICES

The Mach-Zehnder Interferometer (MZI) consists of a structure that splits an optical signal into two separate waveguides, allows their independent propagation over a path of several wavelengths, and recombines them back into a unique waveguide. By inducing a phase delay between the two optical paths, a constructive or destructive interference of the two signals at the recombination point will lead to the controlled modulation of the output signal. The illustration of active PC structures is performed using the shareware version of Translight [14]. Figure 1a shows the simulation of the transmission spectrum of the TE mode in a 2-D planar hexagonal lattice photonic crystal structure along the Γ_k direction. In order to induce a band-gap centered at $\lambda=1.55$ μm, the backbone PC structure consists of a hexagonal arrangement of holes with a lattice constant of d = 650 nm, with dielectric constant of $\varepsilon_1 = 2.9$ and $\varepsilon_2 = 1$. The MZI design will eventually employ tuning of index of refraction to induce phase shifts in the PC waveguides A relatively wide band-gap must therefore be designed to ensure that such shifting of the dielectric constant will not move the band-gap away from the operating wavelength. Figure 1b shows the transmission spectrum under large changes of dielectric constants ($\varepsilon = 2.8{\sim}3.0$), indeed showing that bandgap will easily be maintained under the much smaller shifts expected from normal operating conditions. By proper tuning of the crystal parameters, the bandgap effect has been observed for both TE and TM modes using the hexagonal lattice.

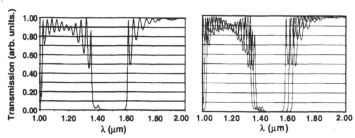

Figure 1: Translight simulations of the transmission spectrum of 2D photonic crystals. a) Using an hexagonal arrangement of holes with lattice d = 650 nm, a bandgap centered at λ = 1.5 um is obtained. b) Similar transmission spectra under large simulated shifts of dielectric constant (ε = 2.8~3.0). The operating wavelength remains within the bandgap.

Figure 2a shows the simulated propagation of the TE mode in an "idle" MZI structure implemented in a square crystal lattice. While more lossy than the hexagonal lattice simulated above, the square lattice was here used due to its simpler implementation with the software employed. The power splitting at the input is clearly occurring, as expected. However, the intensity is somewhat dissipated at the recombination point for two reasons: 1) the material was simulated to have an imaginary component to its index of refraction and is thus leaky, and 2) the E_z component of the electric field is experiencing vector addition at the recombination point because of field components that are flipping orientation as they 'turn the corner'.

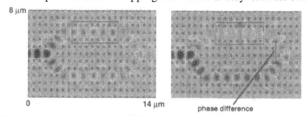

Figure 2: Translight simulations of a MZI photonic crystal structure in electro-optic layers. a) Idle state of device shows clear splitting of the optical field. b) Same structure with shift of dielectric constant of 0.5 of upper waveguide. An output modulation of 0.83 is obtained.

Figure 2b shows the simulation of the same structure, but where the area under the rectangle has had its dielectric constant shifted by 0.5. Such unrealistic shift was employed to offset the unusually short optical paths that have been simulated for easier visualization of the pathways. Eventual designs will rather involve shifts orders of magnitude smaller, and optical paths correspondingly orders of magnitudes longer. Nonetheless, confinement is still accomplished in spite of the resulting bandgap shift. A phase shift is seen between the two paths, and a destructive interference is seen at the recombination point. Just prior to the recombination, the normalized electric field intensity is 1.05 in the lower path, and 0.69 in the upper path (because of its poorer propagation). At the recombination point, the normalized electric field intensity is 0.83.

SYNTHESIS AND CHARACTERIZATION OF ISAM MATERIAL

In this work, a combination of low-molecular-weight molecules and polyelectrolytes are used for such fabrication. Poly(allylamine hydrochloride) (PAH) was used as cation, in combination with either Procion Red MX-5B [12] or poly{1-[4-(3-carboxy-4-hydroxyphenylaz benzensulfonamido]-1,2-ethanediyl, sodium salt} (PCBS) as anion. Second harmonic generatio (SHG) measurements were performed using a Q-switched Nd:YAG 10 Hz pulsed laser with a 1 ns pulse width as light source. The wavelength was $\lambda =1064$ nm and the typical maximum incident energy was 1.7 mJ. A series of PCBS/PAH films was fabricated with 25, 50, 75, 100, 125, and 250 bilayers, with a thickness of ~1 nm/bilayer. Figure 3 a) shows the SHG data such films as a function of the incident angle of the beam with respect to the film normal. Since the films are fabricated by immersion, they form on both sides of the glass substrate. Each film is thin enough to serve as a phase-matched source of SHG. However, as the angle of incidence is increased, the propagation distance between the two films through the substrate is also increase Since the fundamental and second harmonic waves travel at different velocities through the glas the phase between the second harmonic generated in the two films varies with the incident angle yielding the observed fringes. The observed periodicity of 7° corresponds to a coherence length of 21 μm, as expected. The minima of zero also demonstrate that the films are highly homogeneous such that complete destructive interference can occur between the two opposite films. In general, the second harmonic intensity as a function of thickness should follow the Maker fringe equation:

$$I^{2\omega} \propto (l_c \chi_{eff}^{(2)})^2 \sin^2\left(\frac{\pi d}{2l_c}\right) \qquad (1$$

where l is the sample thickness and l_c is the coherence length. Here, the argument of the $\sin^2$ is small since the maximum thickness is 250 nm and the coherence length is >10 μm. It is thus expected that the peak SHG intensity should scale quadratically with the number of bilayers. Figure 3b), which displays the square root of the peak SHG intensity as a function of the numbe of bilayers demonstrates that this is indeed the case. This is an important demonstration that films >250 nm thick can be synthesized with bulk polar order throughout the film.

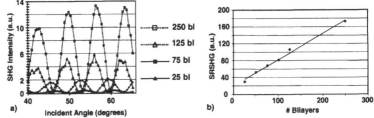

Figure 3: a) Second harmonic intensity of PCBS/PAH ISAM films as a function of the angle of incidence for 25, 75, 125, and 250 bilayers. b) Square root of the peak SHG intensity as a function of the number of bilayers.

DEVICE FABRICATION

The inception of a photonic crystal platform in ISAM materials require the development of a nanofabrication process optimized for the soft polymeric films. Such requirement present a significant departure from the usual process parameters developed in inorganic PCs given the differing chemistries and physical properties involved. We have developed such process through combination of standard electron beam lithography, metal evaporation and lift-off of an etch mask, and dry RIE etching in an oxygen plasma. Oxide films 500 nm thick are first produced by thermal oxidation of Si (100) substrate. The index of refraction of this oxide will insure wave guiding in the vertical direction. A 500 nm thick electro-optic polymer is then cast onto the oxide surface. We initially developed such fabrication process using polymethlymetacrylate (PMMA) as target electro-optic material. While this polymer possesses marginal non-linear properties, its physicochemical properties approach those of the ISAM films to be eventually processed.

A 30 nm thick metal mask is thermally evaporated on the film surface. A 250 nm thick electron beam resist is then spun cast from a 5.5% 495k molecular weight PMMA in anisole solution. The fabrication of optical PBGs must involve the patterning of ~250 nm diameter hole, on a ~400 nm lattice spacing. To that end, optimal exposure doses are 170 fc/dot using a 90 nA beam at 40 kV acceleration voltages. The pattern is then transferred to the metal mask through wet etch in a nitric acid/phosphoric acid solution. The patterned metal then serves as mask for the dry etching of the polymer in oxygen plasma. Photonic crystals consisting of 250 nm diameter holes with 400 nm lattice spacing are routinely produced in 500 nm thick PMMA "optical" films (Figure 4).

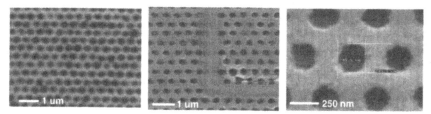

Figure 4: Photonic crystal structures produced in polymeric films. Here, 500 nm-thick PMMA is used as target material. Such process will be amenable to the similar patterning of strongly-active films of similar thicknesses developed by the ISAM method.

Preliminary attempts at the patterning of ISAMs films of similar thickness were also performed. However, the limited morphology of the first films of such thickness produced prevented the production of sub-micron features during those initial attempts. Further work is underway to now apply the fabrication technology to the smoother thick films described above.

CONCLUSIONS

We have presented the development of an active photonic crystal platform in electro-optic materials grown by ionic self-assembly. Preliminary designs and simulations of demonstrated the viability of such platform for the development of phase-modulating devices such as MZI interferometers. The synthesis and characterization of thin ISAM films were discussed. The development of a photonic crystal fabrication process in electro-optic polymers was also reporte Photonic crystals consisting of 250 nm diameter holes with 400 nm lattice spacings are routinely produced in 500 nm thick PMMA films. Such process will be fully amenable to the similar development of photonic crystal circuits in thick ISAM films.

ACKNOWLEDGMENTS

The authors acknowledge funding from AFOSR through the SBIR/STTR program (contract # F49620-02-C-0099).

REFERENCES

1. E. Yablonovitch, *Phys. Rev. Lett.* **58**, 2059(1987).
2. S. John, *Phys. Rev. Lett.* **58**, 2486(1987).
3. E. Chow, S. Y. Lin, S. G Johnson, P. R. Villeneuve, J. D. Joannopoulos, J. R. Wendt, G. A. Vawter, W. Zubrzycki, H. Hou, and A. Alleman, *Nature* **407**, 983(2000).
4. S. Fan, P. R. Villeneuve, J. D. Joannopoulos, *IEEE J. Quantum Elect.* **36**, 1123(2000).
5. S. H. Kwon, H. Y. Ryu, G H. Kim, Y. H. Lee, S. B. Kim, *Appl. Phys. Lett.* **83**, 3870 (2003).
6. K. D. Singer, J. E. Sohn, and S. J. Lalama, *Appl. Phys. Lett.* **49**, 248 (1986).
7. I. R. Girling, N. A. Cade, P. V. Kolinsky, R. J. Jones, I. R. Peterson, M.M. Ahmad, D. B. Neal M. C. Petty, G G Roberts, and W. J. Feast, *J. Opt. Soc. Am. B*. **4**, 950 (1987).
8. H. E. Katz, G Scheller, T. M. Putvinski, M. L. Schilling, W. L. Wilson, and C. E. D. Chidsey *Science*, 254, 1485 (1991).
9. G Decher, *Science* **277**, 1232 (1997).
10. J. R. Heflin, C. Figura, D. Marciu, Y. Liu, R. O. Claus, *Appl. Phys. Lett.* **74**, 495(1999)
11. C. Figura, P. J. Neyman, D. Marciu, C. Brands, M.A. Murray, S. Hair, R.M. Davis, M.B. Miller, and J.R. Heflin, *SPIE Proc.* **3939**, 214 (2000).
12. K. Van Cott, M. Guzy, P. Neyman, C. Brands, J.R. Heflin, H.W. Gibson, R.M. Davis, *Angew. Chem. Int. Ed.* **41**, (2002).
13. K. Van Cott, M. Guzy, P. Neyman, C. Brands, J.R. Heflin, H.W. Gibson, R.M. Davis, unreported.
14. http://www.elec.gla.ac.uk/groups/opto/photoniccrystal/Software/SoftwareMain.htm

Mat. Res. Soc. Symp. Proc. Vol. 817 © 2004 Materials Research Society   L6.12

Structural properties of Si nanoclusters produced by thermal annealing of SiOx films

Simona Boninelli[1], Fabio Iacona[2], Corrado Bongiorno[2], Corrado Spinella[2], and Francesco Priolo[1]
[1]INFM-MATIS and Dipartimento di Fisica e Astronomia dell'Università di Catania, via Santa Sofia 64, 95123 Catania, Italy
[2]CNR-IMM, Sezione di Catania, Stradale Primosole 50, 95121 Catania, Italy

ABSTRACT

The structural properties of Si nanoclusters embedded in SiO_2, produced by high temperature annealing of SiO_x films, have been investigated by energy filtered transmission electron microscopy. The presence of amorphous nanostructures, not detectable by using dark field transmission electron microscopy, has been demonstrated. By taking into account also this contribution, a quantitative description of the evolution of the samples upon thermal annealing has been accomplished. In particular, the nanocluster mean radius and the density of amorphous and crystalline clusters have been determined as a function of the annealing temperature.

INTRODUCTION

Si nanocrystals (nc) embedded in a SiO_2 matrix are currently attracting a great interest as a candidate system to solve the physical inability of bulk Si to act as an efficient light emitter. Indeed, the band gap of Si nc is enlarged with respect to the bulk material due to quantum confinement effects, and an intense visible photoluminescence at room temperature is obtained. Recently, the interest towards this material is greatly increased due to the observation of light amplification in Si nanostructures [1-4], as well as to the demonstration of the feasibility of efficient light emitting devices based on Si nc [5-9]. Indeed, both of these points open the route towards the development of a Si-based optoelectronics.

A key point for a full understanding of the optical properties of this system is the availability of a clear picture of its structural properties and their evolution upon thermal annealing. In this work, the main stages of the thermal evolution of SiO_x films (separation of the Si and SiO_2 phases, formation of amorphous Si clusters and their transition to the crystalline phase) have been investigated in details by energy filtered transmission electron microscopy (EFTEM). The presence of a relevant contribution of amorphous Si nanostructures, not detectable by using the conventional dark field TEM (DFTEM) technique, has been demonstrated. By taking into account also this contribution, the mean size and the density of the Si clusters have been determined as a function of the annealing temperature.

EXPERIMENTAL

SiO_x thin films with a total Si concentration of 46 at.% have been prepared by using a parallel plate plasma enhanced chemical vapor deposition system. The source gases used are high purity (99.99% or higher) SiH_4 and N_2O. Further details about deposition processes can be found elsewhere [10]. After deposition, the SiO_x films have been annealed for 1 h in ultra-pure N_2 at temperatures ranging between 900 and 1250 °C to induce the separation between the Si and the SiO_2 phases and the formation of Si nanoclusters embedded in SiO_2.

The Si nanocluster formation was monitored by using a 200 kV energy filtered transmission electron microscope Jeol JEM 2010F with Gatan Image Filter. This system consists of a conventional TEM coupled with an electron energy loss spectrometer. Si maps have been obtained by using a 4 eV wide energy window centered in correspondence of the Si plasmon loss peak at about 16 eV. The same microscope has been used also to collect dark field TEM images by selecting a small portion (about 10%) of the diffraction ring of the (111) Si planes.

RESULTS AND DISCUSSION

Figure 1 reports the EELS spectra relative to SiO_x films as deposited and annealed at 1250 °C. The EELS spectrum in the 0 - 50 eV range is dominated by the plasmon loss peaks, due to the excitation of collective vibrational modes of valence electrons, and by the zero-loss peak, due to electrons that have been elastically scattered or transmitted without any scattering event. From the analysis of the figure we note that the EELS spectrum relative to as deposited SiO_x presents broad peak centered at about 20.8 eV, in agreement with literature [11]. The energy of this peak is intermediate between those corresponding to pure Si and SiO_2 (about 16.8 and 23.5 eV, respectively). This demonstrates that the as deposited film is characterized by the presence of a homogenous and peculiar phase, consisting of $Si-Si_xO_{4-x}$ (with $0 \leq x \leq 4$) tetrahedra, in agreement with the description of SiO_x films given by the random bonding model (RBM) [12], and it cannot simply be described as a mixture of Si and SiO_2.

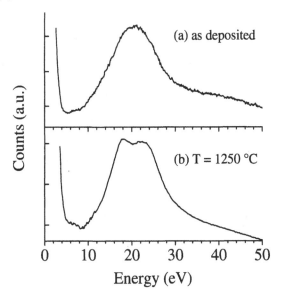

Figure 1. EELS spectra relative to (a) as deposited SiO_x and (b) SiO_x after a thermal annealing at 1250 °C. For all spectra, the zero loss peak is not fully plotted.

On the other hand, the spectrum relative to the sample annealed at 1250 °C presents two well resolved components, whose energies correspond to the Si and SiO_2 plasmon losses. This is a clear evidence of the occurrence of phase separation between Si and SiO_2 induced by the thermal annealing in SiO_x films.

The EFTEM technique allows to generate a micrograph by using only electrons that have lost a specific amount of energy due to the interaction with the sample. This allows to obtain a chemical mapping with the very high spatial resolution typical of TEM and therefore represents a very suitable method to detect all the silicon nanoclusters (both crystalline and amorphous and independently of the crystal orientation) dispersed in a silica matrix. In particular, to map the presence of Si clusters formed inside the SiO_x layer by the annealing process, we have put an energy window 4 eV wide in correspondence of the Si plasmon loss (about 16 eV). The high energy shift between the Si and SiO_2 plasmon peaks (about 7 eV) allows to neatly discriminate the Si contribution from the SiO_2 one.

Figure 2 reports the plan view EFTEM images obtained by using the above described method from SiO_x samples as deposited and annealed at 1000, 1100, and 1250 °C.

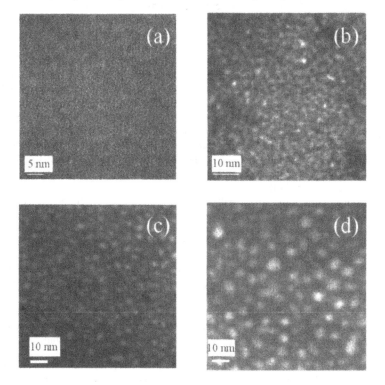

Figure 2. EFTEM plan view images obtained from (a) as deposited SiO_x, and SiO_x annealed at (b) 1000 °C, (c) 1100 °C, and (d) 1250 °C. The bright zones are associated to the presence of Si clusters.

In such kind of images the bright zones are associated to the presence of silicon. The image reported in figure 2(a), showing an uniform background without any appreciable intensity contrast, confirms the absence of any phase separation effects in the as deposited samples. EFTEM data (not shown) demonstrate that clustering effects (formation of a Si network) are already visible at 900 °C. The occurrence of phase separation between Si and SiO_2 becomes much more evident by increasing the annealing temperature, and well defined Si clusters embedded in the oxide matrix are clearly visible in samples annealed at 1000, 1100 and 1250 °C (see figures 2(b) - 2(d)).

To gain a better knowledge on the structural properties of these systems, we have also employed the DFTEM technique. This technique is sensitive to the presence of crystalline planes, and it is therefore able to map the system for the presence of Si nanocrystals. As expected on the basis of our previous data [10], we have not observed any diffraction pattern corresponding to the presence of a crystalline phase for samples annealed at temperatures of 1000 °C of lower. We can therefore conclude that the clusters shown in figure 2(b) are fully amorphous. On the other hand, the samples annealed at temperatures of 1100 °C or higher exhibit the presence of a diffraction pattern mainly consisting of three well distinct rings corresponding to the (111), (220), and (311) planes of crystalline silicon, so that it is possible to conclude that a significant fraction of the clusters shown in figures 2(c) and 2(d) is crystalline.

A first qualitative analysis of the TEM images shown in figure 2 allows to define some aspects of the Si nc nucleation from annealed SiO_x films. As deposited SiO_x films are homogeneous and fully amorphous materials, without any evidence of phase separation between Si and SiO_2. The first stages of the phase separation between Si and SiO_2 become visible at 900 °C, but well defined and amorphous Si clusters are formed only at 1000 °C. At 1100 °C the amorphous nanoclusters (na) begin to become crystalline. We remark here that this general behaviour has to be considered valid for any SiO_x composition; on the other hand, the clustering and crystallization temperatures we have determined (1000 and 1100 °C, respectively) apply in a rigorous way only to the Si concentration used for the experiments reported in the present paper (46 at.%), since lower temperatures are expected when dealing with SiO_x films characterized by higher Si concentrations [10].

More quantitative information on the annealed SiO_x samples can be obtained by measuring the size of the Si nanoclusters. The mean radius of the Si clusters, as obtained by using the EFTEM technique, is reported in figure 3(a) as a function of the annealing temperature. The mean radius of the clusters (both amorphous and crystalline) increases by increasing the annealing temperature from 1.0 (at 1000 °C) to 2.6 nm (at 1250 °C).

For samples in which the amorphous and crystalline phases coexist, we have estimated the crystalline fraction present at a given temperature. In order to do this, the ratio between the number of nc (as detected by DFTEM) and the total number of clusters (as detected by EFTEM) has been evaluated at the different temperatures. The data have been reported in figure 3(b), and a clear trend, showing the progressive increase of the crystalline fraction by increasing the temperature, has been found, clearly demonstrating that in this range the temperature plays a role not only in the cluster growth, but also in extensively promoting the amorphous to crystal transition. In the figure the crystalline fraction at 1250 °C has been set to 100%, since the ratio between the total number of clusters detected by EFTEM, and the nc detected by DFTEM is about 20 : 1, corresponding to the expected value for the fraction of crystals detected by DF in a fully crystalline sample under our experimental conditions [13].

Another information that can be derived from plan view EFTEM images is the number of nanoclusters per unit volume [13]. The values for the Si nanocluster density we have obtained are reported in figure 3(c) as a function of the annealing temperature. The nanocluster density is roughly constant in the 1000 - 1150 °C range (9×10^{17} /cm^3), while it decreases (about 7×10^{17} /cm^3) by increasing the annealing temperature up to 1250 °C. This result, coupled with the continuous increase in cluster size with annealing temperature reported in figure 3(a), suggests that nanocluster growth is not simply due to the inclusion of Si atoms diffusing from the oxide matrix, but also Ostwald ripening effects, leading to the disappearance of small clusters, are probably operating.

Finally, we have used the data on the crystalline fraction reported in figure 3(b) and those on the Si nanocluster density reported in figure 3(c) to calculate the concentration of nc and of na as a function of the annealing temperature. The obtained data are reported in figure 3(d) and demonstrate that the temperature progressively induces the transformation of 9×10^{17} na/cm^3 in 7×10^{17} nc/cm^3, with the loss of about 20% of the clusters present at 1000 °C, due to the occurrence of Ostwald ripening phenomena.

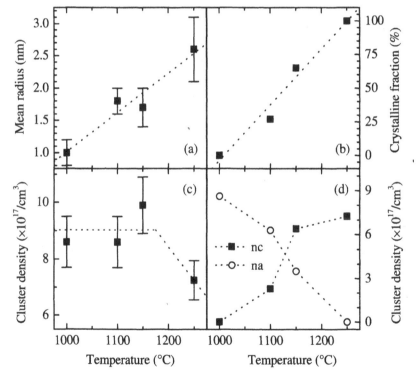

Figure 3. (a) Si nanocluster mean radius, (b) crystalline fraction, (c) density of Si nanoclusters, and (d) density of amorphous (na) and crystalline (nc) nanoclusters, as a function of the annealing temperature. The lines are drawn to guide the eye.

CONCLUSIONS

The comparative analysis of EFTEM and DFTEM images has allowed to elucidate some aspects of the Si nanocluster nucleation in annealed SiO_x films. As deposited SiO_x films are homogeneous and fully amorphous materials, without any evidence of phase separation; well defined amorphous Si clusters are formed only at 1000 °C. At 1100 °C the amorphous clusters start to become crystalline, and the crystalline fraction increases by further increasing the annealing temperature. The capability of the EFTEM technique to detect all the Si clusters has allowed a very reliable determination of the Si nanocluster mean radius and density as a function of the annealing temperature. These data are not available, or can be derived with a high degree of uncertainty, by using dark field or high resolution TEM analyses. Finally, the availability of a quantitative picture allowed the demonstration that amorphous Si clusters constitute a relevant fraction of the overall population in samples annealed at intermediate temperatures. These new evidences constitute a relevant tool for a full comprehension of the properties of Si nanostructures and for their application for the development of a Si-based optoelectronics.

ACKNOWLEDGEMENTS

The authors want to thank G. Franzò, A. Irrera, D. Pacifici, and M. Miritello for collaborating to some of the experiments. This work has been supported by the project FIRB financed by MIUR, and by the IST project SINERGIA financed by the European Commission.

REFERENCES

1. L. Pavesi, L. Dal Negro, C. Mazzoleni, G. Franzò, and F. Priolo, *Nature* **408**, 440 (2000).
2. L. Khriachtchev, M. Rasanen, S. Novikov, and J. Sinkkonen, *Appl. Phys. Lett.* **79**, 1249 (2001).
3. K. Luterova, I. Pelant, I. Mikulskas, R. Tomasiunas, D. Muller, J.-J. Grob, J.-L. Rehspringer, and B. Honerlage, *J. Appl. Phys.* **91**, 2896 (2002).
4. L. Dal Negro, M. Cazzanelli, L. Pavesi, S. Ossicini, D. Pacifici, G. Franzò, F. Priolo, and F. Iacona, *Appl. Phys. Lett.* **82**, 4636 (2003).
5. K.D. Hirschman, L. Tsybeskov, S.P. Duttagupta, and P.M. Fauchet, *Nature* **384**, 338 (1996).
6. N. Lalic and J. Linnros, *J. Lumin.* **80**, 263 (1999).
7. P. Photopoulos and A.G. Nassiopoulou, *Appl. Phys. Lett.* **77**, 1816 (2000).
8. G. Franzò, A. Irrera, E.C. Moreira, M. Miritello, F. Iacona, D. Sanfilippo, G. Di Stefano, P.G. Fallica, and F. Priolo, *Appl. Phys. A: Mater. Sci. Process.* **74**, 1 (2002).
9. A. Irrera, D. Pacifici, M. Miritello, G. Franzò, F. Priolo, F. Iacona, D. Sanfilippo, G. Di Stefano, and P.G. Fallica, *Appl. Phys. Lett.* **81**, 1866 (2002).
10. F. Iacona, G. Franzò, and C. Spinella, *J. Appl. Phys.* **87**, 1295 (2000).
11. M. Catalano, M.J. Kim, R.W. Carpenter, K. Das Chowdhury, and J. Wong, *J. Mater. Res.* **8**, 2893 (1993).
12. H.R. Philipp, *J. Non-Cryst. Solids* **8-10**, 627 (1972).
13. F. Iacona, C. Bongiorno, C. Spinella, S. Boninelli, and F. Priolo, *J. Appl. Phys.* **95**, 3723 (2004).

Mat. Res. Soc. Symp. Proc. Vol. 817 © 2004 Materials Research Society   L6.13

Effects of Annealing Atmosphere on the Characteristics and Optical Properties of SiON Films Prepared Plasma Enhanced Chemical Vapor Deposition

Ki-Jun Yun, Dong-Ryeol Jung, Sung-Kil Hong, Jong-Ha Moon, Jin-Hyeok Kim

Center for Photonic Materials and Devices
Department of Materials Science and Engineering, Chonnam National University
300 Yongbong-Dong, Puk-Gu, Kwangju 500-757, South Korea

ABSTRACT

SiON thin films were deposited by plasma-enhanced chemical vapor deposition method at 350 °C using N_2O/SiH_4 gas mixtures as precursors. As-deposited SiON films were annealed in different gas atmospheres (air, N_2, and O_2) and at different annealing temperatures (800 °C ~ 1100 °C). Effects of annealing atmosphere on the Si-O, Si-N, Si-H, and N-H bonding characteristics in SiON films and their structural and optical properties have been investigated. Cross-sectional and planar microstructures were characterized by scanning electron microscopy and atomic force microscopy, and crystallinity was investigated by X-ray diffraction. Chemical bonding characteristics and optical properties SiON films were studied using fourier transform infrared spectroscopy and prism coupler. X-ray diffractions showed no evidence of any crystals in all SiON films. The deposition rate strongly depended on the processing parameters such as radio frequency (rf) power, N_2O/SiH_4 flow ratio, and SiH_4 flow rate. Deposition rate increased as N_2O/SiH_4 flow ratio increased and SiH_4 flow rate increased. It was possible to obtain SiON films with surface roughness of about 1 nm and a high deposition rate of about 4 μm/h when the processing parameters were optimized as rf power of 200 W, N_2O/SiH_4 flow ratio of 3, SiH_4 flow rate of 100 sccm. It was observed that the intensity and the shift of the Si-O stretch and Si-N peaks depended on the annealing atmosphere as well as the annealing temperature. The intensity of Si-O peaks increased in the samples annealed in oxygen atmosphere, but it decreased in the samples annealed in nitrogen atmosphere. The intensity of Si-N peak decreased in the samples annealed in oxygen atmosphere, but it increased in the samples annealed in nitrogen atmosphere. The position of Si-O peaks shifted from 1030 nm to 1140 nm in the samples annealed both in oxygen and in nitrogen atmosphere. It was also observed that the intensities of Si-H (~2250 cm^{-1}) and N-H (~3550 cm^{-1}) peaks decreased apparently as the annealing temperature increased in all annealed samples.

INTRODUCTION

Planar optical waveguides have been received great interest in the fields of optical communication applications for many years by many companies and universities [1,2]. In the planar optical waveguide, the optically guiding core layer is sandwiched between a lower and an upper cladding layer. Generally, doping with Ge, P, or Ti is done to increase the refractive index of the core layer relative to the cladding layers. The relative difference of the refractive index between the core and the cladding layers is called "index contrast" and is an important parameter for controlling many optical properties of the waveguide [3]. Waveguides with a much higher index contrast are required to decrease the size of a device and to increase the density of devices in a wafer. This has leaded the development of the new core material that has high refractive index in recent years, especially, silicon oxynitride (SiON) [4-5]. The high refractive index contrast waveguide technology based on silicon oxynitride (SiON) makes it possible to fabricate waveguides with considerably higher index contrasts and thus much smaller waveguide bending radii.

Plasma enhanced chemical vapor deposition (PECVD) technique has been used to obtain amorphous non-stoichiometric SiON films from silane and nitrous oxide in recent years [5]. There have been many reports on the preparation of SiON thin films with high deposition rate using PECVD. The fundamental advantage of this technique is that it is easy to control the structural, mechanical and optical properties of the films deposited by adequately adjusting the deposition parameters [6-8]. However, this technique has serious problem of remaining Si-H or N-H bonds, which give detrimental effects on the optical property of the film, inside the film even after the post annealing process. Therefore, it is needed to study systematically the post-annealing effect on the remaining bonding characteristics inside the SiON films.

In this study, we prepared SiON films by PECVD and investigated basic properties of SiON films such as refractive indices and microstructure. Especially, we have focused on the effect of the post-annealing condition that varies by changing the temperature and ambient on the chemical bonding characteristics inside the SiON films.

EXPERIMENT

The SiON films were deposited by the PECVD technique from appropriate gaseous mixtures of silane (10% SiH_4, diluted in N_2) and electronic grade nitrous oxide (99.999% N_2O), in a capacitively coupled reactor. The plasma was activated by 13.56 MHz radio frequency (rf) signal applied via a matching box to two 400-cm^2 parallel grids. The substrate

Table 1. Parameters for the deposition of SiON films using PECVD

Parameters	Value
RF power (W)	100 ~ 200
SiH_4 flow rate (sccm)	50 ~ 200
N_2O/SiH_4 flow ratio	3 ~ 12
Pressure (torr)	0.2 ~ 0.5

holder lies below the grids at 2.35 cm distance, in order to minimize ion-bombardment. All the films were deposited on one-side polished p-type (14-20 Ohm cm) silicon (100) wafers. SiON films with different nitrogen, silicon, and oxygen contents were obtained by varying the process parameters such as, N_2O/SiH_4, SiH_4 flow rate, rf power, and working pressure as shown in table 1. The deposited samples were annealed at different temperatures (800 °C ~ 1100 °C) and in different ambient (Air, N_2, or O_2) for 1 hour.

Crystallinity and microstructure were characterized by X -ray diffraction (XRD) (X'pert-PRO, PHILLIPS, Netherland), scanning electron microscopy (SEM) (S-4700, HITACHI, Japan), and atomic force microscopy (Nanoscope IV, Digital Instrument, USA). A prism coupler (MODEL-1550, Fi-ra, Korea), having a He-Ne laser at a wavelength of 632.8nm as a light source, was employed to obtain the thickness and the refractive index of the samples. The compositional and chemical bonding properties of the as-grown and annealed samples were characterized using Fourier transform infrared (FT-IR) spectrometer (NICOLET 520T, Nicolet instrument, USA)

RESULT AND EXPERIMENT

A number of SiON films were grown in this study by changing process parameters that vary as shown in table 1. While trying to find the best process condition in which the SiON films can be grown with highest deposition rate and without any artifacts, it was found that the deposition rate strongly depends on the process parameters. The deposition rate increased linearly as the silane flow or rf power increased. Little change in the deposition rate was observed as the N_2O/SiH_4 flow ratio or substrate temperature was changed. It was also observed that the refractive index increased linearly as the rf power or N_2O/SiH_4 ratio was increased. It remained almost unchanged as temperature or silane flow rate was changed. Among these experimental conditions, we selected the optimized experimental condition for the deposition of SiON thin films to investigate the post-annealing effect. Therefore, SiON films that have a refractive index of 1.46798 and a thickness of 3.78 μm were deposited with process parameters of 200 mTorr, 200 W, 350 °C, and a N_2O/SiH_4 flow ratio of 3 in this study.

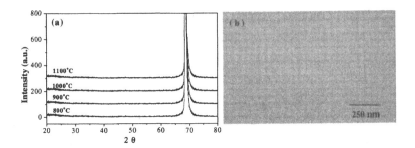

Figure 1. X-ray diffraction patterns of SiON thin films annealed at various temperatures from 800 °C to 1100 °C for 1 h in air ambient (a) and a plan-view SEM micrograph of the SiON film annealed at 1100 °C for 1h in O_2 ambient.

Figure 1 shows XRD patterns of SiON thin films annealed at various temperatures from 800 °C to 1100 °C for 1 h in air ambient (a) and a plan-view SEM micrograph of the SiON film annealed at 1100 °C for 1h. No diffraction peaks except the Si substrate peak are observed in the XRD pattern, which indicates that SiON films are amorphous phase even after the annealing process. Very smooth surface microstructure and no contrast difference that shows the microstructural change is observed in the SEM image (figure 1(b)). This result is well consistent with the XRD result.

Magnified FT-IR spectra shows that peaks for the Si-H bonds of the as-deposited SiON film and of the SiON films annealed at different temperatures and in different ambient such as air (a), O_2 (b), and N_2 (c) are shown in figure 2. It is clearly observed that peak intensities of the Si-H bonding in all samples decrease as the annealing temperature increases. It was also observed that the intensity of peaks for N-H bond decreased as the annealing

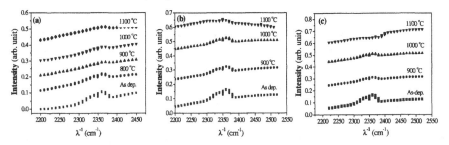

Figure 2. Magnified FT-IR spectra of the as-deposited SiON film and of SiON films annealed at different temperatures and ambient such as air (a), O_2 (b), and N_2 (c).

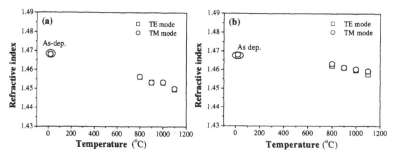

Figure 3. The change in the refractive index of SiON films, annealed in different ambient of O_2 (a) and N_2 (b), as a function of annealing temperature.

temperature increased. (data are not shown here) The effect of the post-annealing conditions on the refractive index of SiON films were also investigated.

Figure 3 shows the change in the refractive index of SiON films, annealed in different ambient of O_2 (a) and N_2 (b), as a function of annealing temperature. Refractive index decreases as the annealing temperature increases in all samples. These results are well consistent with reported else where [9]. It is interesting to see that there is difference in decrease of refractive index between the SiON film annealed in O_2 ambient and the film annealed in N_2 ambient. The refractive index of SiON films annealed in O_2 ambient decreases from 1.46798 to 1.44968 and that in N_2 ambient changed from 1.46798 to 1.45921. This result shows that it is more efficient in keeping the refractive index of SiON films to anneal the SiON film in N_2 ambient rather than in O_2 ambient.

We also investigated the intensity change of the Si-O stretching peaks in SiON films annealed in different ambient as a function of annealing temperature because the Si-O stretching peak may cause mechanical stress inside the film[10]. Figure 4 shows the intensity change of the Si-O stretching peaks in SiON films, annealed in O_2 ambient (a) and in N_2

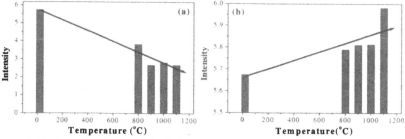

Figure 4. The intensity change of the Si-O stretching peaks in SiON films, annealed in O_2 ambient (a) and in N_2 ambient (b), as a function of annealing temperature.

ambient (b), as a function of annealing temperature. The intensity values were obtained from the FT-IR spectra at around 1050 cm^{-1}. It is observed that the intensity of the Si-O stretching peak increases in SiON films annealed in O_2 ambient and it decreases in N_2 ambient. Details on the relationship between the intensity change of Si-O stretching peak and the mechanical stress remaining inside the film will be studied further.

CONCLUSION

The effects of silicon oxynitride films in different ambient were investigated. Irrespective of the annealing atmosphere, Si-H and N-H bonds are disappeared as increasing the annealing temperature (800 ~ 1100). Also the refractive index of films annealed at atmospheric pressure in a steam-nitrogen ambient shows more stable than in a steam-oxygen and atmosphere ambient. We guess that the annealing ambient and temperature are large effective for the mechanical stress of SiON films.

REFERENCE

1. Y. P. Li, C. H. Henry, *IEE Proc. Optoelectron.* **143,** 263 (1996).
2. N. Takato, K. Jinguji, M. Yasu, H. Toba, M. Kawachi, *IEEE J. Lightwave Technol.* 6, 1003 (1988).
3. E. Voges, H. Bezzaoui, M. Hoffmann, *in Proc. 6th Europ. Conf. on Integr. Optics.* " *ECIO '93,* " *Neuchatel*, Switzerland, P. Roth, Ed., 12.4-12.6 (CSEM, Neuchatel, 1933).
4. H. Albers, L.T.H. Hilderink, E.Szilagyi, F.Paszti, P.V.Lambeck, Th. J. A. Popma, *in proc. IEEE lasers and Electro-Optical Society Annual Meeting "LEOS '95,"* **Vol.2,** 88 (1995).
5. B.J. Pffrein, G.L. Bona, R. Germann, F.Krommendijk, I. Massarek, H.W.M. Salemink, *in proc. 1996 Symp. IEEE/LEOS Benelux Chapter, A. Dressen, R.M. de Ridder*, Eds. **290** (1996).
6. C.F.Lin, W.T.Tseng, M.S.Feng, *J.Appl.Phys.* **37(Pt. 1),** 6364 (1998).
7. C.Dominguez, J.A.Rodriguez, F.J.Munoz, N.Zine, *Vacuum* 52, 395 (1999).
8. J.Viard, E.Beche, D.Perarnau, R.Berjoan and J.Durand, *Journal of the European Ceramic Society,* 2025-2028 (1997).
9. Y.T. Kim, S.M. Cho, Y.G Seo, H.D. Yoon, Y.M. Im, D.H. Yoon, *Surface and Coatings Technology*, (2003).
10. C.M.M.Denisse, K.Z.Troost, F.H.P.M.Habraken, W.F. van der Weg, and M.Hendriks, *J. Appl.Phys.* **60,** 2543 (1986).

New Concepts and
Devices

Mat. Res. Soc. Symp. Proc. Vol. 817 © 2004 Materials Research Society   L5.3

Hybrid Integrated Microphotonics and It's Applications

Suntae Jung *and Taeil Kim
Photonics Solution Laboratory, Telecommunication R&D Center,
Samsung Electronics Co., Ltd. Dong Suwon P.O. Box 105, Suwon City, Korea 442-600.

ABSTRACT

Hybrid Integration of passive and opto-electronic devices is emerging as a key technology of optical component, because it can increase functionality and reliability of optical device module, lowers the packaging cost and enables automated manufacturing. Main applications of hybrid integrated microphotonic devices are FTTH system and metro access network.

There are various technical issues in hybrid integration modules. It requires connecting an active device to a passive waveguide. The embedded functional devices such as grating, coupler, switch and filter are essential elements to improve optical performance of integrated devices. The package solution for low cost and small size is also required.

We have developed SSC LD(Spot Size Converted Laser Diode) and RMF PD(Reflection Mirror Facet Photo Diode) for connecting to planar waveguide. Simple and cost-effective silica PLC platform with terraced-silica, PLC grating and coupler have been developed. Using these technologies, we have made bi-directional diplexer, triplexer and ECL(External Cavity Laser).

INTRODUCTION

As similar to the progressive of electronic device, photonic device has been developed from discrete device to integrated module. Recently, integration of all optical devices is emerging as a key technology of optical component, because it can increase functionality and reliability of optical device module, lowers the packaging cost and enables automated manufacturing[1,2]. Indeed, to widely adopt optics into the consumer market like as FTTH(the fiber-to-the-home), integration technique should be need to reduce the cost of module.

Monolithic integration has been studied for this purpose using compound semiconductor and silicon by several research groups[3,4]. However, laser diode and photo diode are still the hurdle to overcome for telecommunication applications. As a alternative method, hybrid integration has been studied and some devices have been commercialized[1,2].

Hybrid integration of passive and optoelectronic devices requires several techniques for increasing coupling efficiency. We have developed cost-effective silica PLC platform with terraced-silica. For connecting to planar waveguide, SSC LD(Spot Size Converted Laser Diode) and RMF PD(Reflection Mirror Facet Photo Diode) have been also developed.

To increase functionality of the hybrid integration, several passive devices such as direction: coupler, PLC grating, and AWG(arrayed waveguide grating) are used for several applications which are bi-directional transceiver and ECL(External Cavity Laser).

In this paper, we described the more detail results of our R&D activity for hybrid integrated microphotonics.

THE COUPLING EFFICIENCY OF HYBRID INTEGRATION

Hybrid integrated modules use PLC platform to connect an active device to a passive waveguide. Representative PLC platform has used silica-on-terraced-silicon structure [5]. However, this platform requires complicated fabrication processes that make the platform expensive. We proposed a novel fabrication method of silica PLC platforms for hybrid integration. This method is very simple without degradation of performance and does not need additional equipments than conventional silica PLC fabrication except metallization(evaporating plating) equipments. Figure 1. shows the fabrication sequence of the terraced-silica platform for hybrid integration.

Underclad and core silica layer are deposited by FHD on silicon wafer. Waveguides are patterned by dry etching process. Alumina layer is deposited as etch stopper by E-beam evaporator. Alumina have good selectivity against silica during deep dry etching and maintains its integrity after overcladding process. Overclad silica layer is deposited about 25μm. Sintering temperature is above 10008C.

Terraced-silica structure is made by deep dry etching(~30μm). Deep dry etching with OES end-point-detector exposes alumina etch-stop layer and over etching process is continued to make terraced-silica structure. Over etching determines the height of terraced-silica but do not change the reference surface. Electrode patterns are fabricated by evaporation and lift-off metho Solder layer is fabricated by electroplating. Solder height was controlled to 0.5~1μm higher than terraced-silica surface. OE-devices such as LD and PD are bonded using flip chip bonding technique.

The one of the main issues for platform is lateral and vertical position errors. Lateral position error causes from lithography error and flip chip bonding error. The terraced silica structure uses waveguide align marker that is made when making waveguide with same photo-mask. The

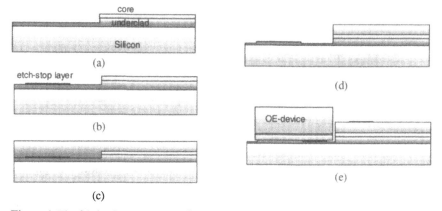

(a)

etch-stop layer

(b)

(c)

(d)

OE-device

(e)

Figure 1. The fabrication sequences of terraced-silica platform. (a) underclad and core deposition, (b)waveguide patterning and etch-stop layer deposition, (c) overclad deposition, (d) deep etch, (e) electrode patterning, soldering and flip chip bonding of OE-devices

bottom surface of OE-devices and top surface of terraced silica was contacted for accurate waveguides alignment. The lateral position error mainly caused from flip chip bonding error. Bonding condition and sensing error of the equipment effect on the lateral position errors. We can control the lateral position error less than 1.5µm which is equal to 1dB tolerance as shown in figure 2(a). For accurate vertical alignment the position of terraced-silica surface relative to the waveguide center must be precisely controlled. The factors causing position error are core silica thickness and waveguides patterning depth. Deep etching process and solder height do not cause position error because etch-stop layer inhibits etching of terraced region and resulting terraced-silica also acts as a mechanical stop. The vertical position can be controlled within 0.5µm.

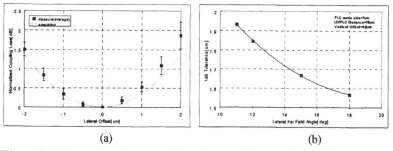

(a)   (b)

Figure 2. (a) the coupling loss according to lateral offset and (b) 1dB tolerance as to lateral far field angle of SSC LD

155

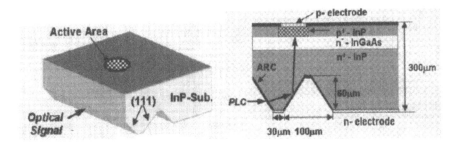

Figure 3. The schematic diagram of reflecting mirror facet photodiode(RMF-PD) [6].

To minimize mode size mismatch between LD and waveguide, the FFA(far field angle) of laser diode should be reduced. The figure 2(b) shows simulation results of 1 dB tolerance as to FFA of SSC LD. We have developed SSC LD with tapering InGaAsP ridge waveguide. The typical FFA is less than 13x15°. This LD shows good reliability and coupling efficiency which is less than 4dB.

We suggested reflecting mirror facet (RMF) photodiode[6]. This is composed of two v-grooves and a conventional vertical PD as shown in figure 3. RMF PD is designed for increasing coupling efficiency from waveguide to photodiode. The two adjacent v-grooves convert a horizontal beam to nearly a vertical beam as like arrows shown in figure 3. The fabrication RMF-PD showed a dark current less than 0.1nA at 5V. The chip capacitance was 0.6 pF which was identical to the conventional PDs with the same layer structure. The responsivity was very high, 0.85 A/W at 1.55 μm light.

PLANAR INTEGRATED CIRCUITE AND INTEGRATED DEVICES

Directional Coupler & Bidirectional Transceiver

We propose a new silica-PLC WDM filter based on double polynomial curve [7] directional couplers (DC's), which shows good performances for the Ethernet-PON (E-PON) applications

The schematic diagram of our PLC WDM filter is described in figure. 4. It is based on double polynomial curve DC's. Its features are as follows. First, in the coupling region (i.e. in coupled waveguides), the core width (Wc) becomes sufficiently narrow (Wc<W=6.5 μm) to support a single mode operation. The single mode waveguides in the coupling region can reduce

156

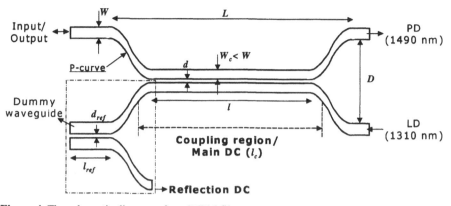

Figure 4. The schematic diagram of our WDM filter.

the unnecessary noise and crosstalk comes from the optical power of higher modes. Higher modes are either excited at the bending region in the DC or can be directly fed in the input waveguide excited at other bending waveguides in the PLC platform. Therefore, the single-mode guiding in the coupling region can be said to make the WDM filter more immune against other platform geometries. Therefore, the design flexibility can be increased. We found that this feature can reduce the length of parallel waveguides in the coupled region by more than 30 % as well as improve the BXT(bidirectional crosstalk) up to 10 dB without additional losses.

Secondly, the bending path was configured with a polynomial curve path design so that single curved waveguide element can perform both tapering and bending functions and thus significantly simplifies the design and drawing processes. Also, the introduction of polynomial curve element reduced about 10 % of connecting waveguides and thus 5 % of entire filter length. Finally, a reflection DC has been introduced to reduce the BXT and improved the level of the BXT by about 5 dB. By adopting these features, that is double polynomial curve DC's in addition to the single mode coupled waveguides, we could make a WDM filter which meets the specification of bidirectional transceiver.

One preliminary schematic design of an entire bidirectional transceiver module is shown in figure 5. Within a chip as small as 8.5 mm × 3.7 mm, an LD, monitor and receive PD's, a TIA (transimpedance amplifier), and the reflection structure (devised to avoid the electrical crosstalk between LD and PD) could be integrated with our WDM filter.

We hope that our filter and especially its design guides can be used in the design and fabrication of not only bidirectional transceiver modules but also other PLC devices in WDM communication systems.

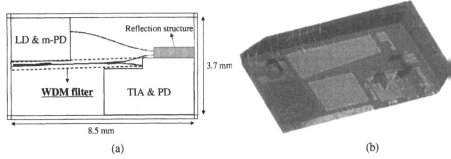

(a)   (b)

Figure 5. A transceiver platform design (a) schematic diagram, (b) OE-devices integrated on platform

PLC grating and ECL

PLC gratings were made using the phase mask technique. The Ge-doped silica planar waveguides were hydrogenated at room temperature under 100atm pressure for 4 days to increase the photosensitivity. As experimental parameter, the laser average fluence per pulse F_p was controlled. According to the coupled-mode theory [8], the peak reflectance R and the bandwidth are expressed by

$$R = \tanh^2[(\eta\pi L\Delta n)/\lambda)] \tag{1}$$

and

$$2\Delta\lambda = (\lambda^2/\pi n_{eff}L)\sqrt{(\eta\pi L\Delta n/\lambda)^2 + \pi^2}, \tag{2}$$

respectively. Here, η is the fraction of the single mode intensity confined in core, which is estimated to be 0.85 for our gratings, Δn is the refractive index modulation and n_{eff} is the effective refractive index of the core. According to the one-photon absorption model and the power law, Δn as a function of t is given by

$$\Delta n = \Delta n_{max}[1 - \exp(-AIt)] \tag{3}$$

and

$$\Delta n = Ct^b, \tag{4}$$

respectively, where Δn_{max} is the maximum refractive index modulation and I is the intensity of the UV light.

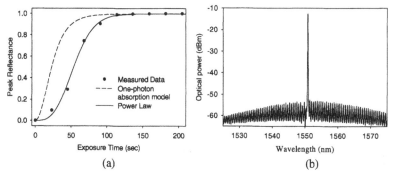

(a)   (b)

Figure 6. (a) Peak reflectance of PLC grating as a function of exposure time. (b) Oscillation spectra of the FP-LD with a PLC grating as an external reflector.

Figure 6(a) shows the peak reflectance of the PLC gratings versus exposure time. The dashed curve is the fit predicted by the one-photon absorption model with $\Delta n_{max}=8.6\times10^{-4}$, which was measured in this work. We obtained the solid curve by fitting Δn to (4) with $C=2.3\times10^{-7}$ and $b=1.53$. While the one-photon absorption model is not suitable for describing the measured data, the power-law can predict well the measured data as shown in figure 6(a).

The bandwidth and reflectance of PLC grating as external reflector for the ECL were designed to be 1.8nm and 0.55, respectively. The measured bandwidth and reflectance were ~1.9nm and ~0.55, respectively.

The FP-LD was spot-size converter integrated LD (SS-LD) having low coupling loss [9] and the cavity length was 600μm. The rear facet of the FP-LD was coated with 80% of reflection and the front facet was antireflection (AR) coated with 1% of reflection to minimize the parasitic reflections [10]. The total external cavity length L_c was ~2.1mm including air gap. Figure 6(b) shows that, a single longitudinal mode oscillation was stabilized at Bragg wavelength with side mode suppression ratio (SMSR) of ~40dB by growing a Bragg grating in a PLC waveguide.

CONCLUSION

We report our recent development of hybrid integration and some applications. The silica terraced platform, SSC LD and RMF PD are developed for increasing coupling efficiency and accuracy of passive alignment.

PLC WDM filters based on double polynomial-curve directional couplers, which will be integrated into the compact and low-cost bidirectional optical transceivers for FTTH systems.

Considering the growth characteristics, we designed and fabricated the PLC gratings as external reflectors for the ECL. We believe the hybrid integration will be a general technology for integrated module.

REFERENCES

1. Y.Yamada, et al., *Optical Engineering* **28**, 1281-1287 (1989).
2. C. H. Henry, G. E. Blonder, R. F. Kazarinov, *IEEE J. Lightwave Technol.* **LT-7**, 1530-1539 (1989).
3. L. Pavesi, *J. Phy.:Condens. Matter* **15**, R1169-R1196 (2003).
4. H. Wong, *PROC. 23rd International Conference on Microelectronics* **1**, 285-292 (2002).
5. Y. Yamada, et al., *Electron. Lett.* **29**, 444-445 (1993).
6. S. Yang, H. Kang, B. Jeon, D. Rhee, Y. Kim, E. Lee, A. Choo, J. Burm and T. Kim, ECOC'03, Rimini, Italy, We4.P.80 (2003).
7. F. Ladouceur and P. Labeye, *IEEE J. Lightwave Technol.* **13**, 481-492 (1995).
8. T. Erdogan, *IEEE J. Lightwave Technol.* **15**, 1277-1294 (1997).
9. Y. Tohmori, Y. Suzaki, H. Oohashi, Y. Sakai, Y. Kondo, H. Okamoto, M. Okamoto, Y. Kadota, O. Mitomi, Y. Itaya, and T. Sugie, *Electron. Lett.* **12**, 1838-1840 (1995).
10. F. L. Gall, S. Mottet, N. Devoldere, and J. Landreau, ECOC'98, Madrid, Spain, 285-286 (1998).

Mat. Res. Soc. Symp. Proc. Vol. 817 © 2004 Materials Research Society   L5.6

Optoelectronic Simulation of the Klein Paradox based on Negative Refraction Phenomenon

Durdu Ö. Güney[1,2], David A. Meyer[1]
[1]Department of Mathematics, University of California, San Diego, 9500 Gilman Drive, La Jolla, California 92093-0112
[2]Department of Electrical & Computer Engineering, University of California, San Diego, 9500 Gilman Drive, La Jolla California 92093-0407

ABSTRACT

Having shown elsewhere [1] that the Klein paradox for the Klein-Gordon (KG) equation of spin-zero particle manifests exactly the same kind of wave propagation and negative refraction phenomena, which also exist in the scattering of TM (transverse-magnetic) –polarized electromagnetic (EM) wave incident on a left handed medium (LHM), we show in this paper that it is possible to simulate the Klein paradox, using this peculiar feature of LHMs. Real time control and processing of certain quantum systems, involving controlled pair production rate and distribution, among others, could be achieved by this optoelectronic simulator using appropriate transformations and approximations.

INTRODUCTION

In 1968, Veselago [2] demonstrated that the refractive index (n) of materials should be negative when permittivity (ε) and permeability (μ) are simultaneously negative. In his paper, Veselago called such materials left-handed materials (LHMs) and treated them purely formally, since no such materials had been discovered or created at that time. Nonetheless, he gave a number of arguments about how one should look for such materials. He pointed out that the Doppler effect, Cerenkov radiation, and even Snell's law are inverted. He also considered some questions related with the physical realization of materials with $\varepsilon < 0$ and $\mu < 0$. His work undoubtedly raised interest in the phenomena of LHMs.

It has been recently demonstrated that an effective LHM, which consists of periodic array of split ring resonators (SRRs) and continous wires manifests $n < 0$ for a certain frequency region in the microwave regime, which then leads to anomalous electromagnetic wave propagation [3-4]. An experimental verification of a negative index has been established by R. A. Shelby et al [5].

Subsequently, we have observed [1] that there is a situation in relativistic quantum theory that also results in similar anomalous wave propagation. The interaction of a relativistic quantum particle with a strong step potential displays a peculiar phenomenon, known as the Klein paradox. One interesting outcome of the paradox is classically forbidden transmission through strong potential barrier, where the height of the barrier is larger than the total energy of the incident particle. Furthermore, negative transmittance is compensated by the reflectance exceeding unity to conserve the probability. This is the essence of the Klein paradox. The paradox is resolved by employing the notion of particle-antiparticle pair production due to the strong potential. The antiparticles create a negative charged current moving away from the interface, while the particles are reflected and combined with the incident beam leading to a positively charged current moving towards the interface. As a result of this fact, group and phase velocities in the potential step become antiparallel, which results in negative refraction in two dimensions [1], analogous to the peculiar refraction phenomenon in LHMs. It is this effect which motivates us to propose the simulator described below. In this work, we show that the spatial profile of the antiparticle wavefunction (or probability density) in the Klein potential could be reproduced in LHM using a TM-polarized electric field in the form of an electric field (or intensity) distribution. We then explain how LHMs could be exploited to simulate systems in which the Klein paradox manifests itself.

THEORY

Following suggestions in Refs [3,6-8] about wave propagation in LHMs, we assume the geometry shown in Fig. 1 in order to study the wave refraction at the interface between two semi-infinite layers of optical media. We consider causal plasmonic forms [7] for ε and μ, which lead to the refractive index for the LHM given by

$$n(\omega) = 1 - \omega_p^2 / [\omega(\omega + i\Gamma)], \tag{1}$$

where ω_p is the plasma frequency. We assume the incident electric field to be a plane wave and linearly polarized along the y-axis. It can be shown by the phase matching condition and causality for the refracted wave that we have

$$Q_L = |K_L|_x x + \sigma\{n^2|K_L|^2 - |K_L|_x^2\}^{1/2} z \tag{2}$$

$\sigma = +1$ for the positive index medium and -1 for the NIM (or LHM) [6]. From the definition of group velocity for isotropic, low loss materials, the refracted beam has group velocity

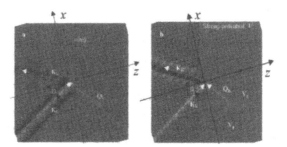

Figure 1. (a) Negative refraction of a beam of photons at the interface between a PIM ($z < 0$) and a NIM ($z > 0$). The white arrows indicate the wavevector for an arbitrary wave (K_L) in the incident flux and its corresponding transmitted (Q_L) and reflected (K_L') wavevector. The angle between the boundary normal and K_L is Θ_i. (b) Negative refraction of a beam of spin-zero particles at the boundary of a strong potential, V, for $z > 0$. White arrows indicate wave vectors for an arbitrary wave (K_K) component in the incident flux and its corresponding transmitted (Q_K) and reflected (K_K') counterparts. The wave vector shown dark grey is not allowed due to the causality and phase matching conditions at the interface [1]. The light grey arrows show the directions of group (V_g) and phase (V_p) velocities, which are antiparallel. The angle between K_K and the normal to the interface is θ_i.

$$V_g = (Q_L / |Q_L|)(\sigma nc/n_g),$$ (3)

where c is the light speed and n_g is the group index, which is by causality always greater than unity [6]. Thus the group velocity is always antiparallel to the phase velocity of the refracted wave for $\sigma = -1$. Since the time-averaged energy flux $\langle S \rangle = \langle u \rangle V_g$ [7], where $\langle u \rangle$ is the time-averaged energy density, it is clear from Eqns. (2) and (3) and by causality that the incident wave should undergo negative refraction.

Now, we turn our attention to the Klein paradox. Although the non-relativistic scattering of a quantum particle is a straightforward problem, the relativistic case displays a quite peculiar phenomenon called the Klein Paradox [9]. Elsewhere [1] we analyze the scattering of a particle from a step potential in two dimensions and derive some of the associated peculiarities, which include transmission through a strong potential barrier, pair production, negative transmission, as well as negative refraction.

Our analysis is for a spin-0 particle, whose dynamics are governed by the Klein-Gordon equation,

$$[E - V(\mathbf{r})]^2 |\Psi\rangle = [-c^2 \mathbf{P}^2 + m^2 c^4] |\Psi\rangle,$$ (4)

where m is the mass of the particle. A particle confined to the xz-plane with energy $E = [(p^2 c^2 + m^2 c^4)]^{1/2}$ is incident on a potential barrier at angle θ_i from the normal to the interface. The potential $V(\mathbf{r})$ vanishes on the incident side and is forms a step, with $V(\mathbf{r}) = V$ for $z > 0$. We have shown under these circumstances [1] that the Klein paradox for the Klein-Gordon equation of spin-zero particle manifests exactly the same kind of wave propagation and negative refraction phenomena which also exist in the scattering of transverse-magnetic polarized electromagnetic wave incident on a left handed medium (see Fig. 1b).

METHOD

Having compared the two problems, we realize that they both have reversed group and phase velocity and manifest negative refraction. Moreover, we have the mathematical expressions above, which connect these two problems, and suggest that one might simulate the other. The question we would like to address now is how to simulate the Klein paradox using LHMs? Although this is possible under many different transformations, here, we only consider the following transformation to simulate the Klein paradox. To be more precise in some results or to solve other specific problems, similar but different transformations based on the same argument could also be applied.

We match the input and output (I/O) in the LHM case to the I/O of the Klein case, respectively, which basically includes the dispersion relation and the phase velocity. This leads to the transformation given by

$$E \rightarrow \hbar\omega$$ (5)

$$V \rightarrow (|n|+1)\hbar\omega$$ (6)

where n is given by Eqn. (1). These arrows should be understood as follows: The left-hand side gives us the physical properties in Klein problem, in terms of the physical properties of the

LHM. This transformation, however, is correct only in the high-energy limit of $mc^2 \to 0$ and $|V_g|$ $\to c$, independent of any group refractive index.

A wave with angular frequency ω incident on an LHM results in exactly the same kind of wave propagation and refraction as that of its equivalent Klein counterpart, which is characterised by the transformations (5) and (6). Since the corresponding parameters $(E, V,$ etc.) are on the order of the energy of a typical photon, they have to be scaled up to describe the Klein system with higher energies. Thus, having obtained the power transmission and reflection coefficients for the LHM, it is a simple matter to find the equivalent pair production probabilities regardless of any scaling factor, since these coefficients are invariant under such scaling.

From the experimental point of view, once we obtain the power coefficients for LHM, their Klein counterparts can be found from

$$|T_K|/T_L \equiv T(\omega,\theta_i) = (1/|\mu|)[\mu s - a]^2/[s-a]^2 \tag{7}$$

$$R_K/R_L \equiv R(\omega,\theta_i) = [(s+a)(\mu s - a)]^2/[(s-a)(\mu s + a)]^2 \tag{8}$$

s and a are defined as

$$s \equiv \sin(\theta_i) \tag{9}$$

$$a \equiv [|n|^2 - \cos^2(\theta_i)]^{1/2} \tag{10}$$

Note that these ratios depend only on the applied frequency and incidence angle. This is remarkable for the physical realization of such a simulator.

Our proposed hardware implementation of this simulator is illustrated at large scale in Fig. 2. (More detailed architecture design issues are beyond the scope of this paper.) Since the reflection coefficient in the Klein paradox is bigger than one, normally it is not possible to obtain such a reflection coefficient using LHM. This problem can be overcome, however, by using an amplitude modulator (AM) in the first block in Fig. 2. AM, which is also driven by a MEMS actuator, modulates the input signal between a peak and minimum value in the form of a square wave. Thus the intensity of the input signal is modulated between $I_0R(\omega,\theta_i)$ and $I_0T(\omega,\theta_i)$ in a square wave manner where I_0 is a constant. If the modulation period is T_m, half of the period the transmitter sends a signal with intensity $I_0T(\omega,\theta_i)$ which results in exactly the same spatial pattern in the LHM as that of antiparticles under the Klein potential, and half of the period it sends a signal which results in reflection greater than one and exactly same spatial pattern as that of the reflected particles due to the Klein step. Since we have divided T_m into two time slots, the sampling time and response speed for the detectors are important.

RESULTS AND CONCLUSION

After the modulation the problem is ready to be simulated by the LHM, whence its output is detected by the pixel arrays in the blue block and *post-processed* by means of analog to digital converters (A/D) and logic circuit to get the output shown in Fig. 3.

The essential point to note in this operation is that although some digital operations take place, the problem is indeed solved by the LHM. More complicated scenarios under these (or other) transformations with (or without) some approximations may also be possible. The

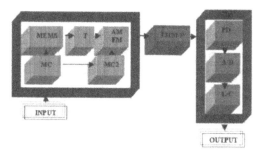

Figure 2. Block diagram for the simulator, which roughly illustrates the components of the simulator. The quantum problem is defined and pre-processed in the first block and then sent to the second block. Having been solved there, it is post-processed in the third block, before obtaining output in fig. 3b. The simulator mainly consists of microcontrollers (MC), microelectromechanical systems (MEMS) actuator, transmitters (T), amplitude (AM) and frequency modulators (FM), LHM, photodetectors, analog to digital converters (A/D) and logic circuits (L/C).

simulations above could also be performed by positive index materials (PIMs). This would, however, require additional circuitry and logic operations. One drawback of the NIM simulation, on the other hand, is a dynamic range problem. For more complex problems, alternative types of NIMs with larger dynamic range would be desirable. Photonic crystals are a more promising technology for creating such NIMs.

Comparing the cost for a typical high energy physics experiment to that of the optoelectronics experiment, such a processor would be priceless provided that the above transformations are extended and/or shaped according to the need. In this paper, we do not consider how to simulate a specific relativistic quantum field theory scenario, but rather present a basic approach for using LHMs to simulate a general class of problems, some of which have (or might have) arisen in connection with the Klein paradox. This kind of simulating device, performing operations based on the mathematical connection between these two systems, could also be helpful in controlling some related quantum processes in real time, such as controlling pair production rate, distribution, etc.

In conclusion, having shown elsewhere [1] that the Klein paradox for the spin-0 KG equation exhibits exactly the same peculiar wave propagation and refraction as occur for a TM-polarized electromagnetic wave incident on an LHM, in this paper we have explained how to simulate the Klein paradox and related problems, as well as discussed possible applications of doing so, using the remarkable properties of LHMs. Our work also highlights an approach to optical computing based on the wave-particle duality in nature. Such optical devices and phenomena could be analog processors and motivating guides for the quantum world, since optical experiments are in general easier and cheaper to realize physically than high energy physics experiments. LHMs and photonic band gap (PBG) materials, which are sometimes called artificial semiconductors for light, are only two kinds of fruits among others harvested from this fertile optics-quantum field.

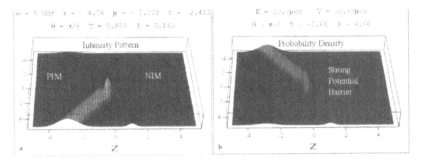

Figure 3. (a) Negative refraction of a gaussian wavepacket with center frequency of 5 GHz, which is incident on a NIM of refractive index -2.412. For an incidence angle of $\pi/6$, the corresponding transmittance T_L and reflectance R_L are 0.855 and 0.145, respectively. The grey-scale show the relative intensity pattern at a random time. An interference fringes occur due to the incident (bottom left) and reflected (top left) wavepackets near the interface. **(b)** Solution to the Klein problem corresponding to the LHM problem in (a)., which is obtained simply by setting $\mu = 1$. In this equivalent problem, the center energy of the wavepacket is 20.7 μeV and the height of the potential barrier is 70.63 μeV. For the same incidence angle of $\pi/6$, the corresponding transmittance T_K and reflectance R_K are -3.66 and 4.66, respectively. Grey-scale show the probability density at the same random time with (a). Here also interference fringes occur near the interface between the incident (bottom left) and reflected (top left) wavepackets.

ACKNOWLEDGEMENTS

This work was supported in part by the National Security Agency (NSA) and Advanced Research and Development Activity (ARDA) under Army Research Office (ARO) grant number DAAD19-01-1-0520.

REFERENCES

1. D. Ö. Güney and D. A. Meyer, to be submitted to Phys. Rev. Lett.
2. V. G. Veselago, *Sov. Phys. Usp.* **10**, 509 (1968).
3. D R Smith, W. J. Padilla, D. C. Vier, S. C. Nemat-Nasser, S. Schultz, *Phys. Rev. Lett.* **84**, 4184 (2000).
4. D. R. Smith, N. Kroll, *Phys. Rev. Lett.* **85**, 2933 (2000).
5. R. A. Shelby, D. R. Smith, S. Schultz, *Science* **292**, 77 (2001).
6. P. M. Valanju, R. M. Walser, A. P. Valanju, *Phys. Rev. Lett.* **88**, 187401 (2002).
7. D. R. Smith, D. Schurig, J. B. Pendry, *App. Phys. Lett.* **81**, 2713 (2002).
8. S. Foteinopoulou, E. N. Economou, C. M. Soukoulis, *Phys. Rev. Lett.* **90**, 107402 (2003).
9. H. Nitta, T. Kudo, H. Minowa, *Am. J. Phys.* **67**, 966 (1999).

Mat. Res. Soc. Symp. Proc. Vol. 817 © 2004 Materials Research Society   L6.23

Carbon Layer as a New Material for Optics

R. Clergereaux[1], D. Escaich[1], S. Martin[2], P. Raynaud[1,a] and F. Gaillard[2]

[1] LGET - Material and Plasma Processes
118, Route de Narbonne 31062 TOULOUSE cedex 04 - France
[2] CEA - LITEN
CEA Grenoble - 17, rue des Martyrs 38054 GRENOBLE cedex 9 - France
[a] raynaud@lget.ups-tlse.fr

ABSTRACT

Plasma enhanced CVD produces carbon layers with various properties which are highly correlated to the different process parameters such as monomer structure, plasma type or plasma power. For example, the modification of monomer (CH_4 to C_4H_{10}) or the plasma source and the increase of plasma power lead to an optical band-gap which runs from 0.9 to 4.3eV, a conductivity from $5 \cdot 10^3$ to $5 \cdot 10^5 S.m^{-1}$ and a refractive index from 1.47 to 2.76. Then, it is able to control the optical and electrical film properties from the external process parameters. This paper will thus be focused on the description of relation between deposition parameters, structural characteristics of material and film properties. These materials can then be used for optoelectronics applications.

INTRODUCTION

Carbon layers (a-CH) are frequently deposited by plasma enhanced chemical vapour processes (PE-CVD). This technology is very promising because of its high control of film quality, its easy integration in current micro-electronics technologies, its low cost, high efficiency and reproducibility. Recently, it has been shown that such layers have various optical and electrical properties [1,2]. Then, as such layers can be easily etched by oxygen-fluorine plasma, they can be used for a great number of optoelectronic components (light source, detector, amplifier, waveguide, optical sensor...).

This paper deals with some correlations using methods such as ellipsometric spectroscopy (ES), Raman scattering and Fourier transform infra-red spectroscopy (FTIR) between the deposition parameters (monomer structure, plasma type and plasma power), the material structure and the film properties (band gap, conductivity and refractive index). The first step is to observe the effect of deposition parameters and, secondly, to highlight some correlations.

EXPERIMENTAL DETAILS

a-CH thin films were deposited on silicon wafer (100, intrinsic) in a capacitively coupled low- (LF, generator 50kHz) or radio-frequency (RF, generator 13.56MHz) plasma reactor. Monomers, in this paper CH_4 or C_4H_{10}, were injected in the chamber at a flow of 100sccm by a shower head gas injector which acts as the anode (spacing between shower and holder electrode of about 20mm). The working pressure was maintained at 1mbar during all the process using a butterfly valve.

Figure 1.a reports the deposition rate evolution with plasma power for the different monomers and plasma types. In contrast with LF condition where the deposition is highly dependant on monomer structure (about 4 times higher for C_4H_{10} than for CH_4), RF plasma leads to the same deposition rate for CH_4 and C_4H_{10}. Moreover, the film characterization by ES reported on figure 1.b shows that both CH_4 and C_4H_{10} based films deposited in RF condition are quite similar compared to the LF one. Then, it appears that plasma types highly influence the monomer dissociation. In RF plasma, CH_4 and C_4H_{10} are completely dissociated (formation of CH_3, CH_2 species and in less quantity CH and C) and then, for both, the deposition mechanisms are the same. On the other hand, in LF plasma, C_4H_{10} is not completely dissociated (C_xH_y species) compared to CH_4 which contains only CH_3 or CH_2 species. Then, these differences play an important role in deposition mechanisms: as the radicals are longer for C_4H_{10} than for CH_4 in LF plasma, it leads to a higher deposition rate.

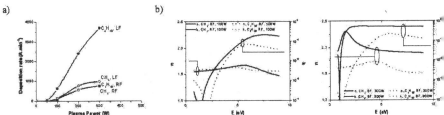

Figure 1: a) deposition rate for CH_4 and C_4H_{10} in LF or RF plasma at different plasma power. b) represents the differences on film properties in RF and LF plasma for CH_4 and C_4H_{10}.

Figure 2 reports the plasma type and plasma power effects on ES measurements. It appears that these two parameters highly affect the films properties. Using Forouhi model [3], it is able to deduce optical and electrical parameters such as the optical band-gap E_g, the conductivity σ and the refractive index n_∞:

$$k(E > E_g) = \frac{A \cdot (E - E_g)^2}{E^2 - BE + C} \qquad \sigma(E) = \frac{nk}{h}E \qquad n(E) = n_\infty + \frac{B_0E + C_0}{E^2 - BE + C}$$

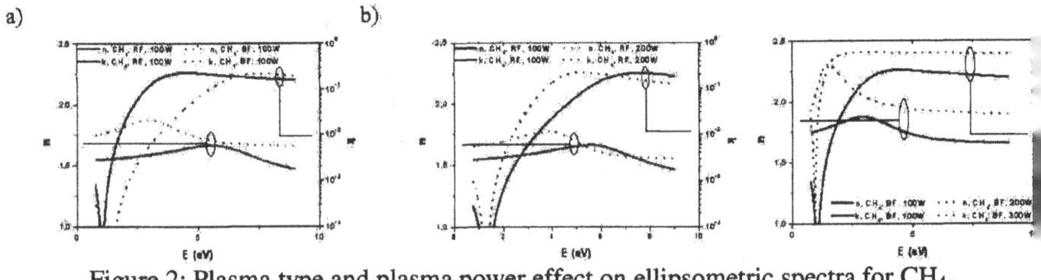

Figure 2: Plasma type and plasma power effect on ellipsometric spectra for CH_4.

From these data, it appears that the optical band-gap runs from 0.9 to 4.3eV, the refractive index from 1.47 to 2.76 and the conductivity from $5 \cdot 10^3$ to $5 \cdot 10^5 S.cm^{-1}$. But what are the relations with the process parameters. The first step is then to characterize the material structure.

RESULTS and DISCUSSIONS

As it is described in the literature [4], it is able to calculate the sp^2 hybridized carbon proportion (Csp^2) from the ES measurements. This method use the comparison of the film effective number of valence electrons per atom taking part in optical transitions n_{eff} at low energies ($\leq 9eV$) with the one, at the same energy, of a pure graphite layer:

$$Csp^2_{a-CH} = \frac{n_{eff,a-CH}(E \leq 9eV)}{n_{eff,graphite}(E \leq 9eV)}$$

For the different processes, Csp^2 runs from 3 to about 30%. It is necessary to use Raman scattering and FTIR to know what the other film components are.

Figure 3.a shows Raman scattering spectra for different films. The classical spectra of carbon films are obtained: the two characteristics bands G and D respectively linked to the graphitic species and the disorder in the material are present. $I(D) / I(G)$ ratio represents the material graphitization. When drawing its power dependency as in figure 3.b, it appears that the films are more and more graphic. The comparison of these values with the Csp^2 shows a good proportionality but a large shift between both values appears. This is due to the fact that a-CH films include a large amount of hydrogen. Furthermore, a large luminescence, generally attributed to the film density, appears too for different processes. The luminescence evolution with plasma power reported on figure 3.c shows that the film density increases with plasma power.

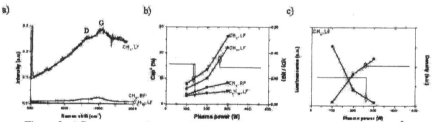

Figure 3: a) Raman scattering on a-CH films. The plasma power evolution of Csp^2 from ES and $I(D) / I(G)$ calculated from these spectra are reported on b). c) represents the luminescence evolution with plasma power.

C-H species of a-CH films are easily shown by FTIR measurements. Figure 4.a reports FTIR spectra where the three peaks characteristics of CH bonds appear. Moreover, the 2956, 2935 and 2875cm^{-1} bands are generally attributed to sp^3 C-H vibrations. These measurements show that a-CH films are highly concentrated in sp^3 hybridized carbon and that these films are highly polymer-like. The hydrogen concentration can be evaluated from FTIR measurements [5] by using the equation:

$$\%H = B \cdot \int_{v_1}^{v_2} \frac{\alpha(v)}{v} \cdot dv$$

where B is a constant which depends on the film thickness. α represents the absorbance and v_1 and v_2 the wave numbers at the edges of the CH absorption peaks. Considering that a-CH films are only composed of sp^2 hybridized carbon – sp^3 hybridized carbon – hydrogen, it is able to found their position in the ternary diagram Csp^2-Csp^3-H. Figure 4.b reports the effects of monomer structure, plasma type and plasma power on the film position in the ternary diagram.

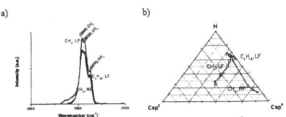

Figure 4: a) represents C-H bonds FTIR spectra. Then from Csp^2 and H concentration calculations, it is able to represent on b) the film structure evolution with plasma power.

It appears that monomer structure, plasma source and plasma power leads to very different materials: as in LF plasma, CH_4 based films tend to graphite-like when increase plasma power, C_4H_{10} based ones stay polymer-like. Moreover, RF plasma deposited films tend to diamond-like materials when increase plasma power. These mechanisms are in good agreement with the differences in monomer dissociation mentioned above: RF plasma leads to CH_2 and CH_3 species for the different monomers but it appears here that the last one is dominant. This behaviour can be attributed too to the ion bombardment which acts in RF condition and not in LF one. It leads then to a highly sp^3 hybridized film. In contrast with RF plasma, LF leads to different dissociation mechanisms as for example C_xH_y polymerisation when using C_4H_{10} which lead to polymer. Then, from the process parameters (monomer structure, plasma source, plasma power), it is able to control the material structure. Furthermore, the optical and electrical properties evolutions with the plasma power, reported on figure 5, shows that they are highly linked to the material structure.

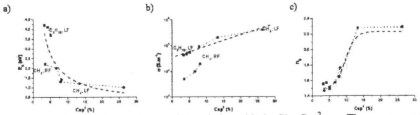

Figure 5: Electrical and optical properties variation with the film Csp^2 rate. They represent the Csp^2 dependency of a) the band-gap, b) the conductivity and c) the refractive index.

According to the increase of localised states by increasing the Csp^2 rate, E_g decreases as it is shown in figure 5.a. This energy is really linked too to the process parameters. For example, it decreases with the plasma power: the more the plasma power is, the more the monomer is dissociated. Then, it forms more CH_2 species which leads to highly sp^2 hybridized material. Furthermore, σ values, shown in figure 5.b with Csp^2, are in good accordance with the ternary diagram: the more sp^2 hybridized it is, the more the film conductivity is high. The conductivity is then related to the chemical structure of the material. Finally, n_∞ seems to be dependent of Csp^2. But, as it is able to deduce from figure 3.b and c, the film density increase with Csp^2. Further experiments will show which of this two different parameters would influence the refractive index.

CONCLUSION

a-CH films can be deposited by LF or RF plasma from different monomers (CH_4, C_4H_{10}). But these different plasma sources lead to different materials. The different monomer dissociation mechanisms explain this behaviour. Then, from these 3 process parameters (monomer structure, plasma source, plasma power), it is able to produce different materials with different optical and electrical properties. From the broad range of film properties and as they can be easily etched with an oxygen-fluorine plasma as it is shown in figure 6, it is possible to realize a great number of optoelectronic components (light source, detector, amplifier, waveguide, optical sensor...).

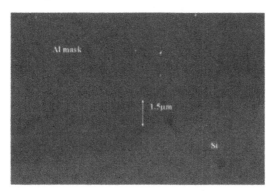

Figure 6: a-CH etched film by oxygen-fluorine plasma.

ACKNOWLEDGMENTS

Thanks to Dr A. ZWICK from Laboratoire de Physique des Solides for his Raman spectra measurement and his precious comments.

REFERENCES

1 A. Grill, Thin Solid Films, **355-356**, 189 (1999)
2 S. Martin, R. Clergereaux, D. Escaich, P. Raynaud, and F. Gaillard, NEMat Conference, Grenoble (2004)
3 A. Canillas, M.C. Polo, J.L. Andujar, J. Sancho, S. Bosch, J. Robertson, and W.I. Milne, Diam. Relat. Mat., **10**, 1132 (2001)
4 N. Savvides, J. Appl. Phys., **59**, 4133 (1986)
5 N. Basu, J. Dutta, S. Chaudhuri, A.K. Pal, and M. Nakayama, Vacuum, **47**, 233 (1996)

Mat. Res. Soc. Symp. Proc. Vol. 817 © 2004 Materials Research Society   L6.27

Ordered Micro-Particle Structures in a Liquid Crystal: Formation and Physical Properties

Ke Zhang, Anatoliy Glushchenko, and John L. West
Liquid Crystal Institute, Kent State University, Kent OH 44242

ABSTRACT

Ordered colloids are of great scientific and practical interests. Liquid crystals offer enhanced ways of producing and stabilizing these complex structures. We therefore studied the rheological and electro-rheological properties of the structured colloids as a means of probing this stabilization. We found that the mechanical properties of the colloids and stability of their 3D structures can be controlled by the particles size and distribution. In addition, when an electric field is applied, we observed an increase in the apparent viscosity with saturation at high electric fields. This effect depends on the shear rate and temperature. The results are also compared with the published data for the viscosity measurements of pure liquid crystals and isotropic colloids. While we are only beginning to understand the details of these complex colloids we expect they will find a wide variety of applications.

INTRODUCTION

We focused our research on micro-particles dispersed in liquid crystalline matrices. The anisotropic liquid crystal phase produces new effects not observed in colloidal suspensions in isotropic liquids, where the interaction between the particles may be explained by a combination of the long-range Coulomb repulsion and short-range Van der Waals attraction [1]. Particles dispersed in the liquid crystal phase produce deviations in the liquid crystal director, n. Because of the surface anchoring of the liquid crystal at the particle surface, these dispersed particles introduce defects in the liquid crystal phase. In order to minimize the free energy of the system, particles will share defects and will tend to reside at defect planes between liquid crystal domains. The particle surfaces can also induce a gradient in the magnitude of the nematic order parameter in their neighborhood, leading to an attractive short-range interaction. There can also be interactions arising from the restriction of thermal fluctuations in the director field by the particle surfaces. The resulting distribution of the particles in an anisotropic fluid depends on the interplay between the described factors.

On the other hand, if a redistribution of the particles is formed somehow, it will impose new physical properties to the system, including electro-optical and mechanical. Terentjev, et al. [2-4] recently explored small (150 and 250nm) polymeric particles dispersed in a nematic liquid crystal. The initial colloid dispersed homogeneously in the isotropic phase formed a cellular structure upon cooling into the nematic phase. The cellular structure is comprised of domains of practically pure nematic phase surrounded by walls of densely packed particles. This long-lived rigid cellular structure enhanced the mechanical properties of the systems, increasing the near-equilibrium (low-frequency) storage modulus by several orders of magnitude, up to $\sim 10^5$ Pa.

In this work we modify the mechanical and electro-optical properties of liquid crystal colloids by forming controlled colloidal structures. Using a combination of optical microscopy and viscoelastic shear measurements, we measured the viscous coefficients of the ordered colloids.

EXPERIMENT

The homogeneous mixture of a liquid crystal and 2 μm plastic particles was sandwiched between two polyimide-covered ITO-glass substrates (Figure 1, a). The polyimides were selecte to provide either planar or homeotropic orientation of the liquid crystal 5CB (4'-pentyl-4-cyano 1,1'-biphenyl). The cell thickness was large enough (20-50 μm) to assure unimpeded movemer of the particles.

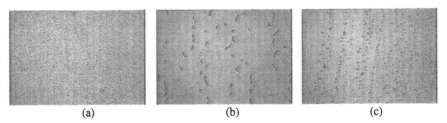

| (a) | (b) | (c) |

Fig. 1 Particles distribution before (a) and after interface sweeping at the speed of 15 degree/min (b) and 30 degree/min (c).

One of the edges of the cell was placed on a temperature controlled hot-stage. This configuration provided a temperature gradient through the entire area of the cell. The distribution of the temperature gradient depended on the balance between the temperature of the hot-stage and room temperature, and it was about 2 deg/cm in our case. This produced a nematic-isotropic transition interface parallel to the heated edge of the cell. Any changes of the hot-stage temperature led to a change in the position of the interface in the cell. The speed of th temperature change, that could be controlled with high precision (0.01 deg/min), determined the speed of the interface. The spatial period of the obtained striped structures (Figure 1,b and 1,c) depended on the cooling rate and the particle size [13]. Increasing the particle size, as well as decreasing the cooling rate, resulted in an increase of the spatial period.

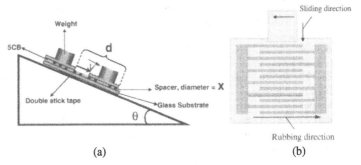

| (a) | (b) |

Fig. 2. Experimental set-up for estimation of the viscosity coefficients of ordered liquid crystal colloids.

There are many physical methods to measure the viscous coefficients in liquid crystal materials. Direct methods determine the viscosity coefficient from first principles. These techniques can provide precise and highly accurate data but require a substantial amount of sample. Alternatively a variety of indirect methods can be used. They explore physical phenomena that are influenced by the visco-elastic behavior of liquid crystals. The viscosity coefficients are then extracted through theoretical considerations of the particular phenomenon. These techniques usually require much less sample and can be used to study properties of heterogeneous systems. However, these techniques yield a viscosity coefficient only when other quantities such as the elastic constants [5-7] or other viscosity coefficients [8-9], the anisotropy of electric or magnetic susceptibility [10-11] or refractive index [12] are known from independent measurements. We found substantial discrepancies among results from the same substance obtained by different techniques reported by other authors, even if the substance is a pure liquid crystal.

Analyzing all the mentioned techniques and taking into account the peculiarity of the ordered colloids [13], we decided to evaluate the apparent viscosity qualitatively by observing the sliding of one cell's substrate relatively to the other (Figure 2,a). The cell, consisting of two glass substrates with an ordered colloidal structure in between, was assembled on a tilt plane with controlled tilt angle θ. In order to control the cell gap and shear stress, a suitable poise (with the weight of 1~10 g) was fixed on the top substrate. Therefore, at appropriate tilt angle and poise weight, the top substrate was able to slide at a quasi-constant speed through a limited distance related to the bottom substrate. The speed, the angle θ, and the weights of the poise and top substrate were recorded to estimate the viscosity of the material.

In order to measure the viscosity of the fluids under in-plane or out-of-plane electric field, the interdigitated electrodes were etched on the bottom substrate before coating alignment film (Figure 2,b). The width of the electrodes was 94 μm and the gap between electrodes was 31 μm.

RESULTS AND DISCUSSIONS

Viscosity of pure 5CB liquid crystal
To check the reliability of our method, we estimated the viscosity of pure 5CB liquid crystal and compared the results with the data from other sources. In our experiment, the viscosity coefficients η were deduced from the equation:

$$G \approx \eta \frac{dV}{dx} = \eta \frac{V}{x} \qquad (1),$$

where the parameter G takes into account all the geometrical factors such as the inclination angle θ of the tilt plane, the area of the substrates, and includes the weight of the poise and top glass substrate; V is the flow speed, and x is the cell gap (as shown in Fig. 2). We assume the velocity is linearly decreasing in the liquid crystal layer and therefore the derivative dV/dx is approximately the same as the value V/x (flow speed difference between the top and bottom liquid crystal layer divided by the cell gap). The flow speed of the top liquid crystal layer, which is the sliding speed of the top substrate, can be calculated by the time and distance of the sliding, while the flow speed of the bottom liquid crystal layer is zero.

As shown in Fig.3, the viscosity of the liquid crystal increases as the applied voltage increases and reaches a plateau at high enough voltages. The saturated viscosity is roughly Miesowicz's viscosity η_2 (when the liquid crystal director is parallel to the velocity gradient), which is ~100 mPa s at 298.2K and corresponds to the data provided in the reference [14]. The viscosity of about 30 mPa s at zero voltage is approximately the same as the apparent viscosity produced by a flow alignment used [15].

An applied electric field tends to align the liquid crystal director along the field which is perpendicular to the substrates in our experimental geometry. On the other hand, shear flow aligns the liquid crystal at some angle to the normal of the substrates. Figure 3 shows how the competition between the electric field and shear flow determines alignment of the liquid crystal. At a smaller shear stress, shear flow alignment can be easily overcome by the alignment imposed by the electric field; therefore the apparent viscosity is saturated at a low voltage, while large voltage is needed when the shear stress is larger.

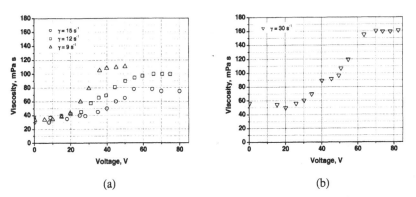

(a)   (b)

Fig. 3 Relationship of the apparent viscosity of the liquid crystal *vs* applied voltage at a constant shear stress. The viscosities at the shear stresses of about 9, 12, and 15 are specified by ▲, ■, and ♦, respectively.

We applied this method to estimate viscosity of liquid crystal colloids when the particles are homogeneously distributed in the liquid crystal or form stripes.

Viscosity of a liquid crystal suspension in an electric field
Addition of even a small amount of particles increases the apparent viscosity of the liquid crystal matrix (Figure 3,b). As we mentioned before, this phenomena was observed and described by many researches. At a high enough electric field (~2.5 V/μm), the viscosity reaches its saturation value. At any applied voltage and at the same shear rate, the viscosity of the suspension is larger than that one for the pure liquid crystal.

Viscosity of liquid crystal suspensions with ordered particle structures
Measurements of the viscosity of the ordered liquid crystal colloids (Figure 1,b and 1,c) revealed a number of interesting phenomena. We found that the viscosity of the ordered structure is higher than the viscosity of the suspension with homogeneous distribution of particles (Table 1). This difference is larger when the sliding direction is perpendicular to the particle stripes. An interesting behavior was observed while repeating the measurements. Sliding of the top substrate disrupts the initially formed particle structures. Observation with optical microscopy showed that repeating the measurements mixes the particles making the distribution approach homogeneous.

Consequently, we obtained the viscosity approach the homogeneous value upon repeated measurements (Figure 4).

Table 1. Viscosity comparison for the pure 5CB liquid crystal, suspension with uniformly distributed particles, and ordered colloids. All the values are brought in [mP s].

Pure 5CB	With uniformly distributed particles	Particle stripes are perpendicular to the sliding direction	Particle stripes are parallel to the sliding direction
30	55	180	75

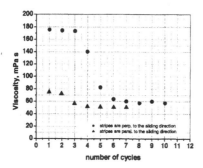

Fig. 4. The graph shows how the measured viscosity changes with each subsequent cycle.

Optical appearance of the structures

Optically all particle containing samples are more opaque than the pure liquid crystal. This is expected because of the enhanced light scattering due to the particle-induced liquid crystal domain structure. However, optical microscopy reveals that even concentrated ordered specimens preserve the underlying birefringence of the liquid crystal host and are therefore optically switchable. We found that completely analogous to dispersed polymer networks, these micro-particle structures affect the switching voltage and speed of the films. They may be used to adjust the optical performance and introduce entirely new effects, like bistability. We are now working on a detailed survey of the electro-optical response of the structured colloids, relating morphology with the viscous properties. The data will be presented in a future paper.

CONCLUSIONS

The mechanical and micro structural properties of an unusual class of heterogeneous soft condensed matter – ordered colloid liquid crystal composites – have been described with a view to explore in qualitative terms how their physical properties depend on internal structure. A strong relationship between viscosity and the particles structures has been revealed. We plan to explore the limits of these materials and develop physical models to explain and predict the phenomena.

ACKNOWLEDGMENTS

The authors would like to thank Yu. Reznikov and D. Andrienko for continuous discussions.

REFERENCES

1. W.B. Russel, D.A. Saville, W.R. Schowalter, *Colloidal Dispersions*, (Cambridge University Press, 1989)
2. S. P. Meeker, W. C. K. Poon, J. Crain, E. M. Terentjev, *Phys. Rev. E*, **61**, R6083 (2000)
3. V. J. Anderson, E. M. Terentjev, S. P. Meeker, J. Crain, W. C. K. Poon, *Eur. Phys. J. E*, **4**, 1 (2001)
4. Anderson, V. J.; Terentjev, E. M. *Eur. Phys. J. E*, **4**, 21 (2001)
5. H. Fellner, W. Franklin, S. Christensen, *Phys. Rev. A*, **11**, 1440 (1975)
6. J.P van der Meulen, R.J.J. Zijlstra, *J. Phys.*, **43**, 411 (1983)
7. G.-P. Chen, H. Takezoe, A. Fukuda, *Liq. Cryst.*, **5**, 341 (1989)
8. K. Skarp, T. Carlsson, *Mol. Cryst. Liq. Cryst. Lett.*, **49**, 75 (1978)
9. M. Imai, H. Naito, M. Okuda, A. Sugimura, *Jpn. J. Appl. Phys.*, **33**, 3482 (1994)
10. F.M. Leslie, G.R. Luckhurst, H.J. Smith, *Chem. Phys. Lett.*, **13**, 368 (1972)
11. P.E. Cladis, *Phys. Rev. Lett.*, **28**, 1629 (1972)
12. K. Skarp, S.T. Lagerwall, B. Stebler, *Mol. Cryst. Liq. Cryst.*, **60**, 215 (1980)
13. J. West, A. Glushchenko, G. Liao, Y. Reznikov, D. Andrienko, M.P. Allen, *Phys. Rev. E*, **66** 012702 (2002)
14. A.G. Chmielewski, *Mol. Cryst. Liq. Cryst.*, **132**, 339 (1986)
15. K. Negita, *J. Chem. Phys.*, **105**, 7837 (1996)

Electro-optic Materials

Mat. Res. Soc. Symp. Proc. Vol. 817 © 2004 Materials Research Society   L7.2

Integrated Optics/Electronics Using Electro-Optic Polymers

Larry R. Dalton
Department of Chemistry, University of Washington
Seattle, WA 98195-1700, U.S.A.

ABSTRACT

Organic electro-optic materials afford the realization of devices with terahertz bandwidths, providing bandwidths are not limited by resistive losses in metal electrodes. Recent realization of electro-optic coefficients (at telecommunication wavelengths) on the order of 200 pm/V permits construction of devices with operating voltage requirements of 1 volt or less. In like manner, substantial progress has been made in both understanding and improving thermal and photostability suggesting that organic electro-optic materials can meet Telacordia standards. However, one of the most intriguing advances afforded by organic materials is their processability including the ability to be integrated with diverse materials. This communication discusses both the systematic improvement, by theoretically-inspired rational design, of relevant material properties and the development of a variety of new processing methodologies, including soft lithography methods, for the fabrication of stripline, cascaded prism, and ring microresonator devices. The fabrication of flexible devices is also discussed.

INTRODUCTION

Electro-optic (EO) phenomena involve the interaction of an optical field with an electrical field through the charge distribution of a material. Typically, a low frequency (dc to 30 THz) electrical field is applied to a material resulting in a perturbation of the charge distribution of the material. This, in turn, results in a change of velocity of light propagating in the material, as a consequence of the electric field component of light interacting with the perturbed charge distribution. Another way of viewing this phenomenon is voltage control of the index of refraction or birefringence of the EO material. The most common mechanisms resulting in electro-optic activity involve electric field induced molecular reorientation of liquid crystalline materials, lattice distortion of inorganic crystalline materials such as lithium niobate, or π-electron redistribution in ordered organic chromophore materials. Of course, the time required to respond to changes in the applied electric field will depend upon the mass that must be displaced. Liquid crystalline materials will, in general, be slow responding materials because significant nuclear mass must be moved; this slow response will translate into a limited bandwidth for devices fabricated from such materials. The response of crystalline materials, such as lithium niobate, is defined by the resistance to lattice displacement and will relate to the inverse piezoelectric effect. Elasto-optic effects will also contribute to index of refraction changes for crystalline materials and thus the frequency dependence of devices fabricated from lithium niobate will be more complex. The "effective" electro-optic activity will be a sum of two terms and will decrease with increasing frequency as elasto-optic effects become unimportant. The EO response of electrically-poled [1-3] or self-assembled [4-6] organic chromophores existing in polymer or macromolecular lattices is defined by π-electron redistribution under the action of the frequency-dependent applied electric field. The response time is fundamentally the phase relaxation time of the conjugated π-electron system of these charge transfer molecules. Pulsed laser studies have shown that such phase relaxation times lie in the femtosecond regime and together with recent terahertz signal generation and detection studies suggest the "intrinsic" bandwidth of organic electro-optic materials lies in the tens of terahertz (THz) range in contrast to tens of gigahertz (GHz) for lithium niobate and a few megahertz (MHz) or less for liquid crystal materials [7,8]. In practice, the bandwidth of stripline (Mach Zehnder amplitude modulators and optical crossbar switches) devices fabricated from organic EO materials are typically limited to 3 dB bandwidths of 200 GHz or less due to resistive losses in metal electrode structures rather than being limited by the fundamental response time of the π-electrons or by velocity mismatch of the propagating electric and optical waves [9-11]. The fact that π-electrons define both dielectric permittivity and refractive indices and that both quantities exhibit low dispersion assures good velocity matching of optical and electrical waves. The advantage of organic electro-optic (OEO) materials for the fabrication of high bandwidth devices is well recognized [9-11].

The electro-optic coefficient of crystalline materials such as lithium niobate is typically fixed (30 pm/V for the device relevant value for lithium niobate). The electro-optic activity of organic chromophore containing EO materials is defined by the molecular first hyperpolarizability of the chromophore and by the degree of acentric (noncentrosymmetric or polar) ordering of that chromophore in a macroscopic lattice. The operational relationship can be considered for many applications to be $r_{33} = 2N\beta f(\omega)<\cos^3\theta>$ where r_{33} is the principal electro-optic coefficient, N is the chromophore number density in the lattice, β is the molecular first hyperpolarizability, $f(\omega)$ takes into account the dielectric nature of the host lattice material, and $<\cos^3\theta>$ is the acentric order parameter. As has been described elsewhere [12-14], optimizing macroscopic EO activity for OEO materials involves optimizing β and optimizing $N<\cos^3\theta>$. Note that because of intermolecular electrostatic interactions, chromophore number density and order parameter are not independent and indeed graphs of EO activity versus N will go through a maximum and the maximum will shift to lower loading with increasing chromophore dipole moment. Theory has suggested how electro-optic activity can be enhanced by control of chromophore shape and covalent interactions that influence chromophore organization [12-14]. Recently, dendronized polymer structures have been used to achieve a factor of two enhancement of electro-optic activity relative to values obtained for the same chromophore in guest-host composite materials [15,16]. Currently, the largest electro-optic coefficients observed for organic materials fall in the range of 180-220 pm/V. We will see in the next section that values are likely to increase as new theoretically inspired chromophores are synthesized and utilized.

The thermal and photostability of organic electro-optic materials is of major concern for commercial application of these materials. A number of researchers [1,17-19] have demonstrated that singlet oxygen chemistry plays a major role in defining photochemical stability of OEO materials. The photochemical figure of merit is defined by B/σ where B is the inverse probability that photodegradation occurs from the charge transfer excited state and σ is the interband electronic absorption coefficient. B/σ is observed to vary by orders of magnitude with the partial exclusion of oxygen, with lattice hardening by crosslinking, and with introduction of chemical and physical quenchers of singlet oxygen. B/σ is observed to increase with increasing separation of the photon energy from the energy of the interband charge transfer transition. Photostability also improves with substitution or steric protection of reactive sites. For a single chromophore type, such as CLD [14], B/σ can vary by four orders of magnitude (e.g., from 10^{32} m^{-2} to 10^{36} m^{-2}) with the aforementioned variations. Packaging will clearly be important for the commercial application of OEO materials in much the same way that it is for OLED materials. Given the high cost of packaging lithium niobate and electro-absorptive modulators, such packaging will not likely be a significant impediment to commercialization. Thermal stability has been extensively investigated. It depends upon crosslink density in hardened materials and upon segmental flexibility of the host lattice. For materials where a glass transition can be defined, thermal stability can be specified in terms of the displacement of the operating temperature from the glass transition temperature. Lattice hardening has been attempted using both thermally and photo-induced crosslinking reactions [20]. Thermally induced crosslinking reactions must be matched to poling temperatures to minimize reduction in poling-induced order by the lattice hardening process. Another concern with condensation reactions is the elimination of water [21]. Recent advances in thermally-induced lattice hardening involve the use of the fluorovinylether moiety [15,16,22] to achieve crosslinking without water elimination and use of reverse (retro) Diels Alder chemistry to achieve additional lattice hardening (at operating temperatures) without crosslinking interfering with poling at higher temperatures [23]. With these latter approaches excellent long-term thermal stability to 85C is obtained. Thermal stability can also be assessed by a dynamic assay wherein second harmonic generation is monitored as a function of increasing temperature (e.g., at a rate of 5°C/minute). Crosslinking typically results in dynamic thermal stabilities in the range 150-230°C.

Organic electro-optic materials can be fabricated into active stripline waveguide [1-3,24-27], cascaded prism [28,29], and ring microresonator [30-32] devices by reactive ion etching. Both vertical and horizontal integration with very large-scale integration (VLSI) semiconductor electronics has been demonstrated [2]. Low loss integration with silica fiber optics has also been demonstrated [33]. The fabrication of three-dimensional (3D) active optical circuitry has been demonstrated [34], including for the realization of active wavelength division multiplexing (WDM) based on electro-optic ring microresonators [30-32]. Recently it has been demonstrated that conformal and flexible devices can be fabricated and the observed performance of these devices is comparable to that for devices fabricated on rigid substrates [35]. The capability to fabricate conformal and flexible devices is particularly relevant to applications such as phased array radar and particularly space-deployed radar antennae. Recently, it has been demonstrated that both active and passive organic optical circuitry can be fabricated by soft lithography techniques and the performance of replicate devices is not significantly different from that of the e-beam etched master devices [36]. This mode of fabrication has the potential of dramatically reducing the cost of production of optical systems based on organic materials. Another advantage of organic electro-optic materials is the capability to be integrated with diverse materials. Recently, such materials have been integrated into silicon photonic circuitry [37] opening the possibility of even greater reduction in circuit dimensions and paving the way for a new generation of sensors including sensors integrated into laboratory-on-a-chip systems. It is becoming increasingly apparent that one of the greatest advantages of organic electro-optic materials is their processability and adaptability. These may lead to dramatic reductions in cost and may greatly increase the range of use of electro-optic materials.

A few comments concerning the relatively neglected topics of optical loss and cladding materials need to be made. Optical loss is a complex topic involving many different loss mechanics ranging from absorption loss, to scattering loss (both from waveguide wall surface roughness and from bending loss in microresonators), to insertion loss (mode size or index of refraction mismatch losses between active and silica waveguides). At telecommunication wavelengths, absorption loss is dominated by hydrogen overtone vibrations and can range from values of 1.3 to 0.2 dB/cm depending on hydrogen content. For typical organic electro-optic materials values of 0.8-1.2 dB/cm are common. Chromophore loss is typically lower than the loss of commercial polymer hosts such as amorphous polycarbonate [38] as expected since unsaturation leads to lower proton density. Partially fluorinated dendrimers with active chromophore cores can lead to exceptional low absorption loss values 0.1-0.3 dB/cm. Scattering losses associated with processing (spin casting, poling, deposition of cladding layers, reactive ion etching) are more serious but can be kept to low values (a few tenths of a dB) with attention to the details of processing. For example, the degree of lattice hardness is an important factor in determining losses induced by deposition of cladding layers and reactive ion etching. A downside of working with organic electro-optic materials is that processing protocols must be optimized for each material and such optimization can be a time-consuming process. Indeed, such processing is clearly the rate-limiting step in the development of new organic electro-optic materials. Use of mode transformers has essentially solved the problem of insertion loss reducing mode mismatch loss to a few tenths of a dB [33,34]. Total insertion loss of 3 dB can be obtained for a two-centimeter stripline device although insertion loss values of 5-8 dB are more common working with new OEO materials. Special device structures, such as ring microresonators and superprism structures, require a higher level of analysis. For example, bending loss is a critical concern for microresonator structures. This contribution to total loss will, in turn, depend on the dimensions of the ring microresonator and the index of refraction difference between the core and cladding materials. Since ring dimensions also define other critical parameters, such as free spectral range (FSR), a thorough analysis is required and the selection of cladding material is critical. The need for detailed analysis and the difficulty of that analysis is even more extreme for superprism and photonic crystal device structures.

To the present time, very little attention has been given to cladding materials used with custom made organic electro-optic materials. Most cladding materials that are commonly used are off-the-shelf commercial materials, which are frequently characterized by high optical loss (. 3 dB/cm) and modest thermal and photochemical stability. Very little attention has been given to properties such as the conductivity of these materials. As the properties of OEO materials have become more carefully defined, performance limitations can increasingly be associated with cladding materials. Clearly, an issue that will be important for the commercialization of organic electro-optic devices will be the development of tailored cladding and packaging materials.

The focus of the research reported here is not to suggest an optimized organic electro-optic material system, but rather to focus attention on several important material advances that illustrate the potential of future (including the relatively near term future) OEO materials. Space limitations prevent inclusion of extensive detail (such as detailed syntheses). The reader is referred elsewhere for this information [39].

THEORY AND EXPERIMENT: EO CHROMOPHORES

Quantum mechanical calculations have clearly guided the improvement OEO chromophores for several decades. Trends relating to the variation of molecular first hyperpolarizability, β, with the chromophore length and bond length alternation of polyene segments are reasonably well understood. Recent attention has turned to the efficacy of quantum mechanical calculations based on density functional theory (DFT) and semi-empirical methods for predicting the change in β with specific systematic variation of the chemical structure of donor, bridge, and acceptor portions of charge-transfer chromophores [1]. While the reliability of theoretical methods for predicting the effect of structural modifications in general has yet to be systematically determined, we present some promising results in this section. Before discussing these results, it is appropriate to note several factors that complicate the comparison of theory and experiment. The first is that theoretical calculations are typically "gas phase" or isolated molecule calculations while experiments are typically conducted on chromophores in condensed phases. Secondly, measurements of molecular hyperpolarizability by Hyper-Rayleigh Scattering (HRS) or by Electric Field Induced Second Harmonic Generation (EFISH) are typically conducted at wavelengths ranging from 1.0 to 1.9 microns while theoretical calculations are most reliably performed for the long wavelength (or zero frequency) limit. Moreover, a number of experimental difficulties, including two-photon effects, can complicate the determination of β values relevant to electro-optic activity. We are sufficiently concerned with such issues that we are reinvestigating the correlation of theory and experiment in a much more fundamental way and have concerns about the quantitative validity of molecular first hyperpolarizability measurements that have been reported to this time. However, qualitative trends may be more reliable particularly if measurements are made over a significant band of wavelengths and concentrations and if the results of various measurements of r_{33}, β, and $<\cos^3\theta>$ are carefully correlated. In this communication, we limit our discussion to qualitative trends.

To illustrate the potential utility of quantum mechanics for guiding structural modifications of chromophores to systematically improve molecular first hyperpolarizability, we discuss the chromophore structures shown in figure 1. Over the past several years, we have investigated several hundred structures using DFT and semi-empirical methods. In this discussion, we focus upon variations in chromophore acceptor structure illustrated by structures "1", "2", and "10" and variation in bridge structure illustrated by structure "150" shown in figure 1. Structure "1" contains the familiar tricyanofuranvinylene (TCF) acceptor of "CLD" and "FTC" (2-dicyanomethylene-3-cyano-4-{2-[E-(4-N,N-di(2-acetoxyethyl)amino)-phenylene-(3,4-dibutyl)thien-5]-E-vinyl}-5,5-dimethyl-2,5-dihydrofuran) type chromophores [1-3,14]. DFT calculations predict a dipole moment, μ, and a zero frequency (long wavelength limit) molecular first hyperpolarizability, β_0, of 11.8 Debye and 23.8 x 10^{-30} esu, respectively. Replacement of one of the methyl groups of the TCF acceptor with a trifluoromethyl group (structure "2") is predicted to result in a factor of 1.45 improvement in β_0 with no significant change in dipole moment (11.9 D). This theoretically predicted trend appears consistent with recent measurements of β and electro-optic coefficients (at telecommunication wavelengths) reported by researchers at Lockheed Martin [40], Corning [41,42], and by our research group. Since dipole moment did not change with this modification, it would be expected that the improvement in β would translate directly to an improvement in r_{33}. This indeed seems to be the case. From a practical standpoint, this simple modification has proven to be of considerable utility leading to electro-optic coefficients on the order of 100 pm/V at telecommunication wavelengths.

Replacement of the TCF acceptor with the tricyanopyrrolinevinylene (TCP) acceptor (structure "10") leads to a predicted factor of 1.8 improvement in β_0 with no significant change in dipole moment (11.7 D). Again, this theoretical prediction appears to be in good agreement with experimental measurements of molecular first hyperpolarizability and macroscopic electro-optic coefficients (measured at 1.3 and 1.55 microns wavelength). For example, the HRS-measured value of β for "10" at 1.3 microns is 3.5 times greater than for "1". The chromophore shown at the bottom of figure 1 (dissolved in amorphous polycarbonate) yields an electro-optic coefficient of 101 pm/V (at 1.55 microns); this is approximately twice the value obtained for the analogous TCF chromophore at the same concentration level. Again, because dipole moment has not changed between structures "1" and "10", improvement in β is expected to translate directly into an improvement in r_{33}.

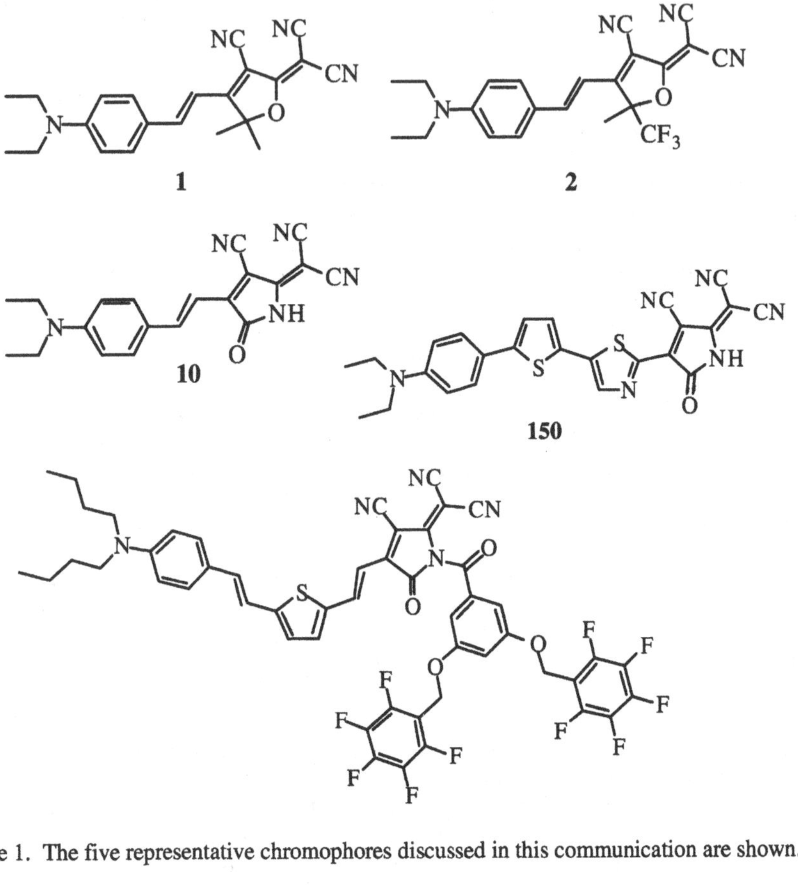

Figure 1. The five representative chromophores discussed in this communication are shown.

Even larger improvements are predicted for other structural modifications. For example, structure "150" is predicted to exhibit dipole moment and molecular first hyperpolarizability values, β_0, of 13.6 D and 188 x 10^{-30} esu respectively. This is a factor of 8 greater than the value predicted for structure "1". The positioning of the thiophene and thiazole moieties in structure "150" is very important. The value of β_0 for structure "150" is 1.55 times greater than for the analogous chromophore with two thiophene rings as the bridge component. Note that the structure of chromophore "150" can be viewed as [strong donor]-[weak donor]-[weak acceptor]-[strong acceptor] and as such the bridge component might be described as a "gradient" bridge.

The above theoretical calculations performed in our laboratory are consistent with previously reported calculations by Mahon and coworkers [43]. Although we have synthesized a number of modifications of structure "150" that should permit a unambiguous test of the validity of theoretical predictions of improved molecular first hyperpolarizability for "gradient" bridge chromophores and extensive characterization of both molecular first hyperpolarizability and electro-optic activity are underway for these materials, sufficient data are not available at this time to comment further on this prediction. The same can be said for a number of other structural modifications including chromophores with multiple donor and acceptor units [44]. These modifications, unlike the modification of TCF and TCP acceptors discussed above, involve significant changes in dipole moment (hence intermolecular electrostatic interactions) and thus statistical mechanical calculations are required to predict the improvement in electro-optic activity that can be expected. Such calculations will be discussed briefly in the next section.

It should be noted that molecular first hyperpolarizability values at telecommunication wavelengths (1.3 and 1.55 microns) are larger than the long wavelength values due to resonance enhancement. The ratio of values for the various structures will also be somewhat larger due to the fact that the structures that exhibit larger β values are somewhat red-shifted in terms of the λ_{max} of their charge transfer transition. The interband transition does not contribute to optical absorption loss, which is typically dominated by hydrogen vibrational (overtone) absorptions.

THEORY AND EXPERIMENT: MACROSCOPIC ORDER

The greatest problem in optimizing macroscopic electro-optic activity of organic materials is that of optimizing $N\langle\cos^3\theta\rangle$. Because of strong intermolecular electrostatic interactions (primarily dipole-dipole interactions), it is difficult to translate even 10% of the electro-optic activity potential of an individual chromophore to macroscopic (and hence device relevant) electro-optic activity. Fortunately, there are two components to the intermolecular electrostatic potential that a chromophore experiences from surrounding chromophores. One component (from chromophores positioned around the equator of the reference chromophore) favors centrosymmetric order. The other component (from chromophores positioned along the poles of the reference chromophore) favors the noncentrosymmetric order necessary for nonzero electro-optic activity. In previous comparisons of theory and experiment [12-14], we demonstrated that chromophore shape has an important influence on the optimization of $N\langle\cos^3\theta\rangle$ with more rotund shapes yielding larger electro-optic activity than chromophores with prolate ellipsoidal shapes. Control of shape can be achieved by simple derivatization of chromophores; however, incorporation of chromophores into dendrimer structures appears to be a systematic way of achieving control of shape (spatial anisotropy of steric interactions).

More recently, it has been demonstrated that the restrictions on chromophore movement and organization inherent in covalent bonds can be systematically used to control the macroscopic ordering of chromophores [1,45]. Multiple chromophores have been incorporated into dendronized polymer structures [15] where covalent bonds prohibit chromophores from achieving centrosymmetric "crystallization". The electro-optic activity for these structures is shown to be approximately twice that of the best value that can be obtained for the same chromophore in a guest host composite [15,45]. By using the acceptors discussed above (CF$_3$-TCF and TCP) electro-optic coefficients in the range 180-220 pm/V are observed.

Recent pseudo-atomistic Monte Carlo statistical mechanical calculations suggest that there are even more desirable supermolecular structures that can be implemented. Pseudo-atomistic Monte Carlo calculations involve treating chromophores in the "united atom" approximation while flexible segments of the supermolecular structure are treated atomistically. As shown in figure 2, a nearly perfect acentric (noncentrosymmetric) chromophore lattice is predicted for some supermolecular structures.

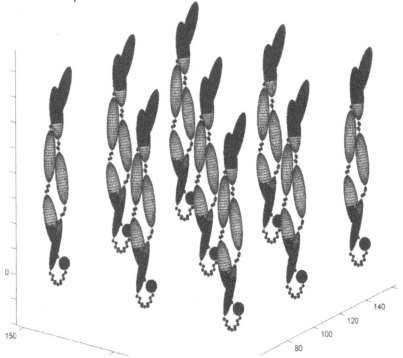

Figure 2. The predicted macroscopic organization of a supermolecular structure consisting of chromophores coupled by flexible segments is shown. The small spheres represent atoms. Chromophores are represented by prolate ellipsoids with embedded dipole moments. The larger spheres represent a rigid assembly of several atoms that serves as the dendritic core of the structure.

In figure 2, chromophores organize in a head-to-tail within a single strand with the two strands (arms) of the dendrimer displaced relative to each other so that chromophores in different strands are positioned at the magic angle (54.7°) with respect to each other. If current chromophores could be incorporated into such dipolar structures, electro-optic coefficients on the order of 100 pm/V would be expected.

Obviously, no such supermolecular structures have been demonstrated to this point in time. However, theoretical calculations suggest a variety of such structures and several of these are currently being synthesized in our laboratory.

PROTOTYPE DEVICE FABRICATION AND TESTING

In this section, we attempt to briefly illustrate recently demonstrated unique advantages of organic electro-optic materials. As has been discussed elsewhere [35], flexible electro-optic devices such as Mach Zehnder modulators can be fabricated by lift-off techniques. In figure 3, we illustrate the impressive performance that can be realized with such devices.

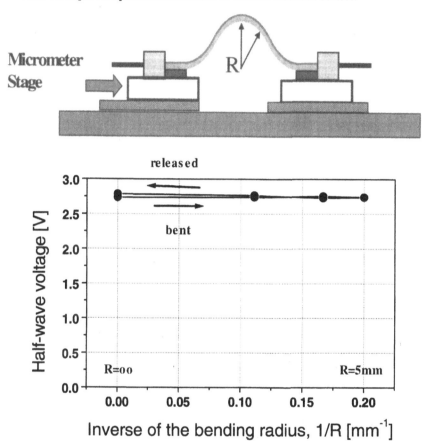

Figure 3. Top: The test bed for measurement of flexing and relaxation of flexible EO Mach Zehnder interferometers is shown. Bottom: The variation of halfwave voltage (V_π) with repeated flexing and relaxation is shown. The electro-optic material used to fabricate devices is a CLD chromophore/Amorphous Polycarbonate (APC) polymer composite [35].

From the data shown in figure 3, it is clear that very little degradation in absolute performance is observed for flexible devices relative to devices fabricated on rigid substrates. Moreover, properties such as halfwave voltage and optical loss are quite immune to extreme and repeated bending. This is a unique feature of organic electro-optic materials and affords important potential for the fabrication of phased array radar antennae structures, particularly structures that can be unrolled and deployed in space.

Another important advance in the fabrication of electro-optic devices from OEO materials is that of using soft lithography [36]. In figure 4, we illustrate recent results.

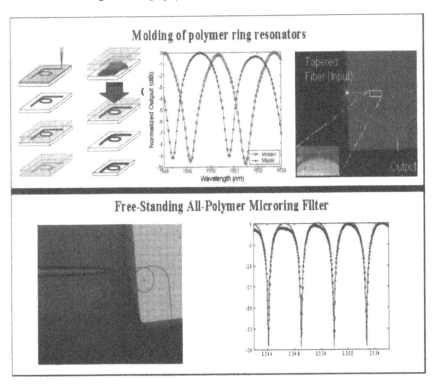

Figure 4. The use of soft lithography to fabricate ring microresonator devices is shown [36]. In the upper left, a schematic flow chart shows electron-beam lithographic fabrication of a microring resonator filter and its soft lithographic replication. In the upper center, a comparison between the transmission spectra of the master microring resonator optical filter and its soft lithographic replica is shown. In the upper right, an optical micrograph is shown with the output of a He-Ne laser injected into the waveguide for demonstration. The inset shows the coupling region. In the lower left, the coupling region is emphasized. In the lower right, comparison of theory and experiment for the transmission spectra of a microring filter is shown.

As with flexible devices, insignificant degradation in performance is observed in going from electron-beam fabricated master to stamped replicate. Soft lithography has also been employed to fabricate free-standing devices.

DISCUSSION

Quantum mechanics and statistical mechanical calculations are likely to permit the realization of OEO materials exhibiting electro-optic coefficients of 300 pm/V in the near future. Values of 1000 pm/V may be obtainable from highly engineered supermolecular materials. Considerable effort will be required to translate these new materials to the fabrication of practical devices. Attention must be paid to development of custom cladding materials and to issues of thermal and photochemical stability and of insertion loss. OEO materials afford unique advantages for the fabrication of devices, particularly for low cost fabrication and the production of conformal and flexible devices.

ACKNOWLEDGMENTS

I wish to thank my graduate students, postdoctoral fellows, and research collaborators (particularly Professors William Steier, Alex Jen, and Bruce Robinson) for participating in the collection of data discussed here and for many helpful discussions regarding interpretation of data. Financial support from the Air Force Office of Scientific Research, the National Science Foundation (DMR-0120967, DMR-0103009, and DMR-0092380) and DARPA (N00014-04-1-0094) is gratefully acknowledged.

REFERENCES
1. L. R. Dalton, *J. Phys. Condens. Mater* **15**, R897 (2003).
2. L. R. Dalton, A. W. Harper, A. Ren, F. Wang, G. Todorova, J. Chen, C. Zhang, and M. Lee, *Ind. Eng. Chem. Res.* **38**, 8 (1999).
3. L. R. Dalton, *Adv. Polym. Sci.* **158**, 1 (2002).
4. G. Evmenenko, M. E. van der Boom, J. Kmetko, S. W. Dogan, T. J. Marks, and P. Dutta, *J. Phys. Chem.* **115**, 6722 (2001).
5. Y. G. Zhao, A. Wu, H. L. Lu, S. Chang, W. K. Lu, S. T. Ho, M. E. van der Boom, and T. J. Marks, *Appl. Phys. Lett.* **79**, 587 (2001).
6. P. Zhu, M. E. van der Boom, H. Kang, G. Evmenenko, P. Dutta, and T. J. Marks, *Chem. Mater.* 14, 4982 (2002).
7. A. M. Sinyukov and L. M. Hayden, *Opt. Lett.* **27**, 55 (2002).
8. L. M. Hayden, A. M. Sinyukov, M. R. Leahy, J. French, P. Lindahl, W. N. Herman, R. J. Twieg, M. He, *J. Polym. Sci., Part B (Polym. Phys.)* **41**, 2492 (2003).
9. M. Lee, H. E. Katz, C. Erben, D. M. Gill, P. Gorpalan, J. D. Heber, and D. J. McGee, *Science* **298**, 140 (2002).
10. D. Chen, H. R. Fetterman, A. Chen, W. H. Steier, L. R. Dalton, W. Wang, Y. Shi, *Appl. Phys. Lett.* **70**, 3335 (1997).
11. D. Chen, D. Bhattacharya, A. Udupa, B. Tsap, H. R. Fetterman, A. Chen, S. S. Lee, J. Chen, W. H. Steier, and L. R. Dalton, *IEEE Photon. Tech. Lett.* **11**, 54 (1999).
12. B. H. Robinson and L. R. Dalton, *J. Phys. Chem.* **104**, 4785 (2000).

13. L. R. Dalton, B. H. Robinson, A. K. Y. Jen, W. H. Steier, and R. Nielsen, *Opt. Mater.* **21**, 1 (2003).

14. Y. Shi, C. Zhang, H. Zhang, J. H. Bechtel, L. R. Dalton, B. H. Robinson, and W. H. Steier, *Science* **288**, 119 (2000).

15. J. Luo, S. Liu, M. Haller, H. Li, T. D. Kim, K. S. Kim, H. Z. Tang, S. H. Kang, S. H. Jang, H. Ma, L. R. Dalton, and A. K. Y. Jen, *Proc. SPIE* **4991**, 520 (2003).

16. J. Luo, T. D. Kim, H. Ma, S. Liu, S. H. Kang, S. Wong, M. A. Haller, S. H Jang, H. Li, R R Barto, C. W. Frank, L. R. Dalton, and A. K. Y. Jen, *Proc. SPIE* **5224**, 104 (2003).

17. A. Galvan-Gonzalez, K. D. Belfield, G. I. Stegeman, M. Canva, S. R. Marder, K. Staub, G. Levina, and R. J. Twieg, *J. Appl. Phys.* **94**, 756 (2003) and references cited therein.

18. M. E. DeRosa, M. He, J. S. Cites, S. M. Garner, and R. Tang, *J. Phys. Chem.* to be published

19. S. M. Garner, J. S. Cites, M. He, and J. Wang, *Appl. Phys. Lett.* to be published.

20. L. R. Dalton, A. W. Harper, R. Ghosn, W. H. Steier, M. Ziari, H. Fetterman, Y. Shi, R. Mustacich, A. K.-Y. Jen, and K. J. Shea, *Chem. Mater.* **7**, 1060 (1995).

21. S. S. H. Mao, Y. Ra, L. Gao, C. Zhang, and L. R. Dalton, *Chem. Mater.* **10**, 146 (1998).

22. S. Suresh, S. Chen, C. M. Topping, J. M. Ballato, and D. W. Smith, Jr., *Proc. SPIE* **4991**, 53 (2003).

23. M. Haller, J. Luo, Hongxian Li, T. D. Kim, Y. Liao, B. Robinson, L. R. Dalton, and A. K. Y Jen, to be published.

24. J. Han, B. J. Seo, H. R. Fetterman, H. Zhang, and C. Zhang, *Proc. SPIE* **4991**, 562 (2003).

25. D. Jin, R. Dinu, T. C. Parker, A. Barlund, L. Bintz, B. Chen, C. Flaherty, H. W. Guan, D. Huang, J. Kressbach, T. Londergan, T. D. Mino, G. Todorova, and S. Yang, *Proc. SPIE* **4991**, 610 (2003).

26. A. Yacoubian, V. Chuyanov, S. M. Garner, W. H. Steier, A. S. Ren, and L. R. Dalton, *IEEE J. Sel. Topics in Quantum Electronics* **6**, 810 (2000).

27. J. H. Bechtel, Y. Shi, H. Zhang, W. H. Steier, C. H. Zhang, and L. R. Dalton, *Proc. SPIE* **4114**, 58 (2000).

28. D. An, Z. Shi, L. Sun, J.M. Taboada, Q. Zhou, X.Lu, R.T. Chen, S. Tang, H. Zhang, W.H. Steier, A. Ren, and L. R. Dalton, *Appl. Phys. Lett.* **76**, 1972 (2000).

29. L. Sun, J. Kim, C. Jang, D. an, X. Lu, Q. Zhou, J.M. Taboada, R.T. Chen, J.J. Maki, S. Tang H. Zhang, W.H. Steier, C. Zhang, and L. R. Dalton, *Opt. Eng.* **40**, 1217 (2001).

30. P. Raibiei, W. H. Steier, C. Zhang, and L. R. Dalton, *Int. Opt. Commun.* **1**, 14 (2002).

31. P. Raibiei, W. H. Steier, C. Zhang, and L. R. Dalton, *OSA Trends in Optics and Photonics. Optical Fiber Commun. Conf.* **70**, 31 (2002).

32. P. Rabiei, W. H. Steier, C. Zhang, and L. R. Dalton, *J. Lightwave Technol.* **20**, 1968 (2002).

33. A. Chen, V. Chuyanov, F. I. Marti-Carrera, S. M. Garner, W. H. Steier, J. Chen, S. S. Sun, and L. R. Dalton, *Opt. Eng.* **39**, 1507 (2000).

34. S.M. Garner, S.S. Lee, V. Chuyanov, A. Chen, A. Yacoubian, W.H. Steier, and L.R. Dalton, *IEEE J. Quant. Electron.* **35**, 1146 (1999).

35. H. C. Song, M. C. Oh, S. W. Ahn, and W. H. Steier, *Appl. Phys. Lett.* **82**, 4432 (2003).

36. Y. Huang, G. T. Paloczi, A. Yariv, C. Zhang, and L. R. Dalton, *J. Phys. Chem.* in press.

37. B. Maune, R. Lawson, C. Gunn, A. Scherer, and L. Dalton, *Appl. Phys. Lett.*, **83**, 4689 (2003) and unpublished results on OEO materials.

38. C. Zhang, H. Zhang, M. C. Oh, L. R. Dalton, and W. H. Steier, *Proc. SPIE* **4491**, 537 (2003)

39. Listing of recent articles and articles in press can be obtained at the following websites: http://depts.washington.edu/eooptic and http://stc-mditr.org

40. S. Ermer, D.G. Girton, L.S. Dries, R.E. Taylor, W. Eades, T.E. Van Eck, A.S. Moss, and W.W. Anderson, *Proc. SPIE* **3949**, 148 (2000).
41. M. He, T. M. Leslie, J. A. Sinicropi. *Chem. Mater.* **14**, 4662 (2002).
42. M. He, T. M. Leslie, J. A. Sinicropi, S. M. Garner, L. D. Reed. *Chem. Mater.* **14**, 4669 (2002).
43. E. M. Breitung, C. F. Shu, and R. J. McMahon, *J. Am. Chem. Soc.* **122**, 1154 (2000).
44. P. A. Sullivan, S. Bhattacharjee, B. E. Eichinger, K. Firestone, B. H. Robinson, and L. R. Dalton, *Proc. SPIE*, **5351**, in press (2004).
45. L. R. Dalton, *Pure and Appl. Chem.*, in press.

Mat. Res. Soc. Symp. Proc. Vol. 817 © 2004 Materials Research Society

Design and Fabrication of All-Polymer Photonic Devices

Claire L. Callender[1], Jia Jiang[1], Chantal Blanchetière[1], Julian P. Noad[1], Robert B. Walker[1], Stephen J. Mihailov[1], Jianfu Ding[2] and Michael Day[2]
[1]Communications Research Centre,
P.O. Box 11490, Station H, Ottawa, ON, Canada K2H 8S2
[2]Institute for Chemical Process and Environmental Technology,
National Research Council of Canada, Ottawa, ON, Canada K1A 0R6

ABSTRACT

High quality optical waveguides have been fabricated from fluorinated poly(arylene ether ketone) materials using a standard photolithographic process. Fabrication of waveguide devices on a polymer substrate is described, including a method of end-facet preparation using excimer laser micromachining. Material issues affecting waveguide birefringence and device performance are discussed.

INTRODUCTION

The material properties required for the fabrication of high performance polymer photonic devices are well known – low optical loss, easy processability, tailorable refractive index, high thermal stability and low birefringence. In the past few years, the required properties have been engineered in several classes of polymer and many photonic devices have been fabricated for demonstration or commercial applications [1,2]. Polymers are particularly attractive for photonic components in metro- and local area networks, where low cost is a key issue and performance specifications are less stringent than those in long-haul WDM networks. However, even with the large (20 nm) channel spacings used in current coarse (C)WDM systems, a birefringence of 10^{-4} or less is desirable for polarization independent operation of wavelength- or phase-control devices.

Birefringence in polymer films for photonic applications has been reported from 8×10^{-3} for highly thermally stable polymers such as polyimides [3], to as low as 10^{-5} in fluorinated acrylates [4]. This property is often difficult to optimize, as decreasing birefringence through molecular engineering often results in a concomitant degradation in the thermal stability of the material. Some birefringence compensation can be achieved through waveguide cross-section and device design, but ultimately, the realization of high performance, thermally stable polymer devices requires a good understanding of the material and process parameters that can be adjusted to minimize birefringence.

Birefringence in polymer films depends on the chemical structure of the polymer chains, and on the stresses built up during the processing into thin film on a substrate. Previous work has investigated several factors influencing birefringence in polymer films for waveguide applications. At the molecular level, designing the polymeric structure with a minimum aromatic content in the polymer network can lower film birefringence by reducing the rigidity of the polymer backbone and minimizing stress-producing orientation effects [5]. In addition, previous studies have demonstrated that careful control of solvent evaporation and polymer cross-linking during film heating and cooling processes are effective in partially relieving internal stress that causes thin film optical anisotropy[6].

A dominant factor in thin film stress and resultant birefringence is the mismatch in coefficient of thermal expansion (CTE) of the polymer film and the substrate material. In this work we focus on substrate effects, and the challenges and advantages of fabricating all-polymer devices. Polymer devices can be fabricated on a variety of substrates, but typically silicon wafers, which are cheap, readily available and can be cleaved or diced using conventional semiconductor processing techniques, are employed. However, the large mismatch in the CTE between polymeric materials and silicon leads to birefringence in the polymer layers, and result in temperature and polarization sensitive devices. Athermal and polarization insensitive polymer devices can be fabricated by employing a plastic substrate with a CTE matching that of the waveguide layers [7],[8]. Polymer substrates offer low cost, and good potential for patterning or chip packaging and alignment features such as fiber attach grooves. A major drawback to the use of plastic substrates is that conventional cutting, polishing and cleaving techniques often cannot be used to dice and prepare facets. We have developed techniques for the fabrication of all-polymer photonic devices, involving end-face preparation and dicing of the devices carried out using excimer laser micromachining.

EXPERIMENTAL

The materials used in this work are pentafluorostyrene modified fluorinated poly(arylene ether ketone)s. Replacing C-H bonds with C-F bonds in the polymers decreases the propagation loss at 1550 nm associated with the C-H vibration overtone. In addition the presence of the pentafluorostyrene crosslinker allows thermal processing at lower temperatures (<200°C). The molecular structure of the waveguide core polymer, FSt-FPAEK, is shown in Figure 1. Details of the synthetic methods used to prepare these materials have been described previously [9]. Cladding layers were produced by mixing the core layer material, FSt-FPAEK with bis (2,3,5,6-tetrafluorostyrol)-1H,1H,6H,6H-perfluoro-1,6-hexanediol ether (BSFHE) also shown in Figure 1.

The substrate material used in this work was CR39-ADC™, which is commercially available in the form of optical flats (3" diameter x 1.5 mm thick) and has a refractive index of 1.483 at 1537 nm. Table I shows the average CTE for the substrate and waveguide materials, over the temperature range of the fabrication process (20-200°C).

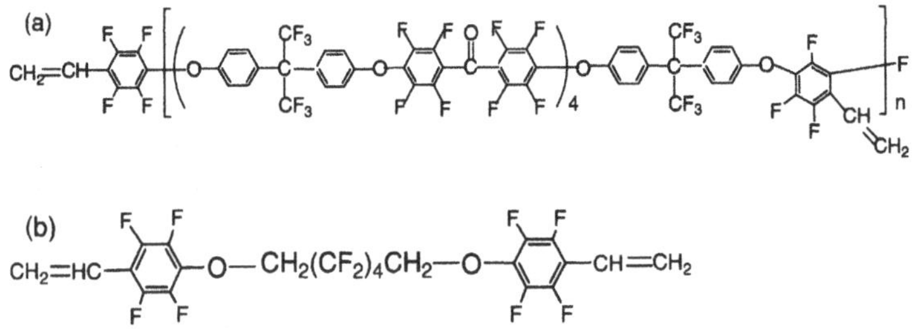

Figure 1: Molecular structures of materials used for waveguide fabrication. (a) (FSt-FPAEK) and (b) (BSFHE).

Table I: CTE values for materials in waveguide structures

Material	CTE(10^{-6}/K) (20 - 200 °C)
Si	2 - 3
SiO$_2$	0.3 - 0.7
ADC	100 - 320
FSt-FPAEK	75 - 185

The refractive indices and birefringence at 1550 nm of the waveguide core and cladding materials were measured using prism coupling. For the waveguides fabricated during this work the core and cladding refractive indices were 1.5033 and 1.4801 respectively. The optical propagation attenuation at 1550 nm was measured using the technique of high index liquid immersion [10]. The slab loss of the core and cladding materials were both around 0.5 dB/cm.

Ridge waveguides were fabricated with FSt-FPAEK polymer on a CR39-ADC™ substrate. A 6 μm thick film of FSt-FPAEK was first deposited by spin coating from a solution of the polymer in cyclohexanone. A standard negative photo-resist lift-off method was used to pattern a nickel mask on the polymer layer, and ridges were then formed using an O$_2$/CHF$_3$ reactive ion etch (RIE) process. A top cladding of FStFPAEK/BSFHE (2:8) co-polymer, with a typical thickness of 10 μm, was deposited over the waveguide ridges by spin coating. The fabrication process is shown in Figure 2. The propagation losses were estimated for similar waveguides fabricated on silica-on-silicon substrate using a cut-back method.

A pulsed ArF excimer laser (λ=193nm) was used for micromachining the all-polymer waveguide devices. The polymer waveguide sample was placed on a motorized translation stage, substrate side toward and normal to the beam. The laser output was apertured and focussed close to the sample. The beam size at the surface of the waveguide sample was typically 0.38 mm x 1.9 mm. In order to minimize problems due to beam non-uniformity and to ensure repeatable

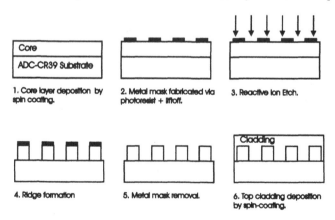

Figure 2: Waveguide fabrication process steps

cuts,the sample was scanned back and forth horizontally relative to the beam using a sweep rate of 0.05 mm/s. The process was divided into two stages: an aggressive and rapid high-fluence ablation through approximately 90% of the substrate followed by a low repetition rate, low-fluence ablation to create a high quality surface on the waveguide end faces. The ablation rate in the substrate material was typically 0.4 μm/pulse. This sequence of cutting optimized the cut quality through both the substrate and waveguide materials, which have widely differing absorption coefficients at 193 nm (26 cm^{-1} and > 1 x 10^4 cm^{-1} respectively). Also, by cutting through the substrate first, redeposition of ablation products on the waveguide layer is minimized. Table II shows typical parameters for the two-stage cutting process.

RESULTS AND DISCUSSION

The birefringence (n_{TE}-n_{TM}) of a 6 μm thick FSt-FPAEK layer on CR39-ADC™ was compared to the value measured for a similar layer on an oxidized (15 μm) silicon substrate. On oxidized silicon the birefringence was +0.0023. On a CR39-ADC substrate the value was -0.0013. The reduction confirms that the closer match in substrate and waveguide CTEs has significantly reduced the stress in the waveguide structure, showing promise for the fabrication of temperature and polarization insensitive devices. The change in sign of (n_{TE}-n_{TM}) suggests that further optimization of birefringence should be possible with a substrate material with an even closer match of CTE. We are currently investigating other substrate materials.

Using the process outlined above, high quality ridge waveguides were fabricated. Propagation losses in 6 μm x 6 μm clad waveguides were typically 0.8-0.9 dB/cm at 1.55 μm. Additional loss of 0.3-0.4 dB compared to the bulk material is typical of ridge waveguides fabricated by photolithographic patterning and can be attributed to sidewall roughness. Waveguides were cut into various lengths using excimer laser cutting. Careful adjustment of cutting conditions was required to avoid thermal degradation of the waveguide ends. Waveguide structures with top cladding were more robust than the bare ridges and typically exhibited much less thermal degradation under similar cutting conditions. The total time of cutting had an effect on the quality of the final endfaces. Immediately after cutting through the waveguide structure, ragged edges, and non-vertical features could be observed; longer cut times resulted in the removal of localized non-vertical features and debris, allowing an optical fiber to be brought into close proximity with the waveguide end. Smooth, vertical faces were easily achieved using the conditions listed in Table II. A laser cut across unclad ridges is shown in Figure 3a, while a typical waveguide profile after laser cutting is shown in Figure 3b, and the end face of clad waveguides is shown in Figure 3c.

Table II: Parameters for excimer laser micromachining of all-polymer waveguide structures

	Initial Cut	Finishing Cut
Fluence	1.5 ± 0.2 J/cm²	0.5 ± 0.07 J/cm²
Repetition Rate	10 Hz	2 Hz
Cut depth	1.35 mm	170 μm
Cut length	9 mm	9 mm
Cut Duration	30 minutes	66 minutes

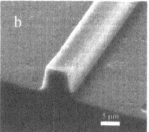

Figure 3: (a), (b) unclad and (c) clad waveguides on CR39-ADC plastic following excimer laser cutting.

The quality of the waveguide facets was assessed by measuring the coupling losses between an optical fiber and the waveguide end face. One piece containing numerous guides was cut into several different lengths, using the parameters listed in Table II. Light from a diode laser at 1550 nm was butt-coupled into each waveguide via a single-mode fiber (SMF-28 9/125) and the transmitted power measured on an optical power meter. Coupling losses were further optimized through the use of index matching fluid between the fiber end and the waveguide facet. The guides support more than one mode, but with careful alignment, the power is propagated predominantly in the lowest order mode. The resulting data are presented in Table III. Theoretical minimum coupling losses were estimated by calculating the overlap between the fiber mode (10.2 μm) and the fundamental mode of each polymer ridge waveguide, using a finite difference method in C2V TempSelene software and are also listed in Table III.

Table III: Experimental and theoretical coupling losses between singlemode optical fiber and all-polymer waveguides.

Waveguide width x height (μm x μm)	5 x 6	6 x 6	7 x 6
Experimental Coupling Loss (dB)	2.6	2.1	1.8
Theoretical Coupling Loss (dB)	1.4	1.2	1.1
Excess Coupling Loss (dB)	1.2	0.9	0.7
Excess Coupling Loss with index matching oil (dB)	0.9	0.4	0.4

CONCLUSIONS

High-quality polymer waveguides have been fabricated in poly(aryl ether ketone) materials using photolithography and reactive ion etching. Propagation losses in straight waveguides are typically 0.8 dB/cm at 1550 nm. A significant reduction in the birefringence of the polymer layers in the waveguide structures has been demonstrated by employing a plastic substrate. All-polymer layer structures offer good potential for the fabrication of polarization insensitive photonic devices. Complete characterization of the birefringence in the ridge structures requires fabrication of photonic devices that are polarization sensitive; work on such devices is currently underway.

Excimer laser micromachining shows excellent promise as a method for preparing endfaces of all-polymer waveguide devices. This technique produces high quality facets, which are often difficult to obtain in polymer materials on semiconductor substrates using cleaving or polishing techniques due to the elastic properties of the polymers and heating effects. The ease with which the plastic substrate can be cut also offers the potential to micromachine alignment grooves for fiber coupling in polymer lightwave circuits fabricated on these substrates. With simple improvements in the beam quality through the use of spatial filters, cylindrical lenses and beam homogenizers, high quality cuts with good depth control could be achieved without sweeping the sample, thus considerably reducing the cut times. This convenient method of preparing high quality waveguide facets will allow more complex all-polymer waveguide devices to be efficiently characterized and fully packaged with optimized insertion losses.

REFERENCES

[1] M. Zhou, "Low-loss polymeric materials for passive waveguide components in fiber optical telecommunication", Opt. Eng., vol. 41(7), pp. 1631-1643, 2002 .

[2] L. Eldada, "Polymer integrated optics: promise vs. practicality", in Proc. SPIE vol. 4642, *Organic Materials and Devices IV*, 2002, pp 11-22.

[3] J. Kobayashi, T. Matsuura, Y. Hida, S. Sasaki and T. Maruno, "Fluorinated polyimide waveguides with low polarization-dependent loss and their applications to thermooptic switches", J. Lightwave Technol. 16(6), 1024-1029, (1998)

[4] L. Eldada, R. Blomquist, M. Maxfield, D. Pant, G. Boudoughian, C. Poga and R.A Norwood, "Thermooptic planar polymer Bragg grating OADM's with broad tuning range" IEEE Photon. Technol. Lett. 11(4), 448-450 (1999).

[5] J.Jiang, C.L. Callender, C. Blanchetière, J.P. Noad, J. Ding, Y. Qi and M. Day, "Optimizing fluorinated poly(arylene ether)s for optical waveguide applications" submitted to Optical Materials.

[6] J. Jiang, C.L. Callender, C. Blanchetière, J.P. Noad, J. Ding and M. Day, "Birefringence Control for Polymer Optical Waveguide Devices", Proc. SPIE vol. 5260, *Applications of Photonic Technology 6* (2003), p. 324-330.

[7] N. Keil, H.H. Yao, C. Zawadzki, J. Bauer, M. Bauer, C. Dreyer and J. Schneider, "Athermal all-polymer arrayed-waveguide grating multiplexer", Electron. Lett., vol. 37(9), pp. 579-580, 2001.

[8] S. Sakaguchi, Y. Moroi, H. Nanai, T. Hayamizu, Y. Yamamoto and K. Maeda, "Fluorinated polyimide for low-loss optical waveguides at 1.55 μm", in Proc. SPIE vol. 4653, *WDM and Photonic Switching Devices for Network Applications III*, 2002, pp 36-44.

[9] J. Ding, F. Liu, M. Li, M.Day and M. Zhou, "J. Polym. Sci. Part A, Polym. Chem., 40, 4205-4216 (2002).

[10] C.-C. Teng, "Precision measurements of the optical attenuation profile along the propagation path in thin-film waveguides", Appl. Opt, 32(7), 1051-1054 (1993).

Mat. Res. Soc. Symp. Proc. Vol. 817 © 2004 Materials Research Society   L7.5.

Dye-Doped, Polymer-Nanoparticle Gain Media for Tunable Solid-State Lasers

F. J. Duarte[1] and R. O. James[2]
[1]Eastman Kodak Company, R& D Laboratories, Rochester, New York 14650, U.S.A.
[2]792 Oakridge Dr., Rochester, NY 14617, U.S.A.

ABSTRACT

Tunable laser action, in the visible spectrum, has been established using dye-doped, polymer-silica nanoparticle gain media. The silica nanoparticles, averaging about 12 nm in diameter, appear to be uniformly dispersed in the polymethyl methacrylate (PMMA) matrix, since the optical homogeneity of the gain medium is maintained. Using rhodamine 6G dye and 30% weight-by-weight (w/w) silica nanoparticles, laser action was established in the 567–603 nm range. At the peak wavelength ($\lambda \sim 580$ nm) laser conversion efficiency is ~63% at a beam divergence of 1.9 mrad (~1.3 times the diffraction limit). The new solid-state nanocomposite gain media also exhibits a reduction in $|\partial n / \partial T|$ because the thermo-optic coefficient of silica is opposite in sign to that of the PMMA polymer-host component.

INTRODUCTION

Tunable solid-state organic lasers using dye-doped polymer gain media have been demonstrated to yield narrow-linewidth [1,2] and single longitudinal-mode (SLM) [3,4] emissions in compact multiple-prism grating oscillator configurations. An extensive review of recent developments of polymeric solid-state dye lasers has been recently reported by Costela et al. [5] The excellent optical homogeneity of these materials is one of the most important features of the dye-doped polymer gain media that allows attainment of TEM$_{00}$ beam profiles and, consequently, the achievement of SLM emission [1,5,6]. Nevertheless, nearly all polymer gain media exhibit relatively high negative $\partial n / \partial T$ values that impose limitations on the pulse-repetition frequency (prf) and, ultimately, on the average power obtainable from these lasers [1,5-9].

One possible avenue available to improve the thermal properties of dye-doped, solid-state gain media is to use a composite organic inorganic matrix where the inorganic part is based on siliceous components (or other oxides with positive values for $\partial n / \partial T$). Examples of such gain media are dye-doped, organically modified silicate (ORMOSIL) [10], tetraethoxysilane [1,11], and silica-polymer composites [12].

As far as narrow-linewidth and SLM-emission is concerned, traditional organic-inorganic composites are not well suited because they have been shown to exhibit internal refractive index inhomogeneities that produce a spatial decomposition of the emission [1,11,12]. This effect is due to internal interference and can be best illustrated by propagating a TEM$_{00}$ laser beam through the material and observing the spatial profile of the beam a few meters from the sample. This problem with the beam quality of hybrid organic-inorganic nanocomposites has been noted in the most recent review by Costela et al. [5] who point out that it should be possible to "*improve the homogeneity of the hybrid materials*" through a range of chemical factors and properties of the gain media components.

In this report, we describe laser emission using a new class of dye-doped, organic-inorganic solid-state gain media that exhibits lower $|\partial n / \partial T|$ values and improved optical homogeneity, as compared to previous composite gain media. Further, this laser emission exhibits lower beam divergence [13,14]. The new nanocomposite gain media consists of dye-doped optics-quality polymethyl methacrylate (PMMA), including dispersed silica nanoparticles. One example of such laser dye-doped polymer-nanoparticle (DDPN) gain media is a rhodamine 6G-doped PMMA matrix containing 30% w/w silica in which the silica content is composed of ~12 nm SiO_2 particles. This particular DDPN gain medium shows conservation of TEM_{00} spatial beam characteristics, approaching that of the DDP gain media, as illustrated in figure 1

Figure 1. Beam profile following propagation of a TEM_{00} beam through a laser DDPN gain medium. As explained in the text, the TEM_{00} characteristics are conserved.

EXPERIMENTAL DETAILS

The laser experiments were performed using nitrogen laser excitation (at 337 nm) and a tunable nitrogen laser-pumped coumarin 152 (at a concentration of 10 mM) prismatic dye laser, incorporating a Glan-Thompson polarizer, as described by Duarte et al. [2] This laser can deliver up to 2 mJ in 3–4 ns pulse (FWHM) and is tunable in the 520–555 nm region. Excitation of the DDPN laser is accomplished longitudinally using the tunable coumarin 152 dye laser [2,13,14]. Beam profiles were recorded photographically. Laser cavities and oscillator architecture for solid-state lasers have been discussed earlier by Duarte [14,15].

Synthesis and methods of fabrication of the DDPN gain media are described in detail elsewhere [13,14,16]. These methods involve mixing colloidally stable SiO_2 nanoparticle dispersions in solvents (organosilicasols) with solutions of optical-grade PMMA resins in solvents that are compatible with the silica sol over a wide range of concentrations. The laser dye is also added as a dissolved component in a compatible solvent. The nanoparticle-polymer, laser-dye composite is formed from this colloidally stable dispersion mixture by slow solvent evaporation from a partially covered mold or container using a solvent-stripping method over a period exceeding one week. An example preparation involves weighting and mixing the following: (1) 54.83 g of silica sol (Nissan Chemical America Corporation), composed of organosilica sol MEK-ST, 12 nm diameter, 30.5% w/w SiO_2 in methyl ethyl ketone (MEK); (2) 97.57 g of PMMA solution (Atoglas, Plexiglas V-series, VLD) at 20.0% w/w solids in MEK; (3) 65.05 g PMMA solution (Plexiglas VLD) at 30.0% w/w in methylene chloride and; (4) 37.17 g of 0.1% w/w solution of rhodamine 6 G (Eastman Kodak Company) in methylene chloride. This dispersion example is 21.92% solids, and the solids are composed of 29.98% SiO_2, 69.95% PMMA, and 0.069% rhodamine 6G. The solvent blend is 41.55% $MeCl_2$ and 58.45% MEK. Samples are placed in containers, and the solvent is slowly stripped away to form a gel, then a solid rigid body. It is possible to vary the solvent mixture ratios in the transparent, colored dispersion and also the SiO_2 particle content of the solid body up to approximately 50% w/w, providing a range of nanoparticle-filled, laser dye-doped polymer gain media.

To investigate the distribution of silica nanoparticles in the gain media, transmission electron microscope nanographs of thin film sections of the composite gain media used in the laser experiments were obtained [14]. The transmission electron microscope used is manufactured by JEOL (Model JEM 100CX II). Its achievable resolution is approximately 0.3 nm. The gain media was sectioned on a Reichert-Jung microtome. A nanograph for the rhodamine 6 G dye-doped PMMA-silica composite gain medium is shown in figure 2.

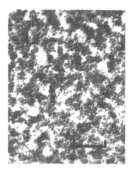

Figure 2. TEM nanograph of a thin section of rhodamine 6G DDPN laser gain medium. The scale shown corresponds to 200 nm.

The depth of field is about 40 nm. Therefore, it is assumed that particles in sharp focus are in a layer thickness of approximately 40 nm. The average diameter of individual silica nanoparticles is about 13 nm, with a range from about 9 to 18 nm. From nanographs such as these, it is estimated that, on average, the silica nanoparticles associated in loose clusters (particle-enriched "islands") leaving the dye-doped polymer depleted of particles ("slits") is approximately in the range of 50–60 nm between the groups of clusters. It appears that the silica particle clusters, or islands, are distributed in a spatial structure perhaps similar to a disorganized fine grating. The electron microscopy image samples were made with 0.3 w/w silica in dye-doped PMMA [13,14]. Using the specific gravity for silica and for PMMA, this corresponds to a volume fraction of 0.19 v/v. The volume of the composite silica nanoparticle islands can be approximated by a disk of about 60 nm in diameter and at least 40 nm in thickness. Using this volume (~113,100 nm^3), the volume of particle (radius ~7 nm) and counting the particles per island (~20 to 25), we make a preliminary estimate of the volume fraction of particles in the nanoclusters to be in the range of 0.25–0.31 v/v or lower. This fairly crude estimated volume fraction for the weak aggregates can be compared with the average 0.19 v/v for the macroscopic gain media.

Gain media with trapezoidal cross sections, at a plane parallel to the plane of propagation, are optically polished using standard techniques. The comparative rhodamine 6G DDP medium is the same used in previous experiments [1-3] and has been characterized in detail by Maslyukov et al. [17] and Popov [18]. The gain length used in these experiments was ~10 mm.

RESULTS AND DISCUSSION

Table I lists laser emission parameters measured for rhodamine 6G DDP and rhodamine 6G DDPN at 30% w/w and 50% w/w silica.

Table I. Performance of solid-state lasers incorporating DDP and DDPN gain matrices

Gain Matrix	C (mM)	λ_a (nm)	Tuning Range (nm)	$\Delta\theta$ (mrad)	% Efficiency
DDP	0.50^a	~525	563–610	2.3	49
DDPN at 30% w/w SiO$_2$	$0.31^{a,b}$	~525	567–603	1.9	63
DDPN at 50% w/w SiO$_2$	$0.31^{a,b}$	~550	570–600	1.6	9

^{a} Initial dye concentration in liquid.
^{b} Dye concentration in the solid gain volume increases to ~1.9 mM in dry solid.

The results in table I indicate that the rhodamine 6G DDPN with 30% w/w SiO$_2$ laser exhibits a conversion efficiency of 63%, with an optimum absorption wavelength (λ_a) and a tuning range similar to that exhibited by the standard rhodamine 6 G DDP gain medium. As the concentration of nanoparticles is increased to 50%, the laser conversion efficiency declines significantly, and the optimum λ_a is red shifted by ~25 nm. A typical cross section of the laser beam achieved with the rhodamine 6G DDPN, at 30% w/w SiO$_2$, is depicted in figure 3.

Figure 3. Laser emission beam profile obtained with the DDPN gain medium at 30% w/w SiO$_2$ using a simple mirror-grating cavity.

In these experiments, we also achieved lasing in the blue-green region of the spectrum using direct nitrogen laser excitation of a coumarin 500 DDPN, at 30% w/w SiO$_2$, gain medium in a simple mirror-mirror cavity.

In addition to the laser performance, the $\partial n / \partial T$ for the new polymer—nanoparticle-optical materials—were measured in a thin film configuration using a commercial prism-coupling device (Metricon 2010). The results, given in table II, clearly illustrate that as the SiO$_2$ content of the optical matrix is increases, the $|\partial n / \partial T|$ value decreases. For a 50% content of SiO$_2$, the $|\partial n / \partial T|$ experiences a decrease by ~33%. Here, it should be mentioned that colloidal and amorphous silica have *positive* $\partial n / \partial T$ coefficients.

Table II. $\partial n / \partial T$ in DDP and DDPN matrices.

Matrix[a]	λ (nm)	$\partial n / \partial T$	Reference
DDP[b]	594	$-1.4 \pm 0.2 \times 10^{-4}$	Duarte et al. [5]
PN at 0% w/w SiO$_2$	632.8	-1.0317×10^{-4}	Duarte and James [13]
PN at 30% w/w SiO$_2$	632.8	-0.8840×10^{-4}	Duarte and James [13]
PN at 50% w/w SiO$_2$	632.8	-0.6484×10^{-4}	Duarte and James [13]

[a] P stands for the PMMA polymer.
[b] Measurement performed using refraction at minimum deviation.
[c] In these measurements, the polymer-nanoparticle matrix was not dye doped.

The results presented here indicate that new rhodamine 6G dye-doped, polymer-nanoparticle gain media, with improved $|\partial n / \partial T|$ coefficients, emit with laser beams exhibiting near-TEM$_{00}$ profiles. For the case of the laser incorporating the matrix containing a 30% w/w of SiO$_2$ nanoparticles, a 63% conversion efficiency was measured at $\Delta\theta \sim 1.9$ mrad, which is ~1.3 times the diffraction limit. It has been shown that for the DDPN described here, even though it does have some non-random nanoclustering in the structure, this structure is small enough that it does not induce significant internal interference that might lead to macro manifestations of laser beam decomposition as was evident in early forms of organic-inorganic composite gain media. The relatively higher conversion efficiency observed with this DDPN matrix is most likely a result of a net higher number of dye molecules in the excitation volume of the gain medium that results from the evaporation of the solvents. We estimate that from the initial concentration of 0.31 mM dye in the liquid, volume contraction increases the dye concentration, in the solid, to approximately 1.9 mM.

The improvement in beam divergence, observed for the laser emission originating from the DDPN gain media appears to be mainly the result of the decrease in thermal lensing as a result of the lower $|\partial n / \partial T|$ values. This reduction in $|\partial n / \partial T|$ is desirable to control thermal lensing in dye-doped polymer gain media, as had been articulated previously [1-3].

An estimate of the effect of scattering by SiO$_2$ nanoparticles on the optical density of gain media with 1cm path length was made using Mie theory calculations. For a wavelength of 590nm in vacuum, taking the refractive index of PMMA as 1.49 and that of the nanoparticle silica as 1.45, the scattering efficiency, Q_{sca}, for 9nm, 12nm and 16nm diameter particles was found to be $2.23*10^{-8}$, $7.05*10^{-8}$, and $2.22*10^{-7}$ respectively. Use of a volume fraction of 0.19 for the SiO$_2$ particles in the PMMA matrix, the computed volume per particle and the geometric cross-section, yields the estimated optical density (OD = $\log_e(I_o/I_t)/2.3026 = N.C_{sca}.l/2.3026$) as

0.0061, 0.0145 and 0.034 respectively. This lends support to the observations that the gain medi■ is transparent, has low haze and allows one to read text placed under the gain media. The results reported in this report may open a new approach toward the development of practical, compact, and inexpensive, tunable lasers emitting in the visible spectrum with attractive conversion efficiencies and lower $|\partial n/\partial T|$ values. Here, it is also important to indicate that promising newly synthesized, dye-doped organic-inorganic gain media, using traditional tetraethoxysilane-based matrices, have been reported by Costela et al. [19,20]

ACKNOWLEDGMENTS

For skillful technical assistance at various stages of this project, the authors are grateful to W. G. Ahearn, A. M. Miller , and L. A. Rowley. The nanographs were recorded by T. R. Vandam.

REFERENCES

1. F. J. Duarte, *Appl. Opt.* **33**, 3857 (1994).
2. F. J. Duarte, A. Costela, I. Garcia-Moreno, R. Sastre, J. J. Ehrlich, and T. S. Taylor, *Opt. Quantum Electron.* **29**, 461 (1997).
3. F. J. Duarte, *Opt. Commun.* **117**, 480 (1995).
4. F. J. Duarte, *Appl. Opt.* **38**, 6347 (1999).
5. A. Costela, I. Garcia-Moreno, and R. Sastre, *Phys. Chem. Chem.Phys.* **5**, 4745 (2003).
6. F. J. Duarte, *Optics and Photonics News* 14(10), 20 (2003).
7. F. J. Duarte, A. Costela, I. Garcia-Moreno, and R. Sastre, *Appl. Opt.* **39**, 6522 (2000).
8. A. Costela, I. Garcia-Moreno, and R. Sastre, in *Handbook of Advanced Electronic and Photonic Material: Liquid Crystals, Display and Laser Materials*, edited by H. S. Nalwa, (Academic, New York, 2001).
9. W. Holzer, H. Gratz, T. Schmitt, A. Penzkofer, A. Costela, I. Garcia-Moreno, R. Sastre, and F. J. Duarte, *Chem. Phys.* **256**, 125 (2000).
10. B. Dunn, J. D. McKenzie, J. I. Zink, and O. M. Stafsudd, *Proc. SPIE* **1328**, 174 (1990).
11. F. J. Duarte, J. J. Ehrlich, W. E. Davenport, T. S. Taylor, and J. C. McDonald, in *Proceedings of the International Conference on Lasers '92*, edited by C. P. Wang, (STS, McLean, 1993), pp. 293-296.
12. F. J. Duarte and E. J. A. Pope, *Ceram. Trans.* **55**, 267 (1995).
13. F.J. Duarte and R.O. James, *Optics Lett.* **28**(21), 2088-2090 (2003)
14. F.J. Duarte and R.O. James, *Appl. Opt.* (2004), *in press.*
15. F.J. Duarte, *Tunable Laser Optics*, (Elsevier Academic, New York, 2003).
16. F. J. Duarte, R. O. James, and L. A. Rowley, U.S. Patent Application (filed on 22 December, 2002).
17. A. Maslyukov, S. Solkolov, M. Kaivola, K. Nyholm, and S. Popov, *Appl. Opt.* **34**, 1516 (1995).
18. S. Popov, *Pure Appl. Opt.* 7, 1379 (1998).
19. A. Costela, I. Garcia-Moreno, C. Gomez, O. Garcia, and R. Sastre, *Appl. Phys. B* **75**, 827 (2002).
20. A. Costela, I. Garcia-Moreno, C. Gomez, O. Garcia, and R. Sastre, *Chem. Phys. Lett.* **369**, 656 (2003).

Mat. Res. Soc. Symp. Proc. Vol. 817 © 2004 Materials Research Society

Enhanced Faraday rotation in garnet films and multilayers

S. Kahl and A. M. Grishin
Condensed Matter Physics, Royal Institute of Technology, S-16440 Stockholm, Sweden

ABSTRACT

Films were prepared by pulsed laser deposition. We investigated or measured crystallinity, morphology, film-substrate interface, cracks, roughness, composition, magnetic coercivity, refractive index and extinction coefficient, and magneto-optical Faraday rotation (FR) and ellipticity. The investigations were partly performed on selected samples, and partly on two series of films on different substrates and of different thicknesses. BIG films were successfully tested for the application of magneto-optical visualization. The effect of annealing in oxygen atmosphere was also investigated - very careful annealing can increase FR by up to 20%.

Periodical BIG-YIG multilayers with up to 25 single layers were designed and prepared with the purpose to enhance FR at a selected wavelength. A central BIG layer was introduced as defect layer into this one-dimensional magneto-optical photonic crystal (MOPC) and generated resonances in optical transmittance and FR at a chosen design wavelength. In a 17-layer structure, at the wavelength of 748 nm, FR was increased from -2.6 deg/μm to -6.3 deg/μm at a small reduction in transmittance from 69% to 58% as compared to a single-layer BIG film of equivalent thickness. In contrast to thick BIG films, the MOPCs did not crack. We were first to report preparation of all-garnet MOPCs and second to experimentally demonstrate the MOPC principle in magneto-optical garnets.

INTRODUCTION

It was realized at the end of the 1960s that bismuth doping strongly enhanced the Faraday rotation (FR) in iron garnets [1]. Since then, iron garnets with higher and higher bismuth contents have been prepared. Thin films made by liquid phase epitaxy could be prepared with higher bismuth content than single crystals [2], but the complete bismuth substitution with three bismuth atoms per chemical formula unit was reported for the first time at the end of the 1980s [3]. Bismuth iron garnet (BIG), $Bi_3Fe_5O_{12}$, is not stable in bulk form and has so far only been prepared by vapour deposition techniques.

FR is a nonreciprocal effect and can therefore be enhanced strongly at a chosen wavelength by placing magneto-optical material inside an optical cavity [4]. This concept was also proposed for iron garnet films where the mirrors of the cavity are highly reflective multilayers composed of two materials with a difference in refractive index [5]. Single layers have quarter-wave thickness in the respective material. The magneto-optical material inside the cavity is an iron garnet film of different thickness: a defect layer inside this one-dimensional magneto-optical photonic crystal (MOPC). More advanced concepts with several defect layers have been proposed as well, in order to improve transmittance [6]. Simple examples of MOPCs have been realized experimentally in two different ways, either by using polycrystalline garnet films sandwiched between dielectric quarter-wave layers [7,8] or by preparation of heteroepitaxial garnet multilayers [9]. The latter have the advantage of better optical quality due to the epitaxial growth and higher FR because a higher bismuth substitution can be achieved on garnet substrates. On the other hand, it is possible to achieve a higher contrast in refractive index between dielectric layers, which reduces the overall lengths of the structures.

Here, we review our results on laser deposited BIG single-layer films and heteroepitaxial garnet multilayers composed of BIG and yttrium iron garnet (YIG) layers.

Bismuth iron garnet films

We prepared BIG films on gadolinium gallium garnet (GGG) substrates by pulsed laser deposition. Details of film preparation can be found in Refs. [10–15]. The results presented in these references can be summarized as follows:

- The deposition parameters of 550 °C substrate temperature and 25 μbar oxygen background pressure give films of a favorable combination of crystalline and optical quality and magneto-optical FR [13].

- BIG films grow in epitaxial quality on all tested garnet substrates [10, 13]. X-ray coherence lengths decrease with increasing film thickness [15].

- 500 to 2500-nm-thick BIG/GGG(111) films have granular structures with grain size of 300 to 500 nm and rms surface roughnesses between 10 and 40 nm [15].

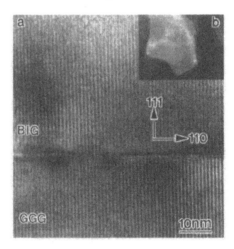

Figure 1. TEM cross section of a BIG film on a GGG(111) substrate (from Ref. [11]). a) High resolution image of the interface. b) Grain of deviating phase.

- The interface between BIG film and GGG(111) substrate is atomically smooth in large parts, but there are also inclusions of a different phase directly at the interface, see Fig. 1 [11].

- Cracks could not be avoided in BIG/GGG(111) films for thicknesses above 1 μm [15].

- We did not obtain stoichiometric transfer from the target at the optimized deposition conditions, the films are all bismuth deficient with concentration ratios $c_{Bi}/c_{Fe} \approx 0.48$ [15]. The target had the correct stoichiometry.

- Depending on film thickness and choice of substrate, the magnetic coercivity H_c varied between 4 kA/m (40 Oe) and 42 kA/m (530 Oe) [13]. Reduction of H_c below 4 kA/m was possible, but only at the cost of reduced optical quality.

- BIG films absorb strongly at wavelengths below 550 nm, but absorption decreases exponentially towards longer wavelengths and the absorption coefficient falls below $\alpha = 1800$ cm^{-1} at $\lambda = 700$ nm [14, 15].

- The refractive index of BIG films falls from approximately 3.05 at 500 nm wavelength to approximately 2.65 at 850 nm [14, 15].

- FR in visible light is very strong with maximum values of more than 20 deg/μm at wavelengths between 530 and 540 nm [11, 14, 15].

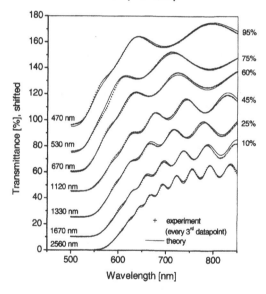

Figure 2. Measured and calculated transmittance spectra for seven BIG/GGG(111) films of thicknesses given in the figure (from Ref. [15]). Each spectrum has been shifted upwards by the percentage indicated.

- Parameters determining transmittance and FR were deduced from the comparison of measurement and theory. Measured spectra were reproduced well by calculations that include the effects of thin film interference, see Figs. 2 and 3 [14, 15].

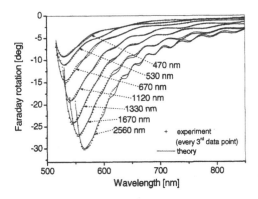

Figure 3. Measured and calculated spectra of Faraday rotation for BIG/GGG(111) film of thicknesses given in the figure (from Ref. [15]).

- The FR spectrum between 500 and 850 nm wavelength can be described by a single diamagnetic line [14, 15].

- The peak of FR is accompanied by strong Faraday ellipticity, which, however, falls off more quickly towards longer wavelengths than FR does [11].

- BIG films can easily be destroyed by annealing. By careful annealing in oxygen, though we could increase the FR angles by 15% to 20% [15].

- BIG films have very competitive magneto-optical figures of merit and optical efficiencies for magneto-optical visualisation [14].

- Test measurements have shown that BIG films can be used as magneto-optical visualizer films [10, 12].

Magneto-optical photonic crystals

This spring, we reported the preparation of magneto-optical heteroepitaxial garnet multi layers with photonic bandgap [9]. The structures comprise BIG and YIG single layers and show promising characteristics. We define a hypothetical BIG layer of equivalent thickness (equal to the sum of all single BIG layer thicknesses) to assess the effect of the multilayer structure on the spectra of transmittance and FR. The obtained results are:

- BIG-YIG MOPCs were grown successfully at the deposition parameters for BIG single layer films.

- All single layers (up to 25) are of epitaxial quality and seem to be exchange coupled.

- No cracks appeared in any MOPC (with up to 2130 nm thickness).

- Properties of the 'best' MOPC with a central BIG layer sandwiched between four (BIG/YIG) double layers on each side, with transmission resonance at 748 nm:
  - Increase in FR angle by 140% as compared with a single layer BIG/GGG of equivalent thickness.
  - Decrease in transmittance by 16% as compared with a single layer BIG/GGG of equivalent thickness.

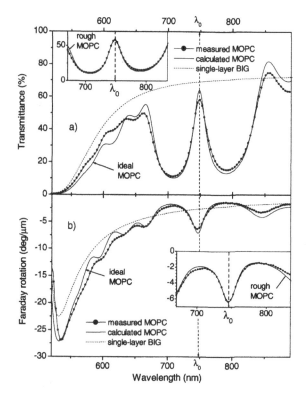

Figure 4. Measured and calculated spectra of the 'best' MOPC (see text), from Ref. [9]: a) of optical transmittance, b) FR. Dotted lines are interference-free transmittance and FR for an equivalent 836-nm-thick single layer BIG/GGG. $\lambda_0 = 748$ nm is wavelength of the transmission resonance. Calculations in the insets were performed for a rough MOPC.

- Transmittance and FR spectra were calculated with good accuracy from the data obtained for single-layer BIG and YIG films, using just two additional fitting parameters: The resonance wavelength λ_0 was adjusted manually, and the oscillator strength of the magneto-optical transition was increased by 25% as compared to the single-layer BIG film in Ref. [15] with the highest FR per film thickness. Measured and calculated spectra are shown in Fig. 4. Calculations in the insets assume a Gaussian distributic of film thicknesses with a standard deviation of 11 nm and homogeneous thicknes variations of all layers.

REFERENCES

1. C. F. Buhrer, *J. Appl. Phys.*, **40**, 4500 (1969).

2. P. Hansen, W. Tolksdorf, and K. Witter, *IEEE. Trans. Magn.*, **MAG-20**, 1099 (1984

3. T. Okuda, N. Koshizuka, H. Hayashi, H. Taniguchi, K. Satoh, and H. Yamamoto, *IEE Translation J. Magn. Jpn.*, **3**, 483 (1988).

4. R. Rosenberg, C. B. Rubinstein, and D. R. Herriott, *Appl. Opt.*, **3**, 1079 (1964).

5. M. Inoue, K. Arai, T. Fujii, and M. Abe, *J. Appl. Phys.*, **83**, 6768 (1998).

6. M. J. Steel, M. Levy, and R. M. Osgood, *IEEE Photon. Techn. Lett.*, **12**, 1171 (2000)

7. M. Inoue, K. Arai, T. Fuji, and M. Abe, *J. Appl. Phys.*, **85**, 5768 (1999).

8. H. Kato, T. Matsushita, A. Takayama, M. Egawa, K. Nishimura, and M. Inoue, *J. App Phys.*, **93**, 3906 (2003).

9. S. Kahl and A. M. Grishin, *Appl. Phys. Lett.*, **84**, 1438 (2004).

10. S. Kahl, A. M. Grishin, S. I. Khartsev, K. Kawano, and J. S. Abell, *IEEE Trans. Magn* 3, 2457 (2001).

11. S. Kahl, S. I. Khartsev, A. M. Grishin, K. Kawano, G. Kong, R. A. Chakalov, and J. ﾟ Abell, *J. Appl. Phys.*, **91**, 9556 (2002).

12. S. Flament, Warsito, L. Mechin, C. Gunther, K. Kawano, G. Kong, R. A. Chakalov, ﾟ S. Abell, S. Kahl, S. I. Khartsev, and A. M. Grishin, *J. Appl. Phys.*, **91**, 9556 (2002).

13. S. Kahl and A. M. Grishin, *J. Appl. Phys.*, **93**, 6945 (2003).

14. S. Kahl and V. Popov and A. M. Grishin, *J. Appl. Phys.*, **94**, 5688 (2003).

15. S. Kahl and A. M. Grishin, *J. Magn. Magn. Mater.*, in press.

Mat. Res. Soc. Symp. Proc. Vol. 817 © 2004 Materials Research Society L8.3

PARTIAL PRESSURE DIFFERENTIAL AND RAPID THERMAL ANNEALING FOR INTEGRATED YTTRIUM IRON GARNET (YIG)

SANG-YEOB SUNG, XIAOYAUN QI, SAMIR K. MONDAL, AND BETHANIE J. H STADLER

In this work, magneto-optical garnets were grown monolithically by low-temperature reactive RF sputtering, followed by an ultra-short (< 15sec) anneal. It was found that in addition to low thermal budgets due to timing, the temperature required (< 750°C) for garnet crystallization was also reduced compared to standard tube furnace annealing (> 1000°C). MgO and fused quartz were used as substrates because they will be useful for future buffer layers and optical claddings. Y-Fe-O films were made with systematically varied compositions and the chemical, structural, and optical properties of the resulting films were analyzed. A solid solution single phase field for YIG was found that spanned a wide range of compositions (30.1 ~ 49.0 atomic % of Fe). The resulting YIG quality was measured by vibrating sample magnetometry (VSM), X-ray diffraction (XRD), and measurements of Faraday rotation (FR). Although the XRD results showed that the films had isotropic crystallinity, the VSM indicated that shape anisotropy dominated the magnetic properties. Out of plane FR measurements yielded up to 0.2°/μm at 632nm rotations. This rotation will be higher in plane. All of these tests demonstrated that the YIG was comparable to YIG grown by standard annealing and also by in-situ crystallization.

Introduction

The critical active element in optical isolators is a magneto-optical garnet. These isolators are required for integrated light sources as they allow extended lifetimes by blocking back-reflected light. However, garnet is difficult to integrate with semiconductors due to the high thermal budget usually required to obtain the garnet crystal structure. For example, current isolator garnets cannot be integrated monolithically into a photonic integrated circuit due to the growth process of liquid phase epitaxy (LPE). Therefore, for semiconductor integration, low-temperature semiconductor-friendly process, such as reactive RF sputtering, is needed. In addition to this, a buffer layer is also needed when using semiconductor platforms. In previous work [1], $Y_3Fe_5O_{12}$ films were successfully fabricated on MgO substrates with single target reactive RF sputtering. MgO is good optical cladding that can also be deposited by sputtering onto semiconductors. However, the conventional gas feed-through of this previous technique led to a slow deposition rate and post-annealing was often required to get the $Y_3Fe_5O_{12}$ garnet phase. Non-reactive sputtering of yttrium-iron-garnet (YIG) target has also been reported to have small deposition rate (5nm/min) [2]. Other research reported a higher deposition rate (10.8nm/min) [3], but very high RF power densities (~4.93W/cm^2) were required. The use of this high power density is likely to lead to cracking of oxide targets. To improve the deposition rate in a less destructive way and to reduce thermal damage from post annealing, new method is needed. In this article previous work with conventional gas feed-through and furnace post-annealing is compared to a new method involving a partial pressure differential and rapid thermal annealing (RTA). The partial pressure differential is defined by a drop in oxygen partial pressure in the sputtering chamber between the substrates and the metallic targets. This differential enables

oxide films to be grown at metallic sputtering rates which are 5 ~ 10 times typical oxid
sputtering rates.

Experimental

Yttrium iron garnet (YIG, $Y_3Fe_5O_{12}$) films were fabricated by two methods. First, Perk
Elmer 2400 RF magnetron sputtering system was used. In this system, the targets cannot k
focused confocally onto a substrate so variations in film composition can only be achieved k
either rotating the substrate between targets, or by using a single target of the desired fil
composition. In the fomer case, the substrate only dwells beneath a target 1/5 of the depositic
time. Since deposition rates of the oxide targets were already very slow, single-target depositic
was used. In order to increase the deposition rate, metallic targets were used in an attempt to c
reactive sputtering. However, gases were fed into the system in a conventional way involvir
one point of entry. This latter feature means that gases mix evenly throughout the sputterir
system and the oxygen partial pressures required to grow oxide films were also sufficient t
oxidize the target. The sputtering rate was therefore similar to that of an oxide target. The targe
was an 8" yttrium target to which Fe foils were attached with 3:1 ratio. Double-side polishe
(100) MgO substrates were used, and other sputtering parameters are listed in Table 1. Th
resulting films were not completely crystallized as deposited. To crystallize the films, a sma
conventional tube furnace was used for post annealing. Varying annealing temperatures wer
used, and the annealing time was fixed at 3hours. A schematic diagram of single target system i
shown in Figure 1.

A second method was adopted in order to improve compositional control and to reduc
the risk of thermal damage for semiconductor substrates. A modified multi-target reactive R
magnetron sputtering system was used to fabricate $Y_3Fe_5O_{12}$. The targets were 2" yttrium an
iron discs. The Ar and O_2 gases were fed into the system at the targets and the substrate
respectively, rather than from one point. The flow rates varied from 20 to 40mTorr and 1 t
4.5mTorr, respectively. Most of films were sputtered 30minutes, and substrate stage was rotate
while sputtering for uniform thickness distribution. Fused quartz and polished MgO were used a
substrates in this work. Other sputter parameters are on table 2. The films made by this techniqu
were all amorphous. To crystallize the films, rapid thermal annealing (RTA) was used for th
post-annealing process. Annealing temperatures and times were varied from 600 to 1000°C an
from 5 to 120 seconds, respectively. Figure 2 is schematic diagram of multi-target system.

Results and Discussion

Most of the films deposited from the single target system were slightly polycrystallin
and had both yttrium oxide and iron oxide phases. After the post annealing process, which used
small conventional tube furnace under air atmosphere, most of films were composed of a mixtur
of three phases: $Y_3Fe_5O_{12}$, $YFeO_3$, and $YO_{1.141}$. The films were crystallized at 1000°C fc
3hours. The exception to this was found only in the films that contained only Y_2O_3 as deposite
These films became single-phase $Y_3Fe_5O_{12}$ after post annealing. X-ray diffraction patterns c
before and after annealing are shown in Figure 3. Since the target was metallic, a high O_2 flov
rate was used to grow oxide films and this caused the target to oxidize, which in turn reduced th
sputter rate. Deposition rate was 1.62~2.11nm/min.

Table 1 Sputtering parameters of single target system.

Sputtering Parameters	Experimental conditions
Target	Fe foils on 8" Y disk target - 3:1 ratio
Substrate material	Polished MgO (100)
RF power density (W/cm^2)	1.37~1.71
Background Pressure (Torr)	2.00E-06
Target distance (mm)	83~103
Process pressure (mTorr)	29~44
Sputtering gas /Flow rate (sccm)	Ar / 20~30
O^2 gas flow rate (sccm)	10~15.4
Deposition rate (nm/min)	1.62~2.11
Deposition time (min)	120
Post Annealing Parameters	Experimental conditions
Method	Conventional Annealing
Temperature (°C)	800~1000
Atmosphere	Air
Time (hr)	3

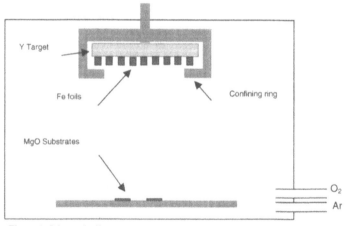

Figure 1 Schematic diagram of the single target RF magnetron sputtering system.

Table 2 Sputtering parameters for multi-target system.

Sputtering Parameters	Experimental conditions
Target	2" Y and Fe
Substrate material	Fused quartz and MgO (100)
RF power density (W/cm^2)	Y: 0.74~1.97 / Fe: 2.47~3.70
Background Pressure (Torr)	4.8~9.6E-6
Target distance (mm)	150
Process pressure (mTorr)	2.6~5.6
Sputtering gas /Flow rate (sccm)	Ar / 20~40
O^2 gas flow rate (sccm)	1~4.5
Deposition rate (nm/min)	4.78~10.85
Deposition time (min)	30~90
Post Annealing Parameters	Experimental conditions
Method	Rapid Thermal Annealing
Temperature (°C)	400~1000
Atmosphere	O^2
Time (sec)	5~120

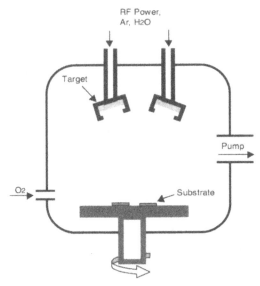

Figure 2 Schematic diagram of the multi-target RF magnetron sputtering system.

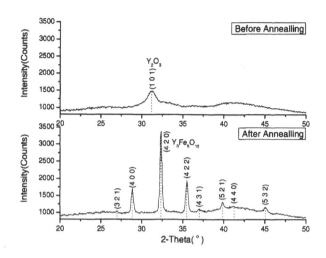

Figure 3 X-ray diffraction pattern of a film deposited on MgO with a single target system. The sample was annealed with conventional furnace at 1000 °C for 3 hours.

The second technique which involved a multi-target system, partial pressure differentials and rapid thermal annealing, produced much better results. To optimize multi-target sputtering system parameters, the reactive sputtering of Y and Fe at varying oxygen flow rates were studied separately at first. In the case of Y, gas flow rates of 1~3sccm O_2 and 20sccm of Ar yielded films that contained the Y_2O_3 phase, but over 3.5sccm of O_2, the Y target was oxidized. In case of Fe only sputtering, magnetite (Fe_3O_4) appeared at 1sccm O_2 and 20sccm Ar and hematite (Fe_2O_3) appeared when 2~3sccm of O_2 was used. At over 4sccm O_2, the Fe target was oxidized. To avoid target oxidation during YIG growth, O_2 flow rates were limited to less than 3sccm with 20sccm of Ar gas. With increasing Ar flow, O_2 flow can be increased. All of the films made with the multi-target system were amorphous as deposited. Previously, it has been shown that oxygen content is important when post annealing Y-Fe-O films if single phase YIG is desired.[4] Here, rapid thermal annealing (RTA) was performed in an O_2 atmosphere instead of air. To find the minimum requirements of annealing temperature and time, the RTA temperature and time was varied. At about 700~725°C, the films started to crystallize and over 750°C, crystallization was finished. The temperature was then set to 750°C, and the time was varied. At 5~10seconds, film crystallization was begun and crystallization was finished by 15seconds. 750°C and 15seconds are both substantial improvements over the previous conventional annealing method. It should be noted that when fused quartz substrates were used, the film surfaces were cracked during anneals over 600°C. Thermal expansion differences between the films and substrates caused these cracks. However, films on MgO substrates did not experience any cracking. The deposition rate was significantly increased compared with conventional method because the oxygen supply was controlled to prevent target oxidation. Deposition rate of multi-target system was 10.85nm/min. X-ray diffraction patterns before and after RTA are shown in Figure 4. All of the $Y_3Fe_5O_{12}$ samples were single phase YIG. Energy dispersive X-ray spectroscopy (EDS) showed a large region of solid solubility for YIG so that a wide range of Fe to Y (1.07 ~ 2.32) deposition ratios all produced single phase YIG. This is encouraging for the manufacturability of this material. The magnetic characteristics of the YIG films were as follows: in-plane anisotropy with coercivity between 17.5 and 20Oe and saturation magnetization about 1200emu/cm³. Faraday rotations as large as 0.2deg/µm (at 830Gauss) were measured at 632nm in the out of plane direction. Waveguide measurements are in progress as ridge waveguide dimensions are being optimized. This rotation will be higher in plane due to the magnetic anisotropy. In addition doping with Bi and Ce is also beginning as future work in order to increase the Faraday rotation of these films.

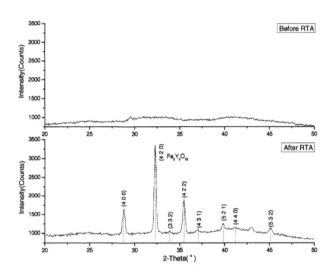

Figure 4 X-ray diffraction pattern of the film deposited on MgO with multi-target system. Sample was annealed with RTA at 800 °C for 120 seconds.

Conclusions

$Y_3Fe_5O_{12}$ films were fabricated on MgO substrate with much shorter times and lower thermal processing than those used in conventional methods. An improved deposition rate of 4.8~10.9nm/min was achieved. Since substrates were not heated during deposition, a post RTA process was the only thermal processing in this $Y_3Fe_5O_{12}$ fabrication method. 750°C and 15seconds were found sufficient for YIG crystallization and they are compatible with semiconductor substrates. MgO is a good optical cladding for YIG and it can be deposited onto semiconductor substrate, also via sputtering. Interestingly, the range of Fe to Y (1.07 ~ 2.32) deposition ratios that yielded single phase YIG was broad, indicating a large region of solid solubility which better reproducibly and less stringent fabrication tolerances will be required.

References

[1] S-Y. Sung, N. Kim and B. Stadler, MRS Symposium - Proceedings, **768**, 111 (2003).
[2] M. Gomi, H. Furuyama and M. Abe, J. Appl. Phys. **70**(11), 7065 (1997).
[3] Y. H. Kim, J. S. Kim, S. I. Kim and M. Levy, J. Kor. Phy. Soc. **43**(3), 400 (2003).
[4] B. Stadler and A. Gopinath, *IEEE Transactions on Magnetics* **36** [6] 3957-3961 (2000).

Mat. Res. Soc. Symp. Proc. Vol. 817 © 2004 Materials Research Society L6.16

Growth and patterning of strontium-titanate-oxide thin films for optical devices applications

M. Gaidi[1], L. Stafford[2], M. Chaker[1], J. Margot[2], & M. Kulishov[3]

[1] INRS-Énergie, Matériaux et Télécommunications, Varennes, Québec, Canada
[2] Département de physique, Université de Montréal, Montréal, Québec, Canada
[3] Adtek Photomask, Montréal, Québec, Canada

ABSTRACT

Strontium-titanate-oxide (STO) thin films have been deposited on silicon substrates by means of a reactive pulsed-laser-deposition technique. The influence of the oxygen deposition pressure on the microstructural properties of the films has been investigated by means of various characterization techniques. It was found that the crystalline quality of the film significantly deteriorates as the oxygen pressure increases. This is accompanied by an increase of the film microporosity. The microstructure of the film is found to directly impact the optical quality of the films. In particular, due to the higher density and crystallinity of the films deposited at lower oxygen pressure, films characterized by lower optical losses can be achieved in such conditions. These films have been used in the context of the development of optical waveguides. For this purpose, patterning of the STO films was investigated using sputter-etching with a high-density argon plasma operated in the very low pressure regime. Highly anisotropic features have been produced with high etch rate and good selectivity over resist. Preliminary results indicate the STO films can be successfully incorporated in functional waveguides.

INTRODUCTION

Strontium-titanate-oxide ($SrTiO_3$ or STO) thin films are of great interest for several applications, including the fabrication of optical devices for telecommunication systems. This is because STO is characterized by a high transparency in the visible and infrared regions. It can also be used as a buffer layer for the growth of electro-optic materials with a perovskite structure such as $(Pb,La)ZrTiO_3$ [1]. The integration of STO layers into optical devices requires understanding of both deposition and patterning processes. This is crucial since the performances of the device are expected to critically depend on film properties and patterning characteristics. For example, STO films can be used as core or cladding layers in optical waveguides : in this case, the attenuation and propagation coefficients of the wave in the waveguide are likely to be strongly influenced by the film microstructural properties and by the etching quality.

STO thin films were grown using a reactive pulsed-laser-deposition (RPLD) technique. In a first step, we investigate the influence of the oxygen buffer gas pressure employed during RPLD on the microstructural properties of the films and we examine how they influence the optical quality. In a second step, sputter-etching of patterned STO layers with a high-density argon plasma is investigated. Specifically, we examine the influence of the argon pressure on the STO etch rate, pattern selectivity and etching anisotropy. Finally, preliminary results on the development of an optical waveguide based on a STO layer is presented.

EXPERIMENTAL DETAILS

Polycrystalline STO layers were grown on (100) silicon substrates by RPLD, using a K$\blacksquare$ excimer laser (wavelength : 248 nm) [2]. Ablation was achieved by focusing the laser bea$\blacksquare$ (energy density of 1.5 J/cm^2 and repetition rate of 50 Hz) on a rotating stoichiometric SrTi$\complement$ target at an incidence angle of 45°. Throughout this work, the substrate temperature was ke$\blacksquare$ constant at 540°C while the oxygen buffer gas pressure P_{O2} was varied from 1 to 100 mTo$\blacksquare$ Since STO has already crystallized right after deposition, it is not necessary to perform pos$\blacksquare$ annealing. The films characterized in the present work have a typical thickness varying from 3$\blacksquare$ to 550 nm. The crystallographic structure of the films was examined using a x-ray diffracti$\blacksquare$ (XRD) spectrometer in the grazing incidence configuration. The surface morphology w$\blacksquare$ analyzed by scanning electron microscopy (SEM) and by atomic force microscopy (AFM). T$\blacksquare$ film composition was characterized using x-ray photo-electron spectroscopy (XPS) a$\blacksquare$ Rutherford backscattering spectroscopy (RBS).

STO samples were patterned by photolithography onto 1 μm thick photoresist with li$\blacksquare$ shape varying from 1 to 10 μm. Dry etching of the patterned STO layers was subsequent. performed using a magnetized high-density plasma sustained by a traveling electromagnet$\blacksquare$ surface wave at 190 MHz [3]. The absorbed power and the magnetic field intensity were set $\blacksquare$ 250 W and 600 Gauss respectively, while the operating argon pressure P_{Ar} was varied from 0.1 $\blacksquare$ 10 mTorr. The energy of the ions impinging onto the substrate was increased by rf-biasing t$\blacksquare$ substrate holder at 13.56 MHz. The DC self-bias voltage was maintained at a constant value $\blacksquare$ -100 V. During etching, the wafer temperature was kept at 10 °C using a water-cooling syste$\blacksquare$ The etch rate ER was determined from He-Ne laser interferometry (λ=632.8 nm) on a$\blacksquare$ unpatterned reference sample placed at about 2 cm from the patterned sample. ER w$\blacksquare$ determined from the interferometry pattern, using the STO refraction index values obtained $\mathfrak{b}$ Variable Angle Spectroscopic Ellipsometry (VASE) [4]. Cross-sectional SEM was employed $\blacksquare$ characterize the etching profiles of the patterned samples after resist strip.

EXPERIMENTAL RESULTS AND DISCUSSION

Characterization of STO thin films

The crystal quality of STO films grown by RPLD is illustrated in Fig. 1 as a function $\blacksquare$ the oxygen pressure. Clearly, in the range of pressures investigated, the diffractograms sho$\blacktriangledown$ well-defined peaks that are correlated to a good crystallinity of the film. As the pressu$\blacksquare$ increases, the XRD peaks becomes broader, thereby indicating a shrinking of the crystallite siz$\blacksquare$ i.e. a degradation of the film crystallinity. Figure 1 also shows that the diffraction peaks shi$\blacksquare$ towards lower angles as the oxygen pressure decreases. The lattice parameter a was calculate$\blacksquare$ from the position of the (110), (111) and (200) peaks, using Bragg's law. It was found that decreases from 0.3930 nm at 1 mTorr to 0.3908 nm at 100 mTorr. These values are close to th$\blacksquare$ of the lattice constant of bulk SrTiO$_3$ ceramic [5]. This dependence of a on the oxygen pressu$\blacksquare$ can be explained by the presence of oxygen vacancies in the crystal lattice at lower depositio$\blacksquare$ pressure [5,6]. The important conductivity of the films deposited at 1 mTorr is worth supporti$\blacksquare$

this idea since it corresponds to a weaker oxidation. In contrast, as the oxygen pressure exceeds 10 mTorr, the film becomes insulating.

Figures 2a and 2b present SEM images illustrating the influence of the oxygen deposition pressure on the film morphology. In Fig. 2a, it can be seen that for an oxygen pressure of 10 mTorr, the surface is very smooth, dense and free of cracks. Such a situation occurs as long as the oxygen pressure remains lower than about 25 mTorr. At higher deposition pressure (Fig. 2b), micropores are observed in the film and droplets (submicron particles) are present on the film surface. The occurrence of droplets is a common feature of laser-ablated oxide thin films deposited at high pressure [6]. In addition, AFM measurements (not shown) indicate that the surface becomes rougher and rougher as the oxygen pressure is increased from 1 to 100 mTorr. In the droplet-free regions, the surface roughness is estimated to vary from 1 to 8 nm as P_{O2} varies from 1 to 100 mTorr. This suggests a relationship between surface roughness and microporosity.

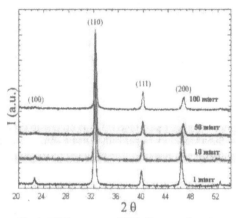 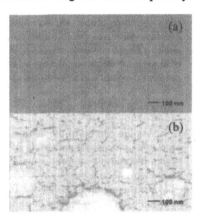

Fig. 1 : XRD patterns of STO films as a function of oxygen deposition pressure.

Fig. 2 : SEM images of STO films for (a) 10 mTorr and (b) 100 mTorr.

In order to characterize the concentration of the various species forming the STO films, we have performed XPS and RBS analysis. RBS measurements were carried out using a 2800 keV ^{4}He^{++} ion beam at normal incidence. The concentration of the various elements was obtained by fitting the RBS data using RUMP simulation [7]. Assuming a uniform distribution in the film, the best fit provides the value of the atomic density of each species, integrated over the film thickness. Absolute density values can thus be calculated by dividing the integrated density by the film thickness. The absolute concentration of Sr, Ti, and O atoms is shown in Fig. 3 as a function of the oxygen deposition pressure. As can be seen, the film density decreases from about 7×10^{22} cm^{-3} to 4×10^{22} cm^{-3} as P_{O2} increases from 1 to 100 mTorr. This decrease agrees very well with the qualitative observations of Fig. 2. A closer analysis of Fig. 3 further indicates that the Sr:Ti ratio remains close to one at any oxygen pressure, while a significant excess of oxygen atoms is observed. Very similar concentration fractions were also observed from XPS measurements.

According to the value of the lattice constant determined from Fig. 1 (close to bulk value one expects an O:Sr(Ti) ratio lower than three (especially at low oxygen pressure). This result in discrepancy with those provided by XPS and RBS measurements. It can however understood if one considers that oxygen is incorporated in the film after air exposure. Since ST is polycrystalline, it is probable that oxygen atoms penetrate through grain spacing and a trapped at the grain boundaries rather than within the grains, as already observed for ZnO [8,9 Thus, this oxygen adsorption should not impact the lattice constant, which is only determined t the crystal structure and stoichiometry.

VASE measurements have shown that increasing the deposition pressure induces degradation of the optical quality of the film [4]. The refraction index n decreases from 2.4 to 1 as the pressure grows from 1 to 100 mTorr, while the extinction coefficient k increas significantly. According to the results presented above, the dependence of n and k on P_{O2} can t understood as resulting from a direct relation between the optical quality of the films and the microstructural properties, in particular film crystallinity (Fig. 1) and density (Figs. 2 and 3). A pressure decreases, the thin film properties approach those of the bulk. As a result, the films a denser, smoother, and possess a higher degree of crystallinity, which yields a better optic quality. In consequence, the achievement of thin films suitable for optical applications requires perform deposition at low oxygen pressure (< 10 mTorr).

Patterning of STO thin films

In the context of the fabrication of optical devices based on STO, it is crucial to develc an etching process for patterning STO layers. Such a study was performed using sputter-etchir with a high-density argon plasma. Figure 4 shows the influence of the etching pressure on bot STO etch rate and selectivity over resist. Recall that selectivity is defined as the ratio of th photoresist etch rate over that of STO. From Fig. 4, it can be seen that the STO etch rate is near constant below 1 mTorr and drastically decreases above this pressure. This can be understood we recall that the sputter-etch rate (ER) of plasma-immersed substrates is given by [10]

$$ER = (1 - RR) J_{+} Y / N_{t},$$ (1)

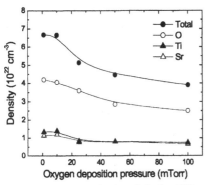

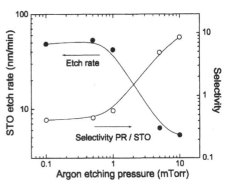

Fig. 3 : O, Ti, Sr and total atomic density of STO films as a function of oxygen deposition pressure.

Fig. 4 : Influence of the argon pressure on the STO etch rate and pattern selectivity. STO films were deposited at P_{O2}= 1 mTorr.

where RR is the redeposition factor resulting from collisions between sputtered species and plasma particles in the sheath, J_+ the positive ion flux impinging onto the substrate, Y the sputtering yield (i.e., the number of atoms ejected per incident ion), and N_t the atomic density of the film. The behavior of the STO etch rate partly results from the opposite variation of the positive ion flux and sheath potential with the argon etching pressure [3]. In addition, since the mean free path of the sputtered species decreases as P_{Ar} increases, the redeposition factor is expected to increase with pressure, resulting in a lower effective etch rate. Also note that redeposition degrades etching quality by forming sidewall residues (fences), as observed for platinum [10]. As further shown in Fig. 4, a good selectivity can be obtained for $P_{Ar} \leq 1$ mTorr. Operation at or below 1 mTorr consequently appear as the most suitable for realizing high quality patterning of STO films.

SEM profiles of patterned samples etched at an argon pressure of 0.5 mTorr are shown in Fig. 6. Clearly, in such conditions, it is possible to achieve highly anisotropic features with no evidence of sidewall residues. Using such a patterned STO film, we have developed an optical waveguide using STO as a core layer. Preliminary results indicate that optical waveguides with relatively low losses (about 1.6 dB/cm for 10 μm patterns) can be obtained.

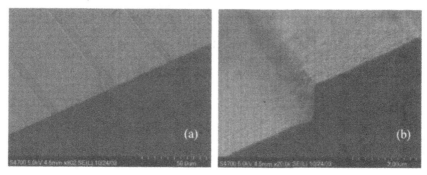

Fig. 5 : SEM images of patterned STO films etched at P_{Ar}= 0.5 mTorr. In fig. 5a, the scale is 50 μm while in Fig. 5b the scale is 2 μm. The film was deposited at P_{O2}= 1 mTorr.

CONCLUSION

In summary, we have demonstrated that the oxygen deposition pressure used during RPLD of STO thin films has a major influence on the film microstructure, in particular on film density and crystalline quality. An increase of the oxygen deposition pressure increases the density of the micropores and decreases the crystallite size. This has a direct impact on the optical characteristics of the film, namely the refraction index and the extinction coefficient. The sputter-etching characteristics of patterned STO films were also investigated as a function of the etching pressure. High etch rate, good selectivity and highly anisotropic features can be achieved provided the etching pressure is maintained at or below 1 mTorr. Finally, capitalizing on our capability of producing STO films with high optical quality and of generating highly anisotropic features, we have been able to implement the STO films into an optical waveguide with a relatively low attenuation coefficient.

ACKNOWLEDGEMENTS

The authors would like to thank A. Amassian & F. Schiettekatte for their technic
assistance and useful discussions during the course of this work. This project is financial
supported by the Natural Science and Engineering Research Council of Canada (NSERC) and tl
Fonds Québécois de la Recherche sur la Nature et les Technologies (FQRNT). M. Chaker
grateful to the Canada Research Chair program for salary support.

REFERENCES

1   F. Sanchez, M. Varela, X. Queralt, R. Aguiar, & J.L. Morenza, Appl. Phys. Lett. **61**, 22⁻
    (1987).
2   M. A. El Khakani, M. Chaker, & E. Gat, Appl. Phys. Lett. **69**, 2027 (1996).
3   L. Stafford, J. Margot, S. Delprat, M. Chaker, & D. Queney, J. Vac. Sci. Technol. A **20**, 5⁻
    (2002).
4   A. Amassian, private communication.
5   M. H. Yeh, K. S. Liu and I. Nan. Lin, Jpn. J. Appl. Phys. **34**, 2247 (1995).
6   M. Hiratani, K. Imagawa, & K. Takagi, Jpn. J. Appl. Phys. **34**, 254 (1995).
7   L. R. Doolittle, Nucl. Instr. Meth. B **15**, 227 (1986).
8   Y. J. Kim & H. J. Kim, Material Lett. **41**, 159 (1999).
9   K. C. Park & K. B. Kim, J. Appl. Phys. **80**, 5674 (1996).
10 S. Delprat, M. Chaker, & J. Margot, J. Appl. Phys. **89**, 29 (2001).

Mat. Res. Soc. Symp. Proc. Vol. 817 © 2004 Materials Research Society

Electro-Optical Properties of $Na_{0.5}K_{0.5}NbO_3$ Films on Si by Free-Space Coupling Technique

Alexander M. Grishin and Sergey I. Khartsev,

Department of Condensed Matter Physics, Royal Institute of Technology,
SE-164 40 Stockholm-Kista, SWEDEN

ABSTRACT

We report electro-optic performance of highly polar axis oriented $Na_{0.5}K_{0.5}NbO_3$ (NKN) films grown directly on $Pt(100nm)/Ti(10nm)/SiO_2/Si(001)$ substrates by rf-magnetron sputtering. Semitransparent gold electrodes (diameter $\varnothing = 2$ mm) were deposited ontop the NKN films by a thermal evaporation through the contact mask. Processing parameters have been specially optimized to obtain "electrosoft" NKN films with a non-linear fatigue-free P-E characteristics: low remnant $P_r = 3.6$ $\mu C/cm^2$ and high induced polarization $P = 26$ $\mu C/cm^2$ @ 522 kV/cm, and the coercive field $E_c = 39$ kV/cm. Electro-optical characterization of NKN/Pt/Si films has been performed using waveguide refractometry: a free-space coupling of a light beam into the thin-film waveguide modes. Intensity of TM- and TE-polarized light of 670 nm laser diode reflected from the free surface of NKN film and Au-cladding NKN/Pt/Si waveguide was recorded at zero and 30 V (100 kV/cm) bias electric field. Extraordinary and ordinary refractive indices as well as electro-optic coefficient have been determined by fitting these experimental data to the Fresnel formulas. Applying 160 V (530 kV/cm) across the parallel plate NKN capacitor ($\varnothing = 2$ mm, thickness 3 μm), modulation of the reflected light as high as 40% was achieved.

INTRODUCTION

Since the first fabrication of Ti indiffused waveguides in $LiNbO_3$ in 1974 [1], this ferroelectric become the leader in the materials arsenal of lightwave systems. Numerous optical devices have been extensively studied which exploit electro-optic, acousto-optic, and nonlinear properties of this material. Integrated optics in $LiNbO_3$ has already reached a stage of maturity. [2] However, such devices suffer from unavailability of low-cost fiber assembly. Retarded development of pigtailing structures and packaging of $LiNbO_3$ chips is one of the reasons for the scarce commercial availability of $LiNbO_3$ integrated optics components. Therefore, an engineering of new materials which enable "true integration" of lasing crystal, waveguides, optical processing circuits and, perhaps, memory all on the same chip remains the challenge in photonics.

Ferroelectric alkaline niobate $Na_xK_{1-x}NbO_3$ (NKN) solid solutions occupy the niche between "electrohard" $LiNbO_3$ and "electrosoft" $Pb(Zr,Ti)O_3$ ceramics. They possess unique combination of functional properties: remnant polization P_r as high as 18 $\mu C/cm^2$, piezoelectric constant d_{33} ~ 160 pC/N, dielectric permittivity ε ~ 400 which promise various applications in the wide frequency range from rf- to the millimeter wave band. Also, clinical tests proved NKN ceramics to be a useful biocompatible implant material. Recently, NKN ceramics attracts revived attention since high performance NKN films have been grown by rf-magnetron sputtering [3-5] and pulse laser deposition [6] techniques with an epitaxial quality on $SrTiO_3$, $LaAlO_3$, and as a highly c axis oriented films on Si, sapphire and quartz wafers. [7] Pulsed laser deposited NKN film exhibit *self-assembling* properties hence grow as highly oriented films on non-matchin substrates (Si, glass). Moreover, their properties can be tailored from the superparaelectric ferroelectric state by control of the oxygen atmosphere. [8] Optical and waveguiding properti of NKN films on sapphire have been recently characterized by a prism-coupling technique. [9]

In this paper we report on electro-optical properties of NKN films grown directly on Pt/ wafer. 40 % of modulation of 670 nm laser light achieved by applying 160 V (530 kV/cm) acro the parallel plate NKN capacitive cell demonstrate feasibility of NKN/Pt/Si as an electro-opt material.

FILMS PROCESSING AND PROPERTIES

NKN films have been grown on $10 \times 10 \text{ mm}^2$ Pt(100nm)/Ti(10nm)/SiO$_2$/Si(001) substrates k rf-magnetron sputtering. 60 W of rf-power was applied to 1 " stoichiometric target of 4.3 g/cn density (95% of theoretical value). The substrate temperature was 650 °C. Ar-O$_2$ gas mixtu (4:1) built up a total pressure of 40 mTorr. Deposition rate was approximately 5.5 Å/s. Film were postannealed at 540 °C in 500 Torr of oxygen atmosphere for 20 min and then slow cooled down to room temperature. Semitransparent gold electrodes ($\varnothing$ = 2 mm) were deposite ontop the NKN films by a thermal evaporation through the contact mask.

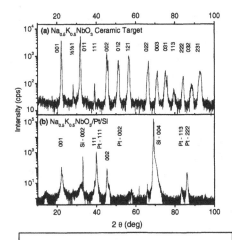

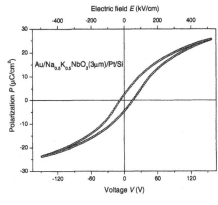

Figure 1. XRD θ-2θ scan of $Na_{0.5}K_{0.5}NbO_3$ (NKN) ceramic target (**a**) and 3 μm thick NKN film deposited on Pt(100 nm)/Ti(10 nm)/SiO$_2$/Si(001) substrate. *hkl* indexes stand for NKN Bragg reflections.

Figure 2. Ferroelectric *P-E* hysteresis loop for the parallel plate Au/Na$_{0.5}$K$_{0.5}$NbO$_3$(3μm)/ Pt(100nm)/Ti(10nm)/SiO$_2$/Si(001) capacitor recorded in continuous triangular 180 Hz wave form regime.

Figure 3. Experimental setup for free-space light-to-film coupling.

Two x-ray diffraction patterns are presented in Fig.1 for comparison: for ceramic $Na_{0.5}K_{0.5}NbO_3$ target used for rf-sputtering (**a**) and NKN film (**b**). It is clear that NKN film grows on Pt/Si as a single phase and exclusively ($00l$) oriented.

Processing parameters have been specially optimized to obtain "electrosoft" NKN film with a strongly non-linear fatigue-free $P\text{-}E$ characteristics (Fig.2): low remnant polarization P_r = 3.6 $\mu C/cm^2$, high induced polarization P = 26 $\mu C/cm^2$ @ 522 kV/cm, and the coercive field as low as 39 kV/cm.

WAVEGUIDE REFRACTOMETRY

Electro-optical characterization of NKN/Pt/Si films has been performed using modernized waveguide refractometry. [10] This method based on a free-space coupling of a light beam into the thin-film waveguide modes. The beam from the solid state (LDM 115-670/3) laser was focused to the spot ($\varnothing$ = 300μm) on the surface of Au/NKN/Pt/Si sample mounted onto a computer-controlled goniometer (Fig.3). A Si photodiode monitored the reflected beam intensity as a function

of incidence angle Θ with a step of 1′. TE and TM polarizations of the incident light have been defined using a polarizer in front of the laser.

Fig.4 shows the interference pattern observed in the light reflected from the free surface of NKN film (symbols ◁) and from Au-cladding NKN/Pt/Si waveguide (symbols ▷). Au-cladding changes the position of the extremes in the reflection spectra. However, the number of waveguided modes which can be excited in 3 μm thick NKN film by 670 nm light at incidence angles Θ from 10 to 90 degree conserves: two TE and three TM-modes. Minima of the intensity of reflected light indicate the excitation of these modes (dark-line spectra).

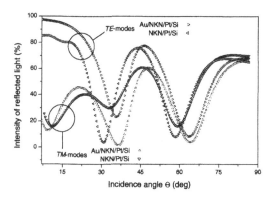

Figure 4. Intensity of TM- and TE-polarized light of 670 nm laser diode reflected from the free surface of NKN film and Au-clad NKN/Pt/Si structure.

227

Reflectivity data for *TE* and *TM* polarizations are shown in Fig.5 at zero and 30 V (100 kV/cm) bias electric field. To determine extraordinary and ordinary refractive indices as well as electro optic coefficient we have used the Fresnel formula for the reflection coefficient:

$$R = \frac{sin^2 \alpha \ (1-\eta)^2}{(1+\eta)^2 \ sin^2 \alpha + 4\eta \cos^2 \alpha} \ .$$

The functions $\alpha(\Theta)$ and $\eta(\Theta)$ are defined differently for *TE* and *TM* modes by the following equations:

$$\alpha = \begin{cases} 2\pi\dfrac{d}{\lambda}\sqrt{\varepsilon_3 - cos^2 \Theta} \ , & TE - mod\,e \\[3mm] 2\pi\dfrac{d}{\lambda}\sqrt{\varepsilon_2 - cos^2 \Theta \dfrac{\varepsilon_2}{\varepsilon_1}} \ , & TM - mod\,e \end{cases}$$

$$\eta = \begin{cases} \dfrac{\varepsilon_3 - cos^2 \Theta}{sin^2 \Theta} \ , & TE - mod\,e \\[3mm] \dfrac{1}{\varepsilon_2} \dfrac{1 - cos^2 \Theta \dfrac{1}{\varepsilon_2}}{sin^2 \Theta} \ , & TM - mod\,e \end{cases}$$

Here ε_1, ε_2 and ε_3 are the components of the dielectric permittivity, the axis "1" is directed perpendicular to the film surface, d is a film thickness, λ is a wavelength in the vacuum and, Θ is an incidence angle. Fitting experimental data from Fig.5 to Fresnel formulas we obtained the ordinary refraction index n_0 : $\sqrt{\varepsilon_3} = 2.247$, $\varepsilon_2 = n_0^2 = 5.049$, $\varepsilon_1 = 4.973$, and electro-optic coefficient $r_{13} = 18.5$ pm/V. Small birefringence $\varepsilon_2/\varepsilon_1 - 1 = 0.01$ observed in zero bias is a natural consequence of the remnant polarization (tetragonal distortions) in ferroelectric NKN film.

We examined Au/NKN/Pt/Si capacitive cell as an electro-optical modulator. The modulation of the intensity of the reflected light as high as 40% was achieved applying 160 V (533 kV/cm) across the parallel plate Au/NKN/ Pt/Si electro-optical cell (Fig.6).

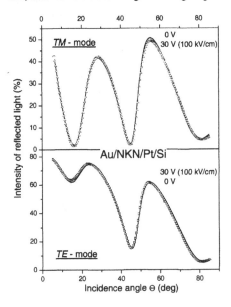

Figure 5 Intensity of *TM*- and *TE*-polarized 670 nm laser diode light reflected from the Au/NKN/Pt/Si electro-optic cell. Data for zero and 30 V bias showed by solid line and triangular symbols, correspondingly.

228

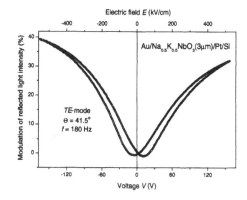

Figure 6. Modulation of polarized laser diode light (670 nm, *TE*-mode) reflected from the Au/NKN/ Pt/Si electro-optic cell. 40 % of light modulation was observed using continuous triangular 180 Hz pulse train.

SUMMARY

Single phase highly c–axis oriented $Na_{0.5}K_{0.5}NbO_3$/Pt/Ti//Si film structures have been shown feasible for electro-optic applications. Free space light-to-film coupling technique was employed to determine Bragg refraction angles. Ordinary n_o = 2.247 and extraordinary n_e ~ 2.23 refractive indices as well as electro-optic coefficient r_{13} = 18.5 pm/V have been determined by fitting experimental data to the Fresnel formulas. The modulation of the intensity of the reflected light as high as 40% was achieved applying 160 V (530kV/cm) across the parallel plate Au/NKN/Pt/Si electro-optic cell.

ACKNOWLEDGMENTS

This research was supported by the Swedish Foundation for Strategic Research (SSF) through the consortium "Functional Ceramics".

REFERENCES

1. R.V. Schmidt, I.P. Kaminow, Appl. Phys. Lett. **25**, 458 (1974).
2. L. Thylen, J. Lightwave Technology **6**, 847 (1988).
3. X. Wang, U. Helmersson, S. Olafsson, S. Rudner, L.-D. Wernlund, S. Gevorgian, Appl. Phys. Lett. **73**, **927** (1998).
4. X. Wang, S. Olafsson, L.D. Madsen, S. Rudner, I.P. Ivanov, A.M. Grishin, U. Helmersson, J. Materials Research **17**, 1183-1191 (2002).
5. M. Blomqvist, J.-H. Koh, S. I. Khartsev, A. M. Grishin, J. Andréasson, Appl. Phys. Lett. **81**, 337 (2002).
6. C.-R. Cho, A.M. Grishin, Appl. Phys. Lett. **75**, 268 (1999).
7. C.-R. Cho, I. Katardjiev, M.A. Grishin, A.M. Grishin, Appl. Phys. Lett. **80**, 3171 (2002).
8. C.-R. Cho, A.M. Grishin, J. Appl. Phys. **87**, 4439 (2000).
9. M. Blomqvist, S.I. Khartsev, A.M. Grishin, A. Petraru, Ch. Buchal, Appl. Phys. Lett. **82**, 439 (2003).
10. B.G. Potter, Jr., M.B. Sinclair, D. Dimos, Appl. Phys. Lett. **63**, 2180 (1993).

Luminescent Materials

Mat. Res. Soc. Symp. Proc. Vol. 817 © 2004 Materials Research Society   L9.1

Reliable InGaAsP/GaAs 40W lasers grown in solid source MBE with phosphorus-cracker

G. K. Kuang, I. Hernandez, M. McElhinney, L. Zeng, B. Caliva, and R. Walker,
Lasertel Inc., 7775 N. Casa Grande Hwy, Tucson, AZ-85743

ABSTRACT

Laser structures with InGaAsP quantum well were grown on GaAs substrates in a solid source MBE system. Threshold current density Jth as low as 290A/cm2 and slope efficiency as high as 0.68W/A per facet were obtained for uncoated laser chips at 25°C. After 857 hours burn-in at 47A (corresponding to around 47W) at room temperature, power degradation rate was measured to be less than 3×10^{-6}/h.

INTRODUCTION

0.8µm lasers have become more and more attractive [1-3] as they have found great application in pumping solid-state lasers and in material processing. Lasers with InGaAsP Al-free active region have shown strong advantages over conventional AlGaAs lasers due to their resistance to dark-line defects [4], and high threshold of catastrophic optical damage (COD) [5]. For 0.8µm high power lasers, reliability is a common concern. In order to improve the laser reliability, a lot of research work has been reported. M. Razeghi et al have reported degradation rate of less than 2×10^{-5}/h for uncoated InGaAsP lasers at facet load of 2 mW/µm in quasi-continuous wave regime [1]. J. Sebastian et al have studied large-optical-cavity (LOC) lasers with tensile strained GaAsP quantum well (QW), and reported degradation rate of less than 1×10^{-5}/h for laser chips at facet load of 30 mW/µm at 25°C. For laser bars with the same epitaxial structure tested at facet load of around 16 mW/µm at 25°C, they obtained an extrapolated lifetime of over 5000 hours [3]. P. Bournes et al from Coherent Inc reported less than 1% degradation for 40W laser bars during 1300 hours operation at facet load of around 14 mW/µm at 25°C, which corresponded to a degradation rate of less than around 8×10^{-6}/h [6]. In this paper we report on degradation rate less than 3×10^{-6}/h obtained from 0.8µm lasers with LOC structure and Al-free active region grown in solid-source MBE with phosphorus-cracker.

EXPERIMENT AND DISCUSSION

The laser structure was grown on n-GaAs substrate at around 500°C. The growth sequence is as follows. At first a GaAs:Si buffer layer was grown mainly with As_2 by using an As cracker cell. Then an AlGaInP:Si cladding layer and a GaInP waveguide layer were deposited, followed by an InGaAsP quantum well (QW). After the QW, similar waveguide layer GaInP and AlGaInP:Be cladding layer were deposited. In the end a GaAs contact layer heavily doped with Be was grown. All the phosphorus-containing layers were grown mainly with P_2 by using a P cracker cell with cracker temperature of 850°C. During the growth, the V/III flux ratio was kept around 15. After growth, the wafer was characterized with photoluminescence (PL), X-ray diffraction and CV-profiler. XRD measurement shows that the lattice-mismatch of the epi-layers is less than 0.04%.

Fig.1 shows the PL spectrum of the wafer at room temperature. It can be seen that the intensity is strong and the full-width at half-maximum is around 24nm which corresponds to 46meV.

Considering the fact that the QW material is quaternary and the PL is measured at room temperature, this PL spectrum is quite narrow and this indicates high crystal quality of the wafer

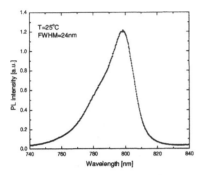

Fig.1, Photoluminescence spectrum of the wafer at room temperature.

After standard processing with wet chemical etching and dry etching, SiO_x deposition with PECVD, polishing, lapping, and metallization, single laser chip and 19-emitters-bars with filling factor of 28.5% were cut and bonded junction-down on Cu heatsink with AuSn. A typical light-current-voltage characteristic of uncoated laser chips with 1.0mm cavity length and 150μm emitter size is shown in Fig.2. It can be seen that threshold current Ith as low as 435mA was obtained, this corresponds to a threshold current density of 290A/cm². Slope efficiency as high as 0.68W/A per facet was measured. The high slope efficiency can be attributed to high crystal quality of the epi-layers and the LOC structure which results in very low optical loss in the laser.

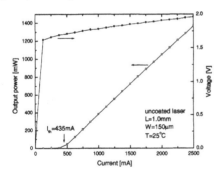

Fig.2, Typical light-current-voltage characteristic of uncoated lasers.

Lifetime testing in CW mode was performed for 2 coated-bars with 19 emitters and the result is shown in Fig.3. After 857 hours of burn-in at 47A (corresponding to around 47W output power) the power degradation rate was measured to be less than $3x10^{-6}$/h. This result is much better than the industrial standard several 10^{-5}/h. To our best knowledge, this is the best reliability result that has ever been reported for 0.8μm laser bars operating at above 40W.

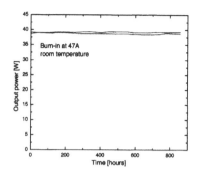

Fig.3, Power at 40A after burn-in at 47A for 2 coated laser bars.

CONCLUSIONS

We have demonstrated excellent performance and reliable result of 0.8μm lasers with InGaAsP QW grown in solid-source MBE with P cracker cell.

REFERENCES

1. M. Razeghi, J. Diaz, I. Eliashevich, X. He, H. Yi, M. Erdtman, E. Kolev, L. Wang, and D. Garbuzov, *14th IEEE International Semiconductor Laser Conference*, 1994, 19-23 Sept. 1994, pp. 159 –160

2. R. F. Nabiev, J. Aarik, H. Asonen, P. Bournes, P. Corvini, F. Fand, M. Finander, M. Jansen, J. Nappi, K. Rakennus, A. Salokatve, *16th IEEE International Semiconductor Laser Conference*, 1998. ISLC 1998 NARA. 4-8 Oct. 1998, pp. 43-44.

3. J. Sebastian, G. Beister, F. Bugge, F. Buhrandt, G. Erbert, H. G. Haensel, R. Huelsewede, A. Knauer, W. Pittroff, R. Staske, M. Schroeder, H. Wenzel, M. Weyers, and G. Traenkle, *IEEE J. Selected Topics in Quantum Electronics*, **7**, 334 (2001).

4. S. L. Yellen, A. H. Shepard, C. M. Harding, J. A. Baumann, R. G. Waters, D. Z. Garbuzov, V. Pjataev, V. Kochergin, and P. S. Zory, *IEEE Photonics Technol. Lett.*, **4**, 1328 (1992).

5. J. K. Wade, L. J. Mawst, D. Botez, R. F. Nabiev, M. Jansen, *Appl. Phys. Lett.*, **71**, 172 (1997).

6. P. Bournes, H. Asonen, F. Fang, M. Finander, M. Jansen, R. F. Nabiev, J. Nappi, K. Rakennus, A. Salokatve, *Lasers and Electro-Optics Society Annual Meeting*, 1998. LEOS '98. IEEE , 1-4 Dec. 1998, pp: 276 -277

Mat. Res. Soc. Symp. Proc. Vol. 817 © 2004 Materials Research Society   L9.4

White-Light Emitting Diode through Ultraviolet InGaN-pumped $Sr_2Si_{1-x}Ge_xO_4$: Eu^{2+} phosphors

J.S. Kim[1], P.E.Jeon[2], W.N.Kim[1], J.C.Choi[2], H.L.Park[1] and G.C.Kim[3]
[1] Institute of Physics and Applied Physics, Yonsei univ., Seoul, 120-749 South Korea.
[2] Department of physics, Yonsei univ., WonJu, 220-719 South Korea.
[3] School of Liberal Arts, Korea University of Technology and Education, Cheonan, 330-708 South Korea.

ABSTRACT

The $Sr_2Si_{1-x}Ge_xO_4$: Eu^{2+} phosphors are formed by means of a new synthesis method. The $Sr_2Si_{1-x}Ge_xO_4$: Eu^{2+} have the mean particle size of 200 nm and the spherical shape. The $Sr_2Si_{1-x}Ge_xO_4$: Eu^{2+} show two emission colors : the blue color of 480 nm and the green color of 525 nm. The intensity of green band with respect to blue band is significantly decreased in comparison with that of solid-reacted bulk sample, described by the hindrance of the energy transfer between the blue and green band by numerous nanoparticle boundaries. As the increase of Ge^{4+} ions, the 480 nm emission intensity of the $Sr_2Si_{1-x}Ge_xO_4$: Eu^{2+} are more dominant. This behavior can be understood in terms of the effect of Ge^{4+} ions on the energy transfer from 480 nm band to 525 nm band. The fabricated white-light emitting diode using ultraviolet InGaN chip with the blue and green emitting $Sr_2Si_{1-x}Ge_xO_4$: Eu^{2+} phosphor shows warm white light of 4300 K and higher color stability against input power in comparison with a commercial blue GaN-pumped YAG:Ce^{3+}.

INTRODUCTION

White light through GaN-based light emitting diode (LED) has a number of advantages over the existing incandescent and halogen lamps in power efficiency, reliability and long lifetime [1]. White LED by the blue GaN-pumped yellow YAG:Ce^{3+} (($Y_{1-a}Gd_a)_3(Al_{1-b}Ga_b)_5O_{12}$:$Ce^{3+}$) phosphor has the following problems; white emitting color change with input power, low color rendering index (CRI) due to two color mixing and low reproducibility due to the strong dependence of white color quality on quantity of phosphor. To solve these problems, the white LED has been fabricated employing a blue, green and red emitting multiphase phosphors excited by a ultraviolet (UV) InGaN chip[2,3]. This type of white LED has the following advantages; high tolerance to UV chip's color variation and excellent color rendering index due to white color generated by phosphors. In white LED, the morphology of phosphors greatly affects the luminescent efficiency because it influences the scattering of incident or emitting light. Phosphors with spherical shape are more effective for forming a good phosphor layer within LED mold rather than irregular shaped one[4].

In this paper, we present a new synthesis route for producing spherical $Sr_2Si_{1-x}Ge_xO_4$: Eu^{2+} phosphors which have fine size and narrow size distribution. We investigate the morphology and optical characteristics. The fabricated white LED through combining ultraviolet chip (λ_{em} = 375 nm) with $Sr_2Si_{1-x}Ge_xO_4$: Eu^{2+} phosphors is presented.

EXPERIMENTAL DETAILS

Strontium nitrate $(Sr(NO_3)_2 \cdot H_2O)$ and Europium nitrate $(Eu(NO_3)_3 \cdot H_2O)$ were used as the precursor of Sr^{2+} and Eu^{2+}, respectively. Colloidal silica and powder GeO_2 were used as the precursor of Si^{4+} and Ge^{4+}, respectively. Those precursors were dissolved in the solution of citric acid and ethylene glycol with weight ratio of 1 : 1. The molar ratio of Eu^{2+} activator was kept as 0.02. The final solution was air dried for 24 hours at 150 °C, and all dried samples were post-treated at 1000 °C for 1 hour under the air environment, and subsequently the reduction was performed at 1000 °C for 1 hour in a reducing atmosphere (a mixture of 25% H_2 and 75 % N_2). Three different sets of samples were synthesized with various amounts of Ge^{4+}, i.e., (a) Sr_2SiO_4:Eu^{2+}, (b) $Sr_2Si_{0.8}Ge_{0.2}O_4$: Eu^{2+}, (c) $Sr_2Si_{0.6}Ge_{0.4}O_4$: Eu^{2+}. The synthesis flow of the samples is summarized in Fig. 1. For comparison, Sr_2SiO_4:Eu^{2+} was also formed through solid state reaction method. The concentration of Eu^{2+} was determined by Inductively Coupled Plasma-Atomic Emission Spectroscopy (ICP-AES, Jobin Yvon 138 Ultrace). Phases of all samples were identified by MXP-3 XRD system (MAC Science. Co., Japan) with Cu_k radiation. The infrared spectrum was recorded on a Fourier-transformation infrared spectrophotometer (FT-IR)(Bruker, IFS 28CS). For the optical investigation, photoluminescence (PL) and photoluminescence excitation (PLE) measurements were obtained with a 200-watt Xe-lamp as an excitation source. The phosphor was coated on an UV LED (Nichia product type= NSHU550A, spectrum peak = 375 nm, optical power = 2 μW). The CIE chromaticity coordinates and the correlated color temperature (T_c) of fabricated white LEDs were measured by PR-705 SpectraScan spectroradiometer. A commercial white LED with blue GaN chip and YAG:Ce phosphor was used as a reference.

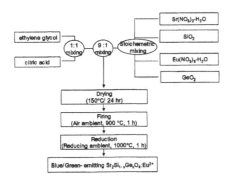

Fig. 1. Synthesis flow chart of $Sr_2Si_{1-x}Ge_xO_4$: Eu^{2+} phosphors.

Fig. 2. SEM image of $Sr_2Si_{0.8}Ge_{0.2}O_4$: Eu^{2+} phosphor

RESULTS AND DISCUSSION

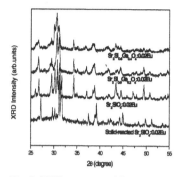

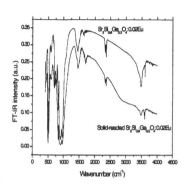

Fig. 3. XRD patterns of $Sr_2Si_{1-x}Ge_xO_4$: Eu^{2+} phosphors.

Fig. 4. FT-IR spectrum of $Sr_2Si_{0.8}Ge_{0.2}O_4$: Eu^{2+} phosphors.

The scanning electron microscopy (SEM) image of $Sr_2Si_{0.8}Ge_{0.2}O_4$: Eu^{2+} is shown in Fig. 2. All prepared particles have spherical shape and their average particle sizes are about 200 nm, whereas the solid-reacted sample shows an irregular shape.

X-ray diffraction (XRD) patterns are shown in Fig. 3. As an increase of larger Ge^{4+} ions, the 2θ values are decreased; the lattice constants are increased. XRD patterns match well with 39-1256 by JCPDF fingerprints. All samples show isostructure and broadening pattern as observed in nanoparticles. The $Sr_2Si_{1-x}Ge_xO_4$:Eu^{2+} phosphors have two different cation sites in the lattice [5, 6]. One site Sr(I) has 10 coordination's and the other site Sr(II) is surrounded by 9 oxygens. Thus, two different sites of Eu^{2+} in $Sr_2Si_{1-x}Ge_xO_4$:Eu^{2+} result in two emission bands.

FT-IR spectra of samples are shown in Fig. 4. Broad peak around 3400 cm^{-1} corresponds to OH stretching. The sharp peaks in the region of 1300 - 1700 cm^{-1} are assigned to stretching or bending of N-O bond and CO_3^{2-} [7]. Thus, FT-IR spectra show that the final $Sr_2Si_{1-x}Ge_xO_4$:Eu^{2+} contains more organic impurities than the sample driven from solid-state reaction.

PLE and PL spectra of $Sr_2Si_{1-x}Ge_xO_4$:Eu^{2+} with varying Ge^{4+} ion concentrations are shown in Fig. 5. The PLE and PL peaks originate from the 4f-5d transition of Eu^{2+} ions. The Eu^{2+} ions have many excited states[4], but they are not resolved. The emission intensity of $Sr_2Si_{0.8}Ge_{0.2}O_4$:Eu^{2+} is maximized, and is 75 % lower than that of solid-state reacted sample due to the more organic impurities. The Eu^{2+} ions occupy the two different cation sites, i.e., Sr(I) and Sr(II), and the PL spectra of $Sr_2Si_{1-x}Ge_xO_4$:Eu^{2+} consist of two bands. In the case of $Sr_2Si_{0.8}Ge_{0.2}O_4$:Eu^{2+}, the blue band is located around 483 nm with a full width at half maximum (FWHM) of 40 nm, and the green band is positioned at 524 nm with a FWHM of 122 nm.

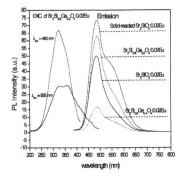

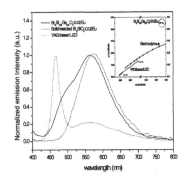

Fig. 5. PLE and PL spectra of Sr₂Si$_{1-x}$Ge$_x$O₄: Eu^{2+} phosphors.

Fig. 6. Emission spectra of 375-pumped white LEDs fabricated with Sr₂Si$_{1-x}$Ge$_x$O₄: Eu^{2+} phosphors. The inlet shows CIE coordinates along with increasing forward-bias.

The integrated intensity ratio of two bands (I_{green}/I_{blue}) is 1.6, whereas that is 1.2 in the case of Sr₂SiO₄:Eu^{2+}, indicating that the energy transfer between the blue and green band is more dominant. More covalent Ge ions lead to the increase of electron delocalization and the more overlapped wave function between the blue and green band [4], and consequently the energy transfer rate increases. [8]. This is supported by the spectral overlap between the blue emission and the green absorption a seen in Fig. 5. Also, I_{green}/I_{blue} of samples synthesized from new methods is significantly decreased in comparison with that of solid-reacted sample (I_{green}/I_{blue} = 2.2). A possible mechanism is the hindrance of the energy transfer between the blue and green band by numerous nanoparticle boundaries

Fig. 6 shows emission spectra of white LEDs formed with Sr₂Si$_{1-x}$Ge$_x$O₄:Eu^{2+} at a forward bias of 20 mA. The emission spectra of our 375 nm-pumped white LEDs are different from 325 nm-excited PL spectrum. This is due to the different excitation sensitivity of excitation energy as seen in Fig. 5. The CRI of our white LED with Sr₂Si$_{0.8}$Ge$_{0.2}$O₄:Eu^{2+} is 62 %, whereas the commercial white LED is 70 %. Also, our white LED show the strong blue emission in comparison with the white LED with solid-reacted Sr₂SiO₄:Eu^{2+} phosphor of which CRI is 60 %

The inlet of Fig. 6 shows the CIE chromaticity coordinates of our white LED with Sr₂Si$_{0.8}$Ge$_{0.2}$O₄:Eu^{2+} under various forward-bias currents (5, 10, 15, 20 and 25 mA). Our white LED shows that the CIE coordinates with the warm white color of 4300 K are unchanged within a standard deviation along with increasing applied currents((x,y) = 0.3930, 0.4767). On the other hand, in YAG:Ce-based white LED with increasing applied currents, the CIE coordinates shift significantly to the blue region and the color temperature increase drastically ((x,y) = 0.2958,

0.3157 → 0.2591, 0.2581). The superior color stability against applied currents can be described in terms of the spectral luminous efficiency function [9]. The CIE coordinates have less sensitivities to the variation of a phosphor-transmitted ultraviolet light than these of the blue light. As a result, our 375 nm-pumped white LEDs have higher color stability than 460 nm-pumped white LED in which white light is generated by a combination of blue and yellow lights.

CONCLUSIONS

The $Sr_2Si_{1-x}Ge_xO_4 : Eu^{2+}$ phosphors are formed through a new synthesis route. The $Sr_2Si_{1-x}Ge_xO_4 : Eu^{2+}$ have the mean particle size of 200 nm and the spherical shape. The $Sr_2Si_{1-x}Ge_xO_4 : Eu^{2+}$ show two emission colors : the blue color of 480 nm and the green color of 525 nm. As the increase of Ge^{4+} ions, the 480 nm emission intensity of the $Sr_2Si_{1-x}Ge_xO_4 : Eu^{2+}$ are more dominant. This behavior can be understood in terms of the increasing energy transfer rate from 480 nm band to 525 nm band due to the increased delocalization resulting from more covalent Ge^{4+} ions. The intensity of green band with respect to the blue band is significantly decreased in comparison with that of solid-reacted bulk sample. This can be described by the hindrance of the energy transfer between the blue and green band by numerous nanoparticle boundaries. The ultraviolet InGaN-pumped white light emitting diode with the blue and green emitting $Sr_2Si_{1-x}Ge_xO_4 : Eu^{2+}$ phosphor shows warm white light of 4300 K and higher color stability.

REFERENCES

1. S. Nakamura and G. Fasol, *The Blue Laser Diode* (Springer, Berlin, 1996).
2. J. S. Kim, P. E. Jeon, J. C. Choi, and H. L. Park, *Appl. Phys. Lett.* **84**, In press (2004)
3. Jong Su Kim, Ji Young Kang, Pyung Eun Jeon, Jin Chul Choi, Hong Lee Park and Tae Whan Kim, *Jpn. J. Appl. Phys.* **43**, 989 (2004)
4. Shigeo Shionoya and William M. Yen, *Phosphor Handbook* (CRC Press, NY, 1998).
5. M.Catti, G. Gazzoni, G. Ivaldi, *Acta Cryst.* **C39**, 29 (1983).
6. B.G.Hyde, J.R.Sellar, L. Stenberg, *Acta Cryst.* **B42**, 423 (1986).
7. P.K.Sharma, M.H.Jilavi, V.K.Varadan, H. Schmidt, *J. Phys. Chem. Solid* **63**, 171 (2002).
8. Zhenggui Wei, Lingdong Sun, Chunsheng Liao, Chunhua Yan, and Shihua Huang, *Appl. Phys. Lett.* **80**, 1447 (2002).
9. Gunter Wyszecki and W.S. Stiles, *Color science* (Wiles-interscience, 1982)

Mat. Res. Soc. Symp. Proc. Vol. 817 © 2004 Materials Research Society   L9.9

The Influence of Processing Parameters on Photoluminescent Properties of Ba_2ZnS_3:Mn Phosphors by Double-Crucible Method

Yu-Feng Lin, Yen-Hwei Chang and Bin-Siang Tsai
Department of Materials Science and Engineering, National Cheng Kung University, Tainan, 701 Taiwan, ROC

ABSTRACT

Red light emitting of Mn^{2+} doped Ba_2ZnS_3 phosphor powders have been synthesized by double-crucible method at different thermal treatments. XRD results indicate that the raw materials are completely sulfurized above 950°C, and the Ba_2ZnS_3: Mn^{2+} powders don't change its orthorhombic crystal structure with increasing doping concentration from 0 to 0.8 mol%. The photoluminescence of Ba_2ZnS_3: Mn^{2+} powders fulfilled the most efficient emission at the excitation wavelength λ_{ex}=358 nm and showed the red emission light with peak wavelength λ_{em}=627nm at the doping concentration of Mn^{2+} ion between 0.2 and 0.8 mol%. The high-luminance red emission results from the $^4T_1$ (^4G) $- {}^6A_1$ (^6S) transition in the Mn^{2+} ion. Ba_2ZnS_3: Mn^{2+} phosphors synthesized by double-crucible method have broad emission spectra (550nm~750nm) with FWHM (full width at half maximum broadband) about 66nm. In our research, the Ba_2ZnS_3 doped with 0.4 mol% Mn^{2+} has the highest luminescent intensity as thermal treatment at 950°C for 16 hours and the CIE coordinate is x=0.66, y=0.33.

INTRODUCTION

Transition metal ions doped wide band gap Ⅱ-Ⅵ compounds such as the zinc chalcogenide ZnS [1-3] and the alkaline-earth chalcogenides (AES) MgS [4-6], CaS [7-9], SrS [10-12] and BaS [6,13-14] are known to be efficient luminescent materials for the realization of multicolor electroluminescent and cathodoluminescent devices. The 3d transition metal ions utilized in commercial powder phosphors have three electrons (in the case of Cr^{3+} and Mn^{4+}) or five electrons (Mn^{2+} and Fe^{3+}) occupying the outermost 3d electron orbitals of the ions. When the 3d ions are incorporated into liquids or solids, spectroscopic properties are considerably changed from those of gaseous free ions. There is considerable interest in Mn-doped wide band gap phosphor materials – ZnS:Mn for example. The high efficiency of ZnS:Mn^{2+} has motivated efforts to shift or broaden the yellow $^4T_1(^4G) \rightarrow {}^6A_1(^6S)$ emission of Mn^{2+} to achieve a wider range of colors for phosphor or filtered white light displays. Hence several Ⅱ-Ⅵ mixed sulfides doped with rare-earth metal ions or transition metal ions have been investigated: $Ca_{1-x}Sr_xS$: Ce [15], $Ca_{1-x}Sr_xS$: Eu [16], $Zn_{1-x}Sr_xS$: Ce [17], $Zn_{1-x}Mg_xS$: Mn [18-19].

To date, barium zinc sulfide was fist prepared by Hoppe [20] by firing zinc oxide and barium oxide in an atmosphere of hydrogen sulfide (H_2S). Subsequently, the solubility of ZnS in BaS has been studied by Megson [21]. It is discovered that the firing of BaS and ZnS (in the molar quantities 2BaS:1ZnS) in an oxygen-free nitrogen atmosphere produced the barium zinc sulfide (Ba_2ZnS_3). The role of the transition metal ion Mn^{2+} doped in barium zinc sulfide has not been completely investigated yet.

In this work, we have synthesized the Mn^{2+} doped barium zinc sulfide phosphor by a double crucible method in conventional solid-state reaction and controlled the doping concentration of Mn^{2+}. Here, we describe the results of powder X-ray diffraction and the morphology and also study on the photoluminescence of $Ba_2ZnS_3:Mn^{2+}$ phosphors and the behavior of cerium as the emitting centre.

EXPERIMENTAL DETAILS

Sample preparations

Samples of composition Ba_2ZnS_3: xMn with $0 \leq x \leq 0.8$ mol%, increment of 0.2 mol%, were prepared via the conventional solid-state reaction by a double-crucible method [22]. The starting ingredients consisted of $BaCO_3$ (NOAH), ZnS (CERAC), S (SHOWA), $MnCl_2 \cdot 4H_2O$ (ALFA) and reducing agent starch (ACS reagent, ACROS). The required amount of $MnCl_2 \cdot 4H_2O$ was first dissolved in an aqueous ethanol medium, then while continuous stirring, the stoichiometric amounts of $BaCO_3$, ZnS and S were added to the ethanol solution and mixed for hours at about 65°C until all ethanol was evaporated. Subsequently, we used two crucibles, one containing the above-mentioned well mixed powders and the other containing proper amounts of starch. These were put into a quartz-tube furnace and fired with different thermal treatment. Starch is a reducing agent to produce a reductive atmosphere and to reduce the raw materials.

Characterization

The structural effect of Mn doping was carefully examined by X-ray powder diffraction using Cu-Kα radiation (Rigaku Dmax-33 X-ray diffractometer) with a source power of 30 KV and 20 mA to identify the possible phases formed after sulfurization-treatment. The surface morphology was examined by high-resolution scanning electron microscopy (HR-SEM, S4200, Hitachi). Photoluminescence spectra were recorded at room temperature with Jobin Yvon HR-320 monochromator and Avantes USB-2000 detector by scanning wavelength region from 200 to 1000 nm under an excitation of UV ray by Perkin Elmer 300 Watts Xenon lamp.

RESULTS and DISCUSSION

Phases in samples

Figure 1 a shows the powder X-ray diffraction patterns of those raw materials with 0.5 mol% Mn^{2+} fired at 750°C~950°C for 30 minutes to be sulfurized in a reductive atmosphere. It is obviously found that the products have a great amount of residual raw materials $BaCO_3$ and ZnS at the fired temperature below 850°C. While the fired temperature above 850°C, the quantity of residual raw materials $BaCO_3$ and ZnS were decreasing and the Ba_2ZnS_3 phase were more distinct with increasing fired temperature. At 950°C fired temperature, there was only a small amount of residual raw materials $BaCO_3$ and ZnS and it suggested that if the fired temperature was extended, we could get the pure Ba_2ZnS_3 powders.

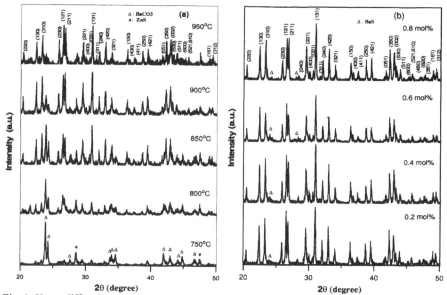

Fig. 1. X-ray diffraction patterns (a) Ba_2ZnS_3:Mn (0.5 mol%) powders with different fired temperatures for 30 minutes (b) Ba_2ZnS_3:xMn (x=0.2, 0.4, 0.6 and 0.8 mol%) powders fired at 950°C for 16 hours.

The powder X-ray diffraction patterns of Ba_2ZnS_3: xMn powders with $0.2 \leq x \leq 0.8$ mol% calcined at 950°C for 16 hours are shown in figure 1 b. All of the peaks were identified to belong to the orthorhombic Ba_2ZnS_3 phase with space group Pnam (No. 62). Only two peaks at $2\theta= 24.16°$ and $27.97°$ showed the existence of a small amount of BaS powders. It was found that the Ba_2ZnS_3: Mn^{2+} powders did not change their orthorhombic crystal structure with increasing doping concentration, from 0.2 to 0.8 mol%. Also they did not show marked differences in crystallinity due to the same thermal treatment procedure.

(a)   (b)

Fig. 2. SEM photograph of (a) 0.2 mol% (b) 0.4 mol% Mn-doped Ba_2ZnS_3 powder prepared at 950℃ for 16 hours.

245

The Ba$_2$ZnS$_3$ powders with a Mn^{2+} ion doping concentration in the range from 0.2 to 0.8 mol% do not have an obvious difference in morphology. It indicates that a trace of Mn^{2+} ions in Ba$_2$ZnS$_3$:Mn^{2+} phosphor does not affect the powder morphology for the same thermal treatment. The SEM images of figure 2. show the Ba$_2$ZnS$_3$ doping with Mn^{2+} ion 0.2 mol% and 0.4 mol% particles aggregated to form a large cluster about several micron meter size which is possibly due to the long thermal treatment time of 16 hours.

Optical properties

Figure 3 shows the PL emission spectra under 358 nm UV ray excitation of Ba$_2$ZnS$_3$: xMn with x=0.1, 0.2, 0.4 and 0.8 mol% measured at room temperature. In these spectra, all samples show a red color photoluminescence originating from the $^4T_1$ (^4G) $-$ $^6A_1$ (^6S) transition of the excited Mn^{2+} ions. The red light emission spectra of Ba$_2$ZnS$_3$: Mn^{2+} have a broad band in the range 550-750 nm with peak around 627 nm and a Stokes shift about 270 nm. Under 358 nm UV ray excitation, the photons absorbed by the Ba$_2$ZnS$_3$:Mn phosphors are quite similar to those in pure Ba$_2$ZnS$_3$ powder. This indicates that the excitation energy absorbed at the band edge of the Ba$_2$ZnS$_3$ host crystal can be efficiently transferred to the unexcited Mn^{2+} ions.

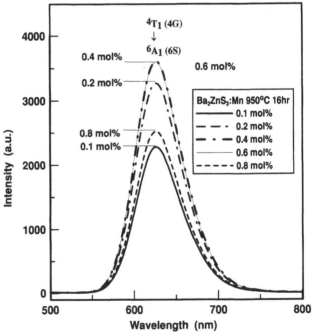

Fig. 3. Photoluminescence spectra of Ba$_2$ZnS$_3$ powders with different concentration of Mn^{2+} doping excited at λ_{ex} = 358 nm.

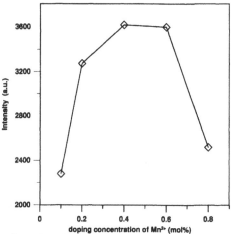

Fig. 4. Photo-luminescent intensity of Ba_2ZnS_3 powders with different concentration of Mn^{2+} doping excited at λ_{ex} = 358 nm.

As shown in Fig. 4, it is found that the luminescence intensity of these phosphor materials depends on the activator concentration. The luminescence intensity of $Ba_2ZnS_3:Mn^{2+}$ powders increased with doping concentration from 0.1 to 0.4 mol% and a maximum intensity emitted around λ_{em}=627 nm occurs for a doping of 0.4 mol% Mn^{2+}. Moreover, the luminescence intensity decreases for increasing doping concentration above 0.4 mol%. While the doping concentration over 0.6 mol%, the luminescence intensity decreased sharply due to concentration quench.

In addition, the colors of the emission light under 358 nm UV ray excitation are red with CIE color coordinates around x=0.66, y=0.33.

CONCLUSIONS

Mn^{2+} doped Ba_2ZnS_3 phosphor powders were successfully synthesized via conventional solid-state reaction by a double-crucible method. Ba_2ZnS_3: xMn powders with $0 \leq x \leq 0.8$ mol% calcined at 950°C for 16 hours display a single orthorhombic structure and give rise to efficient emission at excitation wavelength of λ_{ex}=358 nm. Under the UV ray excitation at λ_{ex}=358 nm, red luminescence was emitted for 0.4 mol% doping and the maximum intensity of the luminescent peak is found at 627 nm with a CIE color coordinate of x=0.66, y=0.33.

ACKNOWLEDGMENTS

The authors would like to thank the National Science Council of the Republic of China for financially supporting this research under Contract No. (NSC 92-2216-E-006-009).

REFERENCE

[1] Lai Qi, Burtrand I. Lee, Jong M. Kim, Jae E. Jang and Jae Y. Choe, J. Lumin. 104 (2003), pp. 261-266.

[2] W. Park, J. S. King, C. W. Neff, C. Liddell and C. J. Summers, Phys. Stat. Sol. (b) 229 (200: pp. 949-960.

[3] A. P. Greeff and H. C. Swart, Surf. Interface Anal. 34 (2002), pp. 593-596.

[4] Ted A. O'Brien, Philip D. Rack, Paul H. Holloway and Michael C. Zerner, J. Lumin, 78 (1998), pp. 245-257.

[5] Nobuhiko Yamashita and Sumitada Asano, J. Phys. Soc. Jpn. 56 (1987), pp. 352-358.

[6] R. P. Rao, J. Mater. Sci. 21 (1986), pp. 3357-3386; J. Mater. Sci. Letters 2 (1983), p 106-110.

[7] Sun Xiaolin, Hong Guangyan, Dong Xinyong, Xiao Dong, Zhang Guilan, Tang Guoqing an Chen Wenju, J. Phys. Chem. Solids 62 (2001), pp. 807-810.

[8] Yasuaki Tamura, Jpn. J. Appl. Phys. 33 (1994), pp. 4640-4646.

[9] Nobuhiko Yamashita, Shigeru Fukumoto, Sumiaki Ibuki and Hideomi Ohnishi, Jpn. J. Appl. Phys. 32 (1993), pp. 3135-3139.

[10] C. J. Summers, B. K. Wagner, W. Tong, W. Park, M. Chaichimansour and Y. B. Xin, Crystal Growth 214-215 (2000), pp. 918-925.

[11] B. Hüttl, K. O. Velthaus, U. Troppenz, R. Herrmann and R. H. Mauch, J. Crystal Growth 159 (1996), pp. 943-946.

[12] Nobuhiko Yamashita, Takaharu Ohira, Hitoshi Mizuochi and Sumitada Asano, J. Phys. Soc Jpn. 53 (1984), pp. 419-426.

[13] Yoshio Kaneko and Takao Koda, J. Crystal Growth 86 (1988), pp. 72-78.

[14] S. Asano, Y. Nakao, N. Yamashita and I. Matsuyama, Phys. Stat. Sol. (b) 133 (1986), pp. 333-344.

[15] B. W. Arterton, J. W. Brightwell, B. Ray, I. V. F. Viney, J. Crystal Growth, 138 (1994) pp. 1051-1054.

[16] Katsuhiro Kato and Fumio Okamoto, Jpn. J. Appl. Phys. 22 (1983), pp. 76-78.

[17] San Tae Lee, Masahiko Kitagawa, Kunio Ichino and Hiroshi Kobayashi, Appl. Surf. Sci. 113-114 (1997), pp. 499-503.

[18] Ryo Inoue, Masahiko Kitagawa, Yoshinori Horii, Takayoshi Nishigaki, Setsuya Kinba, Kunio Ichino, Shosaku Tanaka, Hiroshi Kobayashi, J. Crystal Growth 214/215 (2000) pp. 931-934; J. Luminescence, 87-89 (2000) pp. 1264-1266.

[19] M. K. Jayaraj, Aldrin Antony, P. Deneshan, Thin Solid Films 389 (2001) pp. 284-287.

[20] R. Hoppe, Angew. Chem. 71 (1959), p. 457.

[21] B. H. Megson, (1971) MSc Thesis, Thames Polytechnic, London.

[22] Shigeo Shionoya and William M. Yen, Phosphor Handbook (CRC press, 1999), p. 228.

Mat. Res. Soc. Symp. Proc. Vol. 817 © 2004 Materials Research Society   L6.14

Solvothermal Synthesis and Characterization of Zn(NH$_3$)CO$_3$ Single Crystal

Fushan Wen[1,2,3], Jiesheng Chen[2], Jin Hyeok Kim[1*], Taeun Kim[1] and Wenlian Li[3]

[1] Photonic and Electronic Thin Film Laboratory, Department of Materials Science and Engineering, Chonnam National University, 300 Yongbong-Dong Puk-Gu Kwangju South Korea, 500-757

[2] State Key Laboratory of Inorganic Synthesis & Preparative Chemistry, College of Chemistry, Jilin University, 119 Jiefang Road, Changchun, 130023, P. R. China

[3] Key Laboratory of the Excited States Process, Changchun Institute of Optics, Fine Mechanics and Physics, Chinese Academy of Sciences, Changchun, 130033, People's Republic of China

ABSTRACT

A new 3-dimensional zinc carbonate Zn(NH$_3$)CO$_3$ has been synthesized from a glycol system with urea and zinc acetate as raw materials. The crystal structure and photoluminescent properties have been investigated using X-ray diffraction, smart CCD and FL. The compound had an orthorhombic system with space group of $Pna2_1$ with $M = 142.41$, $a = 9.1449(18)$ Å, $b = 7.5963(15)$ Å, $c = 5.4982(11)$ Å, $V = 381.95(13)$ Å^3, $Z = 4$, $R = 0.0285$ and $R_W = 0.0745$. The NH$_3$ and CO$_3^{2-}$ were connected through the Zn-N bond and Zn-O bond in the symmetric unit. Photoluminescent property was observed in the compound at room temperature and the exited and emission peaks were located at about 350 nm and 426 nm, respectively.

INTRODUCTION

Considerable interest has been focused on the study of luminescent properties of zinc compounds, such as ZnO [1], Zn$_2$SiO$_4$ [2] and other zinc complexes [3,4] because they can be used as photo- and electro- luminescent materials. Most of them were prepared by the traditional high temperature solid-state method. They have disadvantages of high temperature reaction process. There have been a few reports on the synthesis of those compounds by hydrothermal or solvothermal method which have been well known for their ability to get meta-stable compounds and to promote crystal growth [5]. Initially, we just tried to obtain nano-scale ZnO using solvothermal method in this study, but we obtained the Zn(NH$_3$)CO$_3$ single crystal instead of nano-ZnO. To our knowledge, Zn(NH$_3$)CO$_3$ has not been found in other reports. Therefore, we report the formation of the Zn(NH$_3$)CO$_3$ single crystal through the solvothermal method at relatively low temperature using the simple inorganic material, urea and simple organic solvent. Photoluminescent property of the title compound at room temperature will be discussed.

EXPERIMENTAL

To obtain the title compound, zinc acetate (Zn(CH$_3$COO)$_2$·2H$_2$O) and urea (NH$_2$CONH$_2$) were used as the raw materials and the glycol (HOCH$_2$CH$_2$OH) was used as the organic solvent. In a typical synthesis procedure, an initial mixture with a molar composition of

$1Zn(CH_3COO)_2 \cdot 2H_2O$: 4.4 $(NH_2)_2CO$: 53.7 EG was stirred for 12 hours, then sealed in a teflon-lined stainless steel autoclave and heated at 433 K for 108 hours. The reaction product was washed thoroughly with distilled water and dried at room temperature.

The powder X-ray diffraction (XRD) data were collected using a Siemens D50C diffractometer with Cu-K$_\alpha$ radiation (λ=1.5418 Å). The photoluminescent property of the samp. was investigated using a Perkin-Elmer Luminescence Spectrometer-Ls55.

A suitable colorless cubic single crystal of the title compound with dimensions of 0.272×0.231×0.192 mm was glued to a thin glass fiber and mounted on a Bruker Smart CCD diffractometer equipped with a normal-focus, 2.4 kW sealed-tube X-ray source (graphite-monochromated Mo-K$_\alpha$ radiation, λ=0.71073 Å) operating at 50 kV and 40 mA. The data processing was accomplished with the SAINT processing program [6]. Direct methods wer used to solve the structure using the SHELXTL crystallographic software package [7]. The absorption correction was based on the symmetry equivalent reflections using SADABS [8] program. Other effects such as absorption by glass fiber were simultaneously corrected.

RESULTS AND DISCUSSION

The crystal structure was solved using the SHELXTL software [7] and the result showed th. the crystal has an orthorhombic system with space group of $Pna2_1$ with a = 9.1449(18) Å, b 7.5963(15) Å, c = 5.4982(11) Å. Figure 1(a) shows a diffraction pattern of the powder samp. synthesized by solvothermal method. The XRD peaks in figure 1(a) do not correspond to an known phase indicating that the as-synthesized compound should be a structurally new materia Figure 1(b) shows an XRD pattern simulated on the basis of the single crystal structure. Th diffraction peaks in both patterns are well consistent in position, which indicates that th as-synthesized material is a pure phase. (The simulated pattern is a theoretic value, it contains a the peaks. However, not all the peaks are shown in the diffraction of experimental sample due t the experimental factors such as crystal shape, diffraction intensity and time *et al.*.)

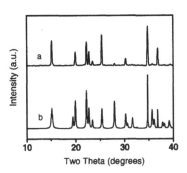

Figure 1. The powder X-ray diffraction patterns of the experimental (a) and the simulated (b) fo $Zn(NH_3)CO_3$.

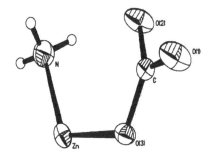

Figure 2. The ORTEP drawing of the asymmetric unit of the proposed Zn(NH₃)CO₃ crystal. Thermal ellipsoids are shown at 50% probability.

The proposed crystal structure of the Zn(NH₃)CO₃ are explained in figure 2 and figure 3. The NH3 group and the CO3 group are connected through the N-Zn bond and the O-Zn bond in the asymmetric unit of the Zn(NH₃)CO₃ crystal as shown in figure 2. There are 3 C-O bonds of which bond lengths vary from 1.245(7) Å to 1.297(6) Å in the CO3 group. The O-C-O bond angle varies from 110.8(5)° to 121.3(5)°. However, bond lengths of all N-H bonds are equal to 0.8900 Å and the H-N-H bond angle is equal to 109.5° in the NH3 group. The bond angle of Zn-N-H and Zn-N bond length is 109.5° and 2.014(6) Å, respectively. The bond angle of Zn-O-C varies from 115.7(3)° to 125.8(3)° and the bond length of Zn-O varies from 1.929(4) Å to 1.989(4) Å.

The Zn atom is coordinated by three O atoms and one N atom and formed ZnO3N tetrahedron. And the three O atoms are connected with other C atoms in three different directions. There exist screw chains along the *a*-direction in the structure as shown in figure 3. The screw chains are connected with C-O bonds and O-Zn bonds then forms 3-D structure. There exist holes formed by C-Zn 8-membered rings and C-Zn 4-membered rings viewed through the *c* axis direction as shown in figure 4.

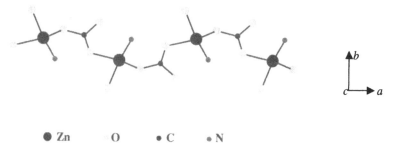

● Zn   O   ● C   ● N

Figure 3. The screw chains along the *a* axis in Zn(NH₃)CO₃.

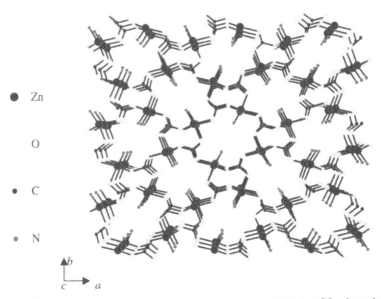

● Zn

O

● C

● N

Figure 4. Schematic drawing showing the packing structure of Zn(NH₃)CO₃ along the c direction.

It was observed that the Zn(NH₃)CO₃ compound has photoluminescent property at room temperature. Figure 5 shows the emission and excitation spectrum. The excitation and emission peak is at around 350 nm and 426 nm, respectively.

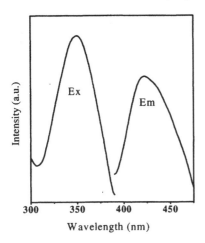

Figure 5. The luminescent spectra of Zn(NH₃)CO₃.

This emission can not be attributed to the ZnO bulk because the emission band is at about 500 nm for the bulk phosphor materials. The coordination of a Zn atom in this compound is shown in figure 6. Three CO3 groups and one NH3 group are connected with a Zn atom. One C atom is connected with three O atoms through 3 δ bonds by sp2-hybridized in every CO3 group. The remaining one electron in a C atom and one electron in the p orbital of an O atom form the delocalized 'extra' π bond. The emission may be due to the electron transfer from the delocalized 'extra' π bonds to the metal Zn centre, which is named as ligand-to-metal charge transfer (LMCT) [9]. The peak of the charge-transfer transition will be at a shorter wavelength than that of d-d transition in the same complex. And the data in the above is agreement with this point. We thought that the NH3 group connected with N atom also had contribution to the luminescent.

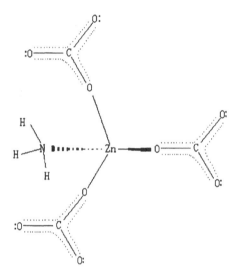

Figure 6. Schematic diagram showing the coordination for the Zn atom.

CONCLUSIONS

$Zn(NH_3)CO_3$ single crystal was synthesized in the solvothermal system. The compound had an orthorhombic system with space group of $Pna2_1$ with $M = 142.41$, $a = 9.1449(18)$ Å, $b = 7.5963(15)$ Å, $c = 5.4982(11)$ Å, $V = 381.95(13)$ Å^3, $Z = 4$, $R = 0.0285$ and $R_W = 0.0745$. It was found that it is a three-dimensional structure with holes. It seems that the emission comes from the ligand-to-metal charge transfer in the compound.

ACKNOWLEDGEMENTS

Dr. Kim is grateful to the financial support from Korea Ministry of Science and Technology through the Center for Photonic Materials and Devices

REFERENCES

1. B. X. Lin., Z. X. Fu., Y. B. Jia., *Appl. Phys. Lett.* **79**, 943 (2001).
2. Q. H. Li, S. Komarneni and R. Roy, *J. Mater. Sci.* **30**, 2358 (1995).
3. C. M. Che., C. W. Wan., K. Y. Ho., Z.Y. Zhou., *New. J. Chem.* **25**, 63 (2001).
4. J. Tao., M. L. Tong., J. X. Shi., X. M. Chen., S. W. Ng., *Chem. Commun.* 20, 2043, (2000).
5. A. Rabenau, *Angew Chem. Int. Ed.* 24, 1017(1985)
6. Software packages *SMART and SAINT,* Siemens Analytical X-ray Instruments a) Inc Madison, WI, (1996).
7. *SHELXTL*, version 5.1; Siemens Industrial Automation, Inc., (1997).
8. G. M. Sheldrick, *"SADABS User's Guide."* University of Gottingen. (1995).
9. R. Bertoncello, M. Bettinelli, M. Cassrin, A.Gulino, E. Tondello and A. Vittadini, *Inorg. Chem.* **31,** 1558 (1992).

Mat. Res. Soc. Symp. Proc. Vol. 817 © 2004 Materials Research Society   L6.15

Hydrothermal Synthesis of Ce^{3+} and Tb^{3+} co-doped Ca$_3$Al$_2$(OH)$_{12}$ Luminescent Material

Fushan Wen[1,2,3], Jiesheng Chen[2], Jin Hyeok Kim[1*], Taeun Kim[1] and Wenlian Li[3]

[1] Center for Photonic Materials and Devices,
Department of Materials Science and Engineering, Chonnam National University
300 Yongbongdong Pukgu Kwangju, South Korea, 500-757
[2] State Key Laboratory of Inorganic Synthesis & Preparative Chemistry, College of Chemistry,
Jilin University, 119 Jiefang Road, Changchun, 130023, P. R. China
[3] Key Laboratory of the Excited States Process, Changchun Institute of Optics, Fine Mechanics
and Physics, Chinese Academy of Sciences, Changchun, 130033, People's Republic of China

ABSTRACT

Ce^{3+} and Tb^{3+} co-doped calcium aluminates luminescent material (Ca$_3$Al$_2$(OH)$_{12}$: Ce^{3+}, Tb^{3+}) was synthesized at 453 K in hydrothermal system without any protective atmospheres. All the reactants used in the system were simple inorganic salts. The crystallinity and luminescent property of as-synthesized compound were investigated using X-ray diffraction and luminescence spectrometer. The compound formula was confirmed to be Ca$_3$Al$_2$(OH)$_{12}$. The emission spectrum showed that only typical Tb^{3+} emission was observed and the emission from Ce^{3+} was almost not observed, which should be attributed to the energy transfer from Ce^{3+} to Tb^{3+} in the compound. The emission peaks at about 486 nm, 498 nm, 540 nm, 549 nm, 582 nm, 595 nm and 623 nm should be assigned to $^5D_4$-$^7F_6$, $^5D_4$-$^7F_5$, $^5D_4$-$^7F_4$, and $^5D_4$-$^7F_3$ transitions of Tb^{3+} ions. The weak emission peak at about 380 nm should be assigned to the emission of Ce^{3+} ions in the compound.

INTRODUCTION

Aluminates phosphors have been widely used for a long time in various luminescent applications such as three-band fluorescence lamps and plasma display panels (PDP) [1,2]. BaMgAl$_{10}$O$_{17}$:Eu^{2+} (BAM) phosphors have been used as a blue component in fluorescence lamps. The green phosphors, Tb^{3+} activated CaGdAlO$_4$, CaGdAl$_3$O$_7$ and GdAlO$_3$ have been synthesized by combinatorial polymerized-complex method [3]. SrAl$_2$O$_4$ phosphors doped with Eu^{2+} and Dy^{3+} are known to exhibit a bright and long-lasting afterglow after irradiation with fluorescent light or sunlight [4-6]. The energy transfer in the Tb^{3+} activated yttrium aluminate was investigated by Kee-Sun Sohn [7]. Ce^{3+} activated Y$_3$Al$_5$O$_{12}$ was also investigated by Baciero et al. [8] which was used as vacuum ultraviolet and x-ray luminescent material.

Even though, aluminates luminescent materials have provoked researchers' considerable interests because of their excellent luminescent properties, all the above luminescent materials were prepared by classical high temperature solid-state method. However, compared with the high temperature solid-state method, hydrothermal method has the advantages as follows: reaction temperature is low and it is easy to control the size of particle.

The purpose of this study is to synthesis Ce^{3+} and Tb^{3+} co-doped calcium aluminates luminescent materials using hydrothermal method and to investigate their luminescent propertie at room temperature.

EXPERIMENT

To obtain the title compound, $Ca(OH)_2$(A.R), $Al(OH)_3$ (A.R.), $NaOH$ (A.R.), Tb_4O_7 (A.R and CeO_2(4N) were used as raw materials. In a typical synthesis procedure, an initial mixture with a molar composition of $1CaO$: $2Al_2O_3$: $2Na_2O$: $0.02Tb_4O_7$: $0.01CeO_2$: $175H_2O$ was stirred till homogeneous, sealed in a teflon-lined stainless steel autoclave and heated at 453 K for 4 day The reaction product was washed thoroughly with distilled water and dried at room temperature

The phase of the powder was identified using the powder X-ray diffraction (XRD) (Siemens D5005) with Cu K_α radiation (λ= 1.5418 Å). The particle morphologies of the powder was observed by scanning electron microscopy using a Hitachi X-650 scanning electron microscope. The luminescent property of the samples was investigated using a Perkin-Elmer Luminescence Spectrometer-Ls55.

RESULTS AND DISCUSSION

Figure 1 shows the XRD pattern of the sample synthesized using the hydrothermal methc and the simulated x-ray diffraction of $Ca_3Al_2(OH)_{12}$ powder from the data of ICSD [9]. Th diffraction peaks of the synthesized powder obtained in our experiment are well consistent wit those of simulated pattern, indicating that $Ca_3Al_2(OH)_{12}$ phase was obtained. The morphology the powder is spherical and the diameter is about 20-50 μm from the SEM picture as shown figure 2.

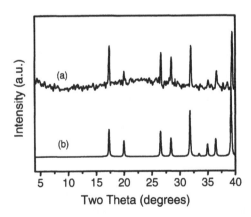

Figure 1. The X-ray diffraction patterns for (a) $Ca_3Al_2(OH)_{12}$ and (b) simulated.

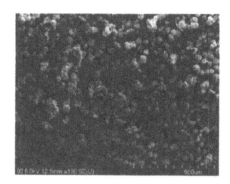

Figure 2. The Scanning electron photograph of $Ca_3Al_2(OH)_{12}$.

Figure 3 shows the excitation and emission spectra of Ce^{3+} and Tb^{3+} co-doped $Ca_3Al_2(OH)_{12}$ sample measured at room temperature. The excitation spectrum is composed of sharp bands at around 234 nm and 265 nm. The photoluminescent emission spectrum was obtained under 234 nm UV irradiation and the excitation spectrum was obtained with emission peak 540 nm. The sharp bands correspond to the absorption of the 4f-4f transition of Tb^{3+} ion [10]. The emission peaks are observed at 486 nm, 498 nm, 540 nm, 549 nm, 582 nm, 595 nm and 623 nm should be assigned to $^5D_4$-$^7F_6$, $^5D_4$-$^7F_5$, $^5D_4$-$^7F_4$, and $^5D_4$-$^7F_3$ transitions of Tb^{3+} ion. The phosphor exhibited a typical, characteristic Tb^{3+} emission and the emission of Tb^{3+} mainly originated from the transitions from $^5D_4$ to ^7F_J (where J = 3; 4; 5; 6; respectively). Since the J value of the ^7F_J ground state multiplets are high, the crystal field splits the levels to many sublevels and gives the spectrum its complicated appearance and fine structures. The emission peaks which are come from the transition of $^5D_4$-$^7F_6$, $^5D_4$-$^7F_5$ and $^5D_4$-$^7F_4$ are split to 486 nm and 498 nm, 540 nm and 549 nm, 582 nm and 595 nm, respectively. However, the split was not found in the emission at 623 nm from figure 2.

The emission from the $^5D_3$ level of Tb^{3+} has not been observed in this compound. The weak emission band at around 380 nm should be attributed to the emission of Ce^{3+} ion in the sample. The emission spectrum of the sample contains both very weak emission of Ce^{3+} and strong green emission of the Tb^{3+}, indicating that there exists the Ce^{3+}-Tb^{3+} energy transfer in this system.

The resource of Tb used in this system is its oxide form Tb_4O_7 where the Tb (+3) and Tb (+4) is co-existent. However, the valence of most Tb in the product is +3 from its emission spectra. The emission was not observed when the raw materials were mixed before the hydrothermal process. It can be inferred that a part of the Tb (+4) should be reduced to Tb (+3) in the hydrothermal system and led to the emission of Tb^{3+} in $Ca_3Al_2(OH)_{12}$.

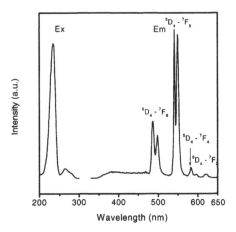

Figure 3. The luminescent spectra of Ce^{3+} and Tb^{3+} doped Ca$_3$Al$_2$(OH)$_{12}$.

CONCLUSIONS

Ce^{3+} and Tb^{3+} co-doped Ca$_3$Al$_2$(OH)$_{12}$ was synthesized at 453 K in hydrothermal system. The photoluminescent property was investigated and found that there exists the energy transfer from Ce^{3+} to Tb^{3+} in the compound. Part of the Tb (+4) was reduced to Tb (+3) because of the reduction atmosphere in the hydrotherm system. And this can be used in the other luminescent materials with changeable valence activators.

ACKNOWLEDGEMENTS

Dr. Kim is grateful to the financial support from Korea Ministry of Science and Technology through the Center for Photonic Materials and Devices.

REFERENCES

1. S. Oshio, K. Kitamura, T. Shigeta, S. Horii, T. Matsuoka, S. Tanaka, and H. Kobayashi, *J. Electrochem. Soc.* **146**, 392 (1999).
2. K. S. Sohn, S. S. Kim and H. D. Park, *Appl. Phys. Lett.* **81**, 1759 (2002).
3. C. H. Kim, S. M. Park, J. K. Park, H. D. Park, K. S. Sohn, and J. T. Park   *J. Electrochem. Soc.*, **149**, 183 (2002).
4. T. Matsuzawa, Y. Aoki, N. Takeuchi, and Y. Murayama, *J. Electrochem. Soc.* **143**,

2670 (1996).

5. M. Ohta and M. Takam, *J. Electrochem. Soc.* **151**, 171 (2004).
6. Y. H. Lin, Z. L. Tang, Z. T. Zhang, and C. W. Nan, *Appl. Phys. Lett.* **81**, 996 (2002).
7. K. S. Sohn and N. Shin, *Electrochem. and Solid-State Lett.* **5**, 21 (2002).
8. A. Baciero, L. Placentino, K. J. McCarthy, L. R. Barquero, A. Ibarra, and B. Zurro, *J. Appl. Phys.* **85**, 6790 (1999).
9. C. Cohen-Addad, P. Ducros and E. F. Bertaut, *Acta. Crystallogr.* **23**, 220 (1967).
10. H. P. You, G. Y. Hong, X. Y. Wu, J. K. Tang and H. P. Hu, *Chem. Mater.* **15**, 2000 (2003).

Mat. Res. Soc. Symp. Proc. Vol. 817 © 2004 Materials Research Society   L6.36

The characteristics of joints with Indium-silver alloy
using diffusion soldering method

Je Yoon Kim[1], Sang Won Park[1], Jee Young Yoon[1], Hwa Young Kim[1], Dae Yeon Lee[1], Gyu Tae Kim[1], Man Young Sung[1,*], and Ey Goo Kang[2]

[1]Dep. of Electrical Engineering, Korea Univ, Anam-dong, Sungbuk-ku, Seoul , Korea
[2]Dep. of Electronic Engineering, Far East Univ, Gamgok-Myun, Chung-buk , Korea
* Email : semicad@korea.ac.kr

Abstract

Bonding process using indium-silver alloy which can withstand high temperature was investigated at relatively low temperature. We used a thermal evaporator and vacuum coater for making indium-silver contact. From the result of experiment, we observed that indium and silver films which have good quality are formed. From phase diagram of In-Ag alloy, we can find that melting point of these compounds increases with the silver content, i.e. eutectic (144 ℃) <$AgIn_2$ (166 ℃) < (300 ℃) < (670 ℃) < (695 ℃). And these compounds are determined by the composition ratio of the source metal. Now we confirmed the thermal characteristics of Indium-Silver alloy is controlled by silver. Consequently we have developed Ag/In/Ag multi-layer composite which has higher melting point than that of normal contact. The melting point of Ag/In/Ag multi-layer is about 700 ℃. The joint cross-sections are studied using SEM(scanning electron microscopy) and EDX(Energy Dispersive X-rays). From these data, we observed that the composition and microstructure of Ag/In/Ag multi-layer were reliable and this bonding procedure is a better technique compared to the conventional structure of quantum well LED and GaN/Si LED structure was made by using sapphire for substrate and might be good for high temperature electronic devices in the future.

Introduction

The primary impediment to gallium nitride (GaN) technology has been the thermal decomposition of this compound at relatively low temperatures to produce metallic Ga and N_2 gas.[1-2] As a result, large-area GaN substrates are difficult to fabricate, necessitating heteroepitaxial growth of GaN thin films onto dissimilar substrates such as sapphire and SiC.[3-4] Despite the growth substrate obstacle, the development of GaN-based light-emitting diodes(LEDs) and laser diodes (LDs) has proceeded rapidly,[5-6] GaN-based high brightness blue LEDs and blue LDs have recently become commercially available. This progress has not only been limited to optoelectronic devices but also includes the rapid development of GaN-based high-power electronics given the large direct GaN band gap (Eg=3.4eV).

The most commonly used growth substrate, sapphire, still imposes constraints on the GaN film quality due to the mismatch of the lattice and thermal-expansion coefficient between the sapphire and GaN. In addition, the sapphire substrate inhibits performance of the LED, LD, and transistor device performance due to its poor thermal and electrical conductivity. For devices processed on sapphire substrates, all contacts must be made form the top side. This configuration complicates contact and packaging schemes, resulting in a spreading-resistance penalty and operating voltages increase. The poor thermal conductivity of sapphire, compared to Si and SiC, also prevents efficient dissipation of the heat generated by GaN-based high-current devices, such as LDs and high-power transistor, consequently inhibits device performance.

Sapphire is widely used material for the substrate for III-nitrides light emitting diode device fabrication due to its high-temperature stability, similar crystal symmetry with the III-nitrides, and relatively low cost, although substrates such as silicon or gallium arsenide would be more ideal due to their thermal and electrical properties along with the increased microsystem functionality through integration, direct deposition of III – nitride-based thin films. A more viable means of material integration for GaN thin films with dissimilar materials is wafer bonding and thin film lift-off techniques.

In this paper, Bonding process using indium-silver alloy which can withstand high temperature was investigated at relatively low temperature and the melting point of Ag/In/Ag multi-layer is above 700 ℃.

Ag-In System

In high temperature electronics, the operating temperature of many electronic devices built on silicon carbide and III-V compound semiconductors has recently risen to 350 ℃ and above[5-10]. Some power devices also need high temperature operation. To support in packages, high temperature joints are required. Conventional joint are fabricated at a process temperature higher than the melting temperature of the joint material. Thus, to produce high temperature joints, conventional methods require high process temperature which tends to induce high stresses when cooled down to room temperature due to thermal expansion mismatch.

To explain the bonding principle, it is appropriate to briefly review the binary phase diagram and intermetallic compounds of the silver-indium.

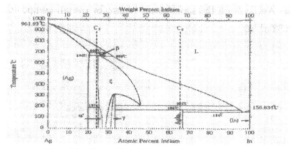

Fig. 1 Indium-silver binary phase diagram

Fig.1 shows the Ag-In phase diagram. There are a total of seven equilibrium phases. Two of them are identified as intermetallic compounds Ag_2In and $AgIn_2$. The terminal solid solution (Ag) can take up to 21 wt.% In over a very wide temperature range. In contrast, the In-rich sold solution (In) has a limited Ag solid solubility of about 1 wt.%. A eutectic reaction occurs at 97 wt.% of In with eutectic point of 144 ℃.

At room temperature, alloy with In composition above 68.1 wt.% is a mixture of $AgIn_2$ intermetallic compound and (In) phase with a solidus temperature of 144 ℃. Upon increasing the temperature above 144 ℃, the alloy converts to a mixture of liquid phase L with $AgIn_2$ compound in the mixture begins decomposing into liquid phase and the γ phase. As a result, the mixture converts into liquid with γ grains until the temperature reaches 205 ℃. The γ phase has an indium composition of 32.4 to 34.9 wt.%, around the stoichiometric value for Ag_2In phase found in some studies. Above 205 ℃, the alloy with indium composition between 47.6 to 92.6 wt.% becomes mixture of the ζ phase and liquid phase. At 205 ℃, the ζ phase decomposes by a metatectic reaction into γ phase and liquid in the composition range 34.4 to 47.6 wt.% In, the phases are not well characterized. However, a β phase was found above 660 ℃ from 26.2 to 31.3 wt.% in. Below 187 ℃ there is a narrow range of α phase around the stoichiometric value for Ag_3In.

Fig. 2 exhibits the In-Ag multilayer structure for Bonding. Chromium, indium, and silver are deposited on a silicon wafer (die) in high vacuum ($10^{-5} \sim 10^{-6}$ Torr) to inhibit indium oxidation. The chromium improves adhesion to the wafer. It was found that silver metal diffuses rapidly into indium via interstitial mechanism and indium diffuses into silver through grain boundaries with a diffusion coefficient D_0 of .24 x 10^{-12} m²/s and activation energy of 0.42 eV The thickness of indium on the die and that of the silver on the substrate depend on the final composition of the joints required. For the present design, the thickness of Ag, In layer on the die are 1.2 um and 1.15 um respectively and the Ag layer on the substrate is 1.2um for overall composition of 75 wt.% Ag and 25 wt.% of In.

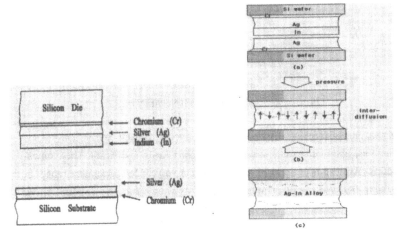

Fig.2 Indium-silver composite design   Fig.3 The bonding principle of In-Ag system

Fig.3 represents the principle of the bonding process for the In-Ag system. This is a two step process. The first step is to produce the joint and the second is an annealing process to increase the melting temperature of the joint. The assembly is loaded into furnace and heated to 220℃ in hydrogen environment for 10 min. Hydrogen is used just to prevent oxygen from getting into the sample, thus inhibiting indium oxidation during the bonding process. If die size and substrate size are same, table 1 present correct rate of calculated weight percent 25% of Indium. Therefore, we cut wafer to same size (1.5 cm by 1.5 cm) and we got following experiment result data

Experimental results and discussions

To confirm the bonding principle and reveal microstructures of the joints fabricated, several specimens are cut into cross-sections. A scanning electron microscope (SEM) with energy dispersive X-ray (EDX) spectroscopic system and an optical microscope are used to examine the microstructures and analyze the element compositions. Thickness of the joints is found to be very uniform along the cross-sections. Fig. 4-5 displays the SEM and EDX image of a specimen produced with the Ag-In design shown Fig.2. The process temperature is 220℃ with dwell time of 30 min. As we can see from the pictures, the bonding layer thickness is 3.2 um.

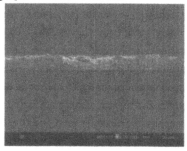

Fig. 4 SEM image of a joint fabrication with indium-silver system

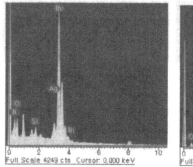 

(a) middle point (b)surface point

Fig.5 Image of Energy Dispersive X-ray

Conclusion

In this paper, we studied that the adhesion layer have stability from low to high temperature using Ag-In systems. We have manufactured the junction that could endure from 250℃ to 700℃. And we observed that indium and silver films which have good quality are formed. The composition of indium-silver contact which is made by diffusion is $AgIn_2$, Ag_2In, Ag_3In, etc. From phase diagram of In-Ag alloy, we can get melting point of these compounds increases with the silver content. And these compounds are determined by the composition of source metal ratio. Now we confirmed the thermal characteristics of Indium-Silver alloy is controlled by silver. Consequently we have developed Ag/In/Ag multi-layer composite which has higher melting point than that of normal contact. The melting point of Ag/In/Ag multi-layer is about 700 ℃. Bonding was executed by annealing and within vacuum atmosphere at a time. But thickness of sample does deposition as is so small, Bonding-force is weak. If we have large thickness on accurate weight percent rate, then Bonding-force might be more than stronger. The joint cross-sections are studied using SEM(scanning electron microscopy) and EDX(Energy Dispersive X-rays). From these data, we observed that the composition and microstructure of Ag/In/Ag multi-layer were reliable and this bonding procedure might be a good technique for high temperature electronic devices in the future and from the conventional structure of quantum well LED, GaN/Si LED structure was made by using sapphire for substrate. And from the InGaN/adhesion/Si structure which has Pd-In adhesion layer, we substituted In-Ag layer for Pd-In adhesion layer. Consequently, the new structure which has In-Ag adhesion layer is going to change the trend of LED market Because anode and cathode electrodes are placed vertically in the experimental LED, the new LED structure is going to be dominant for cost saving and improvement in fabrication.

References

1. Z. A. Munir and a. w. Searcy, J. Chem. Phys. 42, 4233 (1965).

2. N. Newman, J. Ross, and M. Rubin, Appl. Phys. Lett. 62. 1242 (1993)

3. S. Nakamura, M. Senoh, and T. Mukai, Appl. Phys. Lett. 62, 2390 (1993)

4. H. Amano, M. Kito, K. Hiramatsu, and I. Akasaki, Jpn. J. Appl. Phys.,
   Part 2 28, L21 (1989)

5. M. Tomana, R.W. Johnson, R. Jaeger, W. C. Dillard, IEEE Trans. Compon. Hybrids,
   Manuf. Technol. 16 .536 (1993)

6. J. Trew, M. M. Shin, Third Int. Workshop on Integrated Nonlinear Microwave and
   Millimeterwave Circuits, Duisburg, Germany, Oct. 5-7, p. 109 (1994)

7. J. B. Casady, E. D. Luckowski, R. W. Johnson, J. Crofton, J. R. Williams, Proc. Of
   IEEE Electron. Comp., Las Vegas, NV, May 21-24, p.261 (1995)

8. A. K. Agarwal, J. B. Casady, L. B. Rowland, et al., Electron. Device Lett.
   18, 518 (1997)

9. A. Vescan, I. Daumiller, P. Gluche, W. Ebert, E. Kohn, Electron. Device Lett.
   18, 556 (1997)

10. A. K. Agarwal, J. B. Casady, L. B. Rowland, W. F. Valek, M. H. White, C. D. Brandt,
    Electron. Device Lett. 18, 586 (1997)

at. Res. Soc. Symp. Proc. Vol. 817 © 2004 Materials Research Society — L6.40

Structural Characterization of Molecular Beam Epitaxy Grown GaInNAs and GaInNAsSb Quantum Wells by Transmission Electron Microscopy

Tihomir Gugov, Mark Wistey, Homan Yuen, Seth Bank and James S. Harris Jr.
Solid State and Photonics Laboratory, Stanford University, Stanford, CA 94305, U.S.A.

ABSTRACT

In the past decade, the quaternary GaInNAs alloy has emerged as a very promising material for lasers in the 1.2-1.6 μm range with application in telecommunication fiber-optic networks. While most of the challenges in growing high quality laser material with emission wavelength out to 1.3 μm have been successfully resolved, extending the emission beyond 1.3 μm has proven to be quite difficult. Achieving emission out to 1.5 μm requires higher In (up to 40%) and N (up to 2%) compositions. This makes the growth of this thermodynamically unstable alloy quite difficult with phase segregation occurring even at lower growth temperatures. Recently, adding small amounts of antimony has dramatically improved the quality of the material and high luminescence has been demonstrated at wavelengths beyond 1.5 μm. In this study, high-resolution transmission electron microscopy (HRTEM) was used in a novel way in conjunction with dark-field (DF) TEM to elucidate the role of antimony in improving the material quality. The results show that antimony improves the material uniformity via reduction of the local compositional fluctuations of indium.

INTRODUCTION

The GaInNAs(Sb) material system is considered a great candidate for optical sources in fast (gigabit) metro and local area networks. By changing the composition of the alloy, the emission can be tuned to cover the 1.3-1.55 μm range which is the preferred window for telecommunication applications. GaInNAs(Sb) based lasers are expected to have lower cost and better performance compared to the presently employed InGaAsP/InP Bragg grating and distributed feedback (DFB) lasers. Growth on GaAs, an inexpensive and well-developed technology, is sure to greatly reduce the fabrication costs. The large conduction band offset intrinsic to this system ensures better thermal performance and provides an opportunity for the lasers to operate uncooled.

In order to utilize the full potential of this alloy, the foremost prerequisite is the ability to grow high quality material. This alloy is characterized by a large miscibility gap which requires that it be grown at low temperatures in metastable regimes. A lot of the growth challenges have been addressed for 1.3 μm emission alloy (containing up to 30% In and 2% N) and several groups [1-4] have been successful in fabricating lasers at this wavelength. The effort to push emission to 1.55 μm, which requires compositions of up to 40% In, has proven next to impossible due to the tendency of the material to phase segregate in this high In concentration regime [5]. Very recently, the addition of a small amount of Sb has put the race for 1.55 μm material back on track by dramatically improving the optical properties of the alloy [6-8].

In this study, we present a novel way of using high-resolution TEM to investigate the mechanism by which Sb dramatically improves the alloy quality. We use this method to map out the strain variations inside highly-strained quantum well layers. We believe that these strain variations, which are indicative of compositional fluctuations, are to blame for the broad emission spectra, low gain and high threshold currents of lasers based on this material. In addition, we employ dark-field TEM to confirm our findings from the high-resolution analysis.

EXPERIMENT

In this work, we investigated both GaInNAs and GaInNAsSb samples. Structures containing either one or three quantum wells (QWs) were grown by solid source molecular beam epitaxy (SSMBE) on GaAs substrates. All QWs had an 8 nm nominal thickness, with 20 nm barriers in between. GaInNAs samples had a nominal composition of 30% In and 1.6% N. GaInNAsSb samples had a nominal composition of 38% In, 1.6% N and 2% Sb in the wells and 2.3% N in the GaNAs barriers. The substrate temperature was kept low during growth at around 420 °C to prevent phase segregation. Arsenic and antimony were supplied in thermally cracked form and atomic nitrogen was generated from a radio frequency (RF) plasma source. The RF power was 300 W and the nitrogen flow rate was maintained constant at 0.5 sccm.

Cross-sectional TEM samples were prepared in the (110) orientation using the "sandwich" technique. All samples were in the as-grown condition (i.e. not annealed). Thinning was achieved by grinding and polishing followed by 5 keV Ar^+ ion milling at low angles to achieve electron transparency. A final polishing 1 keV low temperature ion milling was performed to achieve very smooth surfaces.

High-resolution TEM lattice images of the QW area were obtained in an 800 kV, 0.15 nm resolution JEOL microscope and recorded on photographic plates. Dark-field images with the chemically sensitive (002) reflection were obtained in a Philips CM-200 microscope equipped with a CCD camera with 1024 x 1024 pixels.

Strain maps were generated from the high resolution lattice images using the DARIP program developed at the National Center for Electron Microscopy (NCEM) at the Lawrence Berkeley National Laboratory (LBNL) in Berkeley, California. DF images were analyzed with Digital Micrograph, a commercial image processing software available from Gatan, Inc.

RESULTS AND DISCUSSION

Figure 1 shows a selection from a HRTEM lattice image of the quantum well area of a GaInNAs sample (a) and the corresponding strain map across the well (b). The image was taken in the [110] zone axis. Performing strain mapping analysis requires very high quality lattice images over large areas of the sample. Therefore, it is critical to perform a final very low energy (1 keV) ion polishing on the sample to achieve very smooth surfaces and large thin areas. Strain mapping is performed on the HRTEM images in several steps. First, the images are filtered to subtract the diffuse background intensity and make the peaks (which can be interpreted as the positions of the atoms) more

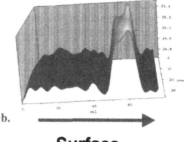

Surface

Figure 1. HRTEM image of the QW area of a GaInNAs sample (**a**) and strain map across the well (**b**).

defined. Then, the DARIP program is used to identify and mark those peaks. Next, a lattice is defined by choosing several peaks and based on this, DARIP calculates a lattice which it locks to these peak positions. The data is then extracted and a strain map is generated, which is essentially a map of the lattice parameter over the selected area. Figure 1 (b) shows this type of strain map for a single QW GaInNAs sample. The well clearly appears as a peak due to the fact that In swells the lattice, being a larger atom than Ga; the higher the In concentration, the greater the lattice distortion. Figure 1 (b) shows clearly that the In concentration profile is very asymmetric. It consists of two bumps with the higher In concentration occurring near the top interface of the well, providing evidence that In segregates near the top of the well. This is very undesirable since there are effectively two different In compositions in the well, which would broaden the emission spectrum. The difference in the rear and leading edge compositions was estimated to be around 8%.

Figure 2 shows the strain map across the three QWs in a GaInNAsSb sample (a) and a GaInNAsSb sample with identical composition but with 5 °C drop in the growth temperature for each successive well (b). The compositional profiles across the wells appear very uniform due to the antimony. The sample in figure 2 (a) exhibits tensile

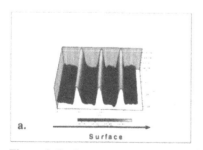

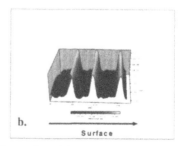

Figure 2. Strain maps across three-QW GaInNAsSb samples with constant growth temperature (**a**) and 5 °C drop in growth temperature for each successive well (**b**).

spikes at the bottom interface of the wells due to the opening of the Sb shutter which increases the incorporation of N. Figure 2 (b) does not show these spikes which indicates that the temperature drop affects the interaction between Sb and N and results in decreased N incorporation at the interface of the well. This should increase the carrier confinement in the well and result in more efficient recombination and increased luminescence efficiency.

Figure 3 shows a dark-field image of a three QW GaInNAs sample taken with the chemically sensitive (002) reflection. The intensity of the (002) reflection has a square dependence on the difference between the composite structure factors of the group III and group V atoms and for the GaInNAs alloy is modeled as shown in equations 1-3 [9-10]:

$$I_{002} = A^2 = k^2 |f_{III} - f_V|^2 \qquad (1)$$

$$f_{III} = x(f_{In} - f_{Ga}) + f_{Ga} \qquad (2)$$

Where I_{002} is the (002) intensity, f designates the structure factors, x and y are the fractions of In and N, respectively, and k is a constant. In figure 3, the wells show up as stripes clearly delineated by dark lines. These occur at the well interfaces because the combination of the fractions x and y are such that the reflection is completely extinguished. Within the wells, as the In concentration increases, the (002) reflection is no longer extinguished and the brightness is recovered. Figure 3 shows that the well interfaces for the GaInNAs sample are quite rough and the deterioration is greatest for the top well. This is in good agreement with figure 1 (b) where the strain map clearly evidenced the tendency of In to phase segregate in a GaInNAs sample.

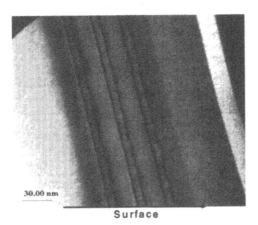

Surface

Figure 3. DF image with the (002) reflection of a GaInNAs sample with 3 QWs.

Figure 4 shows the DF images for the GaInNAsSb samples grown at constant growth temperature (a) and with 5 °C growth temperature drop for successive wells (b). These images show much smoother interfaces compared to the GaInNAs sample in figure 3 and confirm the strain map findings that antimony improves the compositional uniformity of the material. In addition, comparison between the two images in figure 4 reveals that the sample in figure 4 (b) shows slightly smoother interfaces than the one in figure (a) which reinforces the idea that a slight reduction in the growth temperature for successive wells might be beneficial.

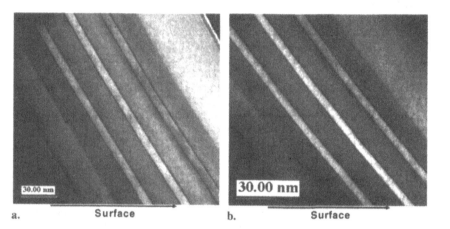

a.   Surface   b.   Surface

Figure 4. DF images with the (002) reflection of three-QW GaInNAsSb samples with constant growth temperature (**a**) and 5 °C drop in growth temperature for each successive well (**b**).

CONCLUSIONS

In this study, high-resolution TEM was used in a novel way to structurally characterize GaInNAs and GaInNAsSb quantum wells. Strain maps were generated from the high resolution images to investigate the compositional fluctuations within the wells. Dark-field images were obtained using the chemically sensitive (002) reflection and used to verify the findings from the strain maps. The results established that: (1) antimony dramatically improves the compositional uniformity within the wells and (2) small reduction of the growth temperature during successive well growth may be beneficial to device performance since it leads to better carrier confinement and additional improvement in the compositional uniformity.

ACKNOWLEDGEMENTS

The authors would like to thank Christian Kisielowski, Cheng Yu Song and Alfredo Tolley at NCEM at LBNL and Kerstin Volz for their valuable help and advice with all aspects of the TEM analysis. We acknowledge DARPA/ONR Optoelectronics Materials Center (contract MDA 972-00-1-0024) and ONR (contract N00014-01-1-0010) for financially supporting this work. T. Gugov acknowledges the Stanford Graduate Fellowship for supporting his studies at Stanford.

REFERENCES

1. S. Sato, Y. Osawa, T. Saitoh and I. Fujimura, *Electron. Lett.* **33**, 1386 (1997).
2. J.S. Harris, *IEEE J. Sel. Top. Quantum Electron.* **6**, 1145 (2000).
3. A. Wagner, C. Ellmers, F. Höhnsdorf, J. Koch, C. Agert, S. Leu, M. Hofmann, W. Stolz and W.W. Rühle, *Appl. Phys. Lett.* **76**, 271 (2000).
4. A. Ramakrishnan, G. Steinle, D. Supper, C. Degen and G. Ebbinghaus, *Electron. Lett.* **38**, 322 (2002).
5. K. Volz, A.K. Schaper, A. Hasse, T. Weirich, F. Höhnsdorf, J. Koch and W. Stolz, *Mater. Res. Soc. Symp. Proc.* **619**, 291 (2000).
6. S. Bank, W. Ha, V. Gambin, M. Wistey, H. Yuen, L. Goddard, S. Kim and J.S. Harris, *J. Crystal Growth* **251**, 367 (2003).
7. S. Bank, M. Wistey, W. Ha, H. Yuen, L. Goddard, and J. S. Harris, *Electron. Lett.* **39**, 1445 (2003).
8. J.S. Harris Jr., *Semicond. Sci. Technol.* **17**, 880 (2002).
9. M. Albrecht, V. Grillo, T. Remmele, H. Strunk, A. Egorov, Gh. Dumitras, H. Riechert, A. Kaschner, R. Heitz and A. Hoffmann, *Appl. Phys. Lett.* **81**, 2719 (2002).
10. J.-M. Chauveau, A. Trampert, M.-A. Pinault, E. Tournie, K. Du and K. Ploog, *J. Crystal Growth* **251**, 383 (2003).

AUTHOR INDEX

SUBJECT INDEX

RTA, 213

second harmonic generation, 133
Si
    nanoclusters, 3, 35
    nanostructures, 139
    QD, 15
silicate phosphor, 237
silicon, 75, 89, 109, 127
    nanocrystals, 121
    nitride film, 121
single crystal, 249
SiN_x, 115
SiON, 27, 145
sol-gel, 21
    method, 83
solid state reaction, 243
sovolthermal synthesis, 249
spectroscopy, 89

strontium titanate, 219

850 nm technology, 95
terbium, 47
transmission electron microscopy, 127, 267
tungsten tellurite glass thin film, 55

viscosity, 173

waveguide(s), 109, 195
    amplifier, 27
white-light emitting diode, 237

YIG, 213
Y_2O_3 thin films, 61

$Zn(NH_3)CO_3$, 249

Printed in the United States
By Bookmasters